QUALITY PLANNING AND ANALYSIS

This senior/graduate-level text explores the quality function and *all* the means used to achieve quality, i.e., the entire collection of activities through which industry creates products and services which can meet the quality requirements of the user. The book is organized so that the early chapters present general principles (both managerial and statistical) and later chapters cover the methodology required during design, development, manufacturing, purchasing, and customer usage of the product. Statistical tools are introduced in separate chapters throughout the book as they are needed, and the large selection of problems and cases reflects the managerial and statistical aspects of the quality function.

SPECIAL FEATURES

- Includes unique treatment of emerging new topics of importance, e.g., concept of fitness for use, effect of quality on manufacturers' income, optimizing users' costs, quantification of systems effectiveness, etc.
- Includes updated treatment of recent developments, e.g., quality cost analysis, reliability techniques, motivation, etc.
- Presents principles in such a way that their applicability to all types of products (e.g., chemicals, food, drugs, electronic and mechanical systems) can be recognized.
- Offers broad coverage of the *entire* quality function.

ont flap

HORS

rsued a varied career
s an engineer, in-
, government admin-
professor, impartial
orporate director,
consultant. Through-
s work has been
rch for the underlying
on to all managerial
rch has produced
ers and ten published
the international
Quality Control Hand-
f degrees in engi-
, Dr. Juran has
200 training seminars
gement subjects in
n numerous countries
. In additon to his
hor and lecturer, Dr.
serves a number of
anies as a member of
ectors or as a

, **Jr.,** holds degrees in
eering and is experi-
nagerial and statistical
ases of quality prob-
sently Professor of
eering at Bradley
as taught at New York
at Rutgers, The State
essor Gryna was also
formerly with the Martin Marietta Corp., where he had major management responsibility for complete quality and reliability programs.

Quality Planning and Analysis

From Product Development through Usage

Dr. J. M. Juran

International lecturer, consultant, and author of

MANAGEMENT OF QUALITY CONTROL
THE CORPORATE DIRECTOR *(with J. K. Louden), also in Spanish*
MANAGERIAL BREAKTHROUGH *also in German, Japanese*
LECTURES IN GENERAL MANAGEMENT *also in Japanese*
LECTURES IN QUALITY CONTROL *also in Japanese*
CASE STUDIES IN INDUSTRIAL MANAGEMENT *(with N. N. Barish)*
MANAGEMENT OF INSPECTION AND QUALITY CONTROL
BUREAUCRACY: A CHALLENGE TO BETTER MANAGEMENT

Editor of:

QUALITY CONTROL HANDBOOK *also in Japanese, Spanish, Hungarian, Russian*

Frank M. Gryna, Jr.

Professor of Industrial Engineering
Bradley University

McGRAW-HILL BOOK COMPANY

New York
St. Louis
San Francisco
Düsseldorf
London
Mexico
Panama
Sydney
Toronto

Quality Planning and Analysis

Library of Congress Catalog Card Number
76-112836
33171

1 2 3 4 5 6 7 8 9 0 MAMM 7 9 8 7 6 5 4 3 2 1 0

This book was set in Caledonia by The Maple Press Company, and printed on permanent paper and bound by The Maple Press Company. The designer was Merrill Haber; the drawings were done by B. Handelman Associates, Inc. The editors were B. J. Clark and Susan Davis. Stuart Levine supervised production.

Preface

This book offers a broad-based educational textbook for preparing people to plan the quality function so that products and services will be fit for use and to analyze quality problems to discover causes and propose remedies.

The need for this book stems from the gathering decision of the human race to organize its life around the quality and availability of manufactured products (and services). To a remarkable and growing degree, human beings now depend on manufactured products for a broad assortment of human necessities and luxuries: nutrition, health, shelter, military security, scientific progress, communication, transport, entertainment, comfort, and so on. Quality failures in these products can and do result in widespread human inconvenience, economic waste, and catastrophic loss of life.

To meet these quality needs of the users requires that all the major functions of the manufacturers take into account the users' needs for quality. Market research must discover just what are these needs of the users; product development must create designs which are responsive to these needs; manufacturing planning must devise processes which are capable of executing the product designs; production must regulate these processes to achieve the desired qualities; inspection and test must prove the adequacy of the product through simulated use; marketing must sell the product for the proper application; customer service must follow the usage to remedy failures and to discover opportunities for improvement.

The quality subactivities present in each of these major company activities are themselves collectively a major activity which has become known as the "*quality function.*" It may be defined as that collection of activities, no matter where performed, through which the company achieves fitness for use. The outlines of this quality function have been emerging clearly. It has now become evident that, in common with other major company functions, successful conduct of the quality function demands much specialized knowledge and many specialized tools, as well as trained specialists to use these tools and apply this knowledge. This book has been designed to enable men to acquire this specialized knowledge and to develop proficiency in use of the tools through which this knowledge is made effective.

Because the quality function interfaces with so many company activities, a book to be used as a basic body of knowledge should examine and expound all these interfaces. In this respect the present book differs significantly from its contemporaries. There are now on the market excellent books for training men in specific tools (or tool kits) essential in the quality function, e.g., statistical methods, reliability engineering, measurement. However, to the knowledge of the authors, the present book is the first textbook which is coextensive in scope with all the critical interfaces between the quality function and the other major functions of the company. It is the belief of the authors that this broader body of knowledge is a necessary response to the broader responsibilities which industry has been giving to the quality planners and analysts.

The book recognizes the important role of statistical techniques in the quality function. To this end a number of separate chapters are devoted to explanation of the nature and use of these statistical techniques as an aid in various aspects of quality planning and analysis. However, the only background assumed for the book is college algebra, and there is no attempt to provide a state of advanced knowledge in statistical methodology.

All chapters include problems, and these problems are so structured as to reflect the "real" world "outside" rather than the more limited world of the classroom. Such problems require the student to face the realities which confront the designers, engineers, marketers, inspectors, users, and others involved in the quality function. The student must make assumptions, estimate the economics, reach conclusions from incomplete facts, and otherwise adapt himself to the imperfect world of the practitioner.

In dealing with the concepts of management of the quality function, some actual case problems have been prepared for study. In addition, there are problems requiring visits to factories, department stores, and still other institutions to learn at first hand of the challenges faced and solutions achieved.

The authors also draw attention to the relationship of *Quality Planning and Analysis* to *Quality Control Handbook,* second edition

(J. M. Juran, editor, McGraw-Hill Book Company, New York, 1962). The handbook is a reference compendium which, through broad sale in the English language plus translation into other languages, has become the standard international reference work on the subject. In preparation of *Quality Planning and Analysis,* the authors have introduced frequent references to the handbook (as well as to other sources) where space limitations placed restriction on detail.

We are indebted to the Literary Executor of the late Sir Ronald A. Fisher, F.R.S., to Dr. Frank Yates, F.R.S., and to Oliver & Boyd Ltd., Edinburgh, for permission to reprint Tables III and IV from their book *Statistical Tables for Biological, Agricultural and Medical Research.*

The authors wish to acknowledge the participation of their respective "teams" in reducing their views to a finished book. For Mrs. J. M. Juran it has been the tenth such collaboration in preparing a book manuscript. Mrs. Marion Nelson carried the principal secretarial burden for Professor Gryna. We are also indebted to Mrs. Virginia Parrett, Mrs. Evelyn Kahrs, Miss Karen Baum, and Miss Nancy Jones for their assistance in the typing. Messrs. Sushil Kumar Suri, Subhash C. Narula, Chester Poremba, and Yogesh Kansal were most helpful in preparing calculations and illustrations.

Professor Gryna owes a special debt of gratitude to his wife Dee, and to Wendy, Derek, and Gary for their encouragement and support.

J. M. Juran
Frank M. Gryna, Jr.

Contents

6

BASIC STATISTICAL CONCEPT OF VARIATION 70

7

ORGANIZATION FOR QUALITY 91

8

DESIGN FOR SYSTEM EFFECTIVENESS 114

19

VENDOR RELATIONS 402

20

STATISTICAL AIDS IN VENDOR RELATIONS 421

21

PROCESS CONTROL 436

22

MOTIVATION FOR QUALITY 453

29
QUALITY DATA SYSTEMS AND QUALITY MANUALS 572

30
DIAGNOSIS TECHNIQUES FOR QUALITY IMPROVEMENT 588

31
QUALITY CONTROL ENGINEERING 623

1 Basic Concepts; Definitions; Terminology

1-1 The Quality Function

All human institutions (industrial companies, schools, hospitals, churches, governments) exist to provide products or services to human beings. An essential aspect of these products or services is that they be fit for use. This phrase "fitness for use" is the basic meaning of the word "quality."

To carry out its mission of providing products or services fit for use, an industrial company engages in a variety of activities, much as does a biological organism. The latter carries out its mission (of life, reproduction, growth) through a number of identifiable systems or functions (e.g., nervous system, circulatory system, skeletal system). The company likewise operates through a number of identifiable systems or functions (e.g., finance, marketing, personnel).

This list of company functions includes a function concerned with quality, or achieving fitness for use. This is a major function. The company's survival depends on the income it gets from selling its products and services, and the ability to sell is based on fitness for use.

The quality function is carried out through a wide assortment of company activities. Figure 1-1 depicts how these activities are related to each other.

Through its field contacts with users, a company determines what qualities are needed by those users. Research and development specialists then create a product concept which can meet these quality needs of the users. Design engineers prepare product and material specifications embodying these needed qualities. Other engineers specify processes and instruments capable of fabricating and measuring these qualities. Purchasing specialists buy materials and components possessing appropriate qualities, which in turn brings the vendors' quality activities into the spiral. Shop operators are trained to use the processes and instruments to put the specified qualities into the product. Inspectors determine whether the resulting product in fact possesses the needed qualities. The sales force, through the distribution chain, urges customers to buy the products possessing these qualities. Customers use the qualities. The experience of use suggests how the product might be improved, thus starting a new turn of the upward spiral.

The spiral (Figure 1-1) is concerned with activities, not with company departments. In tiny companies the entire collection of activities shown on the spiral is conducted by one or a few men, with little specialization. However, in large companies, the activities around the spiral

Figure 1-1 The spiral of progress in quality.

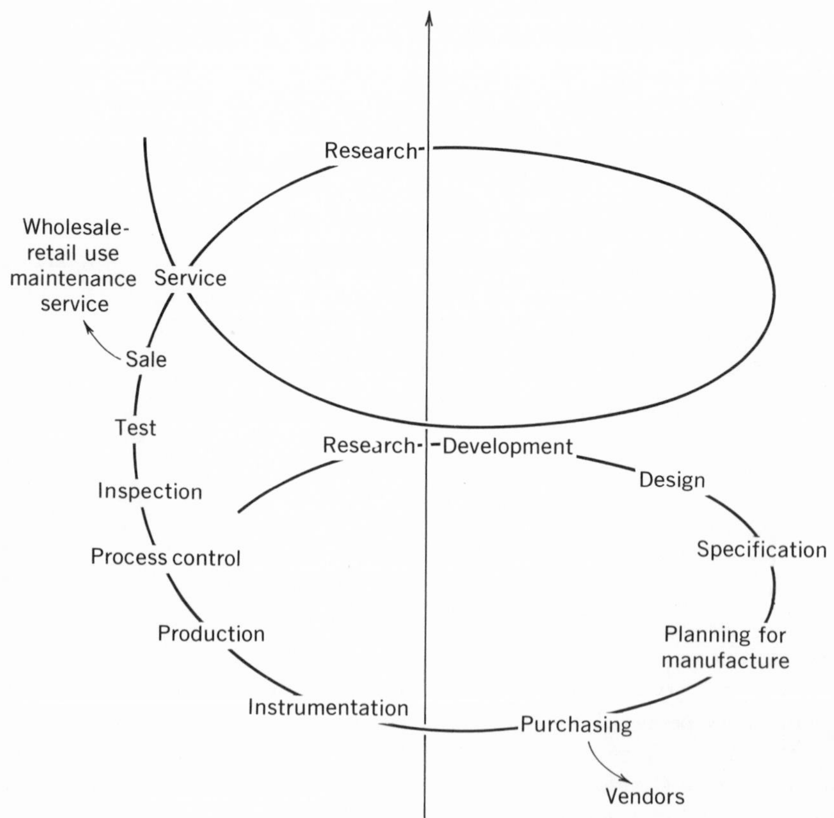

are commonly assigned to specialized departments. Each of these specialized departments is responsible for carrying out the activities inherent in that specialty, including the quality activities. For example, a purchasing department has responsibility for buying materials and components from the right vendors, at the right time, at the right price, at the right terms, *and of the right quality.*

Under such specialization, no single department is "responsible" for the broad quality function. All departments have a role to play. (However, in some companies, a specialized department may be assigned a major role for planning and coordinating all activities which relate to the quality function.)

The spiral emphasizes the broad scope of functions involved in achieving "fitness for use" quality. This is strikingly different from the narrower scope of inspection functions or even conventional "quality control programs" which have generally emphasized adherence to specifications.

In this book, the shorthand label for the collection of activities relating to quality is "quality function." We define the quality function as that *collection of activities through which we achieve fitness for use.*

1-2 The Meanings of "Quality"

The word "quality" has come to have a variety of meanings in addition to "fitness for use." Students and practitioners should understand the more frequently used multiple meanings. Much confusion results when the word "quality" is uttered by a person who has one of these meanings in mind, only to have the word interpreted in a variety of ways by different listeners.

In industrial companies the most usual meanings given to the word "quality" include:

1 Fitness for use. This is the historic meaning. In the ancient marketplaces, where commerce was transacted directly between the one-man producer (tailor, shoemaker, etc.) and the ultimate user, fitness for use was even narrower in meaning, since it referred to the degree to which a specific product or service satisfied the wants of a specific user. In our century, there is still a lot of this face-to-face meeting over fitness for use. Increasingly, however, commerce is transacted through a distribution chain, requiring the concept of a specification, and creating two parameters of fitness for use:
   a Quality of design, or grade
   b Quality of conformance

With the proliferation of complex, long-life products, two added parameters of fitness for use have arisen:

c Availability, i.e., the extent to which the user can secure continuing use of the product

d Customer service, i.e., the extent to which the manufacturer and the distribution chain make good in the event of product failure

2 Grade. This is the degree to which a *class* or category of product possesses satisfactions for people generally. The term "quality of design" is sometimes used as synonymous with grade. The term "brand" is often used to describe a producer's designation for a particular grade.

3 Quality of conformance. This is the degree to which a specific product conforms to a design or specification.

4 Quality characteristic. This is any distinguishing feature of a grade or a product, i.e., appearance, dimension, performance, length of life, dependability, reliability, durability, maintainability, taste, odor, etc.

5 The quality function. This is the name for that area of responsibility in industrial companies through which we achieve fitness for use.

6 A department. Some industrial departments have names containing the word "quality," e.g., quality control department, and these names may be abbreviated to the single word "quality."

In this book, the word "quality," unless otherwise stated, refers to fitness for use.

1-3 The Meanings of "Control"

The word "control" also has a variety of meanings. As used in this book, the word refers to:

1 The planned cycle of activities by means of which we achieve an intended goal, objective, or standard

The word "control" also has other meanings, and it is well to be on the alert for these. The more frequently used meanings include:

2 A device which directs, influences, restrains, commands, verifies, or corrects

3 The act of directing, influencing, etc.

4 The name of a department which conducts control activities

5 A standard of comparison as determined by a statistical test usually referred to as a "state of statistical control"

In this book, the word "control," unless otherwise stated, refers to the cycle of activities by means of which we achieve a goal (definition 1).

1-4 The Meanings of "Quality Control"

This term likewise has a variety of meanings. In this book the term refers to:

1 The entire collection of activities through which we achieve fitness for use, i.e., carry out the company's quality function.

There are other meanings. Of these the most troublesome is:

2 The "partial collection" of quality activities which we assign to that company department which devotes itself primarily to the quality function. Some companies assign most of the quality activities to a centralized department which is mainly concerned with quality. Other companies assign most of the quality activities to other functional departments, e.g., design, marketing, purchasing. Because these organization forms do vary so widely, the above "partial collection" also varies widely. Hence the term "quality control," when used to describe this partial collection, has a decidedly nonuniform meaning from one company to another.

Additional meanings include:

3 The tools, devices, or skills through which quality activities are carried out. (During the 1950s it was common for advocates of use of statistical methodology to use the term "quality control" as synonymous with industrial use of statistical methods.)
4 The name of the department which devotes itself full time to the quality function.

1-5 Concepts and Terminology

While language and dialects differ considerably among industrial companies, the concepts concerning quality are remarkably alike and universal. Communication within and between companies becomes easy or difficult depending on whether one is talking about concepts or about labels for concepts.

For example, a universal concept is that of elimination of chronic quality troubles, i.e., discovering the causes of these troubles and providing a permanent remedy for them. At a meeting of men from a wide assortment of companies there is no difficulty in securing agreement on the usefulness of this concept. The labels in use vary considerably: quality improvement, quality control, defect prevention, "zero defects," etc. The words which make up these labels are, in turn, intricately woven into the fabric of the local dialect. Experience has shown that a hasty change in labels can have unpleasant side effects in unexpected parts of the fabric. This experience makes people cautious about importing labels from other companies.

1-6 Products and Services

The subject matter of commerce includes (1) products, e.g., shoes, automobiles, milk, and (2) services, e.g., electric power, haircuts, bus rides, education.

Human beings acquire products mainly to secure the functional uses which the qualities inherent in these products can provide, e.g., nourishment from fresh milk, transportation from automobiles, shelter from housing, electrical energy from batteries. "People don't really buy aluminum siding. They buy the electro-static baked enamel finish. They buy the 14 decorator colors. They buy the beautiful panel designs. They buy the 20-year guarantee and the paint-free, maintenance-free insulating exterior."[1] Products are also acquired to utilize "nonfunctional" qualities such as appearance or status. ("Services" likewise provide both functional and nonfunctional qualities for use.)

Some products are fully consumed during first usage or during a short period of use, e.g., food, soap. Other products may last for months or years before becoming unfit for use, e.g., clothing, automobiles, cooking stoves, industrial machinery, houses. For such longer-life products the user need not necessarily buy the product in order to utilize its qualities. He may, instead, buy only the services, while ownership of the product rests with someone else, e.g., telephone service, office space, computer time.

There are numerous and ingenious degrees of ownership of products aimed to provide users with optimal services. Automobile tires or batteries are sold based on the number of miles or months of service they provide. Aircraft engines follow a similar arrangement. Some products are sold with a guarantee provision to protect the buyer against losses

[1] *Barron's* p. 3, Mar. 7, 1966.

due to unfitness for use. Other products are now being leased rather than sold outright. For leased products or products sold outright with formal or informal guarantees of dependability, the manufacturer is really selling a *service* rather than a product. This has far-reaching implications because the manufacturer is responsible for the "fitness for use" of the product throughout its full life (not just through a limited warranty period). Thus, the cost of poor quality will directly affect income (leased equipment contracts will be terminated; the notoriety of failure will be harmful to future sales). Such practices are tending to eliminate the distinction between products and services so far as the user is concerned.

The industrial revolution has enormously increased the manufacture and usage of long-life products. For such products, the concept of fitness for use puts great emphasis on the factor of time. To be *effective,* the product must not only be *capable* of performing its intended role; it must be *available* for use, i.e., it must be ready to be used when the user needs it. In turn, this availability depends on whether the product has *reliability* (freedom from failure), along with *maintainabilty* (ease of being restored to service when it does fail). We will later examine all these features in greater detail.

While economists properly distinguish between the manufacturing and service industries, the quality problems have much in common. For example, the schools are an industry supplying a service (education). Yet these schools start with raw material (students), apply a process (teaching), and turn out a finished product (graduates), though with some rejects. There are raw material specifications (minimum entrance requirements) and incoming inspection (entrance examinations). There is a process specification (curriculum, course outlines); process facilities (faculty, laboratories, textbooks); process controls (reports, recitations, quizzes); final product testing (examinations).

1-7 Quality Characteristic

The elemental building block out of which quality is constructed is the *quality characteristic.* Any dimension, chemical property, sensory property (e.g., taste, smell, feel, sound) which contributes to fitness for use is a quality characteristic. Still other characteristics are such properties as length of life, reliability, maintainability.

Every quality characteristic becomes the hub of the continuing spiral of progress in quality (Figure 1-1).

1-8 Quality of Design Distinguished from Quality of Conformance

A difference in specification for the same functional use is a difference in quality of design, often called "grade." The Cadillac and the Chevrolet automobiles serve the same basic functional use. However, they differ in many features of design and are therefore different in quality of design. This results in separate design specifications.

Quality of conformance, on the other hand, relates to the fidelity with which the product *conforms* to the design specification. A Chevrolet which can run and a Chevrolet which cannot run have the same quality of design, but they differ in quality of conformance. Further, both Chevrolet and Cadillac have problems in quality of conformance (i.e., producing Chevrolets and Cadillacs that meet the respective design specifications).

A program for achieving fitness for use quality requires consideration of both quality of design and quality of conformance. One implication of this is the fact that information on quality of design is heavily external to the company while information on quality of conformance is mostly internal. This scope is emphasized by the spiral of quality.

1-9 Quality, Price, and Delivery

There are optimum levels of both quality of design and quality of conformance (see Chapter 4). These optimums are with respect to cost. To the manager, quality is basically a business problem (see section 3-6) and, as with all business problems, the decisions of the management involve "tradeoffs" among cost, schedule, and quality. A poor quality program will result in excessive costs and/or late deliveries. A good quality program *may* cost more initially but will more than pay its way in reducing quality losses and late deliveries due to poor quality. Many of the day-to-day decisions that management must make to resolve a problem require that either cost, schedule, or quality be compromised. A quality program must try to detect quality problems early enough to permit action to be taken in most cases without requiring a compromise in cost, schedule, or quality.[2] Such a lofty objective is stated to emphasize that the business nature of the quality problem necessitates a stress on *prevention* rather than just correction of quality problems.

[2] A concept mentioned in a private communication from Mr. A. M. Carey of North American Rockwell Corp.

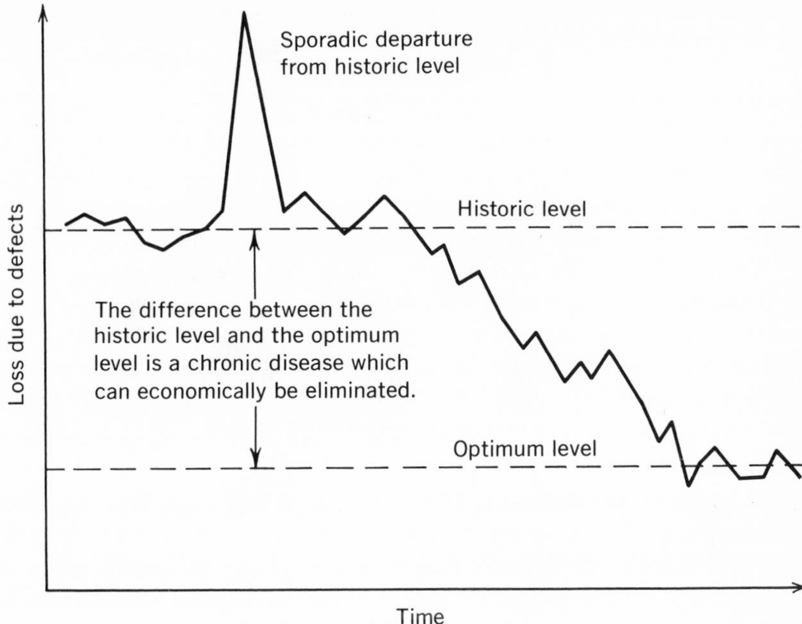

Figure 1-2 Sporadic and chronic quality troubles.

1-10 Sporadic and Chronic Defects

The distinction between sporadic and chronic defects is best understood by reference to Figure 1-2. A sporadic condition is a sudden adverse change in the status quo, requiring remedy through *restoring* the status quo (e.g., changing a worn cutting tool). A chronic condition is a long-standing adverse situation, requiring remedy through *changing* the status quo (e.g., revising a set of unrealistic tolerances).

The aspects of this distinction are shown in Table 1-1.

The dramatic nature of sporadic (yet low loss) defects is exemplified by Schrock.[3] In one plant two sporadic outbursts brought sharp rebukes from top management because these were 240 more defects than usual. Yet the chronic defects on which these two instances were superimposed caused 3,600 defective units per year. The chronic defects were later cut in half for a reduction of 1,800 defective units per year.

Despite the drama of sporadic troubles, the great majority of con-

[3] Edward M. Schrock, paper at Annual Convention of Canadian Ceramic Society, Toronto, February, 1959.

Table 1-1 **Distinction between Sporadic and Chronic Defects**

Aspect	*Nature of Sporadic Defects*	*Nature of Chronic Defects*
Tangible economic loss	Minor	Major
Extent of irritations	Substantial. Sudden nature of trouble attracts supervisory attention	Small. Continuing nature of trouble leads all concerned to accept it as unavoidable
Type of solution required	Restore the status quo	Change the status quo
Types of data needed	Simple data showing trend of quality with respect to one or two variables such as time or lot number	Complex data showing relation of quality to numerous variables
Plan for collectings data	Routine	Specially designed
Data collected	By inspectors, service representatives, etc., in the usual course of their work	Often through special experimental procedures
Frequency of analysis	Very frequent. May require review every hour or every lot	Infrequent. Data may be accumulated for several months before analysis is made
Analysis made by	Line people such as a design or production supervisor	Technical personnel
Type of analysis	Usually simple	Possibly intricate. May require correlation study, analysis of variance, etc.
Action by whom	Usually by line personnel in design, manufacturing, etc.	Usually by personnel other than those responsible for meeting the standard

formance losses are found to exist in a relatively few chronic troubles. This phenomenon arises from the invariable "maldistribution" of defects (see section 5-5).

The distinction between chronic and sporadic problems is essential because there are two different approaches for handling the problems. Chronic problems require the use of principles of "breakthrough" while sporadic problems require the principles of "control." These principles are defined in Chapter 3.

1-11 Quality Improvement

Considerable industrial effort is devoted to improving quality. Much of this improvement has historically taken place in quality of design. Dramatic evidence of this is seen in the modern apparatus for transport or communication as contrasted with the apparatus used in centuries past. But improvement also takes place in quality of conformance. The latter improvements are accomplished by what is vaguely called "defect prevention," but which actually consists of several distinct kinds of activities:

1 The elimination of sporadic causes of departure from historic levels of performance
2 The elimination of chronic causes of difference between historic levels of performance and optimum levels, through quality improvement programs
3 The avoidance of such chronic causes as part of the original planning

Usually the word "prevention" is used without making these distinctions. In consequence, there is confusion as to whether the prevention activity is to be directed at sporadic or chronic troubles, and whether (in the case of chronic troubles) the form of the prevention activity is through a quality improvement program or through planning of future activities.

1-12 The Concept of Self-control

Control has been defined as the cycle of activities by means of which we achieve a goal. An important part of the cycle is the activity of measuring results against the goal and taking action on the difference (this is often called "controlling"). This is the basis for self-regulation or self-control.

In automated control systems, the process is made self-regulating through built-in instrumentation which senses what is going on, compares the measurements with the standard (specification), and energizes a corrective system when the difference exceeds the predetermined tolerance.[4] These automated systems, though highly publicized, control only

[4] See Automation and Quality Planning, section 14-12.

a minority of processes. The majority are regulated by human beings; i.e., part or all of the feedback loop is closed by human sensing and action. The result is an enormous amount of human control effort, which is costly even when well done, and shockingly wasteful when badly done. It should come as no surprise that industrial managers have devoted extensive study to the problem of making all this control work serve as an effective, reliable regulator of industrial activity, whether on the factory floor, in the laboratory, the office, or elsewhere.

Out of this analysis of human control effort has evolved the concept of self-control—the idea that control must be delegated down to the work level, where the action takes place. Failing this, the supervisors, engineers, and managers, though badly needed for planning and break-

Table 1-2 **The Application of the Concept of Self-control**

Should Be Able to	*An Individual in*		
	Product Design	*Manufacturing*	*Field Service*
Know what he is supposed to do	Have complete data on user needs and operating environments with numerical requirements on reliability and maintainability if necessary	Have a clear and complete statement of the desired value and allowable variation for each characteristic	Have complete data on user maintenance skills with a numerical maintainability requirement, if necessary
Know what he is doing	Be provided with feedback on errors in component selection, environmental definition, etc., in sufficient detail to facilitate corrective action	Be provided with feedback on manufacturing errors in sufficient detail to facilitate corrective action	Be provided with feedback on errors in handbooks, maintenance procedures in detail to facilitate action
Take regulatory action	Be provided with information on capability of available materials and components and with funds for development programs when required	Be provided with equipment that is capable of meeting product specifications	Be provided with adequate maintenance skills, equipment, and parts

through activities, become bogged down in an endless succession of control crises. By creating the conditions of self-control at the bottom, the management hierarchy is liberated from the great majority of control tasks and can devote themselves to creative work as well as to the residue of control work which the people at the work level cannot handle.

Creating a state of self-control for a human being requires that we meet several essential criteria. We must provide the man with the means for:

1 Knowing what he is supposed to do
2 Knowing what he is actually doing
3 Taking regulatory action

If *all* these criteria have been met, a state of self-control has been created. (The fact that a state of self-control has been created does not by itself ensure that we will get control. It still requires that the man possess the personal competence and state of mind, i.e., that he be able and willing, to put these faculties to effective use. The nature of these vital needs will be discussed in Chapter 22, Motivation for Quality.)

The concept of self-control is not limited to control of quality. The concept is universal. It applies with equal force to control of cost, delivery, safety, anything. Further, as applied to quality, the concept is not limited to operators in manufacturing. It applies to everyone whose work influences quality. Table 1-2 summarizes the concept as it applies to people in product design, manufacturing, and field service. This is discussed in later chapters.

1-13 The Factual Approach

Anyone who advocates an "improvement" should realize that he is really trying to introduce two very different kinds of change: (1) a change in technology, i.e., a change in the machine, material, tool, tolerance, etc., and (2) a change in the cultural pattern of the people affected by the change in technology, i.e., their habits, beliefs, status, etc.

Normally the advocate of change cannot get his case across by opinion—by saying "I think we should. . . ." The burden of proof is on him, as it should be. In advocacy of a change in the cultural pattern, he is well advised to use the tools of the behavioral scientists. (We will look at this problem in Chapter 31.) In advocacy of a change in technology, he is well advised to use the *factual approach.*

In the quality function there is a continuing procession of questions requiring answers:

Does this lot of product conform to the specification?
Can this process hold the tolerance?
Does this specification reflect consumer needs?

Not only do these questions need answers; the answers should be based on "fact," not "opinion." To illustrate, look in on a conference such as has been held myriads of times in industry. The scene is the office of the works manager.

WORKS MANAGER: I've called this conference to see what we can do about reducing our rejects on the shafts.
PRODUCTION MANAGER: Last month's rejects were 10 percent, which is no different than it's ever been. The machines never could hold those tolerances. And the tolerances have always been too tight.
DESIGN ENGINEER: Whenever we relax tolerances, you fellows just cut down on your controls and slop over the new tolerances. It's happened again and again.
MANUFACTURING ENGINEER: If the machines can't hold the tolerances, how is it that 90 percent of the product is good? You must be doing something different on the other 10 percent.
MACHINE DESIGNER: Our competitors are using the same type of machines and seem to be making out OK. What we need is a real job of maintaining the machines.
CHIEF INSPECTOR: If the operators would check the product more often, and make the adjustments, it would cut the rejections.
PRODUCTION MANAGER: The checks they make would be enough if the machines weren't so erratic and if the gages were more reliable.
SALES MANAGER: You guys ought to get together. Our customers are talking about tightening the tolerances.

Attempts at absolute, foolproof definitions of what is fact and what is opinion are foiled by the limitations of language. Instead, we must regard fact and opinion as the limiting points at opposite ends of a broad spectrum. The distinguishing features of a "fact" are

1 It is sensed from an impersonal source—a unit of product, a machine, a material.
2 The inherent error of the sensing device is negligible in relation to the phenomenon being sensed.

3 It requires no human interpretation or evaluation.
4 It undergoes no transmission between human beings except in terms of the unit of measure.

These requirements are never met 100 percent, and it can be argued philosophically that there is no such thing as a pure fact. However, the foregoing requirements are often met to a degree which satisfies practical men and which is adequate for sound decision. Hence we often encounter information which is accepted as "facts" without ado.

Now we can return to the question "Can the machine hold the tolerance?" A quality control specialist arms himself with a precise gage and measures a series of pieces as they emerge from the machine. He records these measurements and then charts them, in the sequence in which they were made. When these measurements are compared with the tolerance, the matter is settled once and for all. Section 15-2 describes the results of just such a study, conducted in a company in which there had been earnest debate about whether a certain battery of machines could hold the tolerances.

How is it that grown men can for years continue debating an important question for which a factual answer can "easily" be provided? There are many reasons: they are blocked by their axiomatic beliefs; they lack the time or skills to take on the study; they lack the tools but are not aware of this (in the above study a key factor in providing the answer was use of a precise laboratory gage rather than the coarser shop gage); they are afraid the answer might deny what they have said for years. For these and other reasons, it often turns out that the missing facts are so easy to provide that some managers are stunned—"That should have been obvious." At the other extreme, supplying the missing facts requires so heroic an effort that no one thinks of the word "obvious."

The student who is to take on the role of quality control specialist should grasp fully the importance of the factual approach. He becomes an advocate of change in an environment where he may be the least experienced as to the technology. The factual approach is his major tool for raising him to a state of equality in contribution to the problem at hand.

PROBLEMS

1 Consider the discussion under topic 1-6 of the nature of the service rendered by an educational institution, and the process through which this service achieves fitness for use. The discussion identified:

a The product or service rendered
b The specifications established for "raw materials" entering the process
c The nature of the process specifications
d The nature of the "finished product" specifications
e The nature of the quality controls in use during the various stages of progression from raw material to finished product

For each institution listed below, identify the above elements (*a*) through (*e*):

The neighborhood laundry
The U.S. Post Office
The local shoe repair shop
The public bus company
The public library
The local newspaper
The "gas" station
The supermarket
The bank
The telephone company

2 Figure 1-1 identifies the more usual activities through which a manufacturing company achieves fitness for use. Identify, for each institution listed in problem 1, the activities through which it achieves fitness for use. For each of the identified activities, determine who carries them out, i.e., the boss, a specific individual assigned to that activity (full time or part time), a specialized department, an outside service, etc.

3 Visit the assigned institutions listed in problem 1, and determine (*a*) the kinds of sporadic defects they face, (*b*) the kinds of chronic defects they face, and (*c*) their approach to quality improvement.

2 History of Quality Control[1]

2-1 Primitive Man and Quality

In common with all living species, primitive man was concerned with quality. He had to determine whether food was fit to eat, or whether weapons were sound enough to defend him.

In his earliest state, man was a food gatherer; nature did all the manufacture or processing. As man took the first steps toward manufacture (food growing, animal husbandry, cooking of food, dressing of skins), he did so as "usufacture," i.e., manufacture for one's own use.

Usufacture is unique in that the entire cycle of quality activities (Figure 1-1) is performed by the same individual. As we shall see later, these activities can be divided into:

1 *Technological,* relating to the physical, chemical, and other such aspects of the materials, process, and product
2 *Managerial,* relating to setting quality objectives, quality planning, defining responsibilities, training, motivation, etc.

The usufacturer was severely limited in carrying out his technological activities, since technology generally was at a low state. However, his conduct of the managerial activities was excellent, since he was a party to all transactions, and could coordinate these happenings through that superb coordinating device, the human brain.

[1] Some of the material in this chapter is derived from articles by J. M. Juran in *Industrial Quality Control* (*IQC*): "Goals for Quality Management in the Next Decade," *IQC,* May, 1962; also, "The Two Worlds of Quality Control," *IQC,* November, 1964.

2-2 Early Manufacture; the Temple City

The emergence of human communities gave birth to the marketplace, permitting a separation of the user from the maker. Usufacture was supplemented and then largely replaced by *manufacture.* Maker and user met face to face in the marketplace to sell, buy, and barter. Trade was for specific articles, then and there present, which both parties could see, feel, taste, understand. There were no specifications or warranties. Each man was supposed to protect himself by his senses. The rule was caveat emptor—let the buyer beware.

The Temple City constituted one of the earliest forms of a fixed community with a sizable population. A substantial, stable market for goods and services permitted development of specifications for products and processes, and resulted in the evolution of new organization forms. The forms evolved for construction projects differed from those for consumer goods.

The construction projects of the Temple City required great numbers of men and much specialization of work. Design of such projects was entrusted to architects and engineers who had a proven reputation—the forerunners of today's requirements that plans for a project affecting human safety must bear the seal of a licensed engineer or architect. The fact that human life and safety depended on structural integrity resulted in rigid construction specifications. Components were widely standardized, i.e., bricks. Processes were widely standardized, i.e., tempering of clay. Instruments came into enough use to be recorded in the tombs—the square, the level, the plumb bob, and the boning rods for surface flatness. Economics of design was practiced; i.e., stone joints were dressed where they showed, but not to full depth.

The construction inspector was born. Reliefs in a tomb in Thebes (ca. 1450 B.C.) show inspectors checking the quality of stone blocks (Figure 2-1). This was very likely a full-time job classification, since the projects were sizable.

In contrast to the large organizations created for construction projects, the Temple City organization for consumer goods consisted of very small shops.

A typical shop is the textile mill model in the tomb of Meketre (1800 B.C. approximately). It consists of three workers preparing flax, three spinners, two warpers, two weavers, and one overseer. There were process specifications, and inspection was conducted by the overseer. Despite primitive tools, shops such as these performed astounding feats in product quality and uniformity.

Figure 2-1 Stonecutter's dressing blocks (Egyptian, XVIII dynasty, Thebes). Courtesy of The Metropolitan Museum of Art.

2-3 The Growth of Commerce; the Guilds and Quality

The quality problems of the local marketplace could be resolved with comparative ease, since the maker, the user, and the goods were all present simultaneously. (This situation still prevails in the cobbler shop or the hand laundry.) With the growth of commerce, the small shops proliferated, merchants intervened between maker and user, and goods began to move between cities. Now the need arose for specifications, samples, warranties, and other means to provide the equivalent of the face-to-face meeting of maker and user. These needs were met in different societies by various organization forms. One of these forms, the craft guilds, is of special interest.

The European guilds flourished from the thirteenth to the eighteenth centuries. They were monopolies for practicing a given trade (weaver, jeweler, etc.) in a particular city. These monopolies, while exploiting the public through restraint of trade, benefited the public by insisting that guild members adhere to minimum quality standards. Guild regulations governed the quality of the materials used, the nature of the process, and the quality of the finished product. These regulations were often spelled out in great detail. Finished goods were often inspected and sealed by the guild. "Export" of goods to other cities was under particularly strict control, since the reputation of all guild members could be damaged by poor quality shipped by any of the members.

2-4 The Industrial Revolution

The Industrial Revolution made possible an enormous expansion of manufacture and consumption of goods. To meet the needs there arose great manufacturing companies with large and even huge factories. The growth of these institutions aided the solution of some quality problems, but created new problems for which present solutions are still inadequate.

The solution of quality problems was mainly technological. The large companies could appoint full-time specialists to deal with technical problems of materials, processes, measurement, product, etc. The work of these specialists has greatly out-performed the technological achievements of their predecessors.

The new quality problems created were mainly managerial (Table 2-1). In the small shop, the master was able to remain in full command. He was physically present, could personally see and hear all the goings-on, personally issue instructions, and personally see that these instructions were carried out. In contrast, the president of the large company cannot personally do any of these things. Instead, the company must solve its managerial quality problems through use of various managerial tools and skills. As yet, these solutions are substantially short of the mark.

Table 2-1 EVOLUTION OF QUALITY RESPONSIBILITIES

Form of Enterprise	*Who Performs the Quality Activities?* *Technological*	*Managerial*
One man	The one man	The one man
Small shop	The workmen	The master
Large company	The specialized departments	?

2-5 Mass Production and Quality

In ancient and medieval times, the wealthy aristocracy were the prime users of manufactured goods. Quantities were low, machinery was scarce, and handcraftsmanship was widespread. The skills attained by these ancient handcraftsmen were of the highest degree, as attested by exhibits in museums around the globe.

The Industrial Revolution ushered in mass production, made possible by widescale use of power-driven machinery. Attainment of quality became less a matter of handcraftsmanship and more a matter of design, construction, operation, and maintenance of manufacturing processes, and especially the machines and tools which are at the heart of these processes.

Mass production is based on mass consumption. Widespread use of products created new problems of feedback of usage information from many scattered users to permit redesign for quality improvement. The same widespread usage resulted in wide geographical dispersion of use, with resulting variation in environment of use, education of users, etc., again complicating the problems of design.

A vital element in mass production and usage is interchangeability, the ability of any of the components produced by a manufacturing process to function properly in any of the assemblies. For example, automotive vehicles have, as their base, innumerable components which, through interchangeability, permit economic assembly of bearings, carburetors, gearboxes, etc. In turn, these subassemblies, through further interchangeability, assemble into automobiles, trucks, trailers, etc. But the principle of interchangeability does not stop there. It includes the entire system of traffic, with vehicles built by one industry, roads by another, fuel by another, signals by another, etc. And the system of road traffic is but one among the systems of the transport industry. In turn, the transport industry is but one of many industries, i.e., communications, power, machinery, etc.

2-6 Complex Systems and Quality

An early example of a complex system is the telephone system. To permit any subscriber to talk to any other subscriber requires a large array of sensitive terminal apparatus, capable of high fidelity transmission and reception of feeble signals over long distances, plus a network of high precision, interchangeable switching devices, capable of swift interconnection. More recent examples are seen in computers and in modern military apparatus.

The design, construction, test, operation, and maintenance of such systems present extremely complex problems in numerous aspects of quality—precision, interchangeability, reliability, etc. Since systems of such complexity are largely a twentieth-century development, the solution of these quality problems is by no means complete. For example, modern complex systems require that:

1 The time taken for product development, process development, and planning for manufacture be shortened drastically; i.e., one turn of the spiral (Figure 1-1) must now undergo a "time compression."
2 The company which takes on a large project in complex systems must subcontract much of the work to subcontractors, who in turn subcontract to a third layer of subcontractors, etc. These intercompany relationships become very complicated.
3 New technology be developed and used on several fronts, e.g., unprecedented use of electronic circuitry and components; simulation of (literally) unearthly environments and operating conditions; unprecedented precision in measurement and even new forms of instrumentation; extreme miniaturization, requiring levels of cleanliness, purity, uniformity beyond anything dreamed of before this century.
4 Elaborate schemes of checkout and countdown be used to establish the readiness of the system to perform its assigned tasks.

These and other examples illustrate how the new quality problems may become the limiting factor in whether the system is to be produced and whether, if produced, it will provide a high degree of continuous service.

2-7 Human Welfare and Quality

As the benefits of technology have become widespread, men have organized human affairs around these benefits. An incidental result has been to make society increasingly dependent on the successful performance of the quality function. An electrical circuit breaker is defective, and the resulting power failure paralyzes a community. A short-lived product is marketed under a 5-year guarantee, and the company loses millions. A highly publicized missile fails to launch, and a nation is humiliated. A monstrous defect in a drug escaped detection, and thousands of infants are doomed to tragic lives.

As the well-being of society is increasingly bound up with sustained quality performance, the attention devoted to the quality function rises. The heads of great governments publicly discuss the needs for weapons testing. Statutes regulate qualities essential to life and to the health of the economy. Company organization charts give increased status to the quality function.

In addition, the effect of quality on the economics of the user is attracting attention on an unprecedented scale. An institution buying large quantities of machinery, materials, components, etc., can protect itself against poor quality by assigning skilled engineers and buyers to become informed about the technology and the values of these products. However, the householder is largely ignorant of what lies under the hood of the car, behind the switches of the appliances, inside the drug tablets or food packages, etc. Moreover, the relative ignorance of the householder is growing, since products are ever more complex, and he is ever buying more of them as the standard of living rises. In consequence, there is arising a new set of forces which offer to aid the user in choosing the best products and obtaining the best values.

These forces (consumer cooperatives, independent laboratories, government departments, consumer-oriented journals) also concern themselves with truth in advertising, pricing practices of merchandisers, wording of guarantees, and still other matters affecting the consumer.

2-8 Emerging Concepts: Reliability, Maintainability

The dependence of human welfare on manufactured products and services has greatly increased the emphasis on some old concepts. These increases have been so great that new words have been coined to dramatize the changes.

An example of such a change in emphasis is "reliability," defined as "the probability of performing without failure a specified function under given conditions for a specified period of time." We will see, in Chapters 10 and 11, that application of this concept to modern apparatus can be complex indeed.

However, the concept is old. The following record, a clay tablet, was found in the archives of the firm of Murashu Sons of Nippur. The date is the thirty-fifth year of the reign of Artaxerxes I (429 B.C.).

> As concerns the gold ring set with an emerald, we guarantee that for twenty years the emerald will not fall out of the gold ring. If the emerald should fall out of the gold ring before the end of twenty years, we shall pay unto Bel-nadin-shumu an indemnity of ten mana of silver.
>
> Signed by the thumbnail marks of Bel-ah-iddina, Belshunu and Hatin.[2]

[2] Edward C. Bursk, Donald C. Clark, and Ralph W. Hidy, *The World of Business*, Copyright © 1962 by Simon & Schuster, Inc., reprinted by permission of Simon & Schuster, Inc.

Still other such concepts keep coming over the horizon as industrialization proceeds further. For example, electronic, mechanical, and other apparatus is acquired in profusion by myriads of unsophisticated owners. Keeping this apparatus in service is a problem of a new order of magnitude. The word "maintainability" has been coined to state the problem. New language, techniques, etc., are being evolved to deal more effectively with this new level of needs.

Some of the other terms which have emerged include:

1 "Value analysis" (or "value engineering"): study of designs to ensure that the essential function is provided at minimum overall cost of ownership and usage
2 "Producibility": the need for designing apparatus in a way which makes it easier to produce with existing machinery and tooling
3 "Usability": the need for designing apparatus in a way which makes it convenient and foolproof in the hands of the user

There is no end in sight to these magnifications of old problems, though we cannot be sure what will come next. When we look back, the magnifications or "movements" are easy to recognize: proliferation of measuring instruments and technology; growth of inspection departments and test laboratories; use of statistical data to regulate and improve processes.

The past several decades have seen the emergence of several "movements":

1 Statistical quality control. This movement initially emphasized the application of statistical methods to manufacturing problems. It is generally traced back to the creation of the statistical control chart in 1924. The development of statistical sampling plans was another major contribution of the SQC movement.
2 Total quality control. This movement initially emphasized that a quality control program must be comprehensive in scope and include "new design control, incoming material control, product control, and special process studies." This movement emerged in the early 1950s.
3 Reliability. This movement initially emphasized the product design phase of the quality problem, particularly for complex electronic products. Initial emphasis was on the development of techniques for quantifying reliability. This movement emerged in the mid-1950s.
4 Product assurance (product effectiveness). This movement was also associated with complex products and initially emphasized that reliability may need to be supplemented by maintainability or other attributes possibly including cost. This movement emerged in the early 1960s.

5 Zero defects. This movement initially emphasized the motivational aspects of quality during the manufacturing phase. The movement emerged in the early 1960s.

Each of these movements has made a contribution to the quality field; some of these contributions have been major. However, each movement has been surrounded by controversy. *Initially* each movement emphasized some aspect that was not previously stressed. Understandably, the pioneers in each movement wanted to differentiate their contribution from previous quality control efforts, and each therefore coined a new name. These new names did draw attention to the movement. Inevitably, these new movements overlapped with the work of established departments. Simultaneously, some of the new movements tried to enlarge their scope to cover the entire quality function, so that jurisdictional disputes became rampant. The enthusiasm of pioneers in new movements sometimes can easily lead them to underestimate the contributions of the existing similar and allied functions. This causes resentment in the existing functions who then overreact by refusing to recognize that there is anything new or useful in the proposed movement. Actually, "new" movements are almost always a combination of new and old approaches.

Managers and specialists are on the alert to identify such movements. Not only can the company gain by partaking of the movement; the men who pioneer in a successful movement can move more rapidly toward their personal aspirations than can men who are associated with more prosaic activities. The contrast is so great that men are tempted to create pseudomovements as a personal vehicle for achieving higher status. The industrial cemetery is full of the remains of such pseudomovements.

PROBLEM

Conduct a library research to trace the evolution of quality control in any of the following industries (or institutions), or in some other industry acceptable to the instructor:

The pyramids
The medieval cathedral
The practice of medicine
The tanning of leather
Venetian glass
Medieval armor

Muskets
Cotton cloth
Drugs
Bridges
Automobiles
Clerical work
Electronic systems
Highway pavement
Food processing

3 Quality Policy and Objectives

3-1 The Content and Sequence of Administration

Institutions conduct their affairs through the following sequence of activities:

1 Establishing the broad principles which are to guide action; e.g., we will promote from within; we will not be undersold. These principles are, in this book, referred to as "policies."
2 Establishing the quantitative goals or targets for performance, e.g., to cut field failure rates by 30 percent; to set up a vendor rating plan by the first of the year. These goals are, in this book, referred to as "objectives."
3 Defining the list and timetable of deeds which need to be done in order to carry out the objectives, e.g., to find the cause of and the remedy for leaky carburetor valves by July 1; to evaluate performance of suppliers of castings by September 1. Defining the list of deeds is, in this book, referred to as "planning."
4 Defining the organization posts, i.e., jobs which need to be set up so that the planning will be executed. These jobs are known, collectively, as "organization structure." The process of setting them up is known as "organizing."
5 Selecting and training people to man these jobs. This process is known as "manning."
6 Stimulating people to meet the objectives. This is known as "motivating."
7 Reviewing results against objectives, and acting on the differences. This is known as "controlling."

The foregoing is the sequence through which any institution establishes and achieves its objectives. This same sequence is used to establish and achieve quality objectives.

Policies, objectives, and plans are needed at all levels of the company. It is the responsibility of the managers to see that these things are set up. However, much study and analysis of data must commonly precede the setting of policies, objectives, and plans, and the participation of all levels of organization is required. The technical specialists commonly play an extensive, vital role in the data analysis and in the drafting of proposals.

3-2 Establishing Quality Policies

The subject matter of quality policies includes the following:

1 The degree of quality leadership in the marketplace, i.e., whether to go for sole leadership, for sharing leadership, for meeting competition, for quality adequacy. In one company manufacturing containers, there were four competing theories on what policy should guide the outgoing level of quality:
  a A "*capability* theory," i.e., the belief that the plants should keep the available machines going through reasonable maintenance, and that it was then up to sales to sell the resulting product. (This view was held by some of the production supervision.)
  b A "*competitive* theory," held by plant managers who made products for those customers who used multiple sources of supply. The customers made competitive comparisons, and these plant managers felt that they had no choice but to meet the competition.
  c A "*usage* theory," held by plant managers who made the same design of containers for multiple customers. Some customers emphasized appearance, others emphasized dimensions, etc. These plant managers felt that they had no choice but to meet the needs of the customers even though this resulted in several levels of outgoing quality for the same design.
  d An "*excellence* theory," held by the sales function and top management, who wanted the company to be known as a "quality house" but without knowing whether an "excellence" level would cost more or less than the other levels.

2 The pattern of customer relations, including the ethics of advertising truthfully the extent of guarantees of the product, and the extent of rigidity or flexibility in settling customers' claims for defectives.

3 The extent of leadership in adapting to recognize and meet customer quality needs. For example, there has arisen a vast new problem of making goods which are maintenance-free, both for industrial users and for the general public. There is opportunity for taking leadership in solving this new problem.

4 The pattern of vendor relations, e.g., whether to monitor the vendor in a manner similar to that accorded to an in-house department, or to leave the vendor to his own devices.
5 The extent of use of impersonal methods of supervision, i.e., objectives, planning, reports, goals, charts, controls, audits, etc., as against personal supervision.

The foregoing is not a complete list, but it does include some important policy questions which need answers.

A newly appointed quality manager undertook to interview the 32 key managers in his company to secure their views on what were the company's strengths and weaknesses in the quality function. From these interviews emerged a consensus that the company's business had been changing from one of specialty products tailored to meet customer needs to one of standard commodities.

The former was a prescription business in which the company diagnosed the customer's needs, wrote the "prescription," and compounded the needed product. Fitness for use was decided quite as much by the quality of the technical services as by the quality of the product. The price structure was established so that the price per pound included the technical services supplied. Few manufacturers were set up to supply both the technical services and the product, and those that could were able to get good profit margins.

The trend to standardization eliminated the need for the technical services on the standard products. This permitted additional companies to enter the business. These newcomers lacked the technical organization, but were efficient manufacturers and marketers. On the standard products there was no longer a competition in technical services. In addition, the competition in product quality became minimal—all manufacturers had become adequate. By setting prices based solely on costs of manufacture, the newcomers took the market away from the traditionalists whose higher prices were based on services no longer needed in the market.

The consensus discovered by the quality manager made it possible to embark on an up-to-date policy.

There are important advantages to the establishment of *written* policy:

1 Written policy forces those concerned to think out their problem to a depth never before achieved. Before you can write it down you must first think it out. A specification[1] of the National Aeronautics and Space Administration states:

[1] *Quality Program Provisions for Space System Contractor,* NASA Quality Publication NPC 200-2, 1962, par. 3.1.

> The contractor shall demonstrate an organized approach to his quality program which includes:
> Clearly specified policies and objectives. . . .

2 Written policy can be communicated in an authoritative and uniform manner, thus establishing legitimacy as well as minimizing misinterpretation.
3 Written policy provides a basis for management by agreed objective rather than by crisis or opportunism.
4 Written policy facilitates the audit of practice against policy. In one processing company it was the policy to equal or exceed competitor quality. An audit disclosed that whereas the company had extensive knowledge of its own outgoing quality, it had so little knowledge of competitor quality that it did not know whether its own policy was being carried out.

Riordan[2] lists three elements of the quality control policy of the Department of Defense:

1 Quality control plans and procedures must be the outcome of calculated forethought and planning.
2 Quality control operations should be so planned that they serve a preventive function.
3 The Department of Defense must use inspection and test data generated during production operations as a basis for determining product conformance.

3-3 Industrial Quality Objectives

An objective is a specific attainable aim, result, or goal, capable of being so defined as to serve as the basis of a plan for action.

Objectives have little meaning unless they are reduced to writing. Only in this way is there proof that they have been thought out. In addition, objectives should be quantitative, since only in this way are they clearly defined.

There are compelling reasons for quantitative, written objectives:

1 Clearly defined objectives help to unify the thinking of the managers.
2 Clearly defined objectives have the power within themselves to stimulate action.
3 Clearly defined objectives are a necessary prerequisite to running a company on a planned basis rather than on a crisis-to-crisis basis,

[2] John J. Riordan, "Quality Control and Reliability Policy in the Department of Defense," *Electronic Equipment*, vol. 6, pp. 20–23, January, 1958.

4 Only clearly defined objectives permit a subsequent comparison of performance against objective.

It is essential also to distinguish between:

1 Objectives for achieving change, by improving on present performance levels, i.e., breakthrough or changing the status quo
2 Objectives for preventing change by retaining present performance levels, i.e., holding the status quo, or control

There is need for both breakthrough and control objectives. However, the planning, the form of organization, and the method of execution differ widely for the two kinds of objectives.

A great many quality objectives are in the nature of "holding the status quo." Such control objectives imply that current performance is adequate (or that it cannot economically be improved). Under such implications the past level of performance is also the objective for the future. The job of meeting such objectives consists of identifying sporadic departures from past (now standard) levels of performance, discovering the causes of these sporadic changes, and taking action to eliminate these causes so as to restore the status quo.

Objectives for maintaining current levels of performance include setting and holding standards for:

1 Defect levels of purchased materials
2 Yields and defect levels of various processes
3 Levels of quality of finished goods
4 Levels of performance for specific quality characteristics
5 Cost of inspection and testing

In contrast to objectives for holding the status quo are other objectives for change, i.e., improvement. These objectives are of many varieties, but mainly they can be classified as objectives to:

1 Improve the company's income through making the product more acceptable to customers: as by longer life, greater usefulness, freedom from failure, etc.
2 Reduce the company's costs through reduction of the losses due to defects, e.g., lower scrap, less rework, less sorting, fewer customer returns, fewer defects from vendors, etc.

Deciding what should be the quality objectives for a particular company is always a tailor-made job, since companies differ widely as to their levels of performance in the various aspects of quality control. Moreover, since each year brings new accomplishments and new problems, the company's needs change from year to year and therefore quality objectives change. Hence setting objectives is a perennial job.

For example, one company included in its list of objectives:

1 Reduce foundry scrap to 5 percent by the end of the year
2 Reduce rejects on vendors to 8 percent by the end of the year
3 Complete development of the automated gage for the cylinder job by June 30
4 Conduct the training program on nondestructive testing for key production and inspection people during the year

A year later, the list of objectives would change due to the company's having met most of this year's objectives, along with the emergence of new problems requiring solution.

3-4 Securing Unity as to Quality Objectives

It is not difficult to secure unity on meeting departmental objectives for control. The objective for control is a *standard* for quality, cost, productivity, or whatever. The standard is based on analysis of past performance (what has been done) modified by some estimate of what should be done. The departmental supervisor usually participates in the setting of the standard so that it is "his" standard. Usually he can meet the standard because the task is mainly within his department. He is well motivated to meet such standards since his performance as a manager is largely judged by his success in meeting these standards.

> ***Example*** A process has been yielding 98 percent good product. It is concluded that it would not be economic to cut the 2 percent waste down to say 1 percent because the effort to study and remedy the causes of the 2 percent would be greater than the savings. Hence the objective is to hold the former 2 percent. The departmental supervisor is being asked to keep doing what he has already demonstrated he can do. He accepts the standard and (usually) meets it.

In contrast, it is difficult to secure unity on objectives for breakthrough. Such objectives usually involve several departments, and each department head sees the problem differently.

> ***Example*** An important company objective may be to reduce the field failure losses by one-half. No single company department can do this, since field failure losses arise from a long list of causes:

poor engineering design, defective purchased components, inadequate manufacturing planning, production operator mistakes, inadequate test, wrong application, or improper use.

When many departments contribute to a common difficulty, no one of them will act alone. Not only is any department limited in its ability to solve something so widespread; the act of trying to solve it implies a degree of guilt beyond that actually contributed. Hence the departments tend to work on their own internal problems unless an interdepartmental project is defined.

Accordingly, the achievement of unity as to interdepartmental objectives is a complex undertaking. It requires discussion among the various departments, often with the catalytic aid of a staff specialist. This specialist does much of the data collection and analysis which are prerequisite to a meeting of the minds. Once the objectives are agreed on, there remains the interdepartmental problem of planning how to reach the objective, and of executing the plan.

3-5 Planning to Meet Quality Objectives

It takes a lot of activity to meet objectives—especially objectives for breakthrough. These activities are better conducted if they are planned out in advance. This advance planning consists of all the preparations for action: breakdown of objectives into their elements; assigning responsibility; a timetable; selection and training of people; provision of facilities; measurement of progress; etc.

The planning for achieving breakthrough differs remarkably from the planning for control.[3] Breakthrough requires an invariable sequence of events as follows:

1 Breakthrough in attitudes—convincing those responsible that a change in quality level is desirable and feasible
2 Discovery of the vital few projects—determining which quality problem areas are most important
3 Organizing for breakthrough in knowledge—defining the organizational mechanisms for obtaining the knowledge for achieving a breakthrough
4 Creation of the steering arm—defining and staffing a mechanism for directing the investigation
5 Creation of the diagnostic arm—defining and staffing a mechanism for *executing* the technical investigation

[3] For a complete development of this theme, see J. M. Juran, *Managerial Breakthrough,* McGraw-Hill Book Company, New York, 1964.

6 Diagnosis—collecting and analyzing the facts required and recommending the action needed
7 Breakthrough in cultural pattern—determining the effect of proposed changes on the people involved and finding ways to overcome the resistance to change
8 Breakthrough in performance—obtaining agreement to take action
9 Transition to the new level—implementing the change

Thus, breakthrough as used in this book is not due to luck but to the effective execution of the above sequence.

Example An automobile company embarked on a program for a breakthrough in reliability. Management decided it was desirable and feasible to cut the field failure rate in half. (This objective was made primarily to get back a share of the market that had been lost. Thus, product reliability was a marketable item.) A reliability committee was set up to guide the effort, and a staff department was formed to provide the technical assistance. The "vital few" failures were determined for each of the major subsystems of the automobile. Diagnosis enabled symptoms to be traced to causes, and remedies to be provided for these causes. Goals were set for the functional areas responsible for providing these remedies. Actual failure rates were then compared to goals to evaluate the progress of the program.

In contrast, the invariable sequence of events for control consists of:

1 Choosing the control subject—defining the quality characteristic or effort that must be regulated
2 Choosing a unit of measure—defining the terms in which the control subject will be measured
3 Choosing a standard—defining the desired level of performance for the control subject
4 Designing a sensor—creating a method of measuring the control subject
5 Measuring performance—performing the actual measurement
6 Interpreting results—comparing the actual measurement to the standard
7 Decision making—deciding on the action, if any, to be taken on the basis of actual versus standard
8 Action—taking the specific steps to bring performance up to standard

This contrast will be elaborated on in later chapters of this book. However, the foregoing summaries are universal. They apply to breakthrough of any kind and to control of anything. As applied to quality, the *breakthrough* principles are basic to correcting *chronic* quality problems while the *control* principles are basic to correcting *sporadic* quality problems.

Those who administer a quality function are responsible for both breakthrough and control. This point is easy to forget in light of the never-ending list of sporadic problems that arise on a day-to-day basis. The fire fighting on sporadic problems can always serve as an excuse for lack of action on the chronic problems (which are more important in the long range).

3-6 Managerial and Technological Quality Activities

It is most helpful to realize that quality activities are of two main varieties:

1 Entrepreneurial, business, economic, or managerial activities. These include such things as policy formation, setting objectives, planning, organizing, selecting and training people, motivating people, setting up to measure results for control.

2 Technological activities, which include the quality aspects of such things as product design, specification, manufacturing planning, instrumentation, production, inspection and test, selling, and servicing.

In the usual course of events, young engineers, scientists, accountants, salesmen, etc., are brought into the company and assigned to work of a technological nature. As they acquire experience, and expecially if they show managerial aptitude, they are increasingly given responsibilities of a managerial character. In due course they devote

Figure 3-1 Quality activities: managerial and technical.

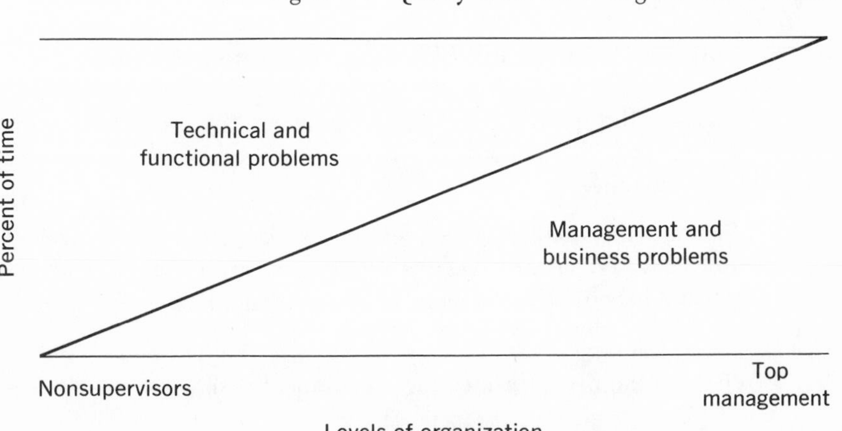

more working time to management problems than to technical problems (see Figure 3-1). This shift from technology to management takes place over the years so gradually that the man may not realize that somewhere along the line he has changed professions, and is now mainly a manager rather than an engineer.

The technological staff in the company exists to help the managers make better decisions and to help them carry out those decisions. But the technology is only one of the inputs to managerial decision making and action. The enterprise has many forces converging on it, and the manager must balance these forces so as to keep the enterprise alive and vigorous. In consequence, the manager often makes decisions which seem illogical to the technologist.

Example A technologist develops an improvement in a process. The manager will not adopt it because the product requiring that process is becoming obsolete. A technologist proposes adoption of a new technique. The manager does nothing because no one has determined the economic significance of the new technique. A technologist asks a manager to order a supervisor to adopt the technologist's recommendation. The manager will not do it because he had delegated that type of decision to that supervisor.

In such ways a technologist may find himself frustrated despite the seeming logic of his position. He may be completely logical from his premises. Yet because his premises do not include the various considerations faced by the managers, his conclusion may still be defective.

PROBLEMS

For any institution acceptable to the instructor, determine:

1 What are the quality policies, and what means are used to communicate these to all concerned? Recommend additional areas that require policy statements.

2 What are the quality objectives for breakthrough and how were they established? What are the quality objectives for control and how were they established? Recommend additional statements of objectives if required.

3 What are the plans for meeting these objectives?

4 Economics of Quality

4-1 Quality and the Basic Business Mission

The needs of society are met by the institutions of society, and meeting these needs is the ultimate justification for the existence of these institutions. These "needs" are met by goods and services created by many varieties of institutions, some organized for private profit, some not.

Collectively, the beneficiaries pay for these goods and services. (The modes of payment vary, i.e., taxes, rent, interest, purchase price.) Because they pay the bills, the beneficiaries decide, in the final analysis (and often in the short run) what goods and services are to be provided and which institutions shall provide them.

The manufacturing company in a consumer-based market is particularly vulnerable to this industrial version of the law of survival of the fittest. When considering quality problems, top managers properly put the first emphasis on the business aspects of quality, i.e.: Will this decision aid or hinder the marketability of our product?

In a competitive form of economy, "marketability" is itself a competitive matter. Companies compete for the purchasing power of the customer by offering better quality, better price, better delivery terms. We saw (section 1-9) that these things are all interrelated through the broad economics which affects them all.

The student will find in this textbook that achieving results in the quality function requires much specialized know-how, and many skills and tools. But the student must never forget that upper management considers quality primarily as a business problem, as a matter of marketability and economics (return on investment, etc.) and only secondarily as a matter of technology, tolerances, statistical sampling, and such. The time will come when the student tries to "sell" something to the upper management. This will require the use of the "breakthrough n attitude" concept. At that time, his likelihood of success will be in direct relation to his ability to present his proposal in terms of its effects on marketability and economics.

4-2 Balance between Cost of Quality and Value of Quality

Quality affects the company's economics in two basic ways:

1 Effect on *income.* With superior quality the company can secure a higher share of market, firmer prices, a higher percentage of successful bids, and still other benefits to income. It is this effect on income which makes quality have *value.*
2 Effect on *cost.* It costs money to build quality, to control it, to pay for the failures.

Finding the correct balance between cost of quality and value of quality is not so easy, since the facts are widely scattered throughout the various company departments, the distribution chain, the customers, the vendors, and still other locations (see Figures 4-1 and 4-2).

The balance to be struck (between cost and value) is not as to quality generally; it applies to each quality characteristic. For the "vital few" characteristics the balance is properly struck in the upper levels of the company. For the "trivial many" characteristics the balance must be struck in the lower levels of organization (see section 7-12).

The facts on cost of quality are often precisely ascertainable. However, facts on value of quality are more nebulous. In particular, while the factors of quality reputation and customer goodwill are conceded to be of great importance, the present methods for evaluating them are quite primitive.

4-3 Economics of Quality of Design

Much confusion arises when the same word "quality" is used indiscriminately for two widely different meanings: quality of design, and quality of conformance.

Grade is a variation in specification for the same functional use, e.g., Cadillac versus Volkswagen. Differences in grade may involve life of the product; appearance; reliability; factor of safety; interchange-

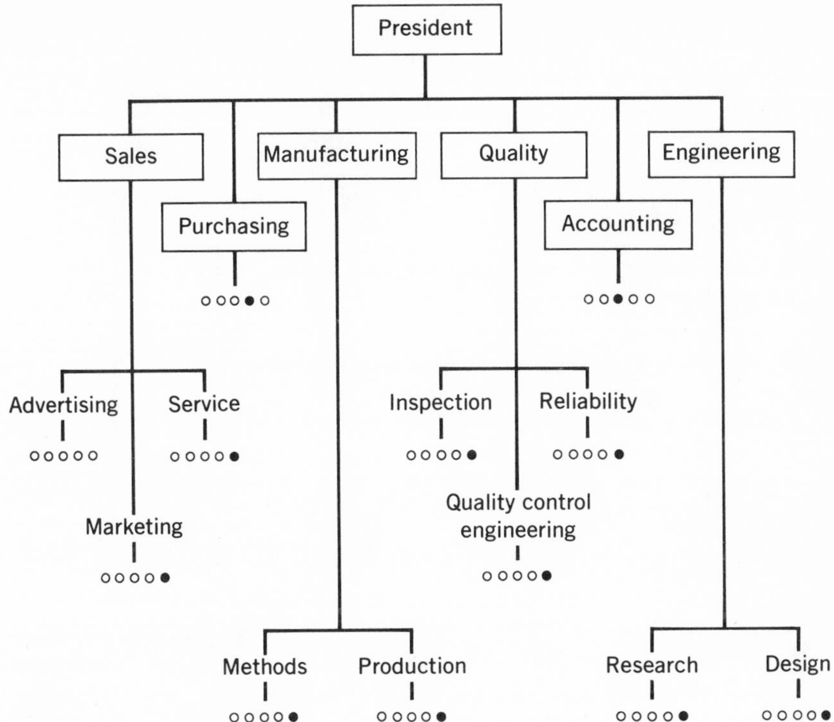

Figure 4-1 Facts as to the cost of quality are widely scattered.

ability features; luxury features; ease of installation, use, or maintenance; still other differences.

As a rule, higher quality of design means higher costs. Quite often it also means higher values (see Figure 4-3). However, human ingenuity often finds ways to make designs both better *and* cheaper.[1] Designs are simplified to use fewer parts, to use less expensive materials, to require fewer operations.[2]

[1] More recently the term "value analysis" has been coined to describe an organized approach to reduction of costs of materials and components. See W. L. Gage, *Value Analysis,* McGraw-Hill Publishing Company, Ltd., London, 1967.

[2] For a summary of some examples, see J. M. Juran (ed.), *Quality Control Handbook* (*QCH*), 2d ed., McGraw-Hill Book Company, New York, 1962, p. 1-12.

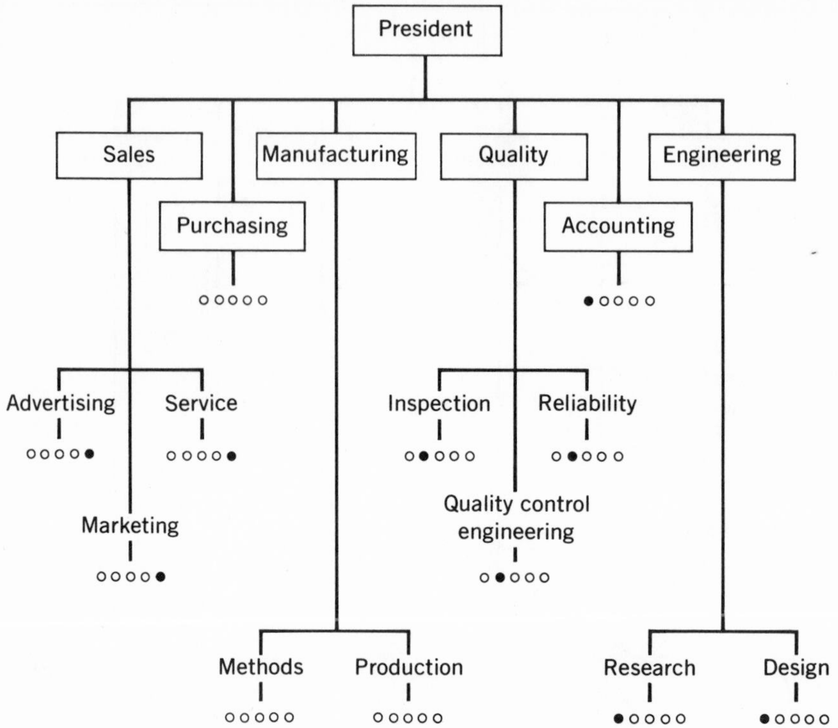

Figure 4-2 Facts as to the value of quality are widely scattered.

For consumer goods, the decision as to quality of design is usually governed by the choice of the market which the company is trying to reach, e.g., luxury, middle class, economy. For capital goods, the decision is usually governed by such considerations as intended life, environmental operating conditions, importance of continuity of service, etc. Either way, it is useful to discover through "market research" the consuming habits of the users, the prices they are willing to pay, the availability of maintenance facilities, and other pertinent factors.

The final responsibility for the market decision rests with the chief executive. To make this decision in the best way requires, on his part, a clear understanding of several implications:

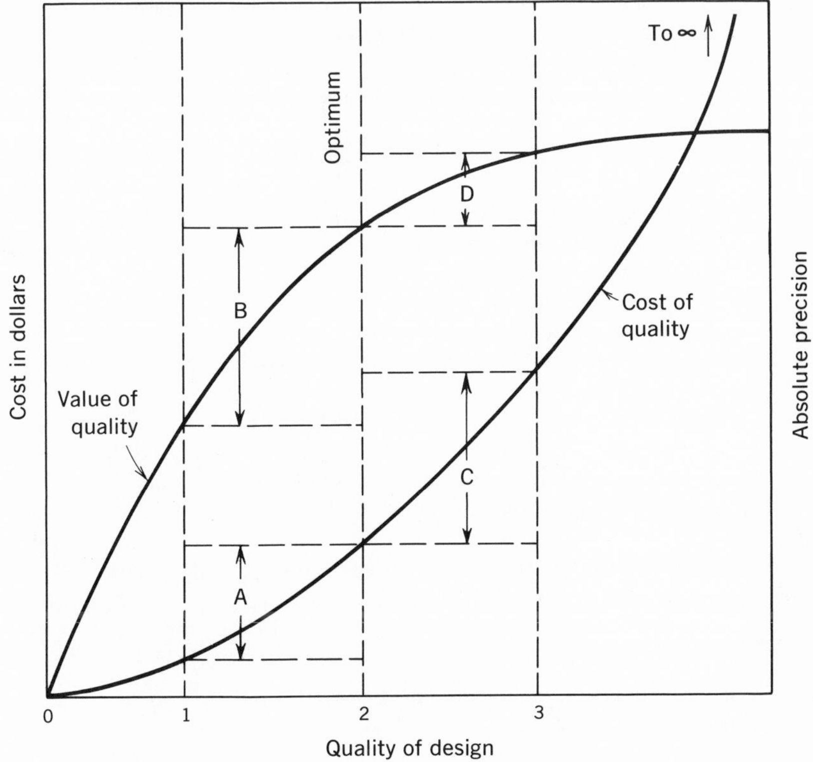

Figure 4-3 Economics of quality of design.

1 That while higher quality of design usually costs more, higher quality of conformance usually costs less
2 That choice of quality of design requires participation by all major departments of the company
3 That this participation should not be left to chance but should specifically be organized by the chief executive

4-4 Competition in Quality

Competition among sellers assumes many aspects—price, service, and still other aspects, *including quality.* Quality of product is thus a weapon of competition.

Quality as a factor in competition can take many forms but has not been so well exploited in industry as has, for example, price or service. The opportunities to exploit quality in competition include:

1 Clear knowledge of the user's service conditions, and of the costs he incurs due to service interruptions as well as costs due to use of the products.
2 Knowledge of "market quality" and use of this knowledge in redesign and in pricing.
3 Design of the product for high customer appeal through function, appearance, life, etc.
4 Development of a positive quality reputation through "invariable" delivery of conforming product. This is an asset of the highest value, as any salesman can testify.
5 Guarantee of quality of product in a way to minimize any losses to the *customer* because of defectives.
6 Advertising the foregoing performance through propaganda, information, etc.
7 Avoidance of any notorious failure which can deal a serious or fatal blow to quality reputation. Companies which have become involved in actual or alleged quality blunders or frauds have been years living down the incidents.

Accomplishment of these objectives involves participation by virtually every department of the company. Quality is a teamwork job.

4-5 Product Design as a Means for Competition in Quality

The product designer is faced with two problems: (1) those in which the consumer is usually "uniformed," i.e., the general public, and (2) those in which the consumer is usually "informed," i.e., other manufacturers.

Product design for the general public places great emphasis on features which can be sensed by an unaided human being. Thus we find such designs emphasizing appearance (through color, shape, luster, etc.), texture, taste, smell, balance, and all the other features to which human senses can respond. The name, the package, the slogans are all part of the marketing process.

Where quality of competing goods differs sharply, such difference is by itself sufficient to drive the inferior goods out of the market. An extensive study[3] of the relationship of consumer quality preferences to "share of market" disclosed that when consumers exhibit a preference,

[3] J. M. Juran, "A Note on Economics of Quality," *Industrial Quality Control (IQC)*, pp. 20–23, February, 1959. The study involved consumer goods. Data were available on competing products showing, for each product, (1) the percent of consumer preference on blindfold tests against competing brands, and (2) the share of market based on actual sales versus competing brands.

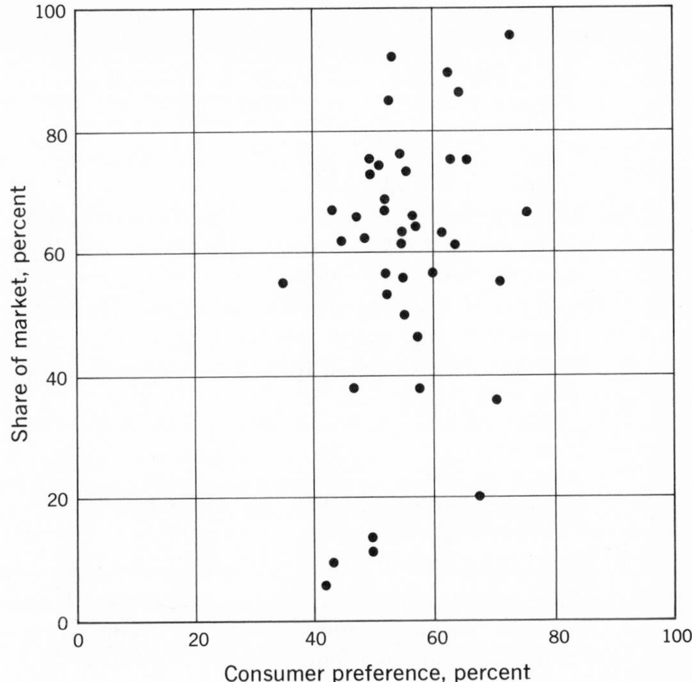

Figure 4-4 Relationship of percent of consumer preference to percent of market sales.

on "blindfold" testing, of 3 to 1 or better, the less favored product loses its market for quality reasons alone. As is seen in Figure 4-4, there were no instances of products remaining in the market in which consumer preference against competing brands was less than 25 to 75.

The same study disclosed that when consumer preference is only slight,[4] such as 55 to 45, any correlation between consumer preference and share of market is obscured by other factors which enter into the overall marketing endeavor, i.e., attractive packaging, sales promotion, prior "franchise," pricing practices, etc.

It follows from this study that companies are well advised, in the case of consumer goods, to put forth substantial effort to secure such quality improvements as will bring their product into the range of acceptability, i.e., above 40 percent or so on customer preference tests. However, once the product has been moved into this range, the effect of quality differences is overshadowed by other marketing factors.

[4] Generally, in this study, the range of statistical error was only several percent, so that such preferences, though slight, were statistically significant.

In the case of products for "informed" consumers, the emphasis is on functional use. The informed buyer normally has access to sound technical advice, and becomes rather insistent about matters such as power consumption, failure rate, interchangeability, fireproofing, and numerous other utility features. Industrial companies have learned to take these things seriously and to direct their design efforts in response to these known needs.

For both kinds of users industrial companies also do much creative design of new features for products. Testing these features "in the market" is comparatively simple when "informed" buyers are involved, as arrangements can be made for tryouts, with feedback of good data. In contrast, a field test on uninformed buyers is difficult to organize, costly to conduct, and frustrating to interpret.

4-6 Knowledge of Market Quality as a Means for Competition in Quality

In a competitive economy, users have a choice as to which products they will buy. In making their choice they compare prices, delivery dates, and other factors, including quality. In this way, there exists a concept of "market quality" similar to the concept of market price.

Industrial companies use a variety of sources of intelligence to discover what is market quality, and to judge their own product relative to the market. These sources include:

1 Use of the field sales force as a field intelligence force. Salesmen report the problems they encounter in selling when a new product or feature is offered by a competitor. They may emphasize this by sending back a sample.

2 Annual, planned comparisons of company product with competitor product. A team, consisting of men from Marketing, Design, and Quality Control, arranges to buy (or study without buying) an adequate sample of competitor products. Comparisons[5] are then made for the principal characteristics, e.g., appearance of the carpets, capacities of the batteries,[6] life of the lamps. These comparisons suggest the action needed, e.g., design in a longer life; ease off the specification to take a cost reduction, etc.

3 Special studies conducted to compare performance of products *under conditions of use*. These are of endless variety. Samples of abrasive cloth from competing manufacturers are compared as to the number of pieces polished per dollar of cloth. Competing brands

[5] See section 27-8.

[6] See *QCH*, 1962, p. 1-17.

of dog food are offered to dogs to measure which brand is preferred. Competing sewing machines are run side by side to see which is the least noisy. Competing razor blades are supplied to users who rate them for various qualities. Competing building materials are tested for resistance to fire, water, wear, etc. Again, the results make clear the extent of quality differences, and become a guide to decision making.

4 Users' data. Increasingly, industrial users are keeping good data on the costs and values associated with the products and services they buy from competing suppliers.[5] To such suppliers these data can be a gold mine of information, since they show the costs and values as seen by the user, who is the decisive force in determining which supplier will thrive in the marketplace. Some astonishing results have been achieved when suppliers have secured such data: redesign of products to reduce users' costs,[7] maintenance, service interruptions; premium pricing for products of superior reliability; new advertising programs to emphasize the qualities which the user regards as vital; elimination of costly frills not needed by the users.

5 Other sources. These include:

a Laboratories of large merchant companies which test competing products before making their purchase commitments. These laboratories commonly limit themselves to test of quality of design and to value analysis.

b "Research" organizations which make comparative tests and publish the findings for the benefit of their clients. These organizations include magazine publishers, consumer cooperatives, and labor unions. In addition to their test results, they publish "ratings" intended to reflect the relative values of the products compared. Usually these laboratories operate on close budgets and are limited to test of quality of design for the main performance features of the product, on a tiny sample consisting of one or a very few units of product.

c Government laboratories. As quality of products increasingly affects human safety, health, and welfare, government agencies have increasingly concerned themselves with quality. Mainly this is in response to legislation affecting safety and health, but there is also a trend for government to become involved in values: conformance

[7] The steam engine provides a striking example of quality improvement due partly to comparative rating of competing products. The early engines were used mainly to pump water out of mines. The mining companies developed a rating or "duty" of engines in terms of millions of foot-pounds of work done per bushel of coal burned. Data on engine performance were exchanged by these companies, and the data became decisive in buying engines. Newcomen's engine (1718) had a "duty" of 4.3 million foot-pounds per bushel of coal, which is 0.5 percent thermal efficiency. By 1816, the Woolf compound engine reached 7.5 percent thermal efficiency, and by 1834 the Cornish engine had reached 17.0 percent. (Charles Singer et al., *A History of Technology*, Oxford University Press, New York, vol. IV, pp. 161–197.)

to label claims, truth in advertising, meaning of guarantees. The findings of these laboratories are a useful input to industrial decision making.

4-7 Costs of Quality

Virtually every company department spends money on some aspect of achieving fitness for use. The activities and costs involved consist mainly of:

1 The market research costs of discovering what the quality needs of users are, and the users' likely responses to new qualities
2 The research and development costs of creating product concepts and proving their technical feasibility
3 The design costs of translating the product concepts into specifications and information adequate to permit manufacture, marketing, and service of the product
4 The manufacturing planning costs needed to provide manufacturing processes and tools able to meet the quality specifications
5 The costs of maintaining the precision of these machines and processes
6 The costs of manning and operating the process controls
7 The costs of marketing the quality aspects of the product: promotion literature, demonstrations, training, etc.
8 The costs of "appraisal," i.e., inspection, gaging, testing, and other forms of product measurement, plus the cost of judging conformance
9 The costs of defect prevention
10 The losses resulting from quality "failures"
11 The work of keeping all hands (including top management) informed on how well the quality function is being carried out

While it is feasible to make these definitions more and more precise, the job of quantifying the "quality costs" can be baffling. A market research study concerns itself with pricing, consumer incomes, territorial growth, etc., as well as with quality. The manufacturing planner is concerned with materials flow, plant safety, power supply, cost reduction, etc., as well as with precision and quality. Many other quality costs are similarly so closely interwoven with "nonquality" costs that separating them is a matter of collective judgment by a team rather than a matter of accounting from the books. However, the appraisal and failure costs (the latter is usually the big one) can be quantified with reasonable precision and this is discussed in Chapter 5.

Of the foregoing 11 classes of costs, (1) through (3) are concerned with determining, creating, and defining fitness for use. The rest are mainly incurred to achieve conformance to specification. (Conformance is usually defined in terms of the fidelity with which the product meets the specification. The real need is, of course, fitness for use. Since most people associated with the materials, components, operations, etc., do not understand how their work relates to fitness for use of the final product, they are asked to conform to specifications which presumably reflect the needs of fitness for use.)

Generally, costs (1) through (7) *must* be incurred if there is to be a product at all and if there is to be any income for the company. Costs (8) through (11) are, in varying degrees, avoidable.

4-8 Economics of Quality of Conformance

Figure 4-5 shows how, over most of the range of conformance, the higher the quality of conformance, the lower the total cost. The optimum is reached when "perfectionism" or "finickiness" begins to set in.

The decisive aspect in economics of quality of conformance is the level of defects which prevails in the various stages of manufacture and in the shipped product. Not only shop economics but also customer relations are strongly, if not critically, affected by this level of defects. It is, accordingly, most useful to know what is the optimum level of defects. On opposite sides of this optimum, opportunities for improved economics exist, either through defect prevention or through relaxing of control effort.

In most situations it is not known with precision what is the optimum. However, in many of these situations it is assumed that past performance, since it represents the accumulated efforts of sincere and able men, must be the optimum level of performance. Accordingly, it is common to find, in industry, that long-standing past performance is regarded as a level which must be endured, since no way has been found for changing it. This long-standing performance will be referred to as the "historic level."

With rare exceptions, the historic levels of defects are found to be on the left of the optimum of Figure 4-5. In consequence, the main opportunity for improved economics of quality of conformance lies, in most companies, in defect prevention. This is developed in Chapter 5.

4-9 Economics of Quality as Viewed by Customers

The user's concern with the "product" is strongly influenced by the needs and economists of the user. To oversimplify, see Table 4-1 on page 49.

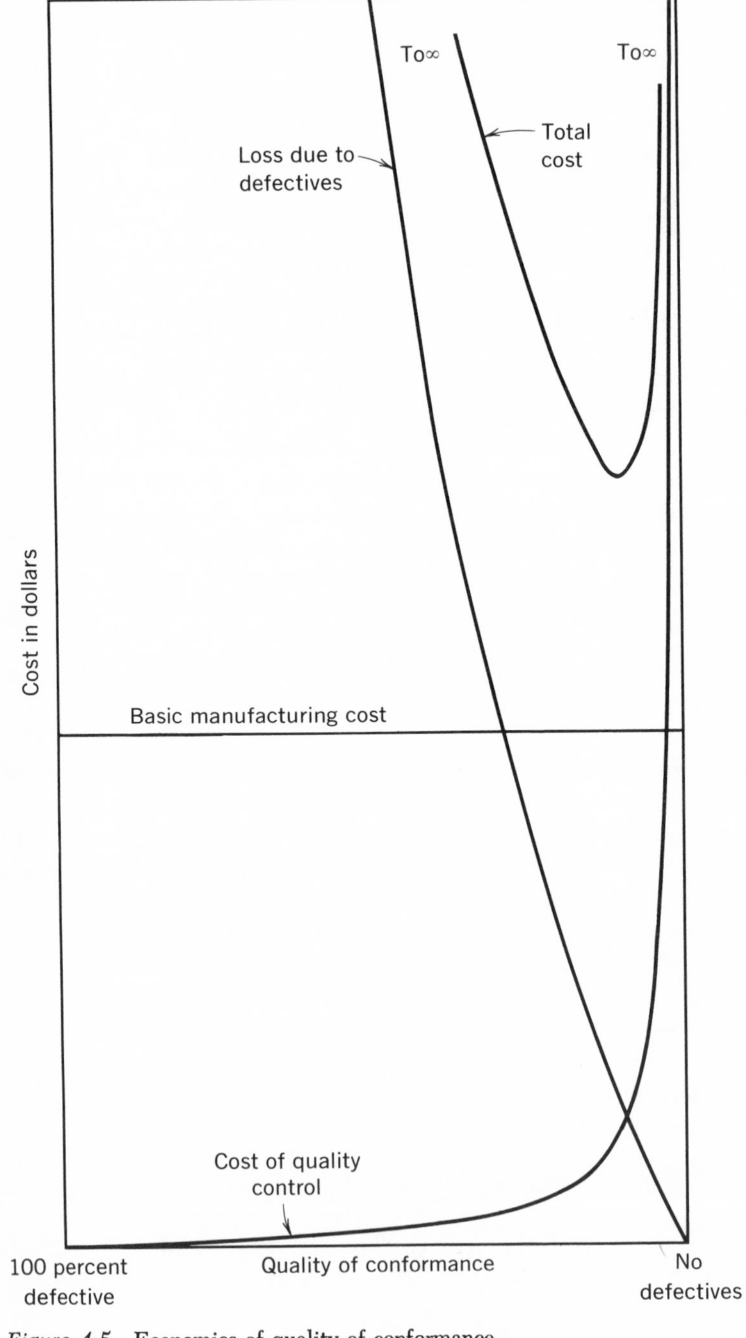

Figure 4-5 Economics of quality of conformance.

Table 4-1 VIEWPOINT OF THE USER

The User Is Not *Interested in*	*The User* Is *Interested in*
The manufacturer's hardware	The service he can get from the hardware. If he could get the service some other way, he might forgo the hardware.
Conformance to specification	Fitness for use. His interest in conformance is only as a means to the end of fitness for use.
Optimizing the manufacturer's costs	Optimizing the user's costs. These include not only the purchase price; they include the costs of maintenance, of loss of service, and still others.

During the latter half of the twentieth century, the gap between these two columns has been widening remarkably. This is in part due to the complexity of products, but mainly due to the proliferation of long-lived products. In earlier centuries, most products were short-lived (food, fuel) or low maintenance (wooden furniture, leather jackets). Either way, the user's total cost over the life of the product was not significantly different from his acquisition cost.

Users have been attempting to reconcile these views by resort to new contract forms (see Products and Services, section 1-6). A more recent development is a formal approach called "life cycle costing." Under this approach, the "life cycle cost" is the sum of (1) the "acquisition cost" (essentially the original purchase price), plus (2) the "logistics cost" (the costs of putting and keeping the product in service).

Because of the historical emphasis on acquisition cost, an analysis of life cycle costs, especially when comparing competitive sources of supply (Manufacturers X, Y, Z) can be illuminating. An example[8] illustrating the analysis as evolved by the military services is seen in Table 4-2.

Note that the life cycle cost (which in this case does not include cost of loss of service) can easily be several times the acquisition cost.

In life cycle costing the term "life" refers to the user's viewpoint, i.e., how long he will retain the product. This is very different from the "warranty life" around which the manufacturer structures so many controls and decisions. To the user, the costs after the warranty period are more important than ever—now he must foot the entire bill. So

[8] *Business Week,* May 13, 1967.

Table 4-2 COMPARISON OF LIFE CYCLE COSTS (THOUSANDS)

	X	*Y*	*Z*
Bid price	$ 42	$ 60	$ 47
5-year maintenance cost	129	116	84
New inventory items	10	20	10
5-year inventory management	45	30	42
New documentation	12	18	12
Operational training	8	8	8
Life cycle cost	$246	$252	$203

the user watches these costs (formally or informally) over the entire life of the product. Sometimes a supplier helps him out! For example, certain suppliers of motor fuels now offer their customers a maintenance data collection and analysis service (at a fee). The customer continually records all maintenance costs (not only fuel costs) on forms provided by the *fuel supplier* who summarizes and analyzes the data on a computer and sends periodic reports to the customer. Note that the service is provided by the fuel supplier and not the manufacturer of the vehicles.

Obviously, the user's future purchase decisions are strongly guided by the total costs as seen by him. The manufacturer is therefore well advised to acquire similar information so as to see his product as the user sees it.

The oportunities for manufacturers to use the concept of optimizing the users' economics are most intriguing because the effect is normally to achieve a large increase in income by a small increase in cost. Two examples[9] may be cited:

Example A company making power tools used by the lumber and construction industries studied the field failure rates of these tools as compared with those of competing tools. The study included an evaluation of the cost of field failures, as seen by the users of the tools. The study found that the company's tools held up significantly better than those of competitors and that downtime and stoppages of such tools were costing the customer far more than had been realized. This deeper understanding of the economics as seen by the users resulted in price increases yielding $500,000 in added annual profit. The cost of the research study was $7,500.

[9] J. M. Juran, "Quality and Profit," *IQC*, vol. 24, no. 1, p. 48, July, 1967.

Example A company selling expendable supplies (abrasive cloth) was able to secure from one of its customers (who also used competitors' cloth) its data on the comparative costs of these competing expendable supplies, i.e., its costs per 100 pieces polished. The information was used to justify a research program to improve the useful life of the supplies. The resulting reduction in users' costs brought an improved share of market to the improved product.

Quality control practitioners have long concentrated on reducing scrap, rework, and other internal costs. A consideration of external costs (i.e., customer costs) may have an equal or even larger potential for increasing company income.

A model of this approach is seen in Figure 4-6.

In the experience of the authors, manufacturers are often operating to the left of the "users' optimum." Quality control practitioners can

Figure 4-6 Economics of life cycle quality.

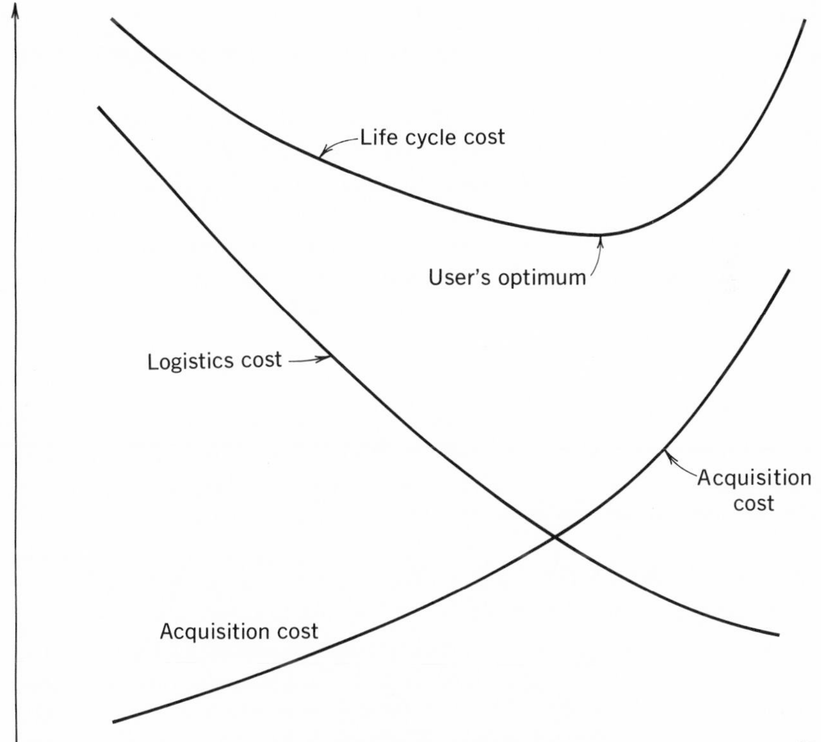

be particularly helpful in putting together the data needed to find and realize these opportunities to improve company income.

The cost of loss of service, once relatively unimportant, has now human beings generally) increasingly plan their activities under an as- risen to dramatic size. The reason is that industrial companies (and sumption of uninterrupted service from today's hardware. When the service fails, the loss of revenue and the wear and tear on the people affected can be shocking. For example:

> British Overseas Airways has estimated that the cost of 1 hour of delay of a jet airliner is in the range of $650 to $900 (aside from the effect on the delayed passengers).[10]
>
> On one large power system, a steam turbine shutdown costs $20,000 a day.[11]

As users dig into the facts and figures, they are often surprised to discover the extent of costs involved in delays, and the multiplying effect of some kinds of failures. (The classic example is the horseshoe nail which lost a kingdom.)

> ***Example*** A reliability engineer[12] carefully recorded all the hours spent taking a new car back to a dealer for the correction of over 100 defects. At the engineer's equivalent hourly pay rate, the defects represented an inconvenience cost of $985.

One effect of the rising costs of loss of service is to drive users to redundancy, i.e., extra facilities to provide continuity of service in the event of failure. For example, one oil refinery provides about 7 percent extra capacity to compensate for downtime on major equipment. This extra capacity involves a $13 million investment and $5,000 per day operating cost.[13]

PROBLEMS

1 Visit a large user of "hardware" to secure information on the economics of quality as viewed by the user. Estimate, as well as you

[10] *Journal of the Royal Aeronautical Society*, p. 397, March, 1966.

[11] T. A. Daly and P. H. Ockerman, "Commercial Reliability Progams," *Proceedings of the Eleventh National Symposium on Reliability and Quality Control*, Miami Beach, January, 1965, Institute of Electrical and Electronics Engineers, Inc.

[12] Private communication to one of the authors.

[13] Private communication to one of the authors.

can, (*a*) the original cost, (*b*) the cost of downtime, (*c*) maintenance labor cost, (*d*) cost of spare parts, (*e*) cost of carrying inventories, and (*f*) other costs.

From these data, draw your conclusions and make your recommendations. Some suggested large users:

Companies (U-drive, U-haul, etc.) which sell the temporary use of automobiles, trailers, trucks, power tools, etc.

Laundromats

Companies using office machines which they own (calculators, typewriters, dictating machines, etc.)

Companies using office machines which they lease (computers, copying machines, etc.)

2 Secure copies of publications of a consumer testing organization which tests competitive products, publishes its results, and makes a judgment on the relative "value" of the various products. Analyze the approach used by the organization to discover the economics of quality as seen by the user. Report your conclusions and recommendations.

3 Prepare a plan for securing data on market quality for a company in any one of the following businesses: (*a*) television sets, (*b*) ball-point pens, (*c*) men's shirts, (*d*) electric typewriters, (*e*) diesel engines, (*f*) office elevators, and (*g*) earth moving equipment.

5 Measurement and Analysis of Quality Costs

5-1 Why Measure Quality Costs?

To manage a large company requires that we set goals, not only for the company, but also for the departments. However, unbalanced pursuit of these departmental goals can actually damage company performance; e.g., pursuit of perfectionism can raise the company's costs to prohibitive levels. To temper this overemphasis on departmental goals, companies conduct interdepartmental studies aimed at improving company effectiveness with respect to various functions, products, markets, etc. One form of such interdepartmental study is that of quality costs.

Generally, the aim is to:

1 Identify all the activities conducted and deeds performed to achieve fitness for use, no matter where in the company these things are done
2 Determine the costs of all these activities and deeds
3 Interpret this information and make it available to all concerned
4 Discover opportunities for optimizing the company's quality costs
5 Provide a running scoreboard of the trends in quality costs

5-2 Defining the Activities and Deeds

Because no two companies are precisely alike, the list of these activities and deeds must be tailor-made for each company. However, there are enough similarities among companies to permit a generalization of the categories of the quality costs.

A quality cost committee[1] of the American Society for Quality Control has recommended that quality costs be defined in four categories:

1 *Prevention:* the costs associated with personnel engaged in designing, implementing, and maintaining the quality system. Maintaining the quality system includes auditing the system.
2 *Appraisal:* the costs associated with the measuring, evaluating, or auditing of products, components, and purchased materials to assure conformance with quality standards and performance requirements.
3 *Internal failures:* the costs associated with defective products, components, and materials that fail to meet quality requirements and result in manufacturing losses.
4 *External failures:* the costs which are generated because of defective products being shipped to customers.

The committee recommends that systems be established to record costs according to the following detailed categories:

Prevention

1 Quality planning and process control planning
   a Quality planning—quality control engineering-type work
   b Process quality control—that portion of compensation and costs associated with implementing the quality plans and procedures
2 Design and development of quality measurement and control equipment
3 Quality planning by functions other than quality control
4 Quality training
5 Other prevention expense

Appraisal

1 Receiving or incoming test and inspection
2 Laboratory acceptance testing
3 Inspection and test
4 Checking labor
5 Setup for inspection and test
6 Inspection and test material
7 Quality audits
8 Outside endorsements or approvals
9 Maintenance and calibration of test and inspection equipment
10 Review of test and inspection data
11 Field testing

[1] *Quality Costs—What and How,* American Society for Quality Control, Quality Cost Technical Committee, Milwaukee, May 31, 1967.

12 Internal testing and release
13 Evaluation of field stock and spare parts

Internal Failures

1 Scrap
2 Rework and repair
3 Troubleshooting
4 Reinspect, retest
5 Scrap and rework—fault of vendor
6 Material review activity
7 Downgrading

External Failures

1 Complaints
2 Product or customer service
3 Returned material processing
4 Returned material repair
5 Warranty replacement
6 Engineering error
7 Factory or installation error

It should be noted that in the foregoing, the category of "Prevention," which also includes quality planning activities, emphasizes the work done by departments called "quality control." As was noted in section 4-7, quality planning is also done by departments called "market research," "product development," etc. A complete set of quality cost categories would properly include, as part of "quality costs," that part of the work of these departments which is done to help achieve fitness for use.

5-3 Determining the Costs

Classifications such as the foregoing are a major step toward securing quality costs. However, "securing the numbers" requires the added steps of (1) defining the activities and deeds in enough detail to make their meaning precise, and (2) working with the accountants and with the supervisors of the various departments to evaluate and estimate the costs associated with these activities and deeds.

An example is seen in Table 5-1, which shows, for one company, the major items of quality costs, the detailed definitions of those items, and the method pursued by the investigator in securing the figures.

It is seen in Table 5-1 that the investigator used a variety of methods to piece together the quality costs:

1 Creation of new reports where none existed, e.g., Scrap
2 Use of existing account totals wherever the accounting system collected these costs on a basis matching the quality cost definitions, e.g., Repairs—inside
3 Combinations of existing account totals in other cases, e.g., Repairs—outside
4 Estimates made jointly by the investigators and departmental supervisors, e.g., Incoming inspection; Engineering
5 New summaries of cost details available from existing accounting records, e.g., 100 percent inspection

The example is quite typical in showing that the quality costs are not secured simply by asking the accounting department for them. "We don't keep the books that way." The investigator must piece things together, using the books where he can, resorting to analysis, estimates, and even creation of new data where he must.

Computing the cost of field failures is even more difficult. Not only is the situation complicated by the difference between user's costs and manufacturer's costs; the effect on the manufacturer's income (through loss of sales, lower prices, etc.) may easily be more important than the entire internal quality cost compilation. To date, quality control specialists have not explored these matters thoroughly, and so there

Table 5-1 **Definitions for Quality Cost Statement**

Item	*Definition*	*Source of Data*
Cost of quality failures	*Losses due to nonconforming parts and assemblies*	*See specific items below*
Scrap	Materials, labor, and burden of nonusable parts	Estimated at $1,000 per week based on first 3 weeks of new scrap reports
Repairs—inside	Labor of making defective parts and assemblies good	Charges to account no. 701 for all departments
Repairs—outside	Material and labor of fixing defective items returned by customers (less the receipts from customers)	Charges to account nos. 6430 and 6451 for all departments, less charges to account no. 6438
Service	Administrative cost of service to customers who receive defective items	Charges to account nos. 6407 and 6485

Table 5-1 DEFINITIONS FOR QUALITY COST STATEMENT (*Continued*)

Item	*Definition*	*Source of Data*
Cost of appraisal	*Costs of evaluating quality, and of identifying and segregating nonconforming parts and assemblies*	*See specific items below*
Incoming inspection	Labor in sampling and sorting of purchased parts	Estimate of costs incurred for this activity based on quality control department charges, accounts nos. 101 and 107
100% inspection	"Direct" labor of 100% inspections and 100% tests to separate "bad" from "good" items	Charges from cost department productive labor records for all departments
Spot-check inspection	"Indirect" labor of roving and spot-check inspections to detect defective lots and prevent production of defects	Charges to account no. 107 for all departments
Outgoing quality rating	Labor and indirect costs of sampling, testing, and rating outgoing product	No costs shown since this activity has just begun
Cost of prevention	*Engineering, technical, and supervisory costs of preventing recurring defects*	*See specific items below*
Quality control engineering	Cost of investigation, analysis, and correction of causes of defects by quality control department	Estimate based on quality control department charges to account nos. 101 and 107
Engineering	Cost of investigation, analysis, and correction of causes of defects by engineering department	Estimate of 30% of costs incurred on engineering expense account no. 6600

is much opportunity for original, creative work to be done in computing the effect of field failures on the economics of quality.

5-4 Interpreting Quality Costs

Standing by itself, the total of quality costs may not excite any attention. The sum may be large, but so is the total payroll, the total cost of purchased materials, etc. When a manager is shown such a total, his instinctive question will be "Is that good or bad?" The need is to answer this question. (If the answer is "bad," there is a further need for proposals as to what to do about it.)

One approach toward interpretation is to look for what is "par"; i.e., how do our quality costs compare with the quality costs of other companies in the same business?

Efforts to discover the extent of quality costs prevailing among many companies in the same industry have usually met with frustration, for a variety of reasons: companies not willing to disclose cost data; differences in dialect in the companies; differences in accounting system; differences in organization form, etc. To date these obstacles have effectively resisted the conduct of really useful studies.

There have been some broader studies, but the resulting concept of "par" is very broad indeed. For example, one of the authors[2] found that annual avoidable quality costs, when divided by the number of production workers, usually ranged between \$500 and \$1,000. In other words, a plant employing 2,000 production workers would normally incur avoidable quality costs ranging between \$1 million and \$2 million per annum.

Another broad study has yielded the following ranges of "par" (Table 5-2).

Such wide ranges have little value to practicing managers, and so other forms of interpretation must be used.

A second major approach in interpretation is to compute what are the "avoidable" costs. This is done by asking "What present costs would disappear if all defects disappeared?" The resulting total, referred to as "the gold in the mine," is a key figure in deciding (1) whether quality of improvement is indicated, and (3) where are the best points of attack. improvement is a major company problem, (2) how big a program

These "avoidable" costs commonly include such elements as:

Internal failures: value of material scrapped due to being defective; cost of repairing defectives; cost of excess production capacity required by low yields; cost of delays and stoppages

[2] J. M. Juran, "Insure Success for Your Quality Control Program," *Factory Management and Maintenance,* vol. 108, no. 10, pp. 106–109, October, 1950.

Table 5-2 "MARKET" FIGURES ON QUALITY COSTS

Industry	*Percent of Sales*
Simple, low tolerance	0.5–2
Normal mechanical process	1–5
Precision industries	2–10
Complex electronic; space	5–25

Breakdown of Quality Costs

Cost Category	*Percent of Total*
Internal failures	25–40
External failures	20–40
Total failures	50–90
Appraisal	10–50
Prevention	0.5–5

External failures: charges to guarantee account; cost of handling customer complaints; loss due to discount on seconds; loss of income due to loss of customers

Appraisal costs: excess costs of appraisal arising from presence of defects

A good deal of accounting technique and especially of common sense must be applied in order to arrive at a figure of "total avoidable quality costs" which will stand up under attack. For example:

1 In normal shop dialect, the term "scrap" includes more than just products so defective that they must be thrown away. The term also includes such "unavoidable" scrap as: chips removed during lathe operations; "skeleton" scrap from press operations; gates, risers, and fins cut from castings.

2 Some planning and appraisal costs are "unavoidable." Even if there were no defects, we would need to design, construct, and maintain testing equipment. We would also need to conduct some minimal level of testing. But the samples would be small, and there would be no resampling, no 100 percent testing of all units, no lot rejections, no delays, etc.

Unless care is taken to keep such unavoidable costs out of the total "gold in the mine" figure, the total will be subject to attack. In such cases the entire purpose of the compilation may be lost in the side debate on some minor element of the total.

Sometimes the total of avoidable costs, when put together for the first time, can stimulate management thought and action. Such a study for a food processing company showed the following:

Annual Quality Loss

Cost of moisture loss	$216,132
Cost of reprocessing	359,685
Cost of dumped (scrapped) material	212,845
Cost of price reduction on downgraded product	46,548
Total loss	$835,210

In this company, the top management had not been aware of the size of this loss. Once aware of it, they were receptive to an improvement program.

In other cases, the result of showing the total to top management is a decision that nothing should be done. For example, in a company making confectionery, the avoidable quality costs were computed to total $44,500 per year. The company executives had numerous other problems which were higher on their priority list and they cleared the air by a decision that they would not undertake a program to reduce these costs.

It should not be assumed that "quality loss" studies are applicable only to manufacturing companies. The following data were put together for a service company—a railroad—to show the loss due to derailments (damage to equipment and tracks, damage to goods, and cost of clearing the derailments):

Year	*Number of Derailments*	*Total Cost*	*Average Cost*
19—	88	$97,604	$1,109
19—	126	86,248	684

A third approach to interpretation is to estimate how much of the quality costs can economically be recovered. In the authors' experience, a reasonable target is to cut in two the *avoidable* quality costs.

It is usually necessary to prepare such information in the language of top management,[3] which is "return on investment." When someone

[3] Two universal languages are spoken in the company. At the "bottom" the language is that of things and deeds: square feet of floor space, schedules of 400 tons per week, rejection rates of 3.6 percent. At the "top" the language is that of money: sales, profits, taxes, investment. The middle managers and the technical specialists *must be bilingual.* They must be able to talk to the "bottom" in the language of things, and to the "top" in the language of money.

proposes to cut in two the avoidable quality costs, he must also propose how he will do it. Actually, to reduce these costs he must spend money, usually in the form of:

1 Diagnosis (to discover the cause of the defects) including:
  a The time of the investigating engineers
  b The time of the participating line supervision
  c The effort of the shop people in segregating trial lots and experiments
  d Supplemental measurement and clerical work as needed
2 Remedy, including:
  a Improved measuring equipment
  b Tool and methods changes
  c Machine and process changes
  d Design changes
3 Holding the new level, including the patrol inspections needed to keep the remedies in force.

At first, only the diagnosis effort can be estimated, since the nature of the remedies depends on what the diagnosis shows. The authors' experience shows that generally the cost of such a program, including remedies, is about 10 to 40 percent of the gains. If return on investment is defined as savings divided by cost, then these figures represent phenomenal returns of from 250 to 1,000 percent.

For example, in a factory making piston rings, an annual scrap loss of $1,200,000 was cut to $550,000 over a 3-year period. To achieve this result required:
1 Nine man-years of engineering time
2 Four man-months of a consultant's time
3 A modest increase in instrumentation
4 Some tool changes
The cost of all was under $200,000, spent once to save $650,000 annually.[4]

5-5 Maldistribution of Quality Losses: The Pareto Principle

In a company making a complex mechanical product *one* component accounted for 10 percent of the *total* company warranty costs and its low quality resulted in a loss of 3 percent of the *total* company sales volume.

[4] Consulting experience of one of the authors.

In a machine tool company, 2 of the 15 departments accounted for over half the loss; 50 percent of the loss was in one intricate type of parts!

A small chemical plant calculated the loss due to scrap, rework, handling complaints, and extra testing. The results are shown in Table 5-3. Note that 4 of the 13 products account for 86 percent of the loss. (The profit from $433,000 in sales would be required to make up for the quality loss on these 4 products.)

In a company making aircraft propellers, 6¼ percent of the piece part numbers accounted for 78.2 percent of the quality losses.

In an automobile parts plant, a department found that the most important defect of 12 codes accounted for 64 percent of the losses. In another department, the first of the 10 defects accounted for 80 percent of the losses.

It is seen from these instances that there is some universal principle which underlies all these cases. The losses are *never uniformly distributed* over the quality characteristics. Rather, the losses are *always*

Table 5-3 ANNUAL QUALITY LOSS IN A SMALL CHEMICAL PLANT

Department A		*Department B*		*Department C*	
Product 1	$ 0	Product 6	$11,384	Product 9	$11,321
2	1,525	7	1,372	10	20,071
3	42	8	160	11	13,243
4	1,735			12	1,473
5	1,122			13	1,790
Totals	$4,424		$12,916		$47,898

maldistributed in such a way that a small percentage, "the vital few," of the quality characteristics always contributes a high percentage of the quality loss (Figure 5-1).[5]

This maldistribution is of great help to managers, for it makes possible an economical attack on the bulk of the quality losses. Once the the "projects." With only a few projects requiring solution, the diagnos- "vital few" defects have been identified, they and they alone become tic and remedial costs are relatively low. Yet since these few projects involve the bulk of the losses, the cost improvement potential remains

[5] Charts of this type were originally used by M. O. Lorenz to depict the maldistribution of wealth. One of the present authors erroneously identified the curves with Vilfredo Pareto. The erroneous name has stuck.

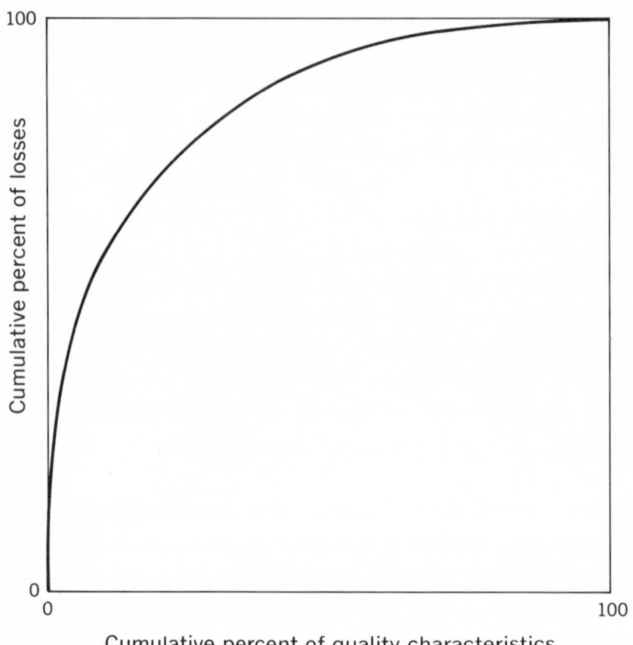

Figure 5-1 Pareto- (Lorenz-) type curves.

high. The "trivial many" defects are not tackled, since for them the cure costs more than the disease.

> ***Example*** In the piston ring company previously cited, the top defect accounted for about 25 percent of all quality losses. In 3 years it was cut from $315,000 to $55,000 annually. This single project, resulting in a reduction of $260,000 annually, accounted for 40 percent of the total reduction of $650,000.

As with any problem in quality costs, applying the Pareto principle requires piecing together information secured from the accounting records, from analysis of details, from estimates, from supervisors, and from creating new data. Moreover, since the dominant quality losses are traceable to specific symptoms (or loosely, diseases) of the product, it is quite important to "price out" the principal diseases of the product; i.e., what losses are encountered because of loose screws, burned out rotors, etc. Since the accountants do not normally "keep the books that way," it becomes necessary for the investigator to dig into the details, analyze, estimate,[6] create new data, etc. However, the resulting summary[7] of the estimated costs of the principal defects is an invaluable aid to managerial decision making.

5-6 The Scoreboard on Quality Costs

The main purpose of putting together a picture of quality costs is *to identify opportunities.* As these opportunities are studied, some quality problems are solved, and the associated losses are reduced. Meanwhile, new problems arise (due to new designs, new processes, new product uses). These new problems create new causes of quality losses. In consequence, the picture of quality costs keeps changing gradually, and needs to be updated periodically.

For the purpose of review of opportunities and for planning new programs of improvement, such updating, though essential, can be infrequent, i.e., annually, or quarterly at most. However, some managers prefer to see a more frequent (quarterly or monthly) report on the status of quality costs, in order to follow progress more closely, and thereby to accelerate action.

Actually, two scoreboards are involved:

1 The score on the projects; i.e., what deeds have been done to date to eliminate the causes of the main quality losses, what remains to be done, and whose move is it? From seeing such a scoreboard (usually provided by minutes of committee meetings, reports of specialists, etc.) the managers can judge whether adequate progress is being made or whether things need to be stirred up.

2 The score on the dollar losses; i.e., what is happening to the figures as a result of some problems being solved and new problems arising?

A good deal of work has been done to make quality cost reports a useful tool for managers. An example of such a report[8] is seen in Table 5-4.

Table 5-4 tries to combine the two purposes of these reports, i.e., identifying opportunities and disclosing trends. The table makes it

[6] To the accountant, the word "estimate" is a "dirty" word.

[7] For an example, see J. M. Juran (ed.), *Quality Control Handbook* (*QCH*), 2d ed., McGraw-Hill Book Company, New York, 1962, p. 1-44.

[8] Derived from training material used by a large manufacturer of electrical products.

easy to see in which categories the costs are concentrated (in the example, the external cost of complaints dominates all else). It also makes clear how small is the prevention effort in relation to the total costs. In Table 5-4, four bases are shown against which to compare the quality costs. These are helpful to managers in interpreting trends and in making comparisons among plants and departments.

Table 5-4 QUALITY COST ANALYSIS $000 BY QUARTERS

	Quarters			
Cost Categories	*1st*	*2d*	*3d*	*4th*
Quality control engineering	5	5	5	5
Tool maintenance	4	5	2	4
Gage control	1	1	1	1
Other	1	—	1	—
Total prevention	11	11	9	10
Inspection	66	72	51	66
Test	24	30	10	18
Test materials	12	22	12	27
Vendor inspection	8	9	6	8
Other	21	20	19	22
Total appraisal	131	153	98	141
Complaints	601	51	668	1,318
Rework	74	98	55	69
Spoilage	58	20	30	35
Other	24	30	22	26
Total failure	757	199	775	1,448
Grand total	899	363	882	1,599
Standard direct labor	226	296	124	138
Net sales billed	4,359	3,557	2,707	1,987
Cost of sales	2,341	2,068	1,646	1,174
Contributed value	1,545	3,171	1,592	1,171

Experience with such reports[9] has disclosed two usual sources of friction among company departments:

1 The "illusion of totality." Some specialists get into heated arguments about whether certain costs, e.g., tool maintenance, properly are a part of quality costs. There is logic on both sides, but the significant point is that it is unimportant whether tool maintenance is included or left out. The managerial decisions on identifying

[9] For another example, see *QCH,* 1962, p. 1-41.

opportunities are hardly affected. The trends of total costs are likewise not affected so long as there is consistency in including or excluding the debatable category.

2 The jurisdictional conflict. The accounting department becomes unhappy whenever anyone undertakes to prepare cost reports. The accountants argue that two sources for such reports inevitably come up with inconsistent figures, create duplication of effort, etc. (The accountants also strongly prefer to have a monopoly on plying their trade within the company.) These arguments are usually well founded. In practice it is well to set matters up so that the accountants prepare the basic data and issue the report. Aside from the accountants' reasons, there is the important fact that top management looks to the accountants for objective financial reports, and suspects all other financial reports as containing biases.

PROBLEMS

1 Identify the quality cost elements associated with any of the institutions listed below. Define each of the elements, and suggest how each might be quantified in practice: (*a*) a university; (*b*) a hospital; (*c*) the local bus line; (*d*) a bank; (*e*) a manufacturer of (any of these): wooden furniture, cotton cloth, men's shoes, glass bottles, castings, drugs; and (*f*) a textbook publisher.

2 Conduct a study of the applicability of the Pareto principle to any of the following: (*a*) causes of human mortality, (*b*) persons committing crimes, (*c*) population of countries, (*d*) population of American cities, and (*e*) sales of American companies.

3 You have been asked to prepare a survey of your machine shop to determine whether a quality control program can be economically justified. The machine shop is a supporting department which makes tools, fixtures, gages, dies, etc., for the manufacturing departments of your plant. The machine shop is necessarily engaged in small lot manufacture, but the variety of items is not large and the same items are produced several times during the year. In other words, it does a large volume of repeat business even though it makes a few pieces of item at any one time.

Your initial tour of the department discloses that all the output of the shop is routed through a group of six inspectors in a central inspection station (reporting to chief inspector who ranks with production superintendent). At the inspection operation, each item is inspected "100 percent" against the drawings. Any item not meeting specifications is tagged and returned to the machine shop foreman for his disposal—either to be repaired (and resubmitted for inspection) or to be scrapped. Defective items discovered before submission to inspection are also scrapped by the foreman.

In addition to the six tool inspectors, there are also two floor inspectors, reporting to the machine shop foreman, who circulate around the shop approving setups, answering questions on interpretation of drawings, and generally acting as technical liaison between the foreman and his machinists. They make no record of their activities.

The inspection records are perused weekly by the machine shop foreman. He notes any unusual rejections and lets individual machinists know if he thinks they are being careless. He estimates that he spends about 2 hours a week in this way.

You also visit the manufacturing departments which use the products of the machine shop. What do they think of the kind of quality the machine shop puts out? The assembly department has no complaints, but the punch press foreman states that several times he has had to scrap large lots of punching because the dies were incorrectly made. He suspects that these dies "slipped through" the machine shop inspectors.

Your next visit is to the cost department, where you obtain the following figures:

Machine Shop—Dept. 692—1969 Charges to Date, Oct. 1, 1969

Direct labor charges	$210,509
Rework labor for repairing rejected items	26,304
Raw materials consumed	69,305
Inventory value of finished goods, October 1	92,500
Burden account:	
Salaries of supervision	15,602
Salaries of floor inspectors	6,295
Heat, light, space, etc.	24,009
Supplies and equipment	49,602
Service charges from other departments	62,560
Depreciation of machinery	122,907
Analysis of scrap tickets:	
Number of tickets	42
Material value of scrapped items	5,607
Salvage value of scrapped items	444
Direct labor charged to scrapped items	12,848

Machine Shop Inspection—Dept. 511—1969 Charges to Date, Oct. 1, 1969

Salaries of tool inspectors (indirect labor)	$ 16,004
Salary of supervisor	5,820
Heat, light, space, etc.	2,260
Supplies and equipment	4,496
Service charges from other departments	2,620
All other burden charges	3,607

Punch Press Department—Dept. 408—1969 Charges to Date, Oct. 1, 1969

Analysis of scrap tickets:	
Number of scrap tickets charged to machine shop	4
Material value of scrapped punchings	4,965
Salvage value of scrapped punchings	575
Direct labor charged to scrapped punchings	1,428
The inspection records kept by the machine shop inspection for 1969 show:	
Total number of items inspected	7,256*
Total number of items rejected to shop	922

* This includes original inspection of new items as well as reinspection of resubmitted items, since no separate records are kept.

Prepare a quality cost summary for the machine shop. Show the buildup of failure cost, appraisal cost, and prevention cost. Put it in such a form that it will be self-explanatory to the machine shop foreman. State whether you believe a quality control program can be economically justified and why.[10]

[10] Problem adapted from course notes of Management of Quality Control Course, U.S. Air Force School of Logistics, 1959.

6 Basic Statistical Concept of Variation

6-1 Statistics as a Science

Statistics is the collection, organization, analysis, interpretation, and presentation of data. The application of statistics to engineering (and other fields) has resulted in the recognition that statistics is a complete field in itself. Unfortunately, some engineers have become so involved in statistics that they have unknowingly become "technique oriented" instead of remaining "problem oriented." The student is warned that statistics is just one of many *tools* necessary to solve quality problems. Statistical methods may or may not be required for the solution of a problem. The objective is to make a quality product and not to promote the statistical method as an end in itself.

6-2 The Concept of Variation

The concept of variation states that no two items will be perfectly identical even if extreme care was taken to make them identical in some respect. Variation is a fact of nature and a fact of industrial life. For example, even twin children vary slightly in height and weight at birth. The net contents of a can of tomato soup varies slightly from can to can. The overall weight of a particular model of automobile varies slightly from unit to unit of the same model. These are all "probabilistic" (as contrasted to "deterministic") situations in that one set of input conditions yielded a range of outputs. To disregard the existence of variation (or to rationalize falsely that it is small) can lead to incorrect decisions on major problems. Statistics helps to analyze data properly and draw conclusions, taking into account the existence of variation.

It may seem that an extreme amount of variation in data makes it impossible to apply statistical methods. However, it is in this situation that statistical methods offer the greatest potential for use.

6-3 Distinction between Variables and Attributes Data

Statistical data can be categorized as either variables data or attributes data. Variables (or continuous) data are data on a characteristic that is measurable and can assume any value over some interval. For example, an impact distance of 6,296 feet for a missile would be classified as one variable measurement.

Attributes (or discrete) data are data on a characteristic that can only assume certain distinct values (e.g., integer values). An example of attributes data would be a statement that a missile fell within a required target area. The only results possible are "in" or "out" of target area. The corresponding variables data would give the exact location of impact for the missile.

The statistical methods for analyzing variables data are different from those required for attributes data. In general, a great deal more information can be obtained when data are in variables form. However, statistical methods are available to analyze the data in either variables or attributes form.

METHODS OF SUMMARIZING DATA

Data summarization can take several forms: tabular, graphical, and numerical indices. Sometimes, one form will provide a useful and complete summarization. In other cases, two or even three forms are needed to assure complete clarity.

6-4 Tabular Method of Summarizing Data: The Frequency Distribution

A frequency distribution is a tabulation of data arranged according to size. The measurements of the electrical resistance of 100 coils are given in Table 6-1. This table presents the raw data from the test data sheet.

Table 6-1 **Resistance (Ohms) of 100 Coils**

3.37	3.34	3.38	3.32	3.33	3.28	3.34	3.31	3.33	3.34
3.29	3.36	3.30	3.31	3.33	3.34	3.34	3.36	3.39	3.34
3.35	3.36	3.30	3.32	3.33	3.35	3.35	3.34	3.32	3.38
3.32	3.37	3.34	3.38	3.36	3.37	3.36	3.31	3.33	3.30
3.35	3.33	3.38	3.37	3.44	3.31	3.36	3.32	3.29	3.35
3.38	3.39	3.34	3.32	3.30	3.39	3.36	3.40	3.32	3.33
3.29	4.41	3.27	3.36	3.41	3.37	3.36	3.37	3.33	3.36
3.31	3.33	3.35	3.34	3.35	3.34	3.31	3.36	3.37	3.35
3.40	3.35	3.37	3.35	3.35	3.36	3.38	3.35	3.31	3.34
3.35	3.36	3.39	3.31	3.31	3.30	3.35	3.33	3.35	3.31

Table 6-2 shows the frequency distribution of these data with all measurements tabulated at their actual values. For example, there were 14 coils each of which had a resistance of 3.35 ohms, there were 5 coils each of which had a resistance of 3.30 ohms, etc. The frequency distribution immediately tells where most of the data are grouped (the data are centered about a resistance of 3.35) and how much variation there

Table 6-2 **Tally of Resistance Values of 100 Coils**

Resistance ohms	Tabulation	Frequency	Cumulative Frequency
3.45			
3.44	|	1	1
3.43			
3.42			
3.41	||	2	3
3.40	||	2	5
3.39	||||	4	9
3.38	卌 |	6	15
3.37	卌 |||	8	23
3.36	卌 卌 |||	13	36
3.35	卌 卌 ||||	14	50
3.34	卌 卌 ||	12	62
3.33	卌 卌	10	72
3.32	卌 ||||	9	81
3.31	卌 ||||	9	90
3.30	卌	5	95
3.29	|||	3	98
3.28	|	1	99
3.27	|	1	100
3.26			
Total		100	

is in the data (resistance runs from 3.27 to 3.44 ohms). The centering and amount of variation in a large set of data would not be readily apparent from an original data sheet. The frequency distribution spotlights this information. Table 6-2 shows the conventional frequency distribution and the cumulative frequency distribution in which the frequency values are accumulated to show the number of coils with resistances equal to or less than a specific value. The particular problem determines whether the conventional or cumulative or both distributions are required.

When there is a large amount of highly variable data, the above frequency distribution can become too large to serve as a summary of the original data. The data may be grouped into cells to provide a better summary. Table 6-3 shows the frequency distribution for this

Table 6-3 **FREQUENCY TABLE OF RESISTANCE VALUES**

Resistance	*Frequency*
3.415–3.445	1
3.385–3.415	8
3.355–3.385	27
3.325–3.355	36
3.295–3.325	23
3.265–3.295	5
	100

data grouped into six cells. The width of each cell is 0.03 ohm. Grouping the data into cells condenses the original data and therefore some detail is lost. However, one can always refer back to the original data if necessary. The literature contains detailed instructions for constructing a frequency distribution. Although the exact frequency distribution which results is usually not of great importance in itself, there are several guidelines that are helpful (but not necessary) for later calculations and analysis:

1 Use between six and twenty cells.
2 Make the cell width a round number and keep it constant throughout the distribution.
3 Make the cell limits to one more decimal place than the original data and end in a "five" (see Table 6-3). This eliminates the problem

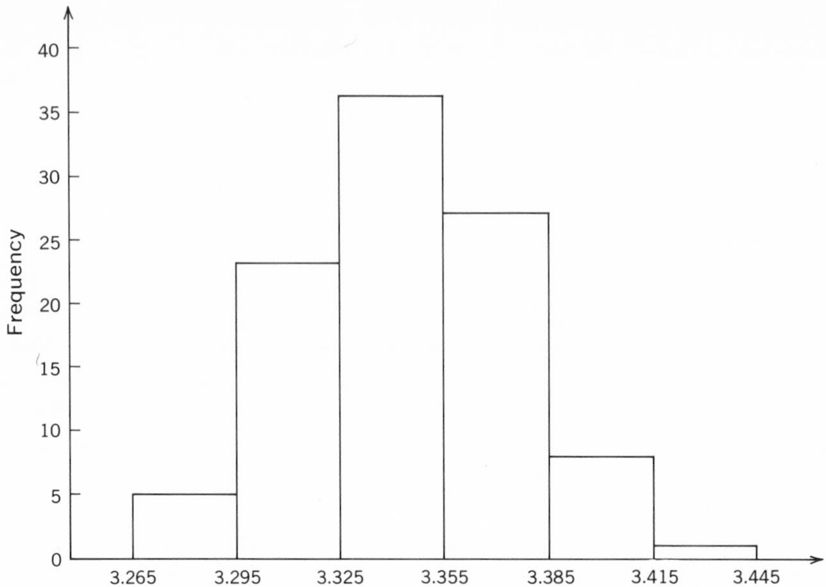

Figure 6-1 Histogram of resistance.

of tallying a value which falls on one of the cell limits. If the cell limits are to one more decimal place than the original data, the problem will never arise.

6-5 Graphical Summarization of Data: The Histogram

A histogram is a vertical bar chart of a frequency distribution. Figure 6-1 shows the histogram for the electrical resistance data. The simplicity of construction and interpretation of the histogram makes it an effective tool in the elementary analysis of data. Complete problems in quality control have been solved with this one elementary tool alone. Section 20-2 demonstrates how the shape of the histogram provides insights into a manufacturing process that would not otherwise be apparent. Note that as in the frequency distribution, the histogram highlights the center and amount of variation in the sample of data.

6-6 Quantitative Methods of Summarizing Data: Numerical Indices

Data can also be summarized by computing a measure of central tendency to indicate where most of the data are centered and a measure of variation to indicate the amount of variation in the data. Often these two measures provide an adequate summary.

The key measure of central tendency is the arithmetic mean or average.[1] The definition of the average is

$$\bar{X} = \frac{\Sigma X}{n}$$

where $\bar{X}$ = sample average
X = individual observations
n = number of observations

There are two measures of variation commonly calculated. When the amount of data is small (10 or fewer observations), the *range* is a useful measure of variation. The range is the difference between the maximum value and the minimum value in the data. As the range is based on only two values, it is not as useful when the number of observations is large.

In general, the *standard deviation* is the most useful measure of variation. Like the average, the definition of the standard deviation is a formula:

$$s = \sqrt{\frac{\Sigma(X - \bar{X})^2}{n - 1}}$$

where s = sample standard deviation

The significance of $(n - 1)$ in the denominator will be explained later. It helps to provide a standard deviation that is most useful for later analyses.

When students first study statistics, there is usually difficulty in understanding the "meaning" of the standard deviation. The only definition is a formula. There is no hidden meaning to the standard deviation, and it is best viewed as an arbitrary index which shows the amount of variation in a set of data. Later applications of the standard deviation to making predictions will help clarify its meaning.

With data in frequency distribution form, shortcut calculations can be employed to find the average and the standard deviation.[2]

A problem that sometimes arises in the summarization of data is that one or more extreme values are far from the rest of the data. A simple (but not necessarily correct) solution is available, i.e., drop such values. The reasoning is that a measurement error or some other

[1] In this book, the terms "mean" and "average" will have the same meaning and will be used interchangeably.
[2] See J. M. Juran (ed.), *Quality Control Handbook*, 2d ed., McGraw-Hill Book Company, New York, 1962, p. 13-11.

(unknown) factor makes the values "unrepresentative." Unfortunately, this may be rationalizing to eliminate an annoying problem of data analysis. The decision to keep or discard extreme values rests with the investigator. However, statistical tests are available to help make the decision.[3]

CONTINUOUS PROBABILITY DISTRIBUTIONS

In statistical analysis, a distinction is made between a sample and a population. A sample is a limited number of measurements taken from a large source. A population is a large source of measurements from which the sample is taken. Many problems are solved by taking a *sample* of measurements and then, based on the sample, making predictions about the defined population containing the sample. It usually is assumed that the sample is a random one; i.e., each possible sample of n measurements has an equal chance of being selected. (The prediction tool ties in with the concept of defect prevention by signaling potential quality problems before many defects occur.) For variables data, continuous probability distributions are used to make predictions.

A probability distribution function is a mathematical formula which relates the values of the characteristic with their probability of occurrence in the *population.* Figure 6-2 summarizes some continuous distributions and their functions. When the characteristic being measured is one which can take on any value (subject only to the fineness of the measuring process), its probability distribution is called a continuous probability distribution. For example, if the diameter of a ball bearing is measured, the probability distribution for this is an example of a continuous probability distribution because the diameter could be 1.69876 or 1.69234 or any value limited only by the fineness of the measuring instrument. Experience has shown that most continuous characteristics follow one of several common probability distributions, i.e., the "normal" distribution, the "exponential" distribution, and the "Weibull" distribution. These distributions find the probabilities associated with occurrences of the *actual values* of the characteristic. Other continuous distributions (such as the t, F, and chi-square) are used to analyze data but are generally not helpful in predicting the probability of occurrence of actual values of the characteristic (see Chapter 11).

[3] Mary Gibbons Natrella, *Experimental Statistics,* National Bureau of Standards Handbook 91, U.S. Department of Commerce, Aug. 1, 1963, pp. 17.1–17.6.

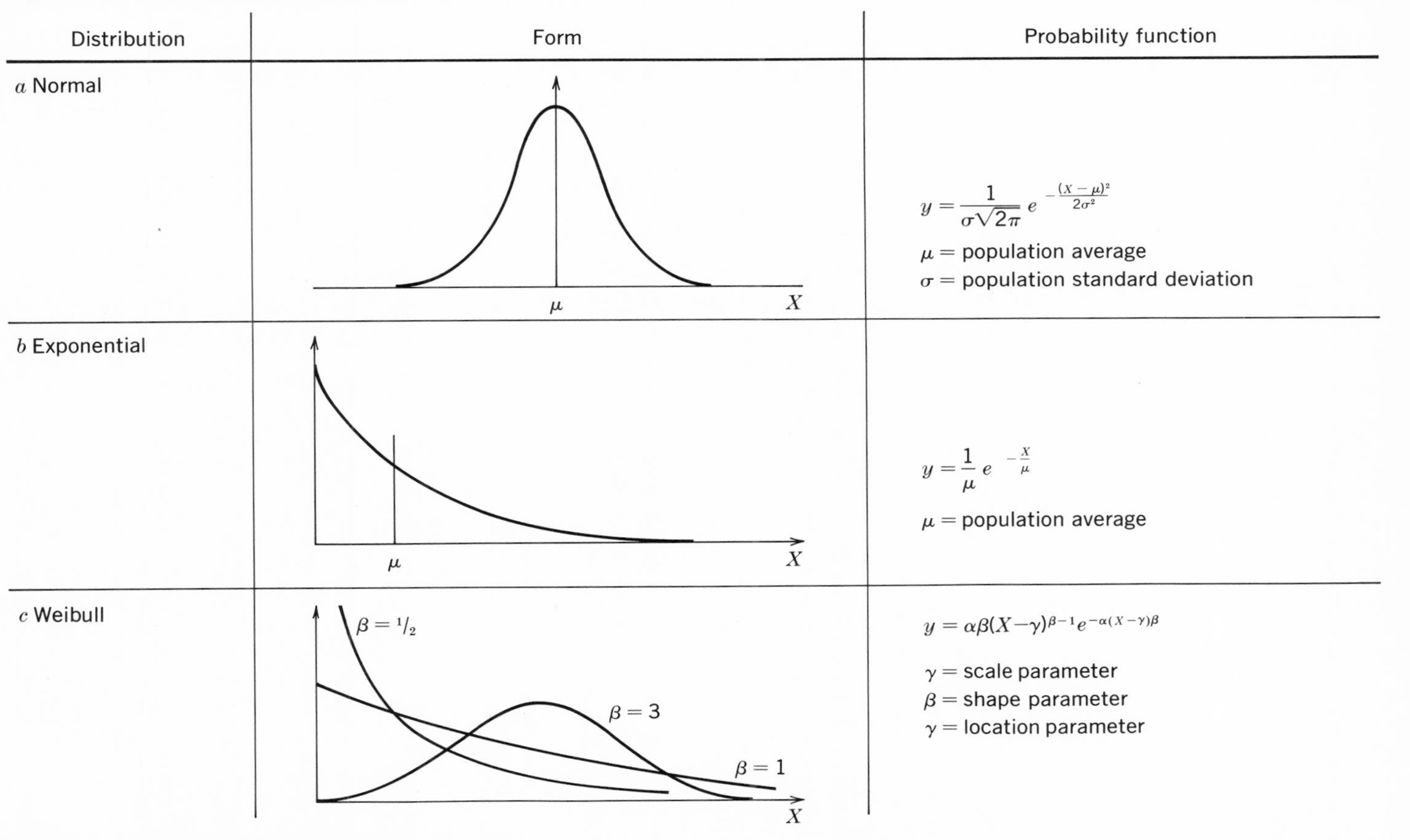

Distribution
Form
Probability function
a Normal
μ
X
$y = \frac{1}{\sigma\sqrt{2\pi}} e^{-\frac{(X-\mu)^2}{2\sigma^2}}$
μ = population average
σ = population standard deviation
b Exponential
μ
X
$y = \frac{1}{\mu} e^{-\frac{X}{\mu}}$
μ = population average
c Weibull
$\beta = 1/2$
$\beta = 3$
$\beta = 1$
X
$y = \alpha\beta(X-\gamma)^{\beta-1} e^{-\alpha(X-\gamma)^{\beta}}$
γ = scale parameter
β = shape parameter
γ = location parameter

Figure 6-2 Three continuous probability distributions.

6-7 The "Normal" Probability Distribution

Many engineering characteristics can be approximated by the normal distribution:

$$y = \frac{1}{\sigma\sqrt{2\pi}} e^{-(X-\mu)^2/2\sigma^2}$$

where $e = 2.718+$
μ = population average
σ = population standard deviation

Note that the formula requires estimates of only the average (μ) and standard deviation (σ) of the *population*[4] in order to make probability predictions about the population. The curve for the normal probability distribution can be related to a frequency distribution and its histogram. As the sample becomes larger and larger, and the width of each cell becomes smaller and smaller, the histogram approaches a smooth curve. If the entire population were measured, and if it were normally distributed, the result would be as shown in Figure 6-2*a*. Thus, the *shape* of a histogram of *sample* data provides some indication of the probability distribution that the population follows. If the histogram resembles[5] the shape shown in Figure 6-2*a*, then there is some basis for assuming that the population follows a normal probability distribution. ("Goodness of fit" tests[6] provide a quantitative means of evaluating any distribution assumption.)

Making Predictions Using the Normal Probability Distribution

To make predictions about a normally distributed *population*, estimates of the average and the standard deviation of the population are used in conjunction with a tabularized probability function. The estimate of the average of a normal population is simply the average of the sample ($\bar{X}$) taken from that population. Thus:

$$\text{Estimate of } \mu = \bar{X} = \frac{\Sigma X}{n}$$

[4] Unless otherwise indicated, Greek symbols will be used for population values and Roman symbols for sample values.

[5] It is *not* necessary that the sample histogram be perfectly normal. The assumption of normality is applied only to the population. Small deviations from normality are expected in random samples.

[6] B. W. Lindgren and G. W. McElrath, *Probability and Statistics,* The Macmillan Company, New York, 1966, pp. 156–157.

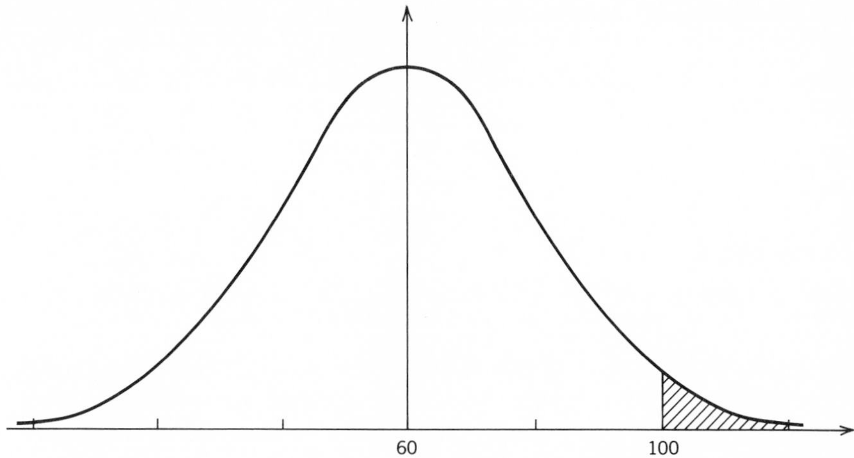

Figure 6-3 Distribution of light bulb life.

The estimate of the standard deviation of a population is obtained by incorporating a correction factor $(\sqrt{n/(n-1)})$ in the formula. Thus:

$$\text{Estimate of } \sigma = s = \sqrt{\frac{\Sigma(X - \overline{X})^2}{n}}\sqrt{\frac{n}{n-1}} = \sqrt{\frac{\Sigma(X - \overline{X})^2}{n-1}}$$

This is the same as the definition formula in section 6-6.

Example From past experience, a manufacturer concludes that the burnout time of a particular light bulb he manufactures is normally distributed. A sample of 50 bulbs has been tested and the average life found to be 60 days, with a standard deviation of 20 days. How many bulbs in the entire population of light bulbs can be expected to require replacement after 100 days of life?

Figure 6-3 shows the problem graphically. The middle of the curve is 60. The problem is to find the area under the curve beyond 100 days. The area under any probability distribution curve represents the probability of values within two stated limits actually occurring. Therefore, the area beyond 100 days is the probability that a bulb will last more than 100 days. To find any area under a normal curve, a fraction is calculated which expresses the difference between a particular value and the average of the curve in units of standard deviation:

$$K = \frac{X - \mu}{\sigma}$$

where K = distance of X from population average (μ) in units of standard deviation (σ)

In this particular problem, $K = (100 - 60) \div 20 = +2.0$. Table A in the Appendix gives the area under the curve between the average and the particular X value. The area under the curve between μ and $+2\sigma$ is 0.4773. Thus, the probability of a bulb lasting between 60 and 100 days is 0.4773. Because the normal curve is symmetrical about the average, the area above the average is 0.5000. Thus, the probability of a bulb lasting more than 100 days is 0.5000 — 0.4773 or 0.0227; that is, 2.27 percent of the bulbs in the population will last more than 100 days. If a characteristic is normally distributed and if estimates of the average and standard deviation of the population are obtained, then a simple calculation will provide an estimate of the probability that the characteristic will fall between any pair of stated values. For example, this method can estimate the total percent of production that will fall within engineering specification limits.

Figure 6-4 shows some representative areas under the normal distribution curve that can be derived from Table A in the Appendix. Thus, 68.26 percent of the *population* will fall between the average

Figure 6-4 Areas of a normal curve.

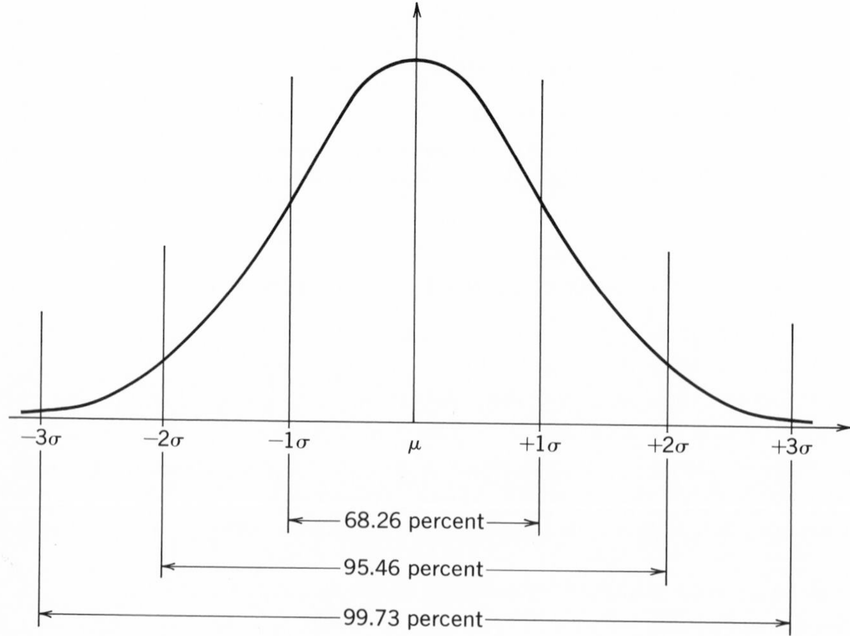

of the population plus or minus 1 standard deviation of the population. 95.46 percent of the poulation will fall between the average plus or minus 2 standard deviations. Finally, plus or minus 3 standard deviations will include 99.73 percent of the population. Remember that these predictions refer to the *population* and not to the sample. The percentage of a sample within a set of limits can be quite different from the percentage within the same limits in the population. This important fact is an underlying principle of testing hypotheses (see Chapter 11).

6-8 The Exponential Probability Distribution

The exponential probability function is

$$y = \frac{1}{\mu} e^{-X/\mu}$$

Figure 6-2*b* shows the shape of an exponential distribution curve. Note the distinctly different shapes of the exponential and the normal distributions. For example, an examination of the tables of areas will reveal that 50 percent of a normally distributed population will have values above the average value and 50 percent will be below. In an exponentially distributed population, 36.8 percent will be above the average and 63.2 percent below the average. This refutes the intuitive idea that the average is always the 50 percent point! This characteristic of a higher percentage below the average sometimes helps to indicate applications of the exponential. For example, the exponential describes the loading pattern for some structural members because smaller loads are more numerous than larger loads. The exponential has also proved useful in describing the distribution of failure times of complex equipments.

Making Predictions Using the Exponential Probability Distribution

Predictions based on an exponentially distributed population require only an estimate of the average of the population. For example, the time between successive failures of a complex piece of equipment is measured and the resulting histogram found to resemble the exponential probability curve. The results of a sample of measurements indicate that the average time between failures (commonly called MTBF—mean time between failures) is found to be 100 hours. What is the probability that the time between two successive failures of this equipment will be at least 5 hours? The problem is one of finding the area under the curve beyond 5 hours (Figure 6-5). Table B in the Appendix gives

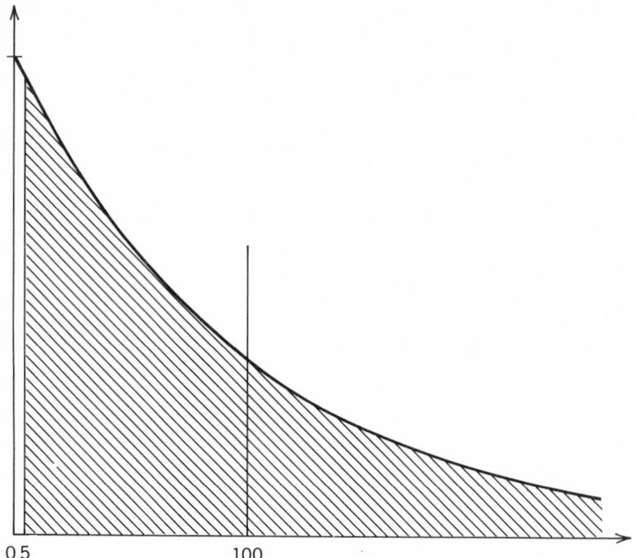

Figure 6-5 Distribution of time between failures.

the area under the curve beyond any particular value X that is substituted in the ratio X/μ.

$$\text{Thus, } \frac{X}{\mu} = \frac{5}{100} = 0.05$$

From Table B in the Appendix the area under the curve beyond 5 hours is 0.9512. That is, the probability that the time between two successive failures is greater than 5 hours is 0.9512; that is, there is a 95.12 percent chance that the equipment will operate without failure continuously for 5 or more hours. Similar calculations would give a probability of 0.9048 that the equipment would operate continuously for 10 or more hours. This probability (frequently called reliability) can also be calculated for the specified mission time of a product.

6-9 The Weibull Probability Distribution

The Weibull distribution is a family of distributions having the general function

$$y = \alpha\beta(X - \gamma)^{\beta-1}e^{-\alpha(X-\gamma)^\beta}$$

where α = scale parameter
β = shape parameter
γ = location parameter

The curve of the function, Figure 6-2*c*, varies greatly depending on the value of the constants. The location parameter is the smallest possible value of X. This is often assumed to be 0, thereby simplifying the above equation. The shape parameter reflects the pattern of the curve. Note that when the shape parameter is 1.0, the Weibull function reduces to the exponential. Also note that when β is about 3.5 (and $\alpha = 1$ and $\gamma = 0$), the Weibull closely approximates the normal distribution. The ability of the Weibull to cover several important distributions has made it increasingly popular in practice because it reduces the problem of examining a set of data and deciding which of the several common distributions (e.g., normal or exponential) would be most appropriate.

Making Predictions Using the Weibull Probability Distribution

For normal and exponential populations, probability predictions are usually done analytically as discussed in the previous paragraphs. However, the analytical approach for the Weibull distribution is cumbersome and therefore the predictions are usually made by the use of special Weibull probability paper (see Figure 6-6). As an example, suppose five heat-treated shafts were subjected to a certain stress test until each of them failed. The fatigue life (in terms of number of cycles to failure) is shown below:

10,000
12,100
16,900
18,000
23,200

The problem is to predict the percentage of the population of all shafts of this type which will fail before any given number of cycles. The solution is to plot the data on Weibull paper, to observe if the points fall approximately in a straight line, and if so, to draw in the straight line. The probability predictions can then be made by reading the graph.

In quality problems, Weibull paper is a plot of a log function of a usage characteristic X (e.g., cycles of life) versus the cumulative percent frequency of X. For example, based on the sample data, it would be estimated that 20 percent of the population would fail before a fatigue life of 10,000 cycles. However, it can be mathematically shown that 20 percent is not the best estimate of the cumulative percent corresponding to a fatigue life of 10,000 cycles if a small sample (20 or less) is involved. Instead of plotting the 10,000 cycles against a cumulative percent of 20 percent, the data point is plotted against the "median

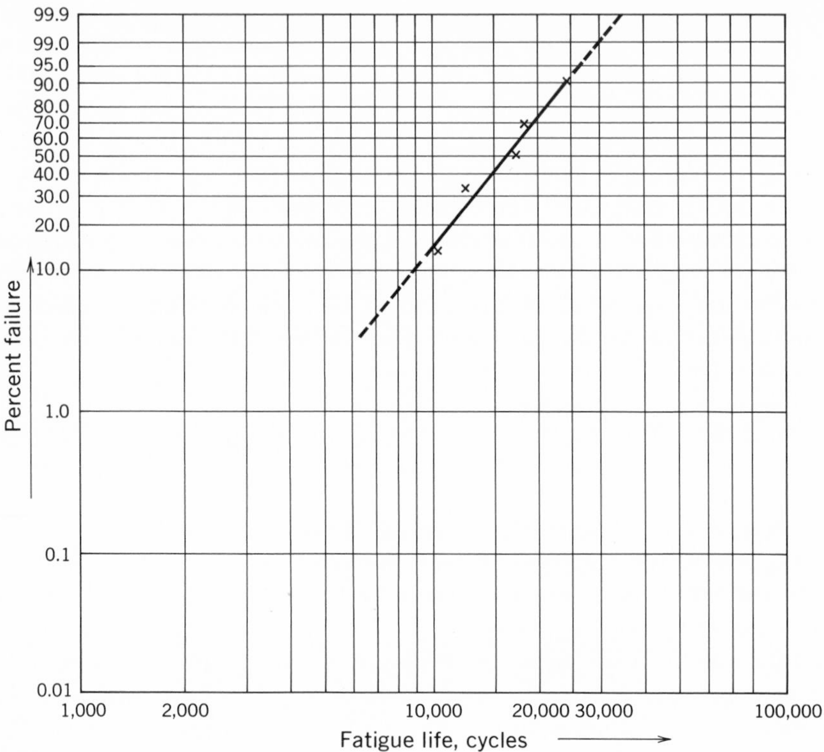

Figure 6-6 Distribution of fatigue life.

rank." Table C in the Appendix gives the values of the median rank for various sample sizes. The median ranks necessary for this particular example are based on a sample size of 5 failures and are as follows:

Failure Number	*Median Rank*
1	0.1294
2	0.3147
3	0.5000
4	0.6853
5	0.8706

The cycles to failure are now plotted on the Weibull graph paper against the corresponding values of the median rank (see Figure 6-6). These points fall approximately in a straight line so that it can be assumed that the Weibull distribution reasonably applies to this set of data. The line has been drawn in on the graph. The vertical axis gives the cumulative percent of failures in the population corresponding to the fatigue life shown on the horizontal axis. For example, about 50 percent of the population of shafts will fail in less than 17,000 cycles. About 90 percent of the population will fail in less than 23,000 cycles. By appropriate subtractions, the percent of failures between any two fatigue life limits can be predicted.

It is tempting to extrapolate on probability paper, particularly when life is the characteristic. For example, suppose the minimum fatigue life were specified as 8,000 cycles and the 5 measurements above were from tests conducted to evaluate the ability of the design to meet the specification. As all 5 tests exceeded the minimum specification, it would seem that the design is adequate and therefore should be released to production. However, extrapolation on the Weibull paper predicts that about 7 percent of the *population* of shafts would fail in less than 8,000 cycles. This suggests a review of the design before release to production. Thus, the small *sample* (all *within* specifications) gave a deceiving result.

Notice that a probability plot of life test data does *not* require that all tests be completed before the plotting starts. As each unit fails, the failure time can be plotted against the median rank. If the early points appear to be following a straight line, then it is tempting to draw in the line *before* all tests are finished. The line can then be extrapolated beyond the actual test data and life predictions made without accumulating a large amount of test time (see Chapter 27). The approach has been applied to predicting, *early in a warranty period,* which components of a complex product will be most troublesome. This provides a technique for predicting the "vital few" early enough to facilitate corrective action. However, extrapolation has dangers. It requires the judicious melding of statistical theory and engineering experience and judgment.

Probability graph paper is available for the normal, exponential, Weibull, and other probability distributions. The mathematical functions and tables will, of course, provide the same information as the graph paper. However, the graph paper can be useful in revealing *relationships* between probabilities and values of X that are not readily apparent from the calculations. For example, the reduction in percent defective in a population as a function of wider and wider tolerance limits can be easily portrayed by the graph paper.

PROBLEMS

Note: Many of the statistical problems in the book have intentionally been stated in industrial language. Thus, the specific statistical technique required will often *not* be specified. Hopefully, the student will then gain some experience in translating the industrial problem to a statistical formulation and then choosing the appropriate statistical technique.

1 The following data consist of 80 potency measurements of the drug streptomycin.

Potency Measurements

4.1	5.0	2.0	2.6	4.5	8.1	5.7	2.5
3.5	6.3	5.5	1.6	6.1	5.9	9.3	4.2
4.9	5.6	3.8	4.4	7.1	4.6	7.4	3.5
4.9	5.1	4.6	6.3	8.3	6.3	8.8	1.0
5.3	5.4	4.4	2.9	7.5	5.7	5.3	3.0
4.2	5.2	7.0	3.7	6.7	5.8	6.9	2.8
6.0	8.2	6.1	7.3	8.2	6.2	4.3	2.2
5.2	5.5	3.5	7.1	7.9	5.6	5.4	3.9
6.8	8.2	4.2	4.2	5.5	6.2	3.5	3.4
6.8	4.7	4.6	4.1	4.7	5.0	3.4	7.1

a Summarize the data in tabular form.
b Summarize the data in graphical form.

2 The data below are 50 measurements on the pitch rating of pulp.

Pitch Rating of Pulp

95	87	110	113	85
78	92	101	115	78
81	81	61	109	103
73	74	122	60	102
101	66	109	77	93
91	84	116	87	107
93	74	123	100	80
102	95	115	81	94
99	124	93	60	93
93	108	90	95	64

a Summarize the data in tabular form.
b Summarize the data in graphical form.

3 The following data were taken in testing the elongation of 24-gage wire before breaking.

Inch/Inch Percent Elongation

21	21	19	16	20
18	17	18	19	18
17	18	20	19	19
19	16	16	18	20
20	17	19	21	16
20	19	20	19	18
23	19	17	23	15
20	18	18	22	22
18	19	19	18	21
20	22	19	21	18
17	18	21	19	19

a Summarize the data in tabular form.
b Summarize the data in graphical form.

4 Compute a measure of central tendency and two measures of variation for the data in problem 1.

5 Compute a measure of central tendency and two measures of variation for the data in problem 2.

6 Compute a measure of central tendency and two measures of variation for the data in problem 3.

7 From past data, a manufacturer of a photographic film base knows that the tensile modulus of the film base follows the normal distribution. Data show an average modulus of 521,000 psi with a standard deviation of 13,000 psi.

a The lower specification limit is 495,000 psi. If there is no upper specification limit, what percent of the film base will meet the specification?
b If the film base needs a modulus of only 550,000 psi, what percent of the film base will exceed this requirement?
c What recommendations would you make?

8 A company has a filling machine for low pressure oxygen shells. Data collected over the past 2 months show an average weight after filling of 1.433 grams with a standard deviation of 0.033 gram. The specification for weight is 1.460 grams ± 0.085 gram.

a What percent will not meet the weight specification?
b Would you suggest a shift in the aim of the filling machine? Why or why not?

9 A company that makes fasteners has government specifications on a self-locking nut. The locking torque has both a maximum and a minimum specified. The offsetting machine used to make these nuts has been producing nuts with an average locking torque of 8.62 inch-pounds, and a variance σ^2 of 4.49 inch-pounds.

a If the upper specification is 13.0 inch-pounds and the lower specification is 2.25 inch-pounds, what percent of these nuts will meet the specification limits?

b Another machine in the offset department can turn out the nuts with an average of 8.91 inch-pounds and a standard deviation of 2.33 inch-pounds. In a lot of 1,000 nuts, how many would have too high a torque?

10 A power company defines service continuity as providing electric power within specified frequency and voltage limits to the customer's service entrance. Interruption of this service may be caused by equipment malfunctions or line outages due to planned maintenance or to unscheduled reasons. Records for the entire city indicate that there were 416 unscheduled interruptions in 1967 and 503 in 1966.

a Calculate the mean time between unscheduled interruptions assuming power is to be supplied continuously.

b What is the chance that power will be supplied to all users without interruption for at least 24 hours? For at least 48 hours? Assume an exponential distribution.

11 An analysis was made of repair time for an electrohydraulic servovalve used in fatigue test equipment. Discussions concluded that about 90 percent of all repairs could be made within 6 hours.

a Assuming an exponential distribution of repair time, calculate the average repair time.

b What is the probability that a repair would take between 3 and 6 hours?

12 Three designs of a certain shaft are to be compared. The information on the designs is summarized as:

	Design I	*Design II*	*Design III*
Material	Medium carbon alloy steel	Medium carbon unalloyed steel	Low carbon special analysis steel
Process	Fully machined before heat treatment, then furnace heated, oil quenched, and tempered	Fully machined before heat treatment, then induction scan heated, water quenched, and tempered	Fully machined before heat treatment, then furnace heated, water quenched, and tempered

	Design I	*Design II*	*Design III*
Equipment cost	Already available	$125,000	$550
Cost of finished shaft	$57	$53	$55

Fatigue tests were run on six shafts of each design with the following results (in units of thousands of cycles to failure):

I	*II*	*III*
180	210	900
240	360	1,400
100	575	1,500
50	330	340
220	130	850
110	575	600

a Rearrange the data in ascending order and make a Weibull plot for each design.

b For each design, estimate the number of cycles at which 10 percent of the population will fail. (This is called the B_{10} life.) Do the same for 50 percent of the population.

c Calculate the average life for each design based on the test results. Then estimate the percentage of the population that will fail within this average life. Note that it is not 50 percent.

d Comment on replacing the current design I with II or III.

13 The following results[7] are the failure ages (in cycles) of nine radiators of Model D design. The design was one of four designs evaluated by a development test program. The failure ages were:

Radiator Number	*Failure Age*
1	7,400
2	13,000
3	35,000
4	42,192
5	56,068
6	73,382
7	90,884
8	106,793
9	112,056

[7] Russel J. Wiesenberg, "Reliability and Life Testing of Automotive Radiators," *General Motors Engineering Journal*, vol. 9, no. 3, pp. 9–13, third quarter, 1962.

Plot the data on Weibull paper and draw in the "best" line by eye. Then graphically estimate:

a The percent failures in the population at 20,000 cycles or less.

b The percent failures in the population at 100,000 cycles or less.

c Median failure age (age at which 50 percent of the population will have failed).

d Suppose, after five failures had occurred, it was necessary to make rough estimates of items (*a*), (*b*), and (*c*) above. Draw in the "best" line (by eye) based on the first five failures and again estimate (*a*), (*b*), and (*c*) above.

Organization for Quality

7-1 The Nature of Organization

The process of organization, as applied to the accomplishment of work by human beings, involves:

1 Division (of the total work to be done) into logical subdivisions, called jobs
2 Definition of the responsibilities and authorities associated with each job
3 Definition of the relationship of each job to other jobs

Organization planners now have available to them some useful tools, chief of these being:

1 The organization chart, which shows the lines of authority and responsibility flowing from one job to another
2 The job description, listing the responsibilities and authorities of the job, and the relationship to other jobs

These two basic tools are extremely useful in defining within-department responsibilities. However, these basic tools are not adequate to define interdepartment responsibilities. These can be defined by interdepartmental tables of responsibilities (discussed later).

7-2 Top-management Responsibility for Organization

The design of an organization plan is always tailor-made for each company, since the objectives, product, processes, skills, traditions, and other factors differ from one company to another. But while the organization plan may well differ between companies, the responsibility of the executive is invariable. *He should see that a plan is designed and put into effect.* If he fails in this, the plan of organization will, by default, be left to his assistants (or their assistants, etc.) to work out as best they can. In some companies these lower levels do generate the informal leadership needed to accomplish what the top executive has failed to do. But in other companies, the informal leadership does not spring up, and the absence of a plan becomes a focal point for confusion and dispute.

Some middle managers avoid becoming involved with organization design on the theory that since final approval rests with top management, the work of preparing proposals likewise rests with top management. This latter view is erroneous. Middle management should aid top management by preparing organization proposals. Such preparation takes off the shoulders of top management much of the detail work of thinking out the alternatives, while at the same time retaining for top management the right to approve, modify, or reject the proposals.

7-3 Evolution of Organization for Quality

Organization forms are one of the means by which company objectives are achieved. Since these objectives keep changing, organization forms must also keep changing.

The manufacturing shops of antiquity were small, seldom numbering over a dozen people. The master presided over all production operations while at the same time carrying on the commercial activities. He instructed the operators, reviewed their work, and made the product acceptance decisions. There were no full-time inspectors.[1]

With the growth of the shops, the production foreman took over the acceptance function, since the owner or master no longer had the time. In due course, the amount of inspection work outgrew the available time of the foreman as well, and full-time inspection jobs were created. The inspectors appointed to these jobs were made responsible to the departmental production foreman, since they performed the work he had done while he had the time (Figure 7-1).

The events which caused inspectors to become numerous also brought other changes:

[1] The first full-time shop inspector (as distinguished from inspectors on construction projects) was likely the proofreader in the printing shop.

1 Inspection increasingly became a technical problem, requiring special technical skills. These special skills were often not possessed by the production supervisor.

2 In many instances, the force supervised by any one production supervisor consisted typically of 20 to 30 production operators plus one or two inspectors. Under such circumstances it was natural for the production supervisor to emphasize the work of the many and to give less attention to the work of the few, so that the selection and training of inspectors was slighted.

3 There was always the temptation, and too often the practice, for a production supervisor to solve a quality problem by ordering the inspector to accept the product, rather than by investigating the cause of the defects.

These and related conditions, plus repeated instances of quality difficulty, led to the movement, early in the twentieth century, to create a central inspection department. This movement was sharply accelerated by the quality experiences of companies during World War I. The result was the creation of an inspection hierarchy headed by a chief inspector, who was also the highest ranking company executive identified with the quality function. The chief inspector reported usually to the official in charge of manufacture. In multiplant or multidivision companies, the chief inspector for a plant usually reported to the works manager (Figure 7-2).

Figure 7-1 Usual organization form before 1914.

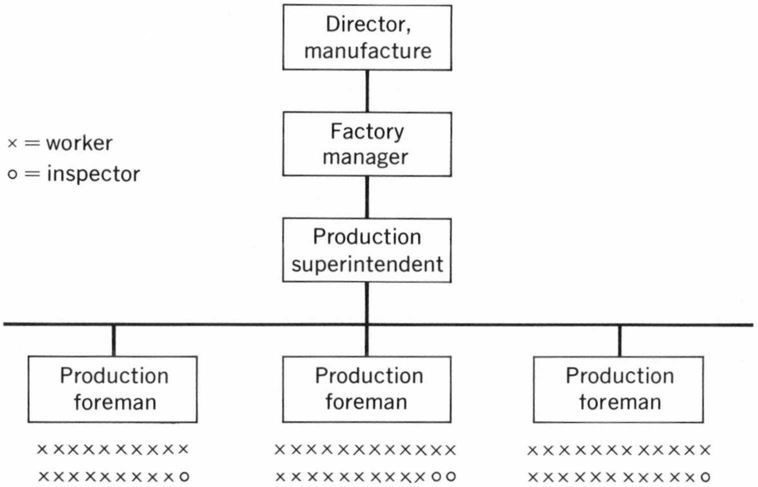

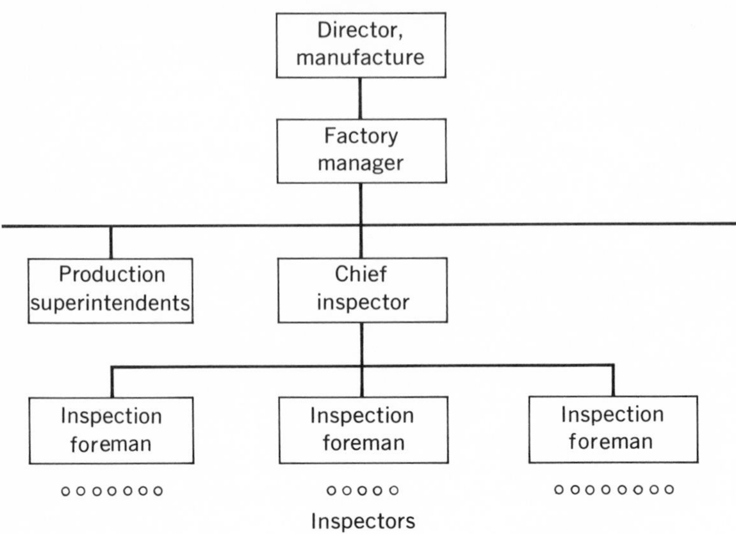

Figure 7-2 Organization form adopted between 1920 and 1940.

As the chief inspectors studied their new role, they took on responsibilities beyond those of simple product acceptance. They developed and refined the services needed by their inspectors: standards, training, record systems, accuracy of measuring instruments. However, a major need was *prevention,* i.e., the work of discovering what caused chronic defects and of providing remedies for these defects. This work required the use of staff specialists to make the necessary factual studies. Guidance of these specialists required imaginative leadership. In consequence, new staff departments were created, under names such as "quality control engineering."

To fit these new departments into the organization charts, a new post of quality manager was created. The new quality manager then presided over (1) the chief inspector (in a multiproduct company there might be several chief inspectors, one for each product line), (2) the common inspection service departments (test set design, measurement laboratories, vendor inspection, process inspection, salvage), and (3) the new quality control engineering department.

Figure 7-3 shows this organization form as adopted by most companies in the mechanical and electrical industries.

In the late 1950s the defense, aerospace, computer, and still other industries faced a new order of magnitude of requirements for products of high reliability[2] and freedom from maintenance. Achievement of

[2] The word "reliability" has a special technical meaning in addition to its dictionary meaning. See Chap. 9.

such new levels of performance depended absolutely on the caliber of the product development and design work. Companies concluded that the traditional product development and design departments could not meet these new levels of performance unless aided by added skills in failure analysis, reliability prediction, environmental testing, etc. To provide these added skills, a new category of specialist, the reliability engineer, was created.

Fitting the reliability engineers into the organization structure turned out to be complicated, and several schemes were tried out (see Chapter 9). The three most usual forms adopted were

1 Creating a new department within the design function
2 Enlarging the duties of the quality control engineering department to include the work of the reliability engineers
3 Creating a new major department of the company, "quality" or similar name, and giving the head of this department, sometimes a vice-president of the company, responsibility for broad coordination of the entire quality function as well as supervision of the departments of inspection, quality control engineering, and reliability engineering (Figure 7-4)

The usual assigned duties of such broad-based departments are shown in Figure 7-5. In some companies a "new" name "quality assurance"[3] has been coined to distinguish this broad list of duties from the narrower list assigned to predecessor departments.

[3] The word "assurance" in department titles has widely different meanings in different companies. One must look to the assigned duties of the department to understand just what is meant by the word "assurance" in that company.

Figure 7-3 Organization form evolved between 1945 and 1955.

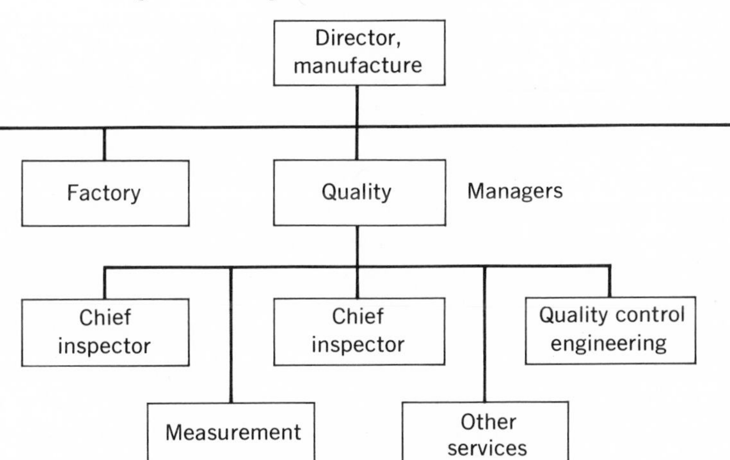

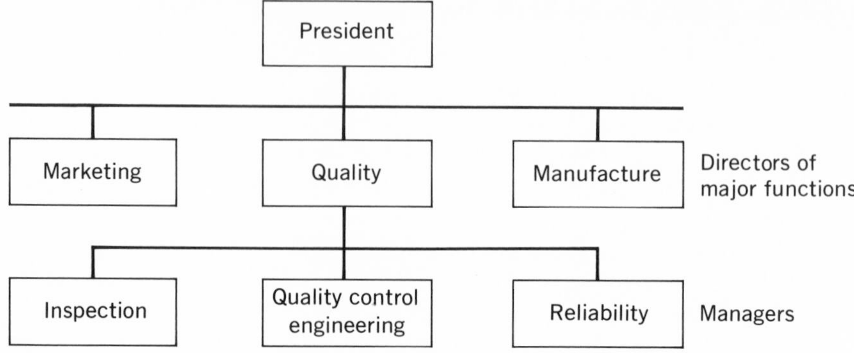

Figure 7-4 Organization trend from 1955 to 1965.

This organization form of "putting quality into the Cabinet" was still a minority form at the beginning of the 1970s. However, it had been widely adopted in the defense and aerospace divisions of the major companies. In addition, it had been put into effect in the major divisions of such companies as General Motors Corporation and General Electric Company, whose views on organization design command respect.

Figure 7-5 Functions under the quality assurance department.

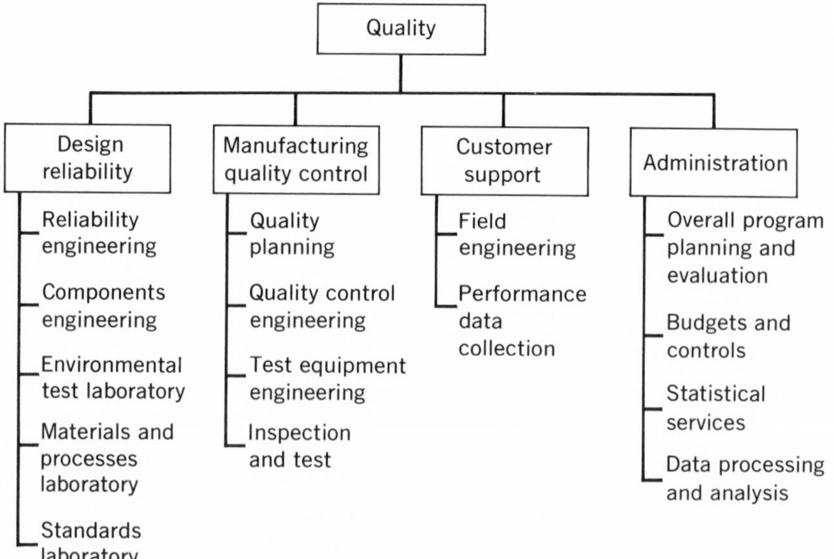

7-4 Organization Forms in the Process Industries

In the process industries, not only the terminology but also the organization form differ from those found in the mechanical industries. The center of product acceptance activity (as well as much process control) is in the "laboratory," variously chemical, metallurgical, physical, etc.

The laboratory generally makes product acceptance decisions across the board, i.e., on purchased materials, on goods in process, and on finished goods. Often the visual examination and sorting of products is done by inspectors responsible to production, but subject to an independent check by the laboratory.

The head of the laboratory is normally responsible to some technical manager, in one of several usual forms of organization (see Figure 7-6).

In the single-plant form, the quality control staff functions may report to the process development department (the usual form) or to an executive who also supervises the laboratory. In the multiplant company, each plant will have its own laboratory. However, at company headquarters there will sometimes be a quality control group which carries out various staff activities for all plants.

In the process industries, the function of setting standards of quality and issuing specifications is often the responsibility of the quality manager. In turn, he assigns it to a standards section, which may be part of the laboratory or may be a separate unit reporting directly to the quality manager.

Figure 7-6 Organization form: process industries, single plant.

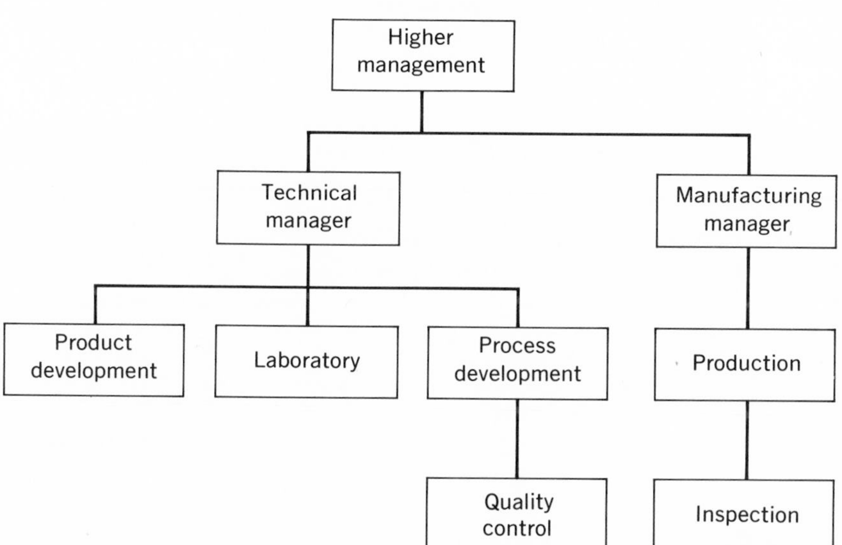

7-5 Organization Forms in the Larger Companies

Large companies are characterized by multiple locations, multiple divisions, international licensees and affiliates, wholly or partially owned subsidiaries, and other patterns. Much thought and experiment have gone into the design of organization structure needed to deal with the quality problems posed by these patterns. The most usual solutions[4] have been as follows:

1 The multiple-factory[5] company. The common organization form is to set up a quality manager at each factory. He reports to the plant manager and carries out the duties of inspection, quality control engineering, and related services. However, a quality control "staff" is set up at company headquarters to carry out the following duties:

a Develop quality policies and objectives.

b Prepare broad quality plans and publish these in the form of manuals, standards, etc.

c Provide the services of reliability engineering and other services closely related to product design and marketing.

d Provide consulting and training services to the factories.

e Audit the quality effectiveness of the factories.

f Publish executive reports on various aspects of quality performance.

g Represent the company in relations with the industry and the government on quality matters.

2 The multiple-division company. Here each "division" is really a subcompany, e.g., the Buick Motor Division of General Motors Corporation. It has the responsibility for meeting a budgeted profit. It is equipped with its own facilities for product design, manufacture, marketing, and still other functions. And it receives a high degree of autonomous authority for conducting operations. (All this applies to subsidiaries as well.)

To a wide extent, this concept of autonomy is applied to the quality function of these divisions. Each division is made self-sufficient as to inspection, inspection services, quality control engineering, reliability engineering, and other quality staff services.

In a minority of the very large companies, there is created a post of corporate quality manager. Where this post is created, it commonly has duties such as developing corporate quality policies and objectives; studying the quality effectiveness of the divisions; providing consulting services to the divisions on organization, manning,

[4] For further elaboration see J. M. Juran (ed.), *Quality Control Handbook* (*QCH*), 2d ed., McGraw-Hill Book Company, New York, 1962, sec. 6.

[5] The term "factory" is not to be confused with the term "division." The "factory" is usually a manufacturing facility, without responsibility for product design or for marketing.

and other aspects of management of the quality function; publishing executive reports serving on industry and government committees. Experience with such corporate quality posts has been mixed, and they have exhibited a high mortality rate. In part this is the result of the inherent difficulty of coordinating so many layers of staff, i.e., corporate, divisional, and plant. But mainly, such a corporate post is difficult to fill. The corporate quality manager, faced with dealing with powerful, autonomous divisional general managers, derives his authority not from his corporate status, but from his professional competence, his managerial skills, his welcome personality. It is a difficult set of criteria to meet.

3 The international company. Here again, there is a wide delegation of management authority to conduct these foreign operations in an autonomous manner. The unique cultural patterns of the various regions are added reasons for autonomy in management. However, there are some restraints in technology, especially if the subsidiaries sell in overlapping markets. In such cases it is common to impose a policy of uniform product specifications, and uniform controls to ensure adherence to specifications. To achieve this uniformity, issuance of specifications is from a central technical staff at company headquarters. In addition, a procedure is designed to audit the performance of the subsidiaries. This consists commonly of the following:

a The foreign subsidiary sends copies of its test reports to company headquarters.

b The foreign subsidiary also sends samples of its product to company headquarters for check testing.

c Company headquarters sends a traveling "auditor" (annually or so) to the foreign locations to review their plan of quality controls and their adherence to plan.

7-6 Organization for Acceptance

The term "acceptance," usually used to describe the activity of making decisions on whether products are acceptable for use, includes the work of inspection,[6] test, and related service activities.

The various stages of progression of the product, i.e., purchased materials, work in process, and finished goods, provide a natural group-

[6] Industrial dialects are not standardized on the meanings of words such as "inspection." "Inspection" usually refers to that part of product examination which is conducted by the human senses, unaided by instruments, e.g., by sight, feel, smell. "Gaging" almost always refers to measurement conducted with mechanical measuring tools. "Testing" usually refers to a verification conducted on the final product, with specially designed proofing equipment, to see if the product performs satisfactorily under conditions of simulated use.

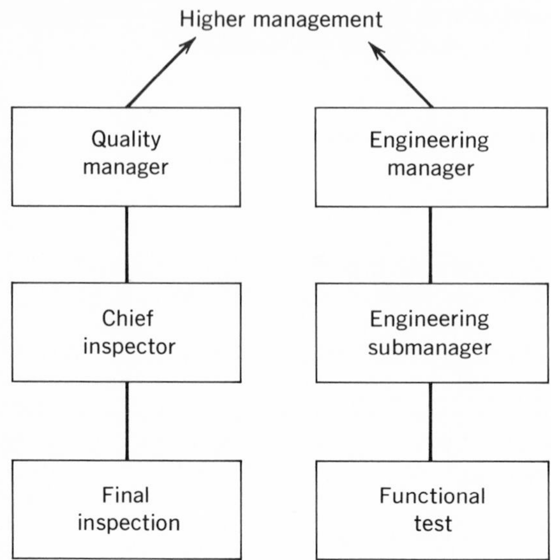

Figure 7-7 Separation of functional test from inspection.

ing of inspectors for organization purposes. In addition, there are supporting activities associated with the work of acceptance (e.g., measurement laboratories; custody and salvage of defective product) which are commonly grouped with inspection for the purpose of organization design. A simple organization form is seen in Figure 7-3.

In some companies there is a separation of inspection from test. The former reports to the chief inspector, but the latter reports to the chief engineer (Figure 7-7).

In the case of products critical to human safety or health (aircraft, safety equipment, drugs) the functional testing commonly reports to top management through some channel other than manufacture.

In some companies, the product as delivered by the process always contains so many defects that a 100 percent inspection or test must be performed. This 100 percent sorting is often conducted by a separate sorting department which is responsible to the production head. However, in such cases it is usual practice to subject the sorted product to a sampling for product acceptance purposes. This sampling is done by personnel who are part of an independent inspection or test department.

It is a common practice to place the work of maintaining the accuracy of gages and other measuring equipment under the jurisdiction of the chief inspector. By "maintaining the accuracy" is meant the system of checking each instrument on a scheduled basis to discover whether it is still within the prescribed limits of accuracy.

7-7 Responsibilities of the Chief Inspector

A typical list includes:

1 Prepare and submit departmental budget for higher approval
2 Approve product and process standards supplemental to engineering specifications
3 Approve manuals and procedures to be used in the department
4 Approve design of gages and test equipment
5 Define the jobs of the department personnel; select and train them; establish standards of performance; motivate and assist them to meet performance standards
6 Render decisions on conformance
7 Collaborate in disposition of nonconforming material
8 Keep informed on new industrial developments in inspection methods, instrumentation, data systems, and other matters affecting the department; make or recommend changes where appropriate
9 Develop the potentialities of subordinate managers
10 Collaborate with other departments in matters requiring a team approach

7-8 Organizing for Prevention

Defect prevention takes three major forms:

1 Planning new products and processes in a way which avoids defects in the first place (see Chapter 14).
2 Eliminating sporadic defects as they arise (restoring the status quo). Table 7-1 shows the activities needed, and the usual assignment of responsibilities.
3 Eliminating chronic defects (changing the status quo). Table 7-2 shows the activities and usual responsibilities. Chapter 30 is devoted to the details of how this is carried out.

Tables 7-1 and 7-2 emphasize the importance of the role of the steering arm and diagnostic arm (see section 3-5). The steering arm:

1 Helps achieve unity of purpose by obtaining agreement on what are the vital few quality problems
2 Provides theories on possible causes of the quality problems
3 Provides the authority to experiment, i.e., assures that those staff people

Table 7-1 **Responsibilities for Diagnosis and Remedy of Sporadic Defects**

Activity	*Line Departments Directly Concerned*	*Quality Control Staff*	*Inspection*
Establish expected levels of performance	X	XX	X
Establish measures of current performance	X	XX	X
Collect data on current performance*	XX		XX
Compare current performance with expected performance	XX	X	X
Take action on differences	XX		X

XX = prime responsibility
X = collateral responsibility
* Responsibility is usually on the production department for process control data, and on the inspection department for product acceptance data.

Table 7-2 **Responsibilities for Diagnosis and Remedy of Chronic Defects**

Activities	*Steering Arm*	*Quality Control Staff*	*Line Departments Directly Concerned*
Decision on priority of problems to be studied	XX		
Draft of plan for diagnosis to discover causes of defects		XX	X
Approval of plan	XX		
Diagnosis to discover causes (data collection and analysis)		XX	X
Diagnosis to provide remedy		X	XX
Agreement on remedy	XX		
Adoption of remedy		X	XX
Holding the gains		X	XX

XX = prime responsibility
X = collateral responsibility

making the investigations receive the complete cooperation of the line functions

4 Provides knowledge and advice about the existing "cultural pattern" and the sources of resistance to change that can be expected

5 Provides the authority to implement the proposed remedies

The diagnostic arm consists of the people (usually staff specialists) who have the time, experimental skills, and objectivity needed to discover the causes and remedy for the chronic quality problems. For problems within a department, the diagnosis can sometimes be done by departmental personnel. However, for problems involving several departments, the diagnostic arm must usually consist of "separate" trained specialists who have the time and objectivity to follow the trail of defects from symptom to cause to remedy (see Table 7-3). If it were otherwise, the line people would have solved the problem long ago.

Table 7-3 WHY A SEPARATE DIAGNOSTIC ARM?

Aspects of the Problems	*Maintaining the Status Quo (Control)*	*Changing the Status Quo (Breakthrough)*
Facts needed	Simple, showing actual performance against standard performance	Usually extensive and complex, to permit deeper understanding of the problem than ever before
Facts collected by	"Regular" scorekeepers, i.e., accountants, inspectors, etc.	Special fact-collecting team, i.e., task force, staff department, etc.
Formality of fact collection	Usually informal. Often not a permanent record	Usually formal. May require special experiments, tests, and formal reports
Analysis of facts by	Line people, i.e., branch manager, foreman, or even nonsupervisory personnel	Specially trained or technical people

7-9 Organization for Coordination

Quality activities are carried out in the many departments involved in the quality spiral (see Figure 1-1). Because company performance rather than departmental performance is to be optimized, there is need for coordination of these activities into a unified whole. This does *not* require that some person or function be given command over all quality activities.

A parallel is seen in the finance function. All departments spend money, and all influence income. Company managements do not leave it to each department to determine for itself how it is to influence income and outgo. Instead, there is a coordinated plan for balancing company income and outgo. This plan is commonly known as the budget. Through use of a common language (chart of accounts) and through coordination by a staff department (controller) the plan is pieced together, talked out, changed, and finally approved by top management. Thereafter the operating departments execute the plan. However, the staff department audits the execution to see how well it conforms to plan, and reports the results to all concerned (see Table 7-4). (Note that the controller does not have command authority over the operating departments.)

Table 7-4 UNIVERSAL APPROACH TO COORDINATION

	Operating Departments	*"Staff" Departments*	
		Planning	*Audit*
Design plan	X	X	
Execute	X		
Audit			X

Generally, industrial companies have not created, for the quality function, the equivalent of the coordination methods used by the finance function. Often, coordination is achieved through precedents established by past actions. Coordination is then predictable but precedents based on the past become obsolete. This leads to other methods of coordination.

7-10 Coordination through the Common Boss

In small companies this is the natural and effective method. The activities are so limited, and the employees so few, that the general manager, by direct contact with the people and the deeds, can know what is taking place and can make intelligent decisions. (Often there is no company budget.) To some extent there are departmental goals. However, the small scale of company operations makes it easy to identify the instances where pursuit of departmental goals will not optimize company performance (and the instances where blind pursuit of departmental goals will damage company performance).

As the company grows, the number of activities requiring coordination grows enormously, to a point that the general manager cannot find the time to coordinate them—or even part of them. In addition, he is now so removed from the action levels that he cannot even identify many of the needs for coordination. In consequence, it is necessary to create additional methods of identifying and solving those problems of coordination which can no longer be handled by the general manager.

7-11 Self-coordination

One of these methods is to leave it to individual supervisors to identify the needs for coordination as they arise, and to convene a meeting of whatever people are needed to solve the problem. In practice, this method is usually effective for dealing with current sporadic problems. For such problems the alarm signals are loud and insistent: the customer is telephoning about the delay in delivery; the process has stopped and the backed-up materials are choking the place. The pain becomes intense and the men feeling pain cannot rest until they get the problem solved.

In contrast, this method is usually ineffective in dealing with chronic problems. These problems can easily go on and on because there is no pain: *the nerves which signal pain have all been killed.*

For example, a process has a chronic level of 10 percent defects. In due course, the procedures are set up to order, always, 10 percent extra raw material to make up for the shrinkage. An extra 10 percent of machinery is provided to supply the extra machine capacity needed. The cost standards are inflated by 10 percent since the loss is "standard." The production manpower assigned is based on incurring "regular" 10 percent losses. Inspectors are provided to sort out the 10 percent defective. As a result, the company is organized to live comfortably, without pain, in the continued presence of 10 percent defectives. The alarm signals have all been defused.

7-12 Coordination through Written Procedure

Written procedure serves some vital purposes:

1 Before something is written out, it must first be thought through. Often the first time a procedure is written out is also the first time it has been thought through.
2 Before a written procedure is approved, there must be a meeting of the minds of the departments involved. Such a meeting of the

minds is better assured from review of a written procedure than from oral understandings.

3 A written procedure guards against drifts of memory, provides for easier training of new people, and serves as a source of reference.

It is seen that a written procedure becomes a perpetual coordination device. It gives impersonal supervision by offering the optimum solution of the repetitive company problem. It has predictability, and tends to engender a sense of law and order.

However, written procedure has its disadvantages:

1 It tends to become static, since many companies fail to provide adequate machinery for review and maintenance of procedure.

2 It provides for the usual situations. When unusual situations arise, they may well be handled in the usual way, with resulting blunders.

3 It can result in an attitude which stresses attention to the procedure rather than to the problems the procedure was designed to solve. In large organizations, whether private or public, this unimaginative devotion to procedure is known as "bureaucracy."

7-13 Coordination through Staff Specialists

Many coordination problems grow out of the interdepartmental nature of projects for change. For example, to get a new product into production requires product development by one department, design by a second, manufacturing planning by a third, etc. Similar collaboration is required on new processes, new standards, new management systems of all sorts.

Successful completion of such projects (fit for use, on time, economical) requires a coordinated plan (specifications, schedules, cost estimates) and a guided execution of the plan. It is common practice to assign "project engineers" to carry out several essential coordinating activities:

1 Draft the plan in the first place and secure the approval of the participating departments.

2 Follow closely the execution of the plan to identify obstacles.

3 Report on the progress of the project and urge action to remove obstacles (expedite).

Where the projects are very large, the practice is to create a post of project manager, to whom are assigned the project engineers needed.

Another form of staff specialist does the detailed work of diagnosing the cause of chronic defects or failures. Commonly these are interdepartmental in nature; i.e., the symptom shows up in one department, but the cause is elsewhere. The quality control engineer or reliability engineer is often cast in such a role, which is really that of the "diagnostic arm" (see section 7-8). This specialist makes his contribution by supplying the time, diagnostic skills, and objectivity not always possessed by the various departmental supervisors.

7-14 Coordination through Staff Departments

A distinction is to be made between coordination on a project by project basis, as described above, and the broad coordination required for such matters as quality policy and quality objectives. At present there is a widespread use of quality control staff departments to supply the specialist manpower needed for diagnosis, for drafting interdepartmental plans, and for other project-by-project coordination. However, there is little clear assignment to these departments to do, for the quality function, what centuries of accountants have evolved for the finance function (see Responsibilities of the Quality Manager, section 7-16).

There is room for much initiative, and opportunity for much original, creative work to be done in developing the means for this broader coordination of the quality function.

7-15 Coordination through Joint Committees

A number of essential activities in the quality function are carried on by committees. The specific duties of the different committees vary and are discussed below. However, the basic responsibilities of industrial committees and the manner of carrying out these responsibilities are generally similar throughout industry.

Generally, a committee has the purpose of facilitating:

1 *Communication.* It is easier for men to communicate with each other while seated around a table than by writing memorandums to each other.
2 *Coordination.* In the same way, it is easier for men in committee to reach common agreement on controversial matters.
3 *Participation.* By providing opportunity for all to be heard when matters are still fluid, there is greater open-mindedness for the ideas advanced. More than this, there is deep psychological satisfaction to all concerned. This satisfaction is derived from being consulted, from being accorded the right to be heard, and from being treated with dignity.

While there are exceptions, the committee normally may not make decisions, issue orders, or tell someone what to do. The committee may only recommend. Other people in the company, including committee members in their noncommittee capacity, decide whether to adopt the recommendations or not.

In practice, because the interested departments are usually represented on committees, an agreement reached in committee means also that each department will carry out its part of the agreement.

Failure of committees to function adequately is usually due to (1) use (really misuse) of the committee as a decision making body instead of as a deliberative body, and (2) lack of skill in use of the machinery of committee operation, i.e., no agenda, chairman unable to keep discussion on the track, no minutes, no follow-up, etc.

The starting point for creating a committee is the existence of problems requiring group deliberation. The character of these problems also suggests the makeup of the committee. The membership is drawn from those departments which will likely have much action to take as a result of the committee's deliberations. The chairman is usually from that department which will have the most to do. The men are chosen not because of their rank, but because of their capacity to contribute to the problems at hand. The secretary is commonly from that department which will do most of the work of analysis ("diagnosis").

It is most desirable for the committee to define its mission carefully. This is done by preparing a written "frame of reference" and then securing the approval of higher authority to legitimize the work of the committee.

The wide variety of quality problems has resulted in creation of several species of committees. Some of the more usual are discussed below. (Other examples[7] are a material review board and a new products committee.)

Quality Improvement Committee

This committee is commonly the "steering arm" for an attack on unsolved, chronic quality problems. Membership is from those departments which can contribute effectively to the list of problems to be solved, but usually includes product design, marketing, manufacturing planning, production, accounting, and quality control. The "charter" of the committee is to:

1 Identify the major unsolved quality problems
2 Agree on the priority of attack

[7] Also see *QCH*, 1962, sec. 6.

3 Discuss possible solutions
4 Arrange for adequate diagnosis to discover causes and remedies
5 Recommend remedial action
6 Follow to see that action is taken and that controls are established to hold the gains

As experience is gained in making quality improvements, it becomes possible to train all levels in how to make such studies and obtain results. The ultimate form is seen in Japanese companies, where work leaders and nonsupervisors have been trained to do the work of both the steering arm and the diagnostic arm.[8]

Quality Motivation Committee

This committee is created to organize and guide a "campaign" for raising the level of quality-mindedness in the company (sometimes the entire company, sometimes only in the plants). Membership is drawn from the departments at which the campaign is directed. Staff assistance is provided by industrial relations, accounting, and quality control. Sometimes a full-time special coordinator is appointed to investigate employee suggestions for "error cause removal," and to help keep the campaign fresh and lively.

Engineering Change Committee

This is a continuing committee, set up to review and expedite the making of engineering changes. Membership usually includes product development, product design, production, quality control, sales service, inventory control, and tool room.

7-16 Responsibilities of the Quality Manager

This official has clear responsibility for the usual duties of managers: setting his departmental policies and goals; getting them approved; getting his budget approved; organizing the work of his department; selecting, training, and motivating his subordinates; collaborating with other managers, etc.

There remains a much broader responsibility which is seldom clearly assigned, whether to the quality manager or to anyone else. This is the responsibility for unifying the activities of the quality function, much as the financial manager does for the finance function.

[8] See J. M. Juran, "The QC Circle Phenomenon," *Industrial Quality Control (IQC)*, vol. 23, no. 7, pp. 329–336, January, 1967. Also see Chap. 22 of the present work.

Centuries of accountants have evolved a unification of the finance function through:

1 Companywide *concepts* such as unified financial planning and control, a common language and units of measure, standards of performance, systematic measurement, summarized reporting, independent audits
2 Standardized *tools* which apply over the entire company (sometimes throughout industry): budgets, charts of accounts, terminology, reports (balance sheet, profit statement, etc.)
3 Standardized *techniques:* cost centers, activity indexes, depreciation methods, double-entry bookkeeping, valuation methods

Quality managers have made much progress in the lower part of this list, i.e., tools and techniques, but little progress in developing of broad concepts. In part this is due to an unwillingness or inability to take the initiative in the absence of a specific assignment from top management. In turn, the top managements have not been fully aware of the need for developing broad policies, etc., and hence have made no clear assignment.

As a result there is a role waiting for an actor.[9]

There is need for someone to take the initiative on such matters as:

1 *Quality policy* on vital questions:
   Quality standards. Shall we go for industry leadership, for competitiveness, or for adequacy?
   Quality costs. Shall we try to optimize our costs or the users' costs?
   Quality motivation. Shall we rely on people or on foolproof systems?
   Vendor relations. Shall we treat the vendor like an in-house department, or on an arm's length basis?
2 *Quality objectives* for such things as failure rates, guarantees, launching failure-free designs, law and order in the tolerance system, quality costs.
3 *Planning* on a companywide basis for facilities, manpower, training.
4 *Organizing* for both quality control and for breakthrough to improve quality. Creating the job descriptions, lines of command, lines of communication, committee machinery, and other interdepartmental forms.
5 *Motivation* to achieve quality. Managers are unified in their belief that motivation is vital, but they are divided in how to achieve motivation, whether by giving people increments of money, information on "why," participation in planning, wider responsibility, threats and penalties, emotional appeals, awards and prizes.

[9] See, generally, J. M. Juran, "The Two Worlds of Quality Control," *IQC*, vol. 21, no. 5, p. 243, November, 1964.

6 *Executive control.* To date the quality specialists have not produced the quality equivalent of the balance sheet and profit statement.

PROBLEMS

1 Conduct a research study and report on how any of the following institutions were organized for quality: (*a*) the Florentine Arte Della Lana (wool guild) of the twelfth, thirteenth, and fourteenth centuries, (*b*) Venetian shipbuilding in the fourteenth century, (*c*) construction of cathedrals in medieval Europe, and (*d*) the Gobelin tapestry industry in the sixteenth and seventeenth centuries.

2 Study the plan of organization for quality for any of the following institutions and report your conclusions: (*a*) a hospital, (*b*) a university, (*c*) a chain supermarket, (*d*) a chain of motels, (*e*) a restaurant, and (*f*) a manufacturing company.

3 You are the manager of a small plant manufacturing precision parts for customers who make various types of assembled units. You have three different processes (e.g., casting, stamping, and machine), each of which is supervised by a foreman. The foremen report to you through a general foreman. In each of these three departments are several inspectors, who report to the department foreman, to inspect the work of the department.

A year ago, you were impelled by the high incidence of customer complaints to institute a final inspection department, where all product now passes before shipment. This is headed up by a chief inspector, reporting to you. The institution of this department has successfully reduced the customer complaint problem, but the losses from scrap and rework are now prohibitive, since nothing much has been done to trace back and find the sources of the defects.

The chief inspector has studied modern quality control methods and believes he can accomplish some reductions. To do so, he urges that all the departmental inspectors now come under his jurisdiction and that he be allowed to allocate one particular man solely to the problem of searching out the causes of defects through capability analysis, control charts, etc.

The general foreman, on the other hand, argues that the main problem is the fact that final inspection is much too strict; he has to spend most of his time arguing with the chief inspector to convince him that certain defects are perfectly acceptable to the customer. He proposes that the chief inspector be brought under him so that decisions can be made more rapidly. If this is done, he would propose to leave the departmental inspectors as they are, but have the chief inspector train them in control chart techniques. This, he believes, will be adequate to cope with the defect problem.

What are the advantages and disadvantages of each of these two plans? Which do you think has the greater prospect of success and why? As plant manager, what would be your decision?[10]

4 Read the Metal Containers, Inc., case in the Appendix and answer the following problem.

You are Lafferty, the new quality control manager for MC. With all the newness, it doesn't look too bad. You do have much know-how which they can use, and you have, on your last two jobs, been in the same situation of an expert from the outside who had to get accepted before he could put his know-how to work. So far, so good.

Aside from making a survey of this can business, you have also made a survey of yourself. You are now 38 years old, with a good record of progress. Each of your previous companies was sorry to see you go—it was just that they couldn't give you a promotion as fast as you found it on the outside. In each case you had a man trained to take over your job, and it was an amiable separation.

You are pretty near the end of the line as a quality control man. In this company, there is no higher quality control job. A few years hence, you might move into a bigger company as quality control manager. But you aren't sure that you want to do that. Quality control has been an exciting specialty, and it has been good to you. But you have been getting interested in becoming a manager. Each year you have been doing more managing and less engineering. You look at what the managers and vice-presidents are doing and it looks good to you. They have interesting, exciting jobs to do, are well paid, and are looked up to. What more can a man want?

Obviously, you have to make good on your present job. You are pretty sure you can do this.

You are aware that making good on your present job requires more than competence in quality control methodology. (Just for dealing with the quality control supervisors it might be enough.) But you suspect that you will find yourself heavily involved with departments you haven't had much to do with—marketing, purchasing, finance, personnel. Being located at the head office (which is in the same fence with the Cleveland plant), you are available to the executive group. You eat in the execu-

[10] Course notes of Management of Quality Control Course, U.S. Air Force School of Logistics, 1959.

tive dining room. So you know vaguely that you need more breadth than you have ever had. On your previous jobs you have participated mainly in technical decisions. Here you are getting involved in business decisions—marketing strategy, return on investment, and the like.

You have always had a good deal of fringe time, both on the job and off. To date, you have used much of this time to go deeper into quality control—statistical methods, quality control engineering, reliability, and the rest. These matters have dictated your choice of conferences, seminars, books, journals, etc. Now you wonder if the time may not have come to put your fringe time into the subject of management—setting objectives, planning, organization, communication, motivation.

You have already learned that managing requires a statement of objectives, followed by planning how to meet these objectives. You decide to go through the exercise of trying to reduce to writing the statement of an objective of "where do I go from here," and then mapping out a plan for getting there.

State your objective and plan.

8 Design for System Effectiveness

8.1 The Life Cycle of a Product

The significance of product design can be appreciated by dividing the life cycle of a product into a number of phases. The type of product influences the breakdown of the life cycle into individual phases and also the time and effort spent in each phase. Although the breakdown may differ widely by product, Figure 8-1 shows the basic phases in the life cycle.

Concept and Feasibility Phase

In this phase, the known or anticipated need for a product is studied in enough detail to determine if it is feasible to design and manufacture a product. An important element of this feasibility is the expected cost to the company and to the customer. This phase culminates in a decision that it is feasible (or not feasible) to continue with the project. This decision recommends one or more design concepts that appear to be promising to explore in the future.

Detailed Design Phase

In this phase, the alternative design concepts are evaluated, the most promising one is selected, and the product is designed in sufficient detail to prepare purchasing specifications for parts and materials, and to plan the manufacture of products for prototype testing. For complex products, the detailed design phase may consist of two major phases:

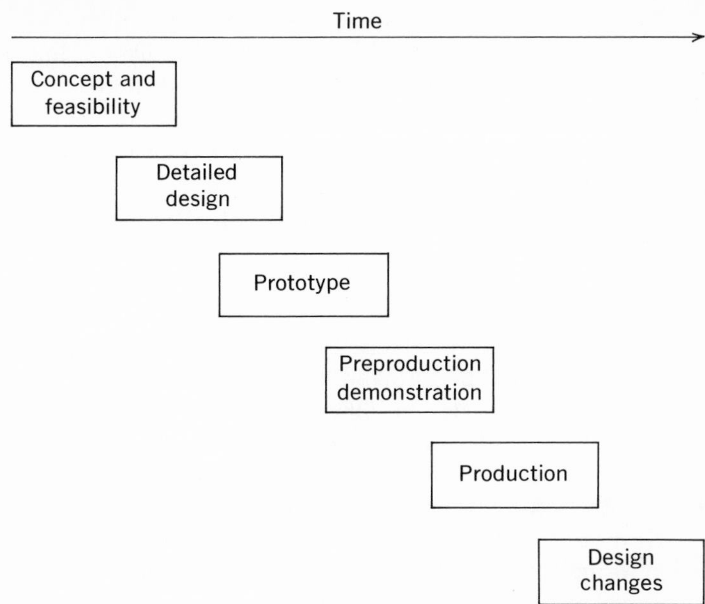

Figure 8-1 Phases in the product life cycle.

(1) a more detailed exploration of several design concepts uncovered in the feasibility stage (these explorations may be conducted simultaneously or sequentially), and (2) selection of a final design concept followed by a full detailed design.

There is always the possibility that during or after the detailed design phase, the design concept selected may prove to be poor and may have to be discarded. The cycle then starts all over. (The detailed design phase may also include major work in new component development and may go beyond and include fabrication of hardware to evaluate the design proposed on paper.)

Prototype Phase

In this phase, the first essentially complete units of the product are built and tested. The tests may evaluate such factors as basic design capability, effects of extreme environments, and reliability for extended periods of operation. The units built may not be complete but are complete enough to test the adequacy of the basic design approach. Hopefully, the prototype tests will indicate that the initial design approach can be adopted and completed in detail. If not, "back to the drawing board."

Ideally, prototype units are built using the manufacturing procedures and equipment planned for full-scale production, but this is sometimes not possible. In such case, the test results may not be completely representative of what the production units will do. However, the prototype phase is a valuable means for the designer to evaluate the adequacy of his design approach ("technical feasibility") and to revise the design prior to setting up manufacturing procedures.

Preproduction Demonstration

In this phase a "production design" is prepared and is evaluated for producibility and performability.

The production design differs from the prototype design due to design changes made for a variety of reasons: simplify manufacturing processes, cut the cost of materials, standardize for greater interchangeability with existing products, utilize present production facilities, etc.

Next, a preproduction lot of product is made to the production design, using the production equipment. This "pilot run" is then tested, sometimes under simulated use environments (proving grounds), sometimes by selected customers under conditions of actual use.

Based on the results on the tests and the experience of manufacture, the design is either released for full-scale production or sent back to be reworked.

Full-scale Production

This is the "regular" production for sale and delivery to customers.

Design Changes

Based on experience gained in production, marketing, and use of the product, design changes are made to improve product performance and to eliminate field failures which showed up under the conditions of use. These design changes go through one or more of the previous phases of the life cycle.

Role of Quality Control Programs

Traditionally, the conduct of the foregoing activities has been delegated to the product development, product design, and manufacturing planning departments. The personnel in these departments were most competent engineers, ingenious designers, resourceful in solving known problems, and to a degree, imaginative in anticipating unforeseen problems. However, they did not develop or accept the modern ways of foreseeing problems through quantifying reliability prediction, measuring process capability, data retrieval on component failure rates, etc.

While these modern ways were generally known, even well known to the quality control departments, they occupied themselves with the later phases of the life cycle of the product: process control, inspection and test, and product conformance to specification. In consequence, the new tools available for use during the early stages of the life cycle of the product were not put to use. The engineers who worked on the early stages did not understand or accept the tools; those who understood them did not participate in these early stages.

All this is changing radically.

The Parameters of Product Quality

The traditional parameters of product quality have long been familiar: the engine shall deliver 360 brake horsepower; the relay shall operate at minimum 55 milliamperes. In response to such parameters, specifications could be unequivocal, test equipment could be designed with confidence, and testing could assure conformance to specification.

All this time, and even back into the beginnings of history, there were other parameters: the product should have long life; during that life it should give service with few failures; if it did fail, it should be easy to restore service. These parameters were almost never quantified. However, they were understood to be present, and earnest men took them seriously, as the historical improvements in products will attest.

What has happened in the last half of the twentieth century is that these parameters must now be quantified. For the first time in history, human safety, health, and convenience depend absolutely on the reliability or uninterrupted service of our vehicles, computers, weapons, drugs, power supply, communications, etc. At the same time, the products have gone into new orders of complexity and precision. Experience has already shown that unless these old parameters are now quantified, the changed needs of humanity cannot be met.

These parameters are not independent of each other—they interact vigorously with each other and with the familiar, quantified parameters. In fact there is a concept that embraces them all—the concept of "system effectiveness."

8-2 System Effectiveness Concept

The term "fitness for use" applies to all products and services. For long-lived and/or complex products, fitness for use is often called "system effectiveness."

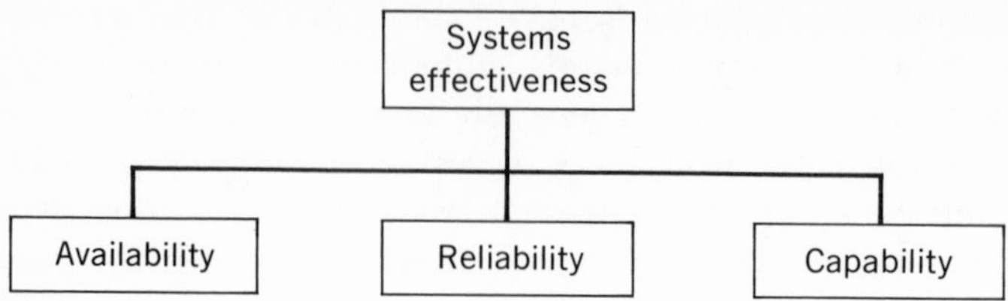

Figure 8-2 Components of system effectiveness.

The following definitions[1] are basic to the concept:

"System effectiveness" is a measure of the extent to which a system may be expected to achieve a set of specific mission requirements and is a function of availability, dependability, and capability.

Availability is the measure of the system condition at the start of a mission and is a function of the relationship between hardware, personnel, and procedures.

Reliability[2] is the measure of the system condition at one or more points during the mission, given the system condition(s) at the start of the mission, and it may be stated as the probability (or probabilities or other suitable mission-oriented measure) that the system (1) will enter and/or occupy any one of its significant states during specified mission and (2) will perform the functions associated with those states.

Capability is a measure of the ability of a system to achieve the mission objectives, given the system condition(s) during the mission, and it specifically accounts for the performance spectrum of the system.

These major components of system effectiveness are shown in Figure 8-2. Von Alven[3] provides a thorough discussion of many other system effectiveness terms.

A more subtle but potentially important advantage of the quantification is the thoroughness that quantifying forces on the analysis. Areas that have previously been treated qualitatively are now examined more closely. Even if these areas are *not* actually quantified, the discussions in attempting to quantify often reveal factors that are overlooked in a qualitative discussion of the problem.

[1] *Final Report of Task Group II—Prediction—Measurement (Concepts, Task Analysis, Principles of Model Construction)*, Weapons System Effectiveness Industry Advisory Committee, U.S. Air Force Systems Command, 1965.

[2] While the report uses the word "dependability," the equivalent term "reliability" has received such universal acceptance that the present authors will use "reliability" in what follows.

[3] ARINC Research Corp., *Reliability Engineering*, Prentice-Hall, Inc., Englewood Cliffs, N.J., 1964.

8-3 Example[4] of System Effectiveness

This example concerns a ground-based radar system used to detect aircraft above the horizon line of sight at ranges up to 200 miles. While aircraft is within this range the radar system is to track (follow) it in range and azimuth (direction) within a defined admissible error and present the information to an operator. The system must operate under climatic conditions present anywhere in the world.

A simple description of the system is shown in Figure 8-3.

If at least one transmitter is working, the radar system will still perform although a failure of one of them will have some effect on the overall effectiveness as discussed later. All other units must work for system success.

An important step in the system effectiveness analysis is the determination of one or more indices to describe product effectiveness. (These indices are sometimes called "figures of merit.") For this product, the following indices have been chosen:

1 Probability of target detection at maximum range at a random point in time
2 Probability of detection at maximum range and continuous tracking within required accuracy during a prescribed time period (assumed to be 30 minutes in this example)
3 Probability of continuous tracking within required accuracy during prescribed time period, given successful detection
4 Curve of probability to detect and track versus range (of special interest if the probability of detection at maximum range is less than the goal)

[4] See footnote 1.

Figure 8-3 Elements in a ground-based radar system. Source: *Final Report of Task Group II—Prediction—Measurement* (*Concepts, Task Analysis, Principles of Model Construction*), U.S. Air Force Systems Command, Weapons System Effectiveness Industry Advisory Committee, 1965.

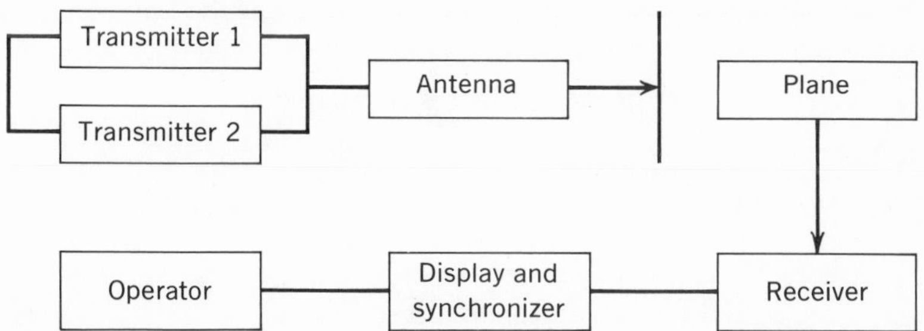

The probability scale will range from 0 (no chance of occurrence) to 1.0 (certainty of occurrence).

The mathematical model will now be developed. The details of obtaining numerical values are omitted here in order to concentrate attention on the overall effectiveness concept. These details will be discussed in later chapters. Further, the presentation here uses simple probability concepts to build the model. The development of models for more complicated examples would be facilitated by the use of mathematical tools such as matrix algebra.

First, it is useful to define a number of system "states." This is simply a list of all possible conditions of the system; for this example:

State	*Definition*
1	All units operating properly
2	One transmitter inoperative, the other transmitter and all other units operating properly
3	System totally inoperative due to inoperative condition of both transmitters or any one of the other units

Expressions for availability, reliability, and capability will be developed to obtain the desired indices of effectiveness.

The first portion of the model to be developed will be availability. Availability will be defined as the probability that the system is in a particular state. Assume the following estimates:

	Mean Time between Failures, Hours	*Average Repair Time, Hours*
Each transmitter T	40	4
Remainder of the system R	100	1

The availability of the product is calculated as

$$P_a = \frac{\text{MTBF}}{\text{MTBF} + \text{MRT}}$$

where MTBF = mean time between failures
MRT = mean repair time

The availability for each transmitter and the remainder of the system can be calculated as

$$P_{at} = \frac{40}{40 + 4} = 0.909$$

$$P_{ar} = \frac{100}{100 + 1} = 0.99$$

These probabilities will be combined according to the following probability theorems:

1 The probability that several independent events will occur is the *product* of their individual probabilities.
2 The probability that either of two events (that cannot occur simultaneously) occurs is the *addition* of their individual probabilities.

The availability of the total radar system for each state can be calculated:

State 1: Availability equals $(0.909)(0.909)(0.99) = 0.818$
State 2: Availability equals $(0.909)(0.091)(0.99) + (0.091)(0.909)(0.99) = 0.164$
State 3: Availability equals $1 - (0.818 + 0.164) = 0.018$

At a given instant there is about an 82 percent chance that all units in the system are operating properly. Also, there is about a 16 percent chance that the system is operating with only one transmitter. Finally, there is about a 2 percent chance that the system is totally inoperative.

The next portion of the effectiveness model is dependability (or "reliability"). It will be assumed that (1) repairs are prohibited during the tracking of an aircraft and (2) the failure time of each unit is exponentially distributed. The reliability will be calculated for each state; i.e., the system could start in State 1 (all units operating) but end in State 1, 2, or 3. Most of the individual terms in the model are simply the reliability calculated from the cumulative exponential distribution; i.e., the reliability of one transmitter is

$$P_t = e^{-X/\mu} = e^{-0.5/40}$$

Note that $1 - P_t$ is the probability of failure during the mission. Table 8-1 summarizes the calculations.

The final portion of the effectiveness model is design capability. The capability problem in this example refers to the ability of the product to (1) detect the aircraft and (2) track the aircraft for 0.5 hour.

The ability to detect is a function of the amount of power available in the radar set and also of the range of the aircraft. The range is specified as 200 miles. However, the amount of power available in

Table 8-1 CALCULATIONS FOR RELIABILITY

State at Start of Mission	State at End of Mission: *1*	*2*	*3*
1	$P = e^{-0.5/40}e^{-0.5/40}e^{-0.5/100}$ $= 0.971$	$P = e^{-0.5/40}(1 - e^{-0.5/40})e^{-0.5/100}$ $+ (1 - e^{-0.5/40})e^{-0.5/40}e^{-0.5/100}$ $= 0.024$	$P = 1 - (0.971 + 0.024)$ $= 0.005$
2	$P = 0$ (Repairs prohibited during mission)	$P = e^{-0.5/40}e^{-0.5/100}$ $= 0.982$	$P = e^{-0.5/40}(1 - e^{-0.5/100})$ $+ (1 - e^{-0.5/40})e^{-0.5/100}$ $= 0.018$
3	$P = 0$ (Repairs prohibited during mission)	$P = 0$ (Repairs prohibited during mission)	$P = 1.0$

the set depends on the transmitters. If both transmitters are operative, the ability to detect is higher than if only one is operative. If neither transmitter is operative, the ability to detect is zero. The relative capability with one or two transmitters operating requires knowledge of radar theory based on noise theory and radar equations. This knowledge provided the following capability figures:

State	*Capability (Probability of Detection at Maximum Range)*
1	0.9
2	0.683
3	0

Thus, if both transmitters are operating, the probability of detection at a 200-mile range is 0.9 while if only one transmitter is operating, the probability is 0.683.

In determining the capability in tracking, theoretical considerations of radar design provided the following estimates:

State	*Capability (Probability of Tracking)*
1	0.98
2	0.98
3	0

The components of effectiveness (availability, reliability, capability) can now be combined to provide numerical estimates for the four indices of effectiveness. The first index was the probability of detection at maximum range. This is the product of availability

0.818 × 0.900 = 0.736. Summarizing for all states:
and capability, e.g., the probability that the system is in State 1 and that the design will detect the aircraft at maximum range is equal to

State	*Probability*
1	(0.818)(0.900) = 0.736
2	(0.164)(0.683) = 0.112
3	(0.033)(0) = 0.000
	0.848

Thus, taking into account that the system may be fully operative, partially operative, or completely inoperative, the overall probability that an aircraft will be detected if it is flying at maximum range is about 85 percent.

The second index of effectiveness was the probability of detecting an aircraft at maximum range and tracking it with required accuracy for 0.5 hour. For a system which starts and ends a mission in State 1, the probability equals the probability (availability) of being in State 1 times the probability (capability) of detection when in State 1, times the probability (reliability) of running continuously for 0.5 hour times the probability (capability) of tracking when in State 1, that is, 0.818 × 0.9 × 0.971 × 0.98. Table 8-2 summarizes for all states.

Table 8-2 CALCULATIONS FOR DETECTION AND TRACKING

		State at End of Mission	
State at Start of Mission	*1*	*2*	*3*
1	(0.818)(0.9)(0.971)(0.98)	(0.818)(0.9)(0.024)(0.98)	(0.818)(0.9)(0.005)(0)
2	$P = 0$	(0.164)(0.683)(0.982)(0.98)	(0.164)(0.683)(0.018)(0)
3	$P = 0$	$P = 0$	(0.018)(0)(1.0)(0)
		$\Sigma = 0.826$	

The overall probability that the system will detect and track the aircraft with prescribed accuracy for a half-hour period is about 83 percent. Notice that this probability is only slightly less than the probability of detection of the aircraft (85 percent).

The third index of performance was the probability of tracking the aircraft. It is easiest to calculate this as the probability of detection and track divided by probability of detection or 0.826/0.848 = 0.974.

The last of the four measures of effectiveness was a curve of the probability to detect and track versus range. Theoretical design capability considerations indicate that the smaller the range the higher the

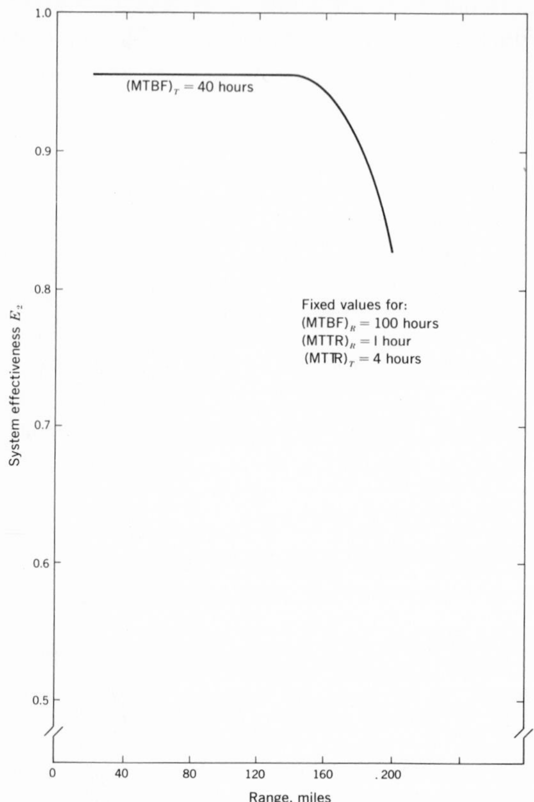

Figure 8-4 System effectiveness versus range as a function of $(\text{MTBF})_T$. Source: *Final Report of Task Group II—Prediction—Measurement* (*Concepts, Task Analysis, Principles of Model Construction*), U.S. Air Force Systems Command, Weapons System Effectiveness Industry Advisory Committee, 1965.

probability of detection. The effect of range on overall effectiveness can be evaluated by making use of the theoretical radar equations to change the capability portion of the overall effectiveness model. The details will be omitted for simplicity, but the results are shown in Figure 8-4. The effectiveness model and the four indices of effectiveness are now complete, and they can be used to investigate various alternative designs of the product. For example, assume that there was an original goal of 90 percent for overall effectiveness. The calculations indicate a predicted effectiveness at maximum range of 82.6 percent.

The real work now begins. What design changes would help meet the requirement? The main problem is in *detecting* the target, since the effectiveness in tracking (assuming detection has already been made) is 0.974 and the effectiveness in detection alone is only 0.848. Further, the calculations for the probability of 0.848 show that even if the availability were 100 percent, the effectiveness in detection could still only be 0.90 due to limitations in design capabilities. Thus, attention has been focused on the inherent *capability* of the design to detect the aircraft as a key problem area. What, then, can be done about capability?

The information gathered in developing the model states that the capability of detection can be increased by decreasing the range at which an aircraft is to be detected or increasing the transmitter power. The effect of changing the range is shown in Figure 8-5. For example, if the range requirement is dropped from 200 miles to slightly over

Figure 8-5 System effectiveness versus $(MTBF)_T$ as a function of range. *Task Analysis, Principles of Model Construction*), U.S. Air Force Systems Command, Weapons System Effectiveness Industry Advisory Committee, 1965.

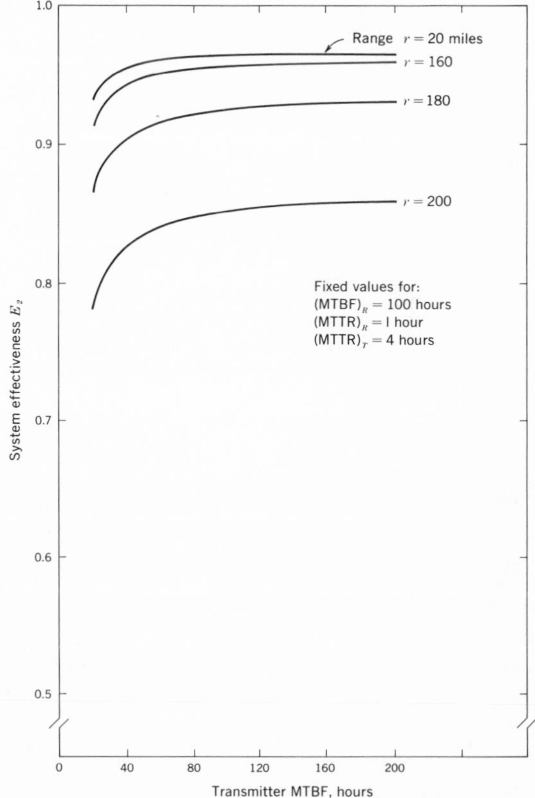

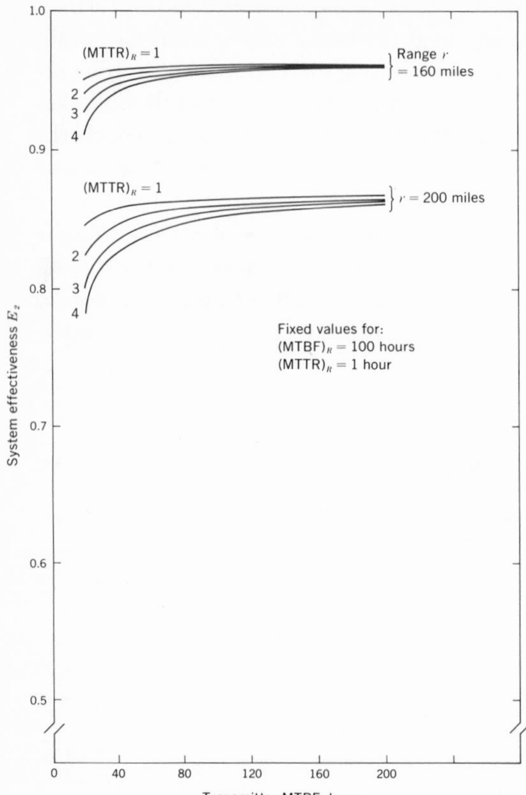

Figure 8-6 System effectiveness versus $(\text{MTBF})_T$ as a function of $(\text{MTTR})_T$ and range. Source: *Final Report of Task Group II—Prediction—Measurement (Concepts, Task Analysis, Principles of Model Construction)*, U.S. Air Force Systems Command, Weapons System Effectiveness Industry Advisory Committee, 1965.

180 miles, the overall effectiveness will increase from about 82 percent to the 90 percent required. This relatively slight decrease in range might be considered satisfactory, particularly in light of the alternative actions that might be necessary if the 200-mile range requirement is considered fixed.

Figure 8-5 also points out some revealing characteristics of the mean time between failures for the transmitter. The overall effectiveness increases rapidly as the MTBF increases up to 100 hours. This means that a program to improve transmitter reliability might have a significant payoff up to an MTBF of 100 hours, but after that, the additional increase in system effectiveness is marginal.

How about the effect of varying the average *repair time* for the transmitter? The lower the repair time, the higher the availability, and, therefore, the higher the overall effectiveness. However, Figure 8-6 indicates that the repair time has a significant effect on overall effectiveness only when the transmitter MTBF is below about 100 hours. Also note that doubling the MTBF from 40 hours to 80 hours has about the same effect as cutting the average repair time in half. This is significant in choosing between higher reliability and better maintainability. Doubling the MTBF and lowering the mean repair time produces the same effect on maintenance manpower costs but not on replacement part costs. Thus, increasing the MTBF is preferable if proportionate changes can be made for the same cost.

Here are some advantages of constructing a model:

1 Overall effectiveness is stated numerically as a function of the vital few product parameters.
2 Alternative designs can be evaluated quantitatively and specific comparisons made to help select the final design.
3 The importance of various parameters of the product can be evaluated by varying these parameters within broad limits and calculating the influence on the overall effectiveness.
4 The quantification provides a means of establishing design goals and predicting actual performance against these goals to detect potential problems and prevent their occurrence.
5 The effect of possible errors in the numerical estimates included in the model can be evaluated by varying the estimates and noting the influence on the overall effectiveness.

8-4 Use of a System Effectiveness Model in Design

A system effectiveness model aims to do *quantitatively* what designers have always done *qualitatively*. The objective is as old as the design function itself. What is new is the emphasis on quantification during the design process instead of after the fact analysis of field data. In the example on a radar set, effectiveness was defined as the product of availability, reliability, and capability. The definition stemmed from an analysis of the function of a radar set. The concept can be applied to any product with effectiveness defined to best describe the function of the product. Thus, definitions of effectiveness would be different for chemical manufacturing equipment, a farm tractor, an automatic turret lathe, and a washing machine. However, the effectiveness of all these products can be expressed in terms of quality parameters.

8-5 Relation of System Effectiveness to Cost and Schedule

System effectiveness quantification means little without a quantification of costs. The recent emphasis on "cost effectiveness" has drawn attention to various ways of defining costs. One concept stresses *total* cost to the customer (see Chapter 4).

Applications of cost effectiveness concepts have mainly involved defense products, but potential applications are broad. For example, the cost per mile of automobile transportation and the cost to move a cubic yard of dirt a given distance with a tractor contain the elements of cost and effectiveness.

A logical extension of the cost effectiveness concept is the addition of schedule and other requirements. Von Alven[5] proposes that "system worth" include cost, effectiveness, schedule, and personnel requirements. All the categories of system worth are influenced by the design of the product and thus must be evaluated from the inception of design.

PROBLEMS

1 Two alternative engines have the following characteristics:

	Engine A	*Engine B*
Capability:	99% sure of generating at least 200 hp and this probability decreases according to the exponential distribution	Essentially 100% sure of generating at least 200 hp and this decreases linearly to 0% at 280 hp
Reliability:	Time between failures (TBF) is exponentially distributed with a mean of 200 hours	TBF is normally distributed with a mean of 300 hours and a standard deviation of 50 hours
Other information:	Mean repair time = 10 hours	Availability = 90%
Investment costs/year:	$2,000	$1,800
Operating costs/year:	$400	$500
Productivity loss when engine is down:	$50/hour	$50/hour

Assume a mission time of 24 hours. Compare the two engines over a 1-year period for a usage that requires 220 hp.

2 For each of the products below, define (in words) one or more measures of system effectiveness. The definitions used in this chapter

[5] ARINC Research Corp., *op. cit.*

may not fully apply to all these products, and so the definitions created should meet the needs of each product.

a Automatic washing machine
b Large digital computer
c Office duplicating machine
d Corn oil refining process equipment
e Large crawler tractor
f Commercial airliner
g Turret lathe
h Rented automobile
i Purchased automobile
j Television set

The measures defined should reflect both quality and cost.

9 Reliability, Maintainability

9-1 The Main Concepts

It was noted in Complex Systems and Quality, section 2-6, that the last half of the twentieth century has brought on a proliferation of new products exhibiting new characteristics (or old characteristics so radically more demanding that the solutions must be new): complex systems; shorter development cycles; growth of subcontracting; new environments; new levels of precision, miniaturization, and other advances in technology.

These changes have contributed to the increased extent to which human beings stake their lives, health, and convenience on man-made products (Human Welfare and Quality, section 2-7). The changes have also affected remarkably the relation of purchase price to cost of keeping the product in service during its life-span (Economics of Quality as Viewed by Customers, section 4-9).

The major response to these causes and needs has been the concept of *quantifying* "reliability." This term was defined in a historic report[1] as "the probability of a product performing without failure a specified function under given conditions for a specified period of time." More simply, reliability is the chance that a product will work. (For "one-shot" products, e.g., explosive bolts on a missile or solid propellant rocket motors, the definition reduces to the probability of performing without

[1] *Reliability of Military Electronic Equipment,* report by Advisory Group on Reliability of Electronic Equipment, Office of the Assistant Secretary of Defense (R&D), June, 1957. (This was the historic "AGREE" report.)

failure under the intended conditions of the mission.) Once quantified, reliability becomes a design parameter just as any other (weight, tensile strength, etc.).

A second major response has been the concept of *planning reliability* as a series of tasks to be assigned to all functions associated with progression of the product from "cradle to grave." This planning breaks the monopoly of the designer by including reviews of the design by other experts.

As a corollary, planning of reliability requires that we meet the key implications of the reliability definition:

1 The quantification of reliability in terms of a probability.
2 A clear statement defining successful product performance.
3 A statement defining the environment in which the equipment must operate.
4 A statement of the required operating time between failures. (Otherwise the probability is a meaningless number for time-oriented products.)

These two concepts, quantifying reliability and planning reliability, are the main subject matter of this chapter.[2] Parallels are also discussed for maintainability.

9-2 The "Gold in the Mine" for Reliability and Maintainability

The reliability and maintainability problem is large: "As a result of less than 100% reliability, United Airlines needs a spare parts inventory of 122 million dollars—about 20% of our total investment in airplanes."[3] "Air Force studies have indicated that annual maintenance costs vary

[2] For some details on the history of the reliability movement, see R. R. Carhart, *A Survey of the Current Status of the Electronic Reliability Problem,* Rand Corporation Research Memorandum RM-1131; *Reliability of Electronic Equipment,* R&D Board *Report* El 200/17, Feb. 18, 1952, vol. I; *Reliability Training Text,* 2d ed., joint publication of Institute of Radio Engineers and American Society for Quality Control, March, 1960; *Reliability Factors for Ground Electronic Equipment,* prepared by Technical Writing Service, McGraw-Hill Book Company, for the Rome Air Development Center, U.S. Air Force, April, 1955; R. G. Fitzgibbons, "Quality Control and the Reliability Engineer," *Industrial Quality Control (IQC),* vol. 14, no. 11, May, 1958.

[3] H. W. Nevin, "Dollars and Cents Aspects of Reliability," *Proceedings of the Third Annual Aerospace Reliability and Maintainability Conference,* Washington, 1964, Society of Automotive Engineers, p. 62.

from 3 to 29 times the original equipment cost."[4] Twenty-four maintenance man-hours per flight hour were required in Navy aircraft in 1949. It was estimated that this would rise to 80 in 1965, primarily because of an increase in electronic equipment complexity from 120 parts in 1949 to 8,900 in 1960 to an estimated 33,000 in 1965.[4]

Although many of the failures were due to workmanship and associated causes, a large percentage of failures was due to improper design and "operational conditions"—inadequate maintenance, storage conditions, etc. (see section 30-16). But, many of the causes that were listed as "operational" eventually required some type of design change to prevent the problem in the future. Thus, attention was focused on the *design* phase of a product as contrasted to conventional quality control emphasis on the manufacturing phase.

Less dramatic as to individual cases but of far greater total economic effect is the enormous network of repair and maintenance shops for automobiles, railroad equipment, household appliances, industrial machinery, electronic sound and video systems, and so on and on. The very evident cost of repairs and the subsurface costs of downtime are so huge a drain on the economy that the signs of user revolt are very much in the air.

9-3 Elements of a Typical Reliability Program

Figure 9-1 lists 20 numbered elements of a reliability program for a hypothetical and sophisticated piece of electronics equipment having a design and development period of 3 years and a manufacturing period of 7 years.[5] Inherent in these elements is a good deal of action beyond that traditionally carried out by the various departments of the company. The need for these actions arises from the fact that new designs are a fertile breeding ground for new causes of unreliability. Unless preventive action is taken to discover these new causes during the design phases, they will not be discovered until several phases (and years) later, when they will have caused a lot of damage.

These elements require specific actions to be taken by all functions which participate in the progression of the product from research and development through use by the consumer. (Some of the elements

[4] E. F. Dertinger, "Funding Reliability Programs," *Proceedings of the Ninth National Symposium on Reliability and Quality Control*, San Francisco, January, 1963, Institute of Electrical and Electronics Engineers, Inc., pp. 16–33.

[5] A more modest project would extend over a shorter time, and would require formal action on only some of these elements.

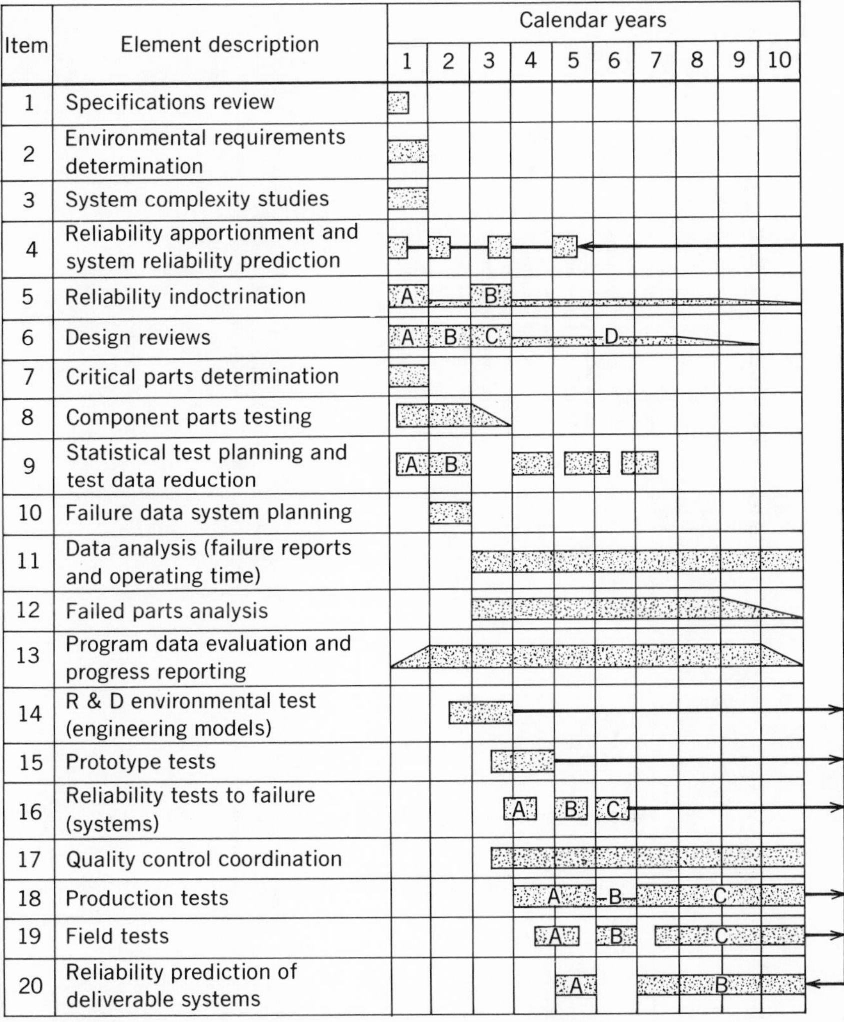

5*A*. Engineering personnel
5*B*. Manufacturing personnel
6*A*. Preliminary design review
6*B*. Interim design review
6*C*. Final design review
6*D*. Reliability-improvement committee reviews
9*A*. Parts tests
9*B*. Systems tests
16*A*. System 1
16*B*. System 2
16*C*. System 3
18*A*. Early production
18*B*. Service-evaluation test phase
18*C*. Full-scale production
19*A*. Contractor-evaluation tests
19*B*. Service-evaluation tests
19*C*. Service-use tests
20*A*. Early production systems
20*B*. Full-scale production systems
→ Reliability calculations and improvement of original prediction

Figure 9-1 Time-phased elements of a typical reliability program.

Table 9-1 EXCERPTS FROM THE MARTIN RELIABILITY MANAGEMENT (MRM) PROGRAM FOR THE ______ PROJECT

Task	Primary Responsibility	Other Functions Concerned	Procedure Approved		Controls	Schedule	
			Date	By		Start	Completion
Perform reliability predictions based on circuit and component analyses and use of reliability handbook. Compare to reliability budgets. If any design revisions are made, review reliability prediction and revise as required	Design engineer	Quality	Completed	Technical director	1 Reliability predictions to be included on page and line schedule 2 Analyses to be reviewed by group engineer 3 All analyses must be submitted to reliability engineer for his approval	7/1/58	10/1/59
Review all designs to determine component part and assembly limitations with respect to reliability requirements using economical processes and procedures	Electrical manufacturing	Quality	8/15/58	Manufacturing manager	1 All drawings must be reviewed by manufacturing representative prior to engineering release	9/1/58	1/1/59
Establish failure reporting and problem resolution system from first development test through service use 1 Establish and maintain failure reporting in test areas and service use 2 Establish and maintain failure reporting in selected vendors' plants	Quality	Engineering, manufacturing, procurement, logistics support	2/15/59	Quality manager	1 Assign specific reliability and assurance engineer and corrective action team to find problems and assure correction 2 Monitor reporting through assigned personnel in reporting areas	3/1/59	Completion of service use

requiring actions will be discussed in the present chapter,[6] others in later chapters.) Securing these actions commonly requires positive, formal planning if the deeds are to be done, and done on time. Moreover, since many departments and individuals are involved, it is useful to set out in detail the tasks to be performed, who is to do them, the timetable, etc. Table 9-1 gives a *small portion* of an example of such a formalized approach.[7] Note the provisions for (1) written procedures to tell each individual what he is expected to perform and (2) controls to assure that tasks have been adequately performed. In military applications, some companies submit a detailed reliability program as part of the proposal to obtain the contract. This "definitization" of tasks in the proposal helps to assure that adequate funds for a reliability program have been planned into the contract.

9-4 Setting Overall Reliability Goals

Designers and managers, as well as users, have long been convinced of the importance of high reliability. Specifications have often required the product to have "high reliability," or "maximum reliability." Designers accepted these statements but usually did not change their design decisions. This was not a matter of the designer preferring "low reliability" any more than a foreman prefers "low quality." The adjectives "high" and "maximum" simply did not permit a meeting of the minds among the parties concerned, any more than in the case of the term "high tensile strength." In addition, the terms "high" and "maximum" had been around a long time, and designers had become shockproof to them. Both authors have witnessed meetings in which a number of designers were asked to write down their numerical target for "high reliability." Not only were the numbers several orders of magnitude apart; the men were not even agreed on the unit of measure, i.e., time period, mean life, availability, etc.

A similar lack of meeting of the minds has existed with respect to operating environments. One product is designed to operate under

[6] 1 Setting overall reliability goals (items 1, 2, 3 in Fig. 9-1)
2 Allocation of the overall reliability goals (item 4)
3 Reliability prediction (item 4)
4 Design review (item 6)
5 Identification of critical parts (item 7)
6 Failure mode/failure effect analysis
7 Role of testing (items 8, 9, 15, 16, 18, 19)
8 Methods of improving reliability
9 Failure reporting and corrective action system (items 10, 11, 12)

[7] F. M. Gryna, Jr., "Total Quality Control through Reliability," *Transactions of the National Convention of the American Society for Quality Control,* 1960, p. 295.

maximum temperatures of 180°; in some applications, the temperatures rise to 200°. Another product is designed to carry maximum loads of 60,000 pounds; some customers load it to 70,000 pounds. The resulting failures have sparked some lively arguments about who is to blame. There are, of course, some glaring cases of "blame," but mostly these failures are traceable to lack of knowledge, by the designer, of the real environments and loads which the product will see in service. (In some cases the user may not have this knowledge either.) One of the contributions of the reliability "movement" has been to force all parties to dig deeper to uncover what will be these real environments.

Hence setting overall reliability goals requires a meeting of the minds on reliability as a number, on the environmental conditions to which that number is to apply, and on a definition of successful product performance. In some cases the customer sets the numerical goal, and the designer is faced with trying to achieve it. In cases where the customer has set no goal, good practice now requires that the manufacturer set the goal, and communicate this to all concerned.

The goal can be expressed in a variety of units of measure. The most usual are

1 In terms of reliability; e.g., the missile shall have 95 percent reliability during a mission time of 1.45 hours
2 In terms of failure rate; e.g., the batteries shall have a failure rate of no more than 1 percent over the 1-year guarantee period
3 In terms of time; e.g., the mean time between failures for the transmitter shall be at least 300 hours

9-5 Reliability Apportionment and Reliability Prediction

The reliability of a total system is a composite of the reliabilities of the various subsystems. The more reliable these subsystems, the more reliable the total system. In turn, if there is a reliability goal for the total system, there must be subsidiary reliability goals for the subsystems. The process of dividing up or "budgeting" of the final reliability goal among the subsystems (and sub-subsystems, etc.) is known as "reliability apportionment."

Our life insurance companies live handsomely on the proceeds of gambling on human mortality (failure rates). Their gambles are backed up by data on racial and family genetics (product design), by data on occupation and personal habits (environments), and by physical examination (test results). These data on specific individuals (components) are supplemented by data on the community mortality rates, long-range trends, new perils (system design). By combining these experience data with cause-and-effect relationships derived from actuarial (statistical) analysis, they *predict* human mortality.

In like manner, "reliability prediction" is the process of estimating quantitatively the probability that a product (component, subsystem, system) will perform its mission.

These two new tools, reliability apportionment and reliability prediction, are closely interrelated, as will be seen from the example in Table 9-2.

Table 9-2 shows both a reliability apportionment and a reliability prediction example. In the top section of the table, an overall reliability requirement of 95 percent for 1.45 hours has been apportioned to the six subsystems of a missile. The second section of the table apportions the budget for the explosive subsystem to the three units within the subsystem. Notice that the allocation for the fusing circuitry is 0.998 or, in terms of mean time between failures, 725 hours. In the final section of the table, the proposed design for the circuitry has been analyzed and a reliability prediction made, using the method of adding failure rates. (Note that the adding of failure rates is analogous to adding standard elemental times or costs to predict the time or cost for an overall manufacturing operation.) As the prediction indicates an MTBF of 1,398 hours as compared to a budget of 725 hours, the proposed design is acceptable. *The prediction technique not only provides a quantitative evaluation of a design or a design change but also can spotlight design areas having the largest potential for reliability improvement. Thus, the "vital few" will be obvious by noting the components with the highest failure rates.* The use of prediction as a tool before design release thus applies the principles of (1) defect prevention and (2) separation of the vital few from the trivial many. A detailed discussion of prediction techniques is included in Chapter 10.

Another example[8] of a reliability apportionment is shown in Table 9-3. A reliability goal was to be set for a new vehicle. Note that the overall goal on the complete vehicle and the apportionment (to the power train, engine, hitch and hydraulics, electrical components, and chassis) was determined by evaluating actual data on a similar vehicle and setting new goals based on expected improvements. Also note that the reliability goal on the complete vehicle was set in terms of average hours per failure and repair cost as a percent of price.

[8] Barrett G. Rich, O. A. Smith, and Lee Korte, *Experience with a Formal Reliability Program,* SAE Paper 670731, Farm, Construction and Industrial Machinery Meeting, 1967.

Table 9-2 ESTABLISHMENT OF RELIABILITY OBJECTIVES*

System Breakdown

Subsystem	*Type of Operation*	*Reliability*†	*Unreliability*	*Failure Rate per Hour*	*Reliability Objective*‡
Air frame	Continuous	0.997	0.003	0.0021	483 hours
Rocket motor	One-shot	0.995	0.005		1/200 operations
Transmitter	Continuous	0.982	0.018	0.0126	80.5 hours
Receiver	Continuous	0.988	0.012	0.0084	121 hours
Control system	Continuous	0.993	0.007	0.0049	207 hours
Explosive system	One-shot	0.995	0.005		1/200 operations
System		0.95	0.05		

Explosive Subsystem Breakdown

Unit	*Operating Mode*	*Reliability*	*Unreliability*	*Reliability Objective*
Fusing circuitry	Continuous	0.998	0.002	725 hours
Safety and arming mechanism	One-shot	0.999	0.001	1/1,000 operations
Warhead	One-shot	0.998	0.002	2/1,000 operations
Explosive subsystem		0.995	0.005	

Unit Breakdown

Fusing Circuitry Component Part Classification	*Number Used* n	*Failure Rate per Part* (λ) *(%/1,000 Hours)*	*Total Part Failure Rate* ($n\lambda$) *(%/1,000 Hours)*
Transistors	93	0.30	27.90
Diodes	87	0.15	13.05
Film resistors	112	0.04	4.48
Wire-wound resistors	29	0.20	5.80
Paper capacitors	63	0.04	2.52
Tantalum capacitors	17	0.50	8.50
Transformers	13	0.20	2.60
Inductors	11	0.14	1.54
Solder joints and wires	512	0.01	5.12
			71.51

$$\text{MTBF} = \frac{1}{\text{failure rate}} = \frac{1}{\Sigma n\lambda} = \frac{1}{0.0007151} = 1{,}398 \text{ hours}$$

* Adapted by F. M. Gryna, Jr., from G. N. Beaton, "Putting the R&D Reliability Dollar to Work," *Proceedings of the Fifth National Symposium on Reliability and Quality Control*, 1959, Institute of Electrical and Electronics Engineers, Inc., p. 65.
† In terms of probability of success for a given time period
‡ For a mission time of 1.45 hours

Table 9-3 EXAMPLE OF RELIABILITY APPORTIONMENT

Development of Reliability Goal for Complete Vehicle

	Hours/Failure (MTBF)	*Repair Cost (Percent of Price)*
Past experience	245	1.8
Past experience adjusted for improvements	305	1.0
Goal for new design	300	1.0

Development of Reliability Apportionment Hours/Failure (MTBF)

	Past Experience	*Past Experience Adjusted*	*New Goal*
Power train	1,200	1,960	1,650
Engine	900	1,260	1,250
Hitch and hydraulics	836	870	900
Electrical components	1,260	1,460	1,500
Chassis	6,580	6,580	6,500

9-6 Stages of Reliability Prediction

Reliability prediction is a continuous process starting with "paper *predictions*" based on a design and past failure rate information and ending with reliability *measurement* based on data from customer use of the product. Table 9-4 lists some characteristics of the various phases.

The quantification and prediction of reliability are receiving increased emphasis even in products under development. Sosdian[9] describes the extent of reliability requirements in a sample of recent development programs sponsored by the U.S. Army Electronics Command: 5 out of 14 exploratory development contracts, 16 of 33 advanced development, and 33 of 38 engineering development contracts required reliability predictions.

However, the numerical prediction procedure must not be viewed as an end in itself. The process of reliability prediction is justifiable *only if it proves useful in yielding a more reliable end product.*

The process of prediction may be as important as the resulting numbers. This is so because the prediction cannot be made without

[9] J. P. Sosdian, "U.S. Army Electronics Command's Reliability Programs," *Proceedings of the Spring Seminar on Reliability Techniques,* 1966, Institute of Electrical and Electronics Engineers, Inc., Boston section, p. 19.

Table 9-4 **Stages of Reliability Prediction and Measurement**

	1 *Start of Design*	*2* *During Detailed Design*	*3* *At Final Design*	*4* *From System Tests*	*5* *From Customer Usage*
Basis	Prediction based on approximate part counts and part failure rates from previous product usage; little knowledge of stress levels, redundancy, etc.	Prediction based on quantities and types of parts, redundancies, stress levels, etc.	Prediction based on types and quantities of parts failure rates for expected stress levels, redundancies, external environments, special maintenance practices, special effects of system complexity, cycling effects, etc.	Measurement based on the results of tests of the complete system. Appropriate reliability indices are calculated from the number of failures and operating time	Same as (4) except calculations are based on customer usage data
Primary uses	**1** Evaluate feasibility of meeting a proposed numerical requirement **2** Help in establishing a reliability goal for design	**1** Evaluate overall reliability **2** Define problem areas	**1** Evaluate overall reliability **2** Define problem areas	**1** Evaluate overall reliability **2** Define problem areas	**1** Measure achieved reliability **2** Define problems areas **3** Obtain data for future designs

Note: System tests in (4) and/or (5) may reveal problems that result in a revision of the "final" design. Such changes can be evaluated by repeating steps 3, 4, 5.

obtaining rather detailed information on product functions, environments, critical component histories, etc. Securing this information often gives the designer new knowledge previously not available to him. Inability to secure this information identifies the areas of ignorance in which the designer is forced to work.

The approach (in complex products) of adding failure rates to predict system reliability is analogous to the control of weight in aircraft structures where a running record is kept of weight as various parts are added to the design. Another analogy is the continuous record that is kept of the increased costs as complexity is added to a design. The reliability engineer, then, is asking that a continuous record be kept of the increasing failure rate as complexity and higher performance requirements are demanded of products.

To date, experience with reliability apportionment and reliability production has been that:

1 The *process* of using these tools is usually most helpful in identifying design weaknesses and strengths, and reaching a meeting of the minds among the men concerned.

2 The precision of reliability prediction still has a long way to go. A study[10] was made on 110 ground electronic equipments to compare the predicted and observed mean time between failures:
a The ratio of predicted to observed MTBF ranged from 0.09 to 5.0.
b Seventy percent of the predictions were pessimistic (the predicted was an average of 45 percent below the observed).
c The optimistic predictions were an average of 224 percent above the observed.

9-7 Design Reviews

Design review, in its broadest sense, is a mechanism for complete review of a proposed design to assure that the design can be fabricated at the lowest possible cost and perform successfully under end use conditions. Design review is not new but in the past has been informally done with relatively little preplanning and follow-up. Complex products often require a more formal program.[11] Formal programs are rela-

[10] Anthony J. Feduccia and Jerome Klion, "How Accurate Are Reliability Predictions?" *Proceedings of the Annual Symposium on Reliability,* 1968, Institute of Electrical and Electronics Engineers, Inc., pp. 280–287.
[11] R. M. Jacobs and H. D. Hulme, "Commercial Design Review and Data Analysis Program," *Transactions of the Nineteenth Annual Conference of the American Society for Quality Control,* 1965, p. 229.

tively new and have often been spearheaded by reliability engineers. A formal design review recognizes that no man can possibly know all the answers to achieve an optimum design and that a design review team may be necessary to break the monopoly of the original designer.

A formal program uses these principles:

1 Emphasis is placed on the review team consisting of the best available technical experts, internal or external to the company.
2 The review includes participation of personnel who had no connection with the proposed design.
3 The review generally includes not only reliability but also producibility, maintainability, and other pertinent parameters.
4 The review is generally made against defined criteria such as specification requirements and checklists of good and bad practices. Preparation for the review meetings requires that all participants thoroughly study the specification and associated documents beforehand and contribute specific constructive comments during the meetings. This is in contrast to reviews that bring a number of people together for a short meeting and ask them to react to a design that they have seen for the first time.
5 All potential problems uncovered by the design review are documented and responsibility is assigned to individuals for further investigation. The final decision on whether a design revision is required usually rests with the individual responsible for the design rather than with the team. The team provides constructive criticism on the design, but it does not relieve the design supervisor of ultimate responsibility for creating a design which best meets the valid criticisms.

Is a formal design review effort worth the time spent? Two surveys summarize some experience.[12]

The following two questions and the resulting answers are particularly important: "What proportion of designs reviewed result in change recommendations?"

Proportion of Design Reviews Resulting in Recommendations	*Percent Replying*
0–25%	25
26–50%	7
51–75%	8
76–100%	49
Unknown	13

[12] J. Y. McClure and E. S. Winlund, "Design Review: A Philosophy, Survey, and Policy," *Proceedings of the Ninth National Symposium on Reliability and Quality Control*, 1963, Institute of Electrical and Electronics Engineers, Inc., pp. 298–299.

"What proportion of design review recommendations are put into effect?"

Proportion of Recommendations Put into Effect	*Percent Replying*
0–25%	10
26–50%	16
51–75%	25
76–100%	36
Unknown	10

6 Reviews are held at several stages of design starting with the early review of requirements and the design approach for the total product down to detailed evaluations of designs on individual parts.[13]

9-8 Critical Components Program

This program aims to identify the "vital" few components which contribute decisively to total system reliability. Generally, a component is considered critical under any of the following conditions:

It is used in the equipment in large numbers.
It is obtainable from only a single source.
It must function to tight limits.
It has no proven record of high reliability.

Often, the individual designer is not aware of the existence of all these conditions. He may not even be able to correctly identify the components with a history of poor reliability. One of the writers has conducted an experiment to see if practicing engineers could identify the poor components (in a proposed design). The engineers were asked to state which type of component was most likely to be troublesome. Usually the *consensus* of the group is *not* the component with the highest failure rate in the reliability handbooks. Also, opinions vary widely on which component will cause the most trouble. This happens even when the engineers are from the same company and have dealt with the same basic product. All this means that what an individual engineer thinks will be the main reliability problem is based on *his* limited personal experience. Component failure rates in reliability handbooks are based on all available experience. The raw data from such failure rates

[13] For further elaboration, see J. M. Juran (ed.), *Quality Control Handbook* (*QCH*), 2d ed., McGraw-Hill Book Company, New York, 1962, p. 20-10.

are sometimes questionable, but the failure rates are probably more dependable than the recollections of one engineer in defining the vital few components. We all tend to be influenced by the sporadic (but dramatic) failures that make the headlines. However, the big potential lies with the everyday chronic failures: these are the "vital few."

A formal list[14] of the critical components is often prepared defining the critical features and plans to determine and/or improve reliability. Such a list serves these purposes: (1) plan test programs for evaluating components, (2) highlight failure causes that require preventive action by the designer if the component is to be used, and (3) highlight failure causes that require special handling procedures during manufacturing and usage of the component.

9-9 Failure Mode/Failure Effect Analysis

This preventive technique studies the causes and effects of failures before a design is finalized. Its only uniqueness is that it provides the designer with a methodical way to examine his design. In essence, a product (at the system and/or lower levels) is examined for all the ways in which a failure may occur. The effect of the failure is then evaluated and a review made of the action already taken or planned to minimize the probability of failure or to minimize the effect of the failure. The following elements may be included in the analysis:

1 Component name
2 Failure mechanism (the cause of failure)
3 Failure mode—the reaction of the component to the failure mechanism
4 Means by which failure is indicated and/or detected
5 Likelihood of failure
6 Immediate effect of component failure
7 Ultimate effect of the component failure on the system performance
8 Productivity effect—whether the system must be stopped and immediately repaired or repaired later during an off-duty cycle
9 Hardware items which must be removed to gain access to the failed component
10 Special tools (other than those ordinarily available) needed to repair or replace the component
11 Estimated repair time
12 Comments and/or recommendations
   a Recommended design changes
   b Requirements to be placed in specifications to minimize likelihood of failure

[14] For an example, see *QCH*, 1962, p. 20-12.

c Instructions to be placed in inspection, maintenance, or operation manuals to minimize likelihood of failure
d Tests to be conducted to more fully evaluate questionable modes of failure

The analysis is usually documented in a table. Each hardware item is listed on a separate line and entries are made in columns for information on elements such as those listed above. The designer will not have answers for all these elements but the analysis forces him to find the answers. Furthermore the information on the failure modes and effects on one item is helpful to the designers of other items in the system. An example of an analysis is shown in Figure 9-2.

The value of this analysis is difficult to grasp because it appears to duplicate the usual thinking of any designer and would, therefore, not reveal anything new. This will sometimes be the case, but for products of moderate complexity and particularly where the design function is divided into units responsible for small portions of the product, the technique deserves serious consideration. Its value can only be appreciated by trying it on a design in one's own company.

A practicing engineer[15] (not a reliability specialist) made a mode of failure analysis on an item of test equipment. He comments as follows:

> You requested that I comment on how the failure mode analysis for the hydraulic fatigue test equipment, upon which I performed a reliability analysis, caused me to realize that I didn't have a complete understanding of the system and helped me to reach that understanding.
>
> First of all, the choice to use a vane type pump rather than piston type was a basic one and was made by considering certain factors without the knowledge of how much effect higher pressures would have on particles coming off the vanes and affecting the electrohydraulic servo valve. This knowledge acquired during the failure mode analysis actually puts the decision to use vane type pumps on the border between right and wrong. In fact, if I had it to do over again I believe I would take the conservative approach and use piston type pumps.
>
> Another item was the directional valve. We have used these many times before but in this application its use was different (actually we had a lower cycle rate in this case) causing its failure rate to be greatly reduced. This points out the fact that you must know how an item functions to be able to choose a reasonable failure rate for the analysis.
>
> Although not important in this case, I completely misunderstood

[15] William E. Dearlove, then Manager of Applied Mechanics, Caterpillar Tractor Co.

PROJECT NO. X101

SYSTEM Planetary group

P = probability of
D = likelihood of
S = seriousness of

1 = very low or none (<1 in 10)   2 = low or minor (3 in 10)   3 = medium or (50-50)

COMPONENT (PART NO.)	POSSIBLE FAILURE	CAUSE OF FAILURE (FAILURE MECHANISM)
Gear, Hub Part No.	Grooved external spline teeth	Wear, case crushing
Plate, Reaction Part No.	Warped	Not made flat
		Excessive heat, slippage
	Worn or smeared	Lack of lube
Disc Assembly Part No.	Warped	Excessive heat, slippage
	Loss of friction material	Bond failure
Spring Part No.	Broken	Fatigue
		Improper assembly

Figure 9-2 Failure mode and effect analysis. Courtesy of Caterpillar Tractor C

the function of the unloading valve. The failure mode analysis straightened me out on this.

I think that the failure mode analysis is like an assembled jigsaw puzzle. Until you put each piece together and see how they are related one to another and get all the pieces interlocked, you cannot

PERSON MAKING ANALYSIS D. R. Longabach

currence

mage to surrounding components

lure to the system

DATE 10/1/69

nificant 4 = high (7 in 10) 5 = very high or catastrophic (>9 in 10)

D	S	EFFECT OF FAILURE ON SYSTEM	HOW CAN FAILURE BE ELIMINATED OR REDUCED
3	5	Will not transmit power	Heat treat splines
2	4	Clutch slippage	Provide straightening
2	4	Clutch slippage	Increase engaging force
2	4	Clutch slippage	Increase lube oil
		Clutch slippage	Increase lube oil
3	5	Clutch slippage	Develop better bonding
2	3	Lack of plate separation	Design for lower stress
2	3	Lack of plate separation	Provide assembly instruction

get the whole picture. I almost believe that you also need to do the failure mode analysis yourself—or at least go over each item in detail to gain the complete understanding. As I mentioned to you before, I did not have this opinion before I made up a failure mode analysis.

9-10 Reliability Testing

As used here, reliability testing means the tests conducted to verify that a product will work for a given time period. This ties in with the definition stated in section 9-1: reliability is the probability of a product performing without failure under given conditions for a specified period of time. The problems of reliability testing focus, therefore, on three elements: performance requirements, environmental conditions during usage, and time requirements.

Performance requirements are defined uniquely for each product, e.g., the horsepower of a motor, the gain of an amplifier. The stress that a product must withstand is a good example for discussion here. The difficulty is that this stress is usually not constant. It will vary between applications of the same product, i.e., the stress on a tractor working in sandy soil versus one working in rocky soil. The stress may vary during 1 day of usage of the same product. Thus, stress must be viewed in terms of a distribution rather than a single value. (Similarly, strength varies and must be viewed as a distribution. The analysis of these two distributions to predict reliability is discussed in Chapter 10.) The problem is to obtain the data necessary to construct the distribution. Such data are costly and there is often reluctance to make the investment because "there are too many variables among the customers." The answer lies in *sampling* the actual use of the product by typical customers. An alternative is to simulate customer usage (in a company "proving ground") and record the stress levels met. Once this information is available, tests must then be conducted to verify that the product can withstand the expected stresses for the required time period.

Environmental conditions (temperature, humidity, vibration, etc.) are critical to many products. The problem is twofold: determining the expected environmental levels and then testing to verify that the product can meet them. Figure 9-3 summarizes the testing on a guidance system. The six systems are first subjected to an acceptance inspection and then divided into two groups of three each. One group is subjected to the transportation and handling environments indicated; the other group does not undergo these tests. This indicates the effect of transportation and handling on later operations. All six systems are then run through the countdown and flight test environments listed. Two systems are run at the expected flight test environment; two are run at a level 10 percent higher than the expected; the remaining two are run at a level 20 percent higher than the expected. (When feasible, raising the levels until failure occurs will provide information on the margin of safety inherent in the design.)

The third element of reliability testing is time. Chapter 24 presents

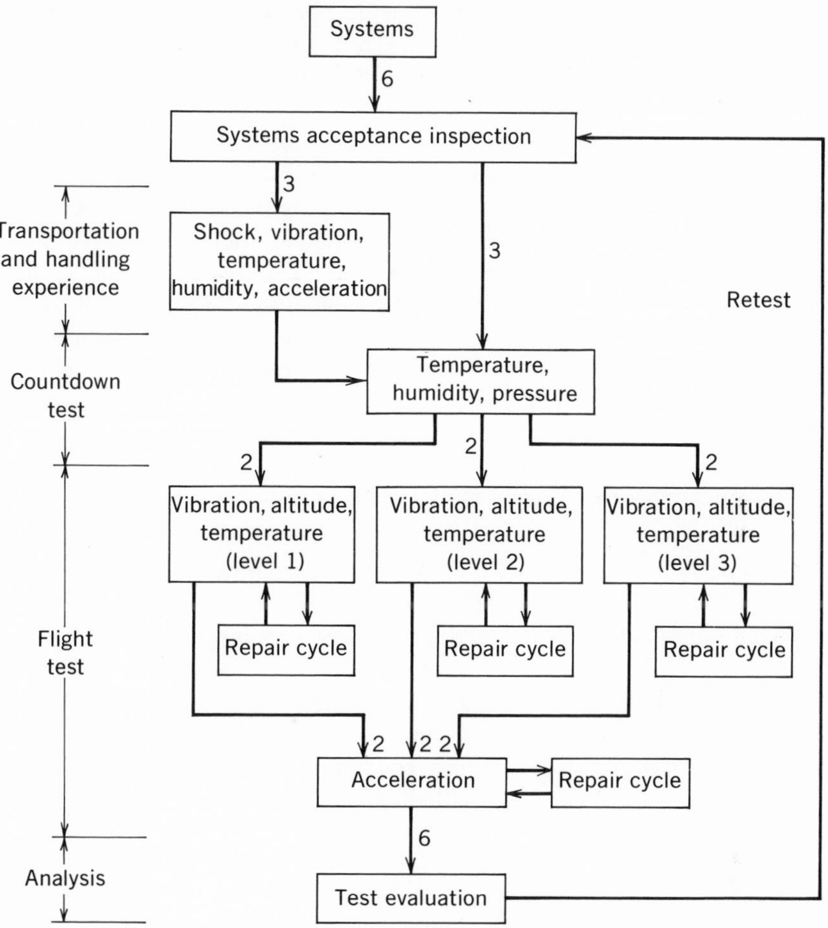

Figure 9-3 Reliability qualification of systems.

examples of sampling procedures which define the number of units to be tested, the test time, and the allowable number of failures. However, even with sampling procedures, there may still be a problem of a test program taking too long. Two approaches for overcoming this are:

1 Run a larger number of units simultaneously for a shorter length of time. (This requires certain statistical assumptions.)

2 "Accelerate" the testing by running the units at more severe stress and/or environmental levels to make them fail sooner. Extrapolations are then used to convert life under severe conditions to life under expected conditions.

Accelerated testing applies to a wide spectrum of products. An electric kitchen mixer is life tested in a thick liquid and the results extrapolated to predict the life when used on normal foods and liquids. Cheseborough[16] reports how 25 new transmissions were placed in New York City taxicabs to accumulate rough miles quickly. Figure 9-4 illustrates[17] how a "test-severity ratio" was established between proving ground testing and customer usage. Proving ground data and warranty data were separately plotted on Weibull paper and the mileage from each plot read for the same percent defects. At the 1 percent defect rate, the ratio of customer miles to proving ground miles is 3.3 to 1.

The reliability testing described here overlaps with conventional engineering testing of a new design and therefore the two must be planned concurrently.

9-11 Methods for Improving Design Reliability

Design reliability is improved by the same breakthrough sequence as was described in Planning to Meet Quality Objectives, section 3-5. In using this sequence, the distinction between diagnosis and action is vital.

Diagnosis to identify symptoms of design weakness and to trace these symptoms to their likely causes can be aided greatly by the tools of quantifying reliability and reliability prediction discussed in this text. At present the men skilled in use of these tools are not the designers, but the "reliability engineers."

Action to change designs to improve their reliability is best taken by the designer himself. It is the designer who understands best the uniqueness of the product and the engineering principles involved. The following actions indicate some approaches used by the designer working jointly with the reliability engineer to improve a design.

1 Review the index selected to define product reliability to be sure it reflects customer needs. For example, "availability" is sometimes more meaningful than "reliability." If so, a strong maintenance program can help to achieve the required availability and thereby help to ease the reliability problem.
2 Question the *function* of the unreliable parts with a view of *eliminating* them entirely if the function is found to be unnecessary.

[16] Harry E. Cheseborough, "Quality through Competition," *IQC*, vol. 20, no. 11, pp. 31–32, May, 1964.
[17] Bruce H. Simpson, "Reliability and Maintainability, Part 2—Automotive: The Ford Reliability Program," *Mechanical Engineering*, pp. 47–51, March, 1966.

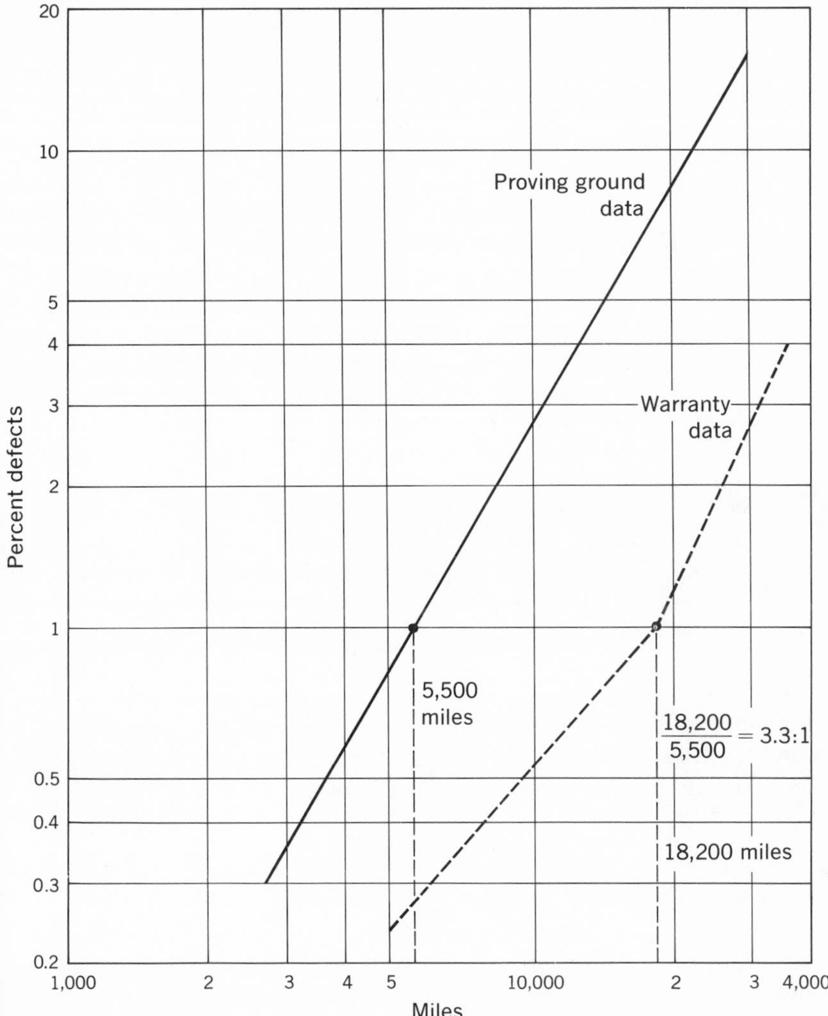

Figure 9-4 Proving ground to customer: subsystems correlation technique. On the one hand, proving ground durability; on the other, warranty data. Source: Bruce H. Simpson, "Reliability and Maintainability Part 2—Automotive: The Ford Reliability Program," *Mechanical Engineering*, March, 1966, pp. 47–51.

3 Review the selection of any parts which are relatively new and unproven. Use standard parts whose reliability has been proven by actual field usage. (However, be sure that the conditions of previous use are applicable to the new product.)

4 Conduct a research and development program to increase the reliability of the parts which are contributing most to the unreliability of the equipment.

5 Specify conservative replacement times for unreliable parts and replace the parts before they fail. Review the need for "burn-in" or "debugging" tests to eliminate "infant mortality" failures.
6 Select parts which will be subjected to stresses which are lower than the parts can normally withstand. This is called "derating." It is the equivalent of using a high safety factor. For example, a pump designed for 3,000 psi is used in a 2,000-psi application; a capacitor rated at 300 volts is selected for a 150-volt application. For some components, data are available showing failure rate as a function of the level of stress imposed on a part (see Chapter 10).
7 Control the operating environment so that a part will be operating under conditions which yield a lower failure rate. For example, a cooling system to reduce operating temperature or the potting of parts for shock and climate protection will reduce the failure rate for some parts.
8 Use redundancy so that if one unit fails a redundant unit will be available to do the job.
9 Consider possible tradeoffs of reliability with functional performance, weight, or other parameters. (Although the reduction of functional performance has disadvantages, the overall product effectiveness for the customer may be higher if a higher reliability level can be provided even at the expense of some functional performance.)

Some of these actions can be taken by the designer himself. Others require management consideration. As he is concerned with the total product, the reliability engineer can play an important role in problems that are beyond the area of an individual designer.

9-12 Corrective Action System

Most of the early efforts in reliability programs were concentrated on documenting failures, preparing and analyzing summaries of performance, and following up to obtain any corrective action necessary. These "failure reporting and corrective action systems" have three basic objectives:

1 Detect and assure resolution of problems affecting reliability
2 Assemble and disseminate failure history and other reference data for use in preventing similar failures in product modifications and in future products
3 Keep management aware of the status of problems affecting the product

Figure 9-5 shows a breakdown of the corrective action process based on concepts developed by A. M. Carey. This breakdown places special

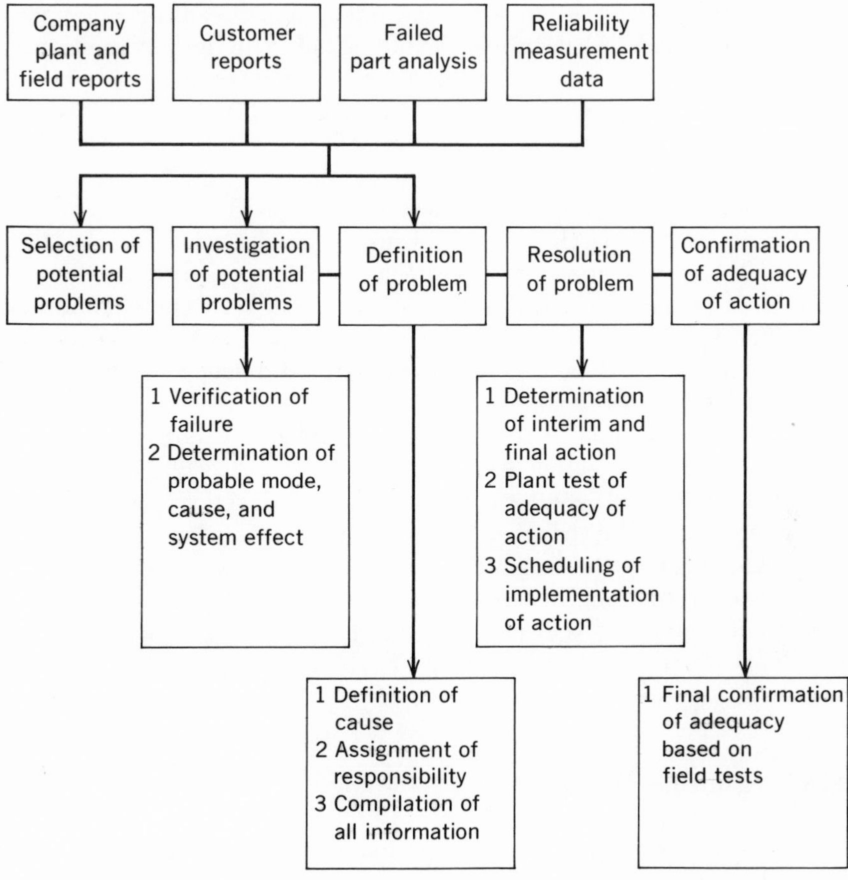

Figure 9-5 Corrective action system.

emphasis on the proper selection and definition of problem areas (see Chapter 30). The corrective action system follows the principles of the control cycle to hold the reliability improvements achieved by the breakthrough cycle previously discussed.

Corrective Action Coordination

The effort to collect and process failure data will be valueless unless means are provided for action to remedy the problem areas disclosed. Assignment of responsibility for corrective action, and follow through to see that the action is completed and becomes effective, are therefore of top importance.

Because the needed corrective action may lie in one (or more than one) of several areas within the organization, and because it is usually not easy to decide what action should be taken, it is important that *joint* analysis of the data be the basis for assignment of responsibility for action. The most effective method is to form a "reliability improvement committee" or "corrective action group."

These committees are usually chaired by the cognizant product-oriented reliability engineer and meet on a regularly scheduled basis. Attendance is dictated by the problems to be discussed and by the trend of the investigation for a particular problem. Generally, the responsible design engineer and a quality control engineer attend all meetings. For specific corrective action requirements, respresentatives of purchasing, product design, production control, components engineering, or other similar activities are invited.

A problem agenda is sent out by the reliability engineer several days ahead of the meeting date. Minutes of the meetings, including a problem-status log, are formally recorded, and actions to be taken are documented and distributed to all concerned. This log[18] summarizes the problem before the committee and shows start and completion of action dates, responsiblility, and action taken. It gives project management an indication of major problems and the status of corrective efforts. Additional effort may then be placed on troublesome areas as deemed necessary.

9-13 Organizing for Reliability: General

The size and complexity of organization for reliability is greatly influenced by the size and complexity of the project at hand. In addition, the organization design is influenced by forces which have their origin in national historical and cultural patterns. In the USA, the creation and training of a highly specialized category of "reliability engineer" has proceeded further than in any other industrialized country. In other countries, the historical developments in patterns of responsibility and of training are different, with resulting differences in the approach to organizing for reliability.

Ideally, the following policies[19] should be recognized in organizing for reliability.

1 A reliability program must start in the conceptual phase of a project and continue throughout design and development, production, test, field evaluation, and service use. This means that the

[18] For an example, see *QCH*, 1962, p. 20-37.
[19] Gryna, *op. cit.*

program cannot be restricted to any one organizational unit but must cover all units that affect the final field reliability.

2 Adequate funds must be provided for a reliability program and these must be determined during the proposal phase. This means that a complete reliability program must be developed in sufficient detail during the proposal effort to permit adequate costing.

3 The execution of a reliability program involves both technical tasks and a management task. The technical tasks consist of the efforts to design reliability into the product and maintain this reliability throughout production and field use of the product with minimum degradation. The management task consists of integrating all of the technical efforts, and controlling these efforts to ensure that all necessary steps are being taken to achieve the required reliability.

4 Reliability *results* can be achieved only by actions taken by the line organization—the designer, the production man, the procurement man, etc. The reliability specialist provides guidance and assistance to the line personnel in executing their fundamental reliability tasks.

5 The program for each project must provide a written plan and must specify responsibility, procedures, and schedule.

6 The program must include controls which will detect and report to management all deviations between plans and actual performance.

7 The program must include suppliers as well as internal company operations.

8 The overall integration and evaluation of the reliability program must be performed by an organization which is independent of those who are responsible for taking the detailed steps necessary to achieve the required reliability.

This ideal is too elaborate for many products in the lower half of the spectrum of size and complexity of products. However, in the upper half of the spectrum it becomes necessary to work close to the ideal.

For a component manufacturer, the formal reliability effort will probably be corrective in nature and concentrate on the collection and analysis of field performance data to measure reliability and define problem areas. This effort is usually located in an engineering, service, or quality function of the company. The efforts to achieve the required reliability will be performed by the line function concerned as part of their normal responsibilities.

For a manufacturer of assembly products, a more formal structure may be required. The program will be both preventive and corrective and will involve some of the specific techniques discussed in this chapter. Reliability specialists from design, manufacturing, and other functions are assigned to each product on a full-time or part-time basis. The reliability specialists perform their specific tasks. Informal coordination may suffice but as products become more complex and stringent quantified reliability requirements are imposed, the reliability organization requires a more formalized structure.

9-14 Organization Forms in the High Reliability Industries

In recent years, the defense, aerospace, computer, and still other industries have faced a new order of magnitude of requirements for products of high reliability. To meet this new requirement has required participation from all functional areas of the company, plus the creation of a new category of specialist, the reliability engineer. As these specialists were appointed, it was necessary to fit them in the scheme of organization design.

Some common organization forms are shown in Figure 9-6 and may be described as follows: (1) a reliability staff department reporting to the director of engineering, (2) a reliability department which is

Figure 9-6 Alternative locations of the reliability function.

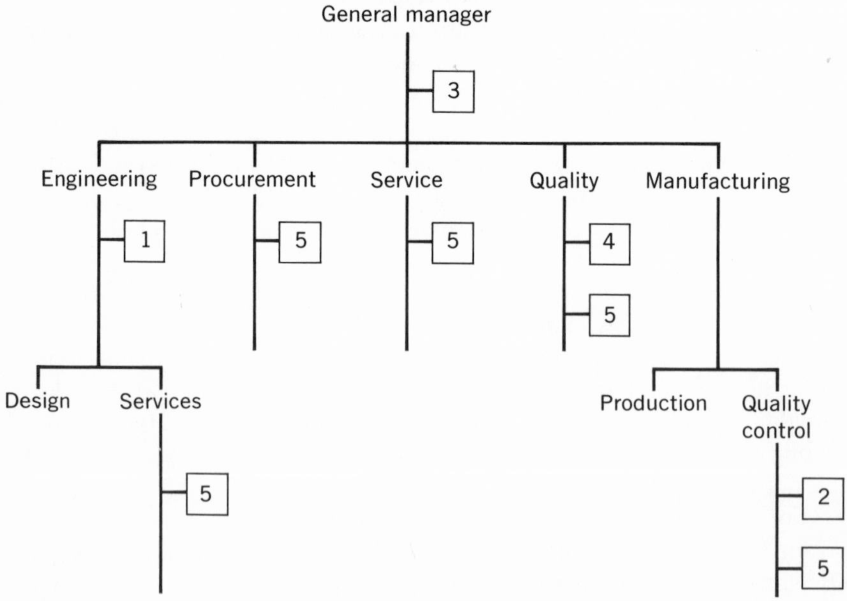

part of a quality control function reporting to a manager of manufacturing, (3) a reliability staff department reporting to the general manager, and (4) a separate department parallel to engineering and manufacturing, and including the manufacturing quality control and customer support functions. These first four forms have all the reliability activities centralized in one organization. In (5), the activities are decentralized and performed by the appropriate line department with the quality function responsible for auditing and coordinating the activities.

The advantages and disadvantages of each of these methods are listed in Table 9-5.

Table 9-5 ATTRIBUTES OF ALTERNATIVE ORGANIZATIONS

Organization Form	*Advantages*	*Disadvantages*
1 Reporting to director of engineering	Easy to coordinate with design activities	Difficult to coordinate with manufacturing and quality control activities
2 Reporting to quality control manager	Easy to coordinate with manufacturing quality problems	Difficult to coordinate with design activities (often not accepted by designers)
3 Reporting to general manager	Can influence policy formation and objectives for reliability and quality control	Receives little effective supervision; too far removed from the action level, especially in large organizations
4 Separate major company department	On policy making and goal setting level; can influence work along entire "spiral." Simplifies coordination among quality specialists	Difficult to coordinate with work of line departments unless they have been well trained in the specialties
5 Decentralized among line departments, coordinated by a separate quality staff department	Access to policy making and goal setting in all departments; access to department heads for dealing with unreliable products; greater likelihood of acceptance by line departments	Difficult to coordinate the scattered reliability activities into a coherent plan; risk that specialists will be assigned to sporadic rather than chronic problems

9-15 Maintainability

Field failures will be around for a long time, and so means must be provided for restoring service. The problem of restoring service is an old one, and it has received much attention from manufacturers and users, though on the vague basis of "easy to maintain." Now, with human safety, health, and convenience depending so largely on continuity of service, the quantification of "maintainability" has become a reality.

This quantification consists of creating two new parameters:

1 A time parameter to express the time during which service should be restored, as for example, maximum downtime.

2 A probability parameter to express the probability of meeting the time parameter. This is also the *definition of "maintainability"*: "the probability that a failed system is restored to operable condition in a specified downtime when maintenance is performed under stated conditions."[20]

As with reliability, these parameters are often combined into one such as "mean time to repair" (MTTR).

Once these parameters are quantified, it becomes possible (as in the case of quantified reliability) to secure a meeting of the minds among all concerned as to what are the goals, how to plan to meet the goals, and how to judge whether the goals have been met.

At present, procedures and organization forms for setting and meeting maintainability goals are in the early stages of development. Techniques for predicting and allocating maintainability are being evolved. The tedious job of collecting information on downtime, time to repair, cost to repair, etc., is underway. (This is a very difficult problem because the repair activities are widely scattered among many users, and because maintenance shops are strongly resistant to record keeping.)

For many complex products, the designer receives little feedback on the adequacy of his design from the viewpoint of maintainability. He will hear if there is a *major* problem with his design from the viewpoint of maintainability, but the feedback loop is usually informal and incomplete.

Neither does the designer always know the effect of a failure in his subsystem on overall system performance. A number of people in one company were asked to classify certain failures as to whether (1) the failure required that the machine be shut down and the repair made immediately or whether (2) the failure could be repaired later when the machine was in an off-duty cycle. They could not agree on many

[20] ARINC Research Corp., *Reliability Engineering*, Prentice-Hall, Inc., Englewood Cliffs, N.J., 1964, p. 346.

of the failures. It became apparent that for complex systems the effect of a component failure is not obvious unless a person has a clear picture of how that component fits into the remainder of the design. Unfortunately, the designer of a component for a complex unit is *not* always knowledgeable *in detail* on the effect of his component on the total system from the viewpoint of maintainability. This means that (1) he may not know if a failure of his component will immediately shut down the machine, and (2) he may not realize the difficulty in repairing his component due to a problem such as gaining access to his component in order to remove and repair it, or repairing it with the minimum of special skills and tools. Note that in the mode of failure analysis mentioned in section 9-9, items of this nature are included.

Finally, although every designer wants his design to be easily maintained, his emphasis on this characteristic will be influenced by the importance that management gives to maintainability. For example, if a designer is not given reasonably clear objectives (quantitative or qualitative) for maintainability, it is only natural for him to conclude that it is not as important as other characteristics that have been defined more clearly.

9-16 Quantification of Maintainability

An overall numerical maintainability goal can be set on a system. The overall goal is apportioned to determine goals for each subsystem. Maintainability can be predicted on paper during the design phase, and it can finally be measured by timing the period required for certain maintenance actions on hardware. (This is analogous to goal setting, apportionment, prediction, and measurement for reliability.)

The methods of maintainability allocation and prediction are still being developed. The methods[21] differ widely. For example, one prediction method extrapolates past experience on maintainability of similar equipment to make predictions on a new design. Another method breaks down the maintenance task into elemental maintenance tasks and uses past data on these elemental tasks to build up a predicted maintainability time for a total repair. This is analogous to predicting the time required for manufacturing operation by working with standard elemental times from a time study handbook. Still another method of maintainability prediction uses a checklist to score significant features of a system and then uses the scores in a regression equation to predict average repair time.

[21] Michael G. Harring and Lyle R. Greenman, *Mantainability Engineering,* Martin-Marietta Corp., Orlando, Fla., 1965.

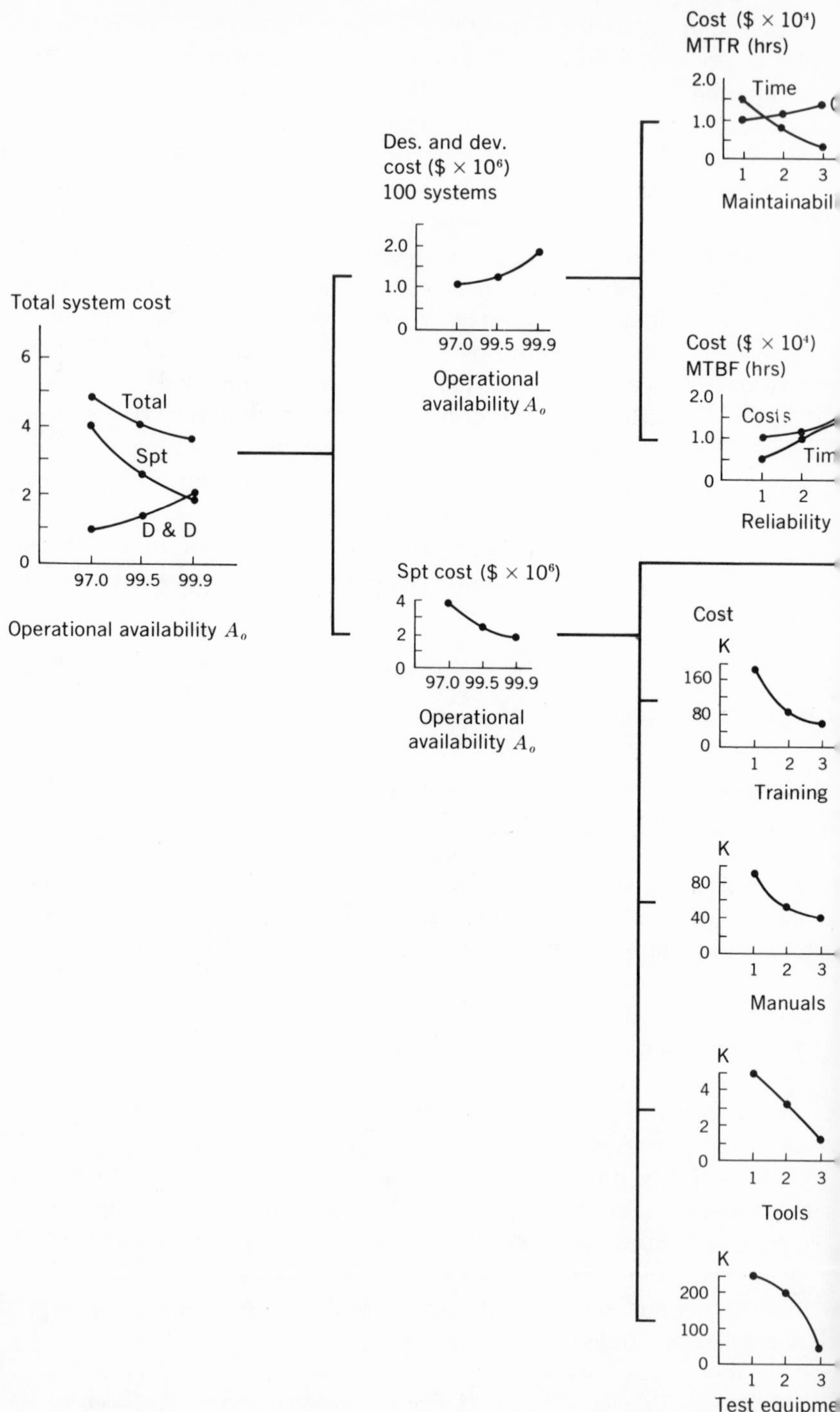

Figure 9-7 Information for a tradeoff analysis. Source: Michael G. Harring and Fla., 1965.

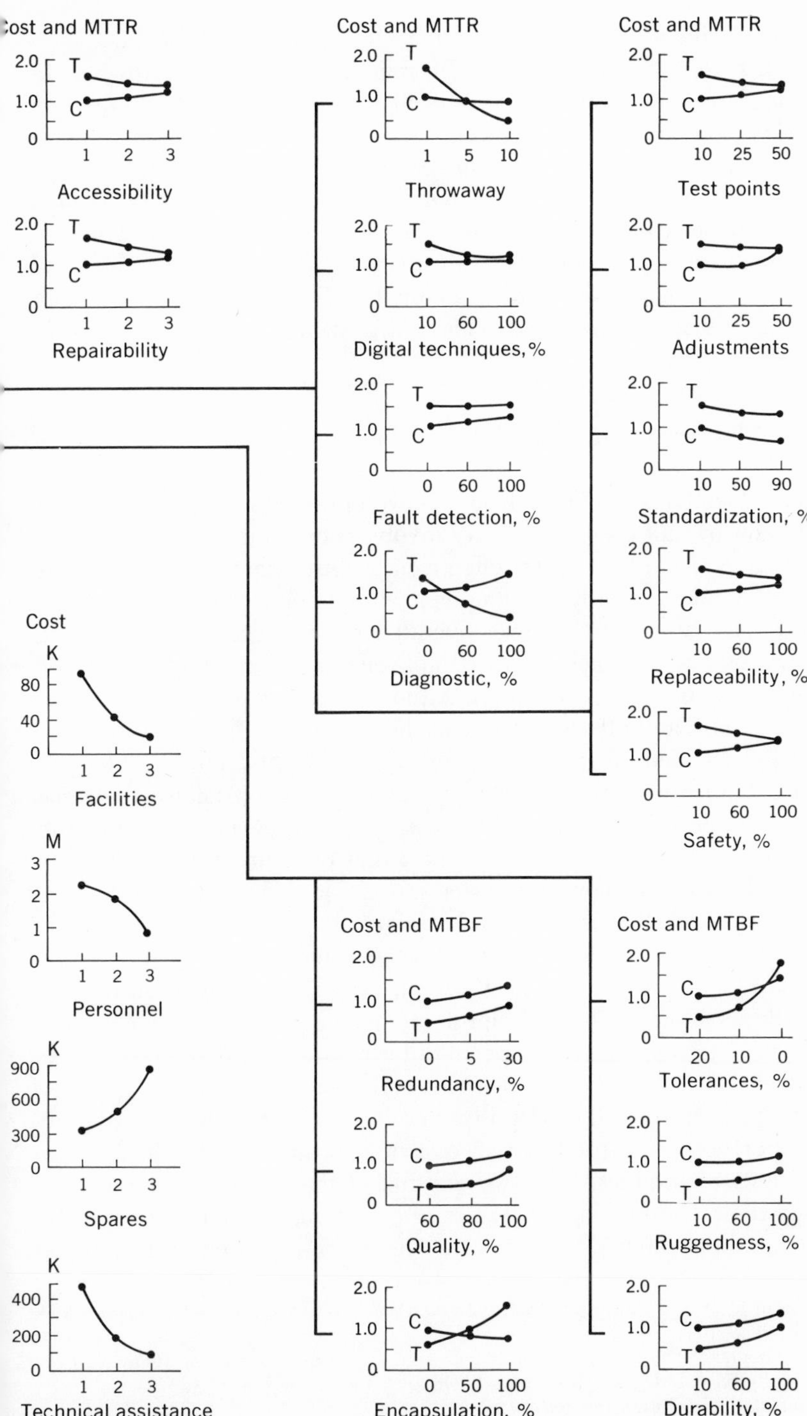

Lyle R. Greenman, *Maintainability Engineering*, Martin-Marietta Corp., Orlando.

Efforts[22] have been made to demonstrate maintainability in a manner similar to the demonstration of reliability. Thus failures are induced into an equipment and repairs made, using standard maintenance procedures and average skill levels. The times for the repairs are recorded on an acceptance/rejection scheme used to decide if the maintainability requirement has been met. An earlier approach to demonstration can be achieved at the mock-up stage of a product. Thomas[23] describes the use of mock-ups to evaluate maintainability on the DC-9 airplane. The mock-ups were used on the DC-9 to prove maintainability before final approval and release of drawings.

9-17 Tradeoffs among Parameters

The real objective in industry is to achieve optimum overall effectiveness at optimum overall cost. This may involve increasing cost in one phase to reduce cost and increase effectiveness elsewhere. Self-lubricating bearings, stainless steel, and redundant circuits all involve greater original cost, but can result in lower overall costs and greater effectiveness. This process of balancing costs and effects is known as *tradeoff*. Through this concept we manipulate the parameters of reliability, maintainability, cost, and others to arrive at the optimum.

Designers have always considered tradeoffs between variables. As system complexity grows, informal qualitative tradeoff analysis becomes difficult. Attention has been given recently to the development of quantitative approaches. It should be stressed that the many interrelationships between variables in complex systems make it difficult to develop any tradeoff approach that is completely quantitative. As with all quantitative techniques in product effectiveness, it aims to narrow the area of uncertainty and provide the decision maker with an evaluation of alternative designs that will be helpful in making final judgements.

Figure 9-7 depicts the type of information included in a tradeoff analysis.[24] Three alternative designs were evaluated and the graph on the left indicates that the proposed design 3 is best with an availability of 99.9 percent and a total system cost of $3.6 million over the life-span for all units of product. The top portion of the figure shows the effect on MTTR of various characteristics affecting maintainability-accessibility, repairability, level of hardware thrown away, digital techniques,

[22] For example, see *Maintainability Demonstration,* MIL-STD-471, Government Printing Office, Feb. 15, 1966.
[23] F. C. Thomas, "DC-9 Maintainability Design Features," *Annals of Reliability and Maintainability,* vol. 4, July, 1965, Spartan Books, Inc., New York.
[24] Harring and Greenman, *op. cit.*

fault detection, ease of failure diagnosis, test points, adjustments stand-dardization, replaceability, and safety.

Each of the three designs is quantitatively evaluated for the effect of each characteristic on design and development costs and MTTR. For example, the accessibility graph shows that design 2 would have more accessibility for repairs (and therefore a lower MTTR) than design 1, but the cost would be somewhat higher. A corresponding "tree" is shown for the characteristics affecting reliability-redundancy, quality encapsulation of hardware, tolerances, ruggedness, and durability. The bottom portion of the graph shows the effect of support characteristics on support costs.

9-18 Organization for Maintainability

A formal maintainability program for a complex product requires that tasks be performed by several functions such as design, manufacturing, and customer service. As with reliability, many problems require *design* action and, therefore, many companies have set up a maintainability organization within the design function. (The effort is sometimes combined with design reliability, value engineering, and associated efforts.) However, as with reliability, other organizational units have long been concerned with maintainability—particularily a customer service function. Therefore, the organization alternatives previously discussed for reliability also apply for maintainability.

PROBLEMS

1 Read the RPM case in the Appendix and answer the following problem. (Your instructor will set up the committee required by the problem.)

You are a committee which has been appointed to come up with recommendations on how to organize for new product control. In your group are members from the following departments: chief engineer, reliability, quality control engineering, quality control, manufacturing engineering, purchasing, production, and customer service.

To date there has been plenty of discussion on how to set up new product control, but the managers have been pretty far apart. The general manager's staff would like to see the differences shaken down better, and the differences more clearly defined, before they get into it. Your committee has the responsibility of working up recommendations for the general manager's meeting. The discussions to date have reached about the following state:

Design review The atlernatives offered so far have been (*a*) appointment of reliability engineers who would be attached to the various engineering design sections, (*b*) appointment of a reliability engineering group which would be part of the chief engineer's department, and (*c*) appointment of a reliability group which would be part of the reliability manager's department.

There is also an argument on how to conduct design reviews. The opposing schools of thought favor: (*a*) a formal study and report by a reliability engineer, (*b*) an informal review by a reliability engineer, and (*c*) a committee review, including the design engineer, a reliability engineer, and others.

Planning and evaluation of prototype tests The argument here has been whether the reliability engineer (wherever located) should be a part of the original determination, or should only review what has been done by the designers; and if done by review, whether to do this formally, informally, or by committee.

Planning of reliability data systems The argument has centered on three alternatives:

a Leave the present responsibilities unchanged, but provide a review by the reliability manager. The present responsibility is about as follows:

Planning of Data System for	*Planning Done by*
Prototype testing	Design engineers
Vendor quality control	Quality control engineering and vendor quality control
Shop losses (scrap, rework, etc.)	Quality control engineering and accounting
Process and final inspection, and test	Quality control engineering and quality control superintendents
Field failures	Quality control engineering and customer service

b Use a committee approach, including reliability.

c Shift the prime responsibility to the reliability manager.

Preparing overall quality plans during manufacture The present responsibility is about as follows:

Planning Activity	*Prepared by*
Determination of production operations to be performed	Manufacturing engineering (master mechanic) and production supervision
Determination of gaging and test to be performed by production	Manufacturing engineering and production supervision
Planning of inspection and acceptance testing	Quality control engineering and quality control superintendents
Design of gages and test equipment	Quality control superintendents

Again there are three alternatives, similar to those available for planning of reliability data systems.

Collect and analyze failure data The reliability manager feels that he should take over this responsibility from the quality control engineering manager, who does not think so.

Problem: Assign to each man a committee member role. Discuss the foregoing, and prepare recommendations on who should do what.

2 Read the RPM case in the Appendix and answer the following problem.

You are Engblom, the quality manager. You have analyzed the needs of the car division for reliability activities. You have also talked with the other managers about how the car division might best carry out these activities. You now have tabulated the activities you consider necessary, showing, for each, the present status:

Necessary Activities	*Status*
1 Establish reliability goals for car and for subsystems	Not being done. Present departments lack the skill to do this
2 Predict reliability in various phases (design, prototype testing, manufacture, use)	Not being done. Present departments lack skill to do this
3 Review designs for reliability	Not being done in "modern" fashion. But the chief engineer points out that his checkers and supervision now check the designs

Necessary Activities	*Status*
4 Identify critical parts and characteristics	Some of this is now being done as the result of field failure analysis (by the service department or quality control engineering); as the result of shop troubles (by accounting and quality control engineering)
5 Choose tests for prototypes	Now being done by chief engineer's department
6 Evaluate test results	Now being done by chief engineer's department. Not using modern methods of analysis
7 Plan reliability data systems	Done by chief engineer for prototype tests; by quality control engineering for shop and vendor failures; by service department and quality control engineering for field failures
8 Conduct reliability indoctrination	Not being done
9 Plan for such gaging and test as are to be performed by production departments	Now done by manufacturing engineering and the production supervisors
10 Plan inspection and acceptance testing; design gages and test equipment	Now done by quality control engineering and quality control superintendents.
11 Conduct vendor surveys and surveillance of vendor reliability programs	Not being done. Control is through incoming inspection
12 Conduct incoming inspection, process inspection, final inspection, and test	Now being done by quality control line departments
13 Collect and analyze field failure data	Now being done by quality control engineering
14 Follow up to correct designs causing failures	Now done by quality control engineering
15 Follow up to correct vendor performance	Done by vendor quality control manager and purchasing
16 Follow up to correct shop processes	Done by quality control line departments
17 Surveillance of dealer contribution to field failures	Not being done
18 Reports to top management on reliability	Not being done

Problem: For the activities listed, determine to which of your departments you will assign each.

3 Visit a local plant and determine whether any formal or informal numerical reliability and maintainability goals are issued to the design function for their guidance in designing new products.

4 Obtain a schematic diagram on a product for which you can also obtain a list of the components that fail most frequently. Show the diagram to a group of engineering students most closely associated with the product (e.g., a mechanical product would be shown to mechanical engineering students). Have the students *independently* write their opinion of the most likely components to fail by ranking the top three. (*a*) Summarize the results and comment on the agreement among the students and (*b*) comment on the student opinions versus actual product history.

5 Examine the warranty statements of two competing brands of one of the following products: (*a*) a product acceptable to the instructor, (*b*) a passenger automobile, (*c*) an automatic washing machine, and (*d*) a phonograph.

Comment on the adequacy of the statements from the viewpoint of the manufacturer and user.

6 Make a failure mode/failure effect analysis for one of the following products: (*a*) a product acceptable to the instructor, (*b*) a flashlight, (*c*) a toaster, and (*d*) a vacuum cleaner.

7 Speak with some practicing design engineers and learn the extent of feedback of field information to them on their own design work.

8 Section 1-12 defined the concept of self-control. For a design department with which you are familiar, state whether the engineers "know what they are supposed to do, know how they are doing, and have the means to regulate."

9 Consider the specific techniques mentioned in this chapter. Prepare a table consisting of three columns. The first column should list the name of the technique. The second column should list the objective of the technique. In the third column, list an effectiveness rating (strong, adequate, weak) for accomplishing the objectives in a company with which you are familiar.

10 A reliability prediction can be made by the designer of a product or by an engineer in a staff reliability department which might be a part of the design function. One advantage of the designer making the prediction is that his knowledge of the design will likely enable him to do a more thorough and faster job. (*a*) What is one other advantage in having the designer make the prediction? (*b*) Are there any disadvantages to having the designer make the prediction?

11 Figure 9-5 lists five main elements of a corrective action system. For a company with which you are familiar, rate the effectiveness (strong, adequate, weak) in executing these elements.

12 Prepare a formal presentation to gain adoption of one of the following: (*a*) quantification of reliability goals, apportionment, and prediction, (*b*) formal design reviews, (*c*) failure mode/failure effect analysis, and (*d*) critical components program.

You will make the presentation to one or more people who will be invited into the classroom by the instructor. These people may be from industry or may be other students or faculty. (The instructor will announce time and other limitations on your presentation.)

13 Outline a reliability test for one of the following products: (*a*) a product acceptable to the instructor, (*b*) a household clothes dryer, (*c*) a motor for a windshield wiper, (*d*) an electric food mixer, and (*e*) an automobile spark plug.

The testing must cover performance, environmental, and time aspects.

10 Statistical Aids to Reliability

10-1 Failure Patterns for Complex Products

Complex products often follow a familiar pattern of failure. Consider the data in Table 10-1. Assume that one unit was started on test and the time recorded when it failed. The unit was repaired, again placed on test, and the time of the next failure recorded. The "failure rate" for the unit can be calculated for equal time intervals as the number of failures per unit time. When the failure rate is plotted against a continuous time scale, the resulting chart (Figure 10-1), known as the "bathtub curve," exhibits three distinct periods or zones. These zones differ from each other in frequency of failure and in the failure causation pattern as follows:

1 The infant mortality period. This is characterized by high failure rates which show up early in usage. Commonly these failures are the result of blunders in design, manufacture, or use or of misapplication and other identifiable causes. Sometimes it is possible to "debug" the product by simulated use test or by overstressing (in electronics this is known as burn in). The weak units still fail, but the failure takes place in the test rig rather than in service.

2 The constant failure rate period. Here the failures result from the limitations inherent in the design plus accidents caused by usage or poor maintenance. The latter can be held down by good control on operating and maintenance procedures. However, a reduction in the basic failure rate requires a basic redesign.

3 The wear out period. These are failures due to old age; e.g., the metal becomes embrittled; the insulation dries out. A reduction in failure

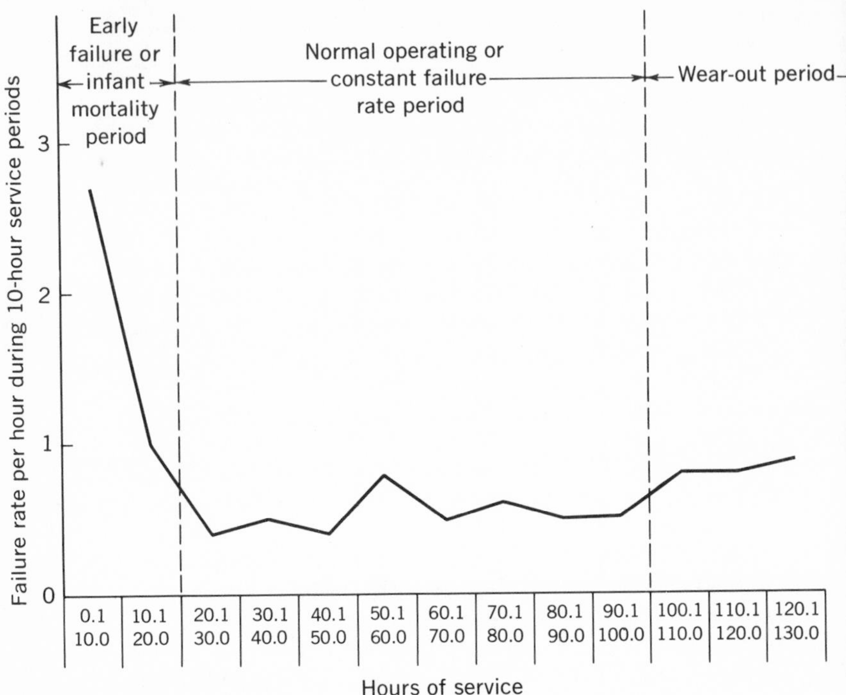

Figure 10-1 Failure rate versus time.

rates requires preventive replacement of these dying components before they result in catastrophic failure.

The Distribution of Time between Failures (TBF)

Users are concerned with the length of time that a product will run without failure. For repairable products, this means that the "time between failures" is a critical characteristic. (The corresponding characteristic for nonrepairable products is usually called the "time to failure.") The variation in time between failures can be studied statistically (Figure 10-2).

When the failure rate is constant, the distribution of time between failures is distributed exponentially. Consider the 42 failure times in the constant failure rate portion of Table 10-1. The time between failures for successive failures can be tallied, and the 41 resulting TBFs can then be formed into the frequency distribution shown in Figure 10-2*a*. The distribution is roughly exponential in shape, indicating that when the failure rate is constant the distribution of time between failures (not

Table 10-1 FAILURE HISTORY FOR A UNIT OF ELECTRONIC GROUND SUPPORT EQUIPMENT

Time of Failure, Infant Mortality Period		*Time of Failure, Constant Failure Rate Period*		*Time of Failure, Wear Out Period*	
1	7.2	28.1	60.2	100.8	125.8
1.2	7.9	28.2	63.7	102.6	126.6
1.3	8.3	29.0	64.6	103.2	127.7
2.0	8.7	29.9	65.3	104.0	128.4
2.4	9.2	30.6	66.2	104.3	129.2
2.9	9.8	32.4	70.1	105.0	
3.0	10.2	33.0	71.0	105.8	
3.1	10.4	35.3	75.1	106.5	
3.3	11.9	36.1	75.6	110.7	
3.5	13.8	40.1	78.4	112.6	
3.8	14.4	42.8	79.2	113.5	
4.3	15.6	43.7	84.1	114.8	
4.6	16.2	44.5	86.0	115.1	
4.7	17.0	50.4	87.9	117.4	
4.8	17.5	51.2	88.4	118.3	
5.2	19.2	52.0	89.9	119.7	
5.4		53.3	90.8	120.6	
5.9		54.2	91.1	121.0	
6.4		55.6	91.5	122.9	
6.8		56.4	92.1	123.3	
6.9		58.3	97.9	124.5	

mean time between failures) is exponential. This is the basis of the exponential formula for reliability.

10-2 The Exponential Formula for Reliability

The distribution of time between failures indicates the chance of failure-free operation for the specified time period. The chance of obtaining failure-free operation for a specified time period *or longer* can be shown by changing the TBF distribution to a distribution showing the number of intervals equal to or greater than a specified time length, Figure 10-2*b*. If the frequencies are expressed as relative frequencies, they become estimates of the probability of survival. When the failure rate is constant, the probability of survival (or reliability) is

$$P_s = R = e^{-t/\mu} = e^{-t\lambda}$$

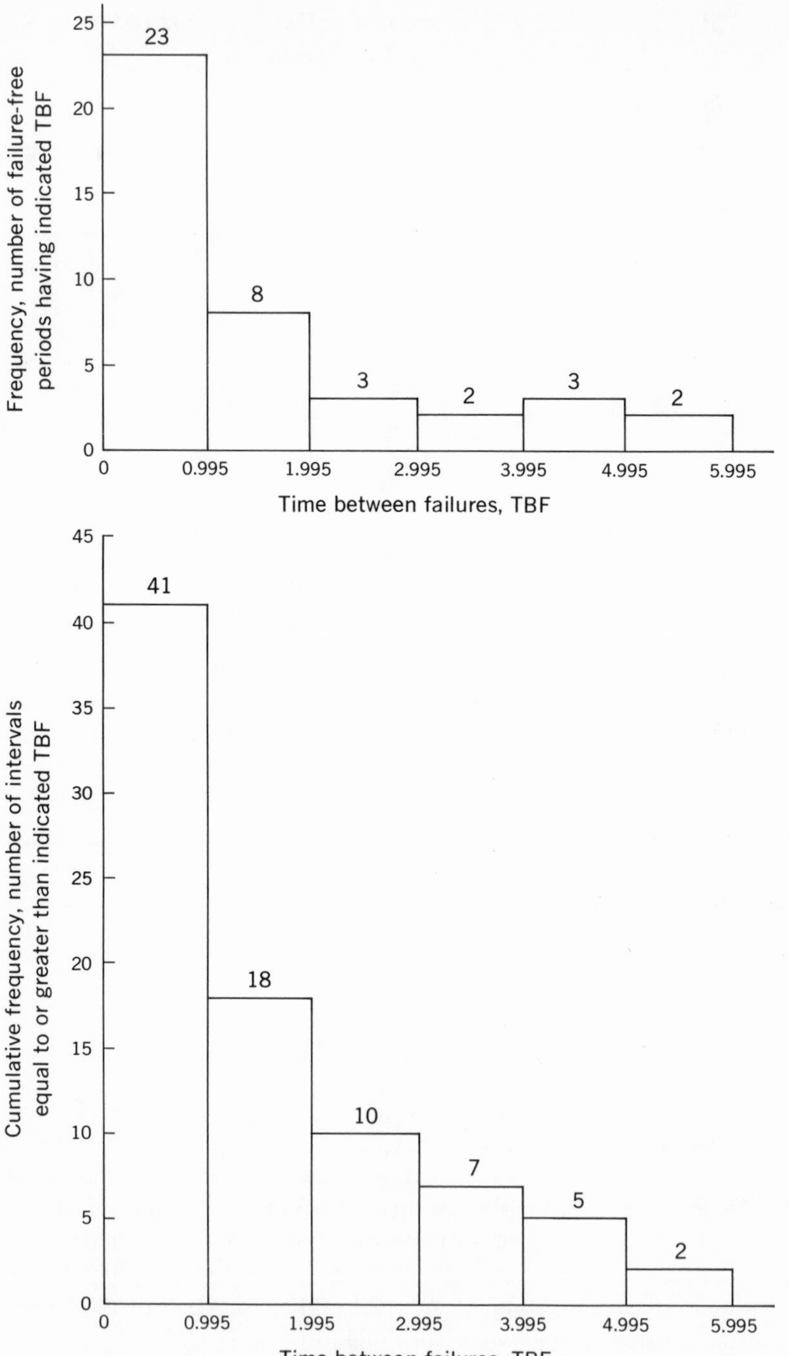

Figure 10-2a Histogram of time between failures. *b* Cumulative histogram of time between failures.

where $P_s = R =$ probability of failure-free operation for a time period equal to or greater than t
$e = 2.718$
$t =$ a specified period of failure-free operation
$\mu =$ mean time between failures, or MTBF (the mean of the TBF distribution)
$\lambda =$ failure rate (the reciprocal of μ)

Note that this formula is simply the exponential probability distribution (Chapter 6) rewritten in terms of reliability.

Example Previous experience shows that the mean time between failures of a radar set is 240 hours. Assuming a constant failure rate, what is the chance of running the set for 24 hours without failure?

$$R = e^{-t/\mu}$$
$$R = e^{-24/240} = 0.90$$

There is a 90 percent chance of obtaining 24 hours or more of failure-free operation.

The assumption of a constant failure is rightly questioned. However, published data[1] suggest that the assumption is a fair one to make. (More fundamental than arguing the validity of the assumption is the need to take design actions to yield a constant failure rate. For example, the careful determination of burn-in periods and replacement periods for both parts and systems are good design practices that will help achieve a constant failure rate.)

The Meaning of "Mean Time between Failures"

Confusion surrounds the meaning of mean time between failures (MTBF). Further explanation is warranted:

1 The MTBF is the mean (or average) time between successive failures of a product. This definition assumes that the product in question can be repaired and placed back in operation after each failure. These conditions are not always met. Furthermore, the MTBF is not always the most appropriate index for a product. For example, some other index such as percent uptime[2] is more pertinent when a product operates continuously as compared with the product needed for specific mission times (e.g., petroleum refinery equipment versus a guidance system for a missile).

[1] D. J. Davis, "An Analysis of Some Failure Data," *Journal of the American Statistical Association*, vol. 27, pp. 113–150, June, 1952.
[2] Uptime is equivalent to percent operative (as contrasted with downtime).

2 If the failure rate is constant, the probability that a product will operate without failure for a time equal to or greater than its MTBF is only 37 percent. (R is equal to 0.37 when t is equal to the MTBF.) This is contrary to the intuitive feeling that there is a 50–50 chance of exceeding an MTBF.

3 MTBF is not the same as "operating life," "service life," or other indices which generally connote overhaul or replacement time.

4 An increase in an MTBF does not result in a proportional increase in reliability (the probability of survival). If $t = 1$ hour, the following table shows the mean time between failures required in order to obtain various reliabilities:

MTBF	*R*
5	0.82
10	0.90
20	0.95
100	0.99

A fivefold increase in MTBF from 20 to 100 hours is necessary to increase the reliability by 4 percentage points as compared with a doubling of the MTBF from 5 to 10 hours to get 8 percentage points' increase in reliability. This is important because MTBF is often used as the criterion for making important decisions affecting reliability, whereas the probability of survival for a specified time t may be the more important index to the user.

Alternate methods of denoting reliability are sometimes used to avoid the serious consequence of misinterpreting the meaning of MTBF. One alternative is the reciprocal of MTBF, i.e., the failure rate. This eliminates the confusion with "service life" or "operating life." Another recognizes that MTBF is really just a substitute for the reliability percent R and its associated time t. Reliability is then stated as (1) the percent reliability required and (2) the mission time, instead of condensing these two numbers into one number (MTBF).

10-3 The Relationship between Part and System Reliability

It is often assumed that system reliability (i.e., probability of survival P_S) is the product of the individual reliabilities of the n parts within the system:

$$P_S = P_1 P_2 \cdots P_n$$

This is known as the product rule. The formula assumes (1) that the failure of any part will cause failure of the system and (2) that the reliabilities of the parts are independent[3] of each other, i.e., that the reliability of one part is not dependent on the reliability of another part. A set of lights in series on a Christmas tree demonstrates the product rule. These assumptions are usually not 100 percent correct. However, the formula is a convenient approximation that should be refined as information becomes available on the interrelationships of parts and their relationship to the system. (The redundancy formula discussed below is an example of this.) The following illustrates the product rule:

Example The following reliability requirements have been set on the subsystems of a communications system:

Subsystem	*Reliability (for a 4-hour Period)*
Receiver	0.970
Control system	0.989
Power supply	0.995
Antenna	0.996

What is the expected reliability of the overall system if the above requirements are met?

$$P_S = (0.970)(0.989)(0.995)(0.996) = 0.951$$

The chance that the overall system will perform its function without failure for a 4-hour period is 95 percent.

If it can be assumed that each part follows the exponential distribution, then

$$P_S = e^{-t_1\lambda_1}e^{-t_2\lambda_2} \cdots e^{-t_n\lambda_n}$$

Further, if t is the same for each part:

$$P_S = e^{-t\Sigma\lambda}$$

Thus, when the failure rate is constant (and therefore the exponential distribution applies), a "reliability prediction" of a system can be made based on the addition of the part failure rates. This is illustrated in the next section.

[3] For a good discussion of this and other assumptions in reliability calculations, see Ralph A. Evans, "Problems in Probability," *Proceedings of the Annual Symposium on Reliability*, 1966, Institute of Electrical and Electronics Engineers, Inc., p. 347.

10-4 Predicting Reliability during Design Based on the Exponential Distribution

Section 9-5 introduced the reliability prediction method based on the addition of failure rates. Two additional examples will now be presented. The steps in reliability prediction can be summarized as follows:

1 Describe how the product works, in sufficient detail to construct the reliability prediction equation.
2 Construct the equation stating system reliability as a function of the reliability of the lower-level units.
3 Obtain basic failure rates based on past experience with parts and components similar to those used in the proposed design. A number of reliability handbooks are now available[4] giving such failure rates. Figure 10-3 shows an example of a page from such a handbook. The stress ratio is operating voltage/rated voltage. The basic failure rate depends on the stress ratio and ambient temperature.
4 Select modification factors to the basic failure rates to reflect product characteristics such as type of stress levels, type of operation, etc. The specific type of modification factors depends on the handbook used.
5 Combine all the above to obtain the numerical reliability prediction.

Table 10-2 shows a reliability prediction[5] for a proposed logic gate (a special type of circuit). All the components must operate in order for the logic gate to function. Thus, the prediction equation is simply the product of the reliabilities of the components. Assuming a constant failure rate, the failure rate for the logic gate is then the sum of the component failure rates. Table 10-2 shows how "derating" (Section 9-11) was used in selecting the components. For example, the stress ratio of 0.2 for the diode means that the diode selected has a rated voltage such that the ratio of operating to rated voltage is 0.2, or the resistor was "derated" by 80 percent. With an expected operating temperature of 40°C and a stress ratio of 0.2, Figure 10-3 shows a failure rate of 3.0 failures per million hours. The K_A factor (obtained from the failure rate handbook) is the application factor for a diode used in a vehicle-mounted ground environment. This modification factor adjusts the basic failure rate to reflect other stresses such as vibration, shock, and humidity. The total failure rate for the circuit is 76.461 failures per million hours.

For more complex products, the prediction can be aided by special

[4] *Reliability Stress and Failure Rate Data for Electronic Equipment,* MIL-HDBK-217A, U.S. Department of Defense, 1965.
[5] *Reliability for the Engineer, Book Three: Using Failure Rate Data,* Martin-Marietta Corp., Orlando, Fla., 1965.

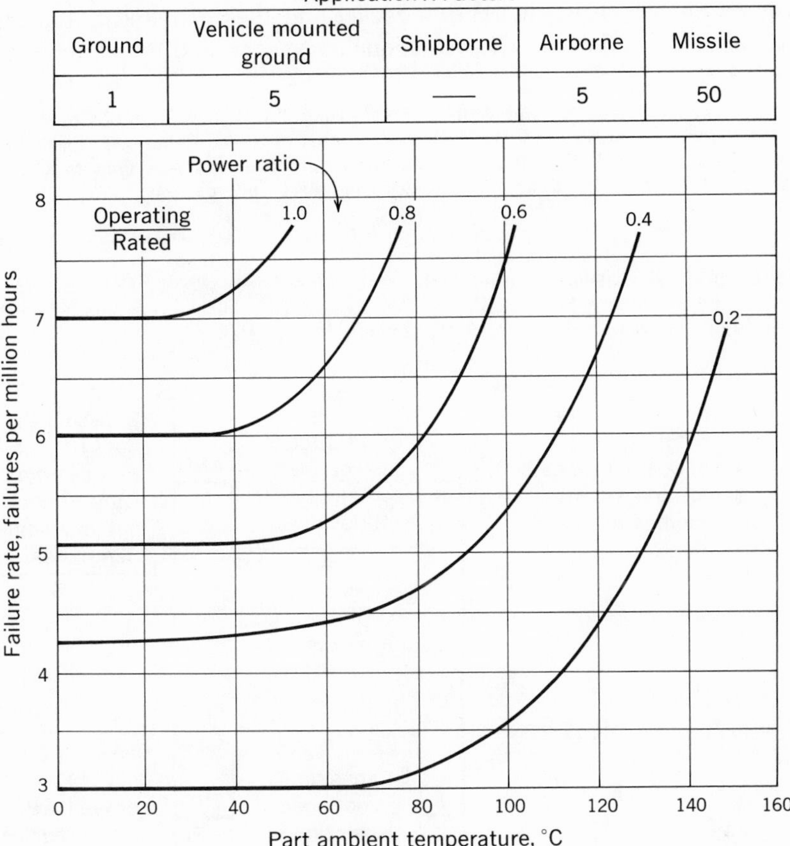

Figure 10-3 Failure rates for MIL-S-19500, microwave diodes, detector applications. Source: *Reliability Stress and Failure Rate Data for Electronic Equipment,* MIL-HDBK-217A, U.S. Department of Defense, 1965.

Table 10-2 Failure Rate Calculation of Proposed Logic Gate

Component	*Stress Ratio*	*No. Used N*	*Failure Rate*	*K_A for Vehicle Ground Mounted*	$\lambda_T = N\lambda_{\text{derate}}K_A$
Resistor, film (1.3K)	0.8	1	$0.19/10^6$	0.3	$0.057/10^6$
Resistor, film (3.32K)	0.2	1	$0.14/10^6$	0.3	$0.042/10^6$
Resistor, film (46.6K)	0.2	1	$0.14/10^6$	0.3	$0.042/10^6$
Transistor, silicon NPN	<1 watt 0.15 normalized junction temperature	1	$0.165/10^6$	8	$1.320/10^6$
Diode, 1N31A	0.2	5	$3.0/10^6$	5	$75.000/10^6$
					Total $= \lambda_e = \Sigma\lambda_T = 76.461/10^6$

diagrams. The approach will be illustrated[6] for a rocket-borne photoflash system which is carried in a rocket and then actuated to deliver a flash of light. The overall system includes a ground-based subsystem

[6] C. Weldon Holtzman, Jr., and William E. Marshall, "A New Method of Communication between Engineer and Mathematician Aids System Reliability Prediction," *Proceedings of the Sixth National Symposium on Reliability and Quality Control*, 1960, Institute of Electrical and Electronics Engineers, Inc., p. 403.

Figure 10-4 Rocket-borne photoflash unit (functional diagram). Source: C. between Engineer and Mathematician Aids System Reliability Prediction," *Proceed-* Institute of Electrical and Electronics Engineers, Inc., p. 403.

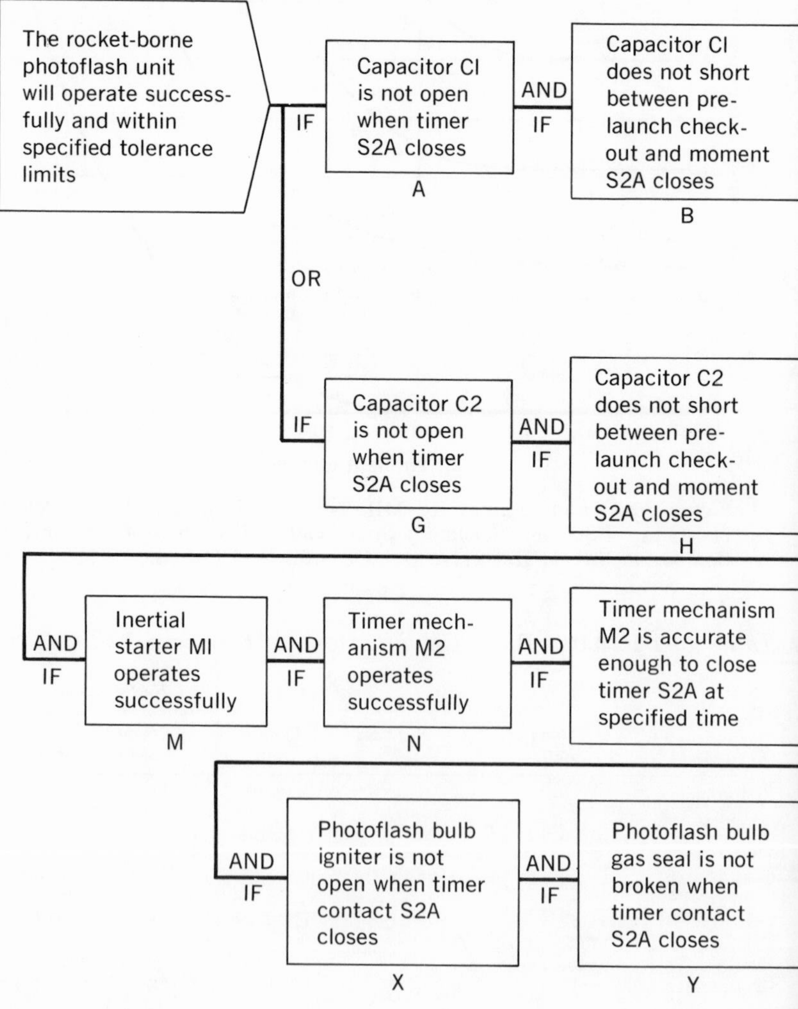

which checks out the rocket-borne subsystem and prepares it for firing. Batteries in the ground subsystem charge the capacitors in the rocket subsystem and an electrostatic voltmeter on the ground monitors the charges in the capacitors. As the rocket is fired, an inertial mechanism starts a timer. When the timer runs down, the capacitor charge is delivered to the photoflash bulb. The functional diagram (Figure 10-4) developed from a schematic diagram shows the sequence of events nec-

Weldon Holtzman, Jr., and William E. Marshall, "A New Method of Communications *ings of the Sixth National Symposium on Reliability and Quality Control,* 1960,

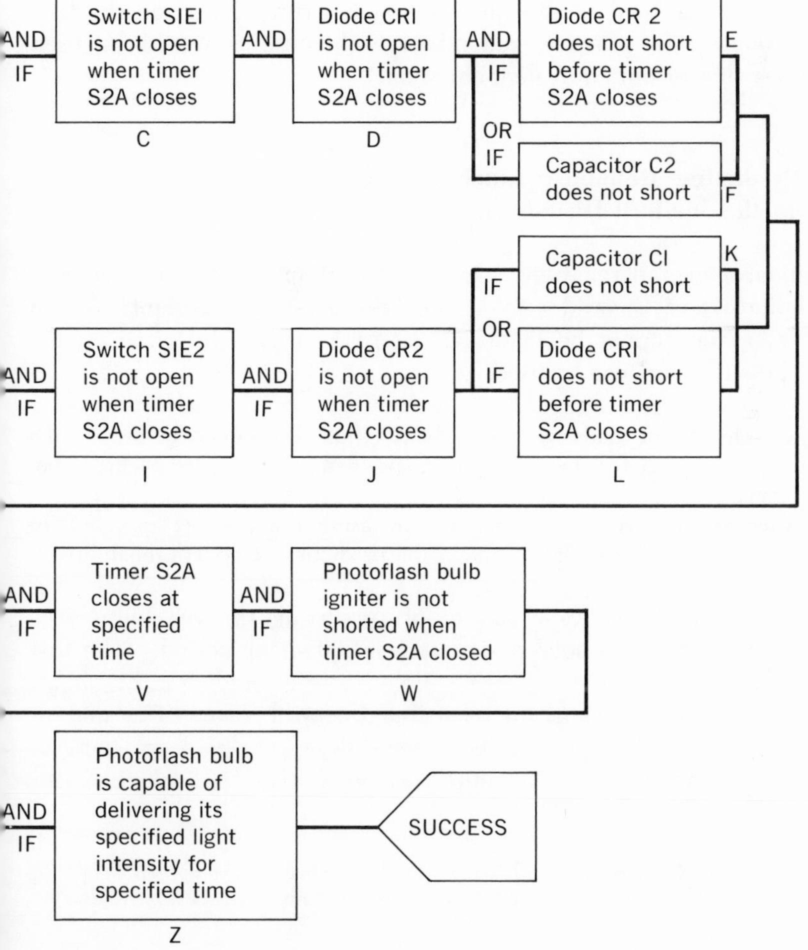

essary for system success. The events are stated in whatever way will facilitate the construction of the reliability equation. It is not unusual for questions to arise during the construction of the functional diagram. Competent designers may disagree on exactly what events must occur for the product to work successfully. In such cases, the diagram alerts all concerned that the design is not thoroughly understood. A proponent of this diagram states: "The author must confess that his preparation of such diagrams for systems of his own design has given him additional understanding of his system which he previously believed he understood completely!"[7] Figure 10-4 can be further simplified, if desired, into a diagram (Figure 10-5) showing the probability relationships among the events. Figure 10-5 also shows the four combinations of events that would produce system success. Note that events M through Z are common to all combinations. For clarity of the probabilities involved, the final prediction equation is written based on the four combinations. (In practice, redundancy relationships would be used to derive a shorter but equivalent equation.)

10-5 Predicting Reliability during Design Based on the Weibull Distribution

Prediction of overall reliability based on the simple addition of component failure rates is valid only if the failure rate is constant. When this assumption cannot be made, an alternate approach, based on the Weibull distribution, can be used.

1 Graphically or analytically use the Weibull distribution to predict the reliability R for the time period specified. Do this for each component.

2 Combine the component reliabilities using the product rule and/or redundancy formulas to obtain the prediction of system reliability.

Table 10-3 shows an example[8] of predicting the reliability of a tractor pump for a 500-hour period and a 1,500-hour period. The first analysis was on a proposed design. Note how the prediction highlights the "vital few" parts such as the oil seal and control valve. The analysis after some design changes on these parts shows a significant increase in reliability and decrease in failure cost per tractor. (Table 10-3 lists

[7] *Ibid.*

[8] Barrett G. Rich, O. A. Smith, and Lee Korte, *Experience with a Formal Reliability Program,* SAE Paper 670731, Farm, Construction and Industrial Machinery Meeting, 1967.

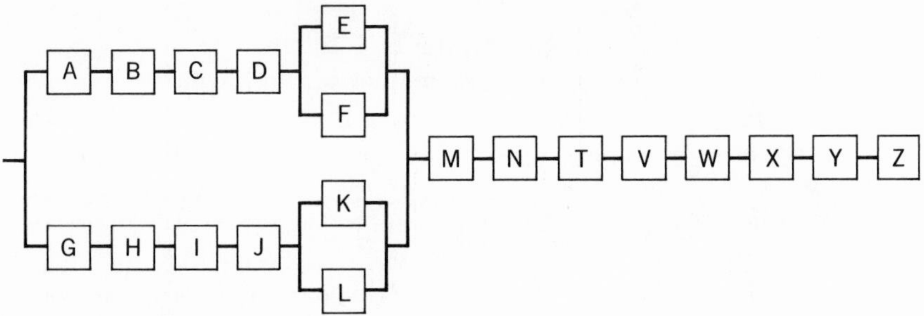

Event combinations for system success

1	A	B	C	D	E	M	N	T	V	W	X	Y	Z
2	A	B	C	D	F	M	N	T	V	W	X	Y	Z
3	G	H	I	J	K	M	N	T	V	W	X	Y	Z
4	G	H	I	J	L	M	N	T	V	W	X	Y	Z

$$R = P_S = [P_A P_B P_C P_D P_E + P_A P_B P_C P_D P_F + P_G P_H P_I P_J P_K + P_G P_H P_I P_J P_L](P_M P_N P_T P_V P_W P_X P_Y P_Z)$$

Figure 10-5 Development of prediction equation.

data for important items only and then summarizes reliability and cost for the complete assembly of 68 parts.)

10-6 Redundancy

Redundancy is the existence of more than one element for accomplishing a given task, where all elements must fail before there is an overall failure to the system. In parallel redundancy (one of several types of redundancy), two or more elements operate at the same time to accomplish the task, and any single element is capable of handling the job itself in case of failure of the other elements. When parallel redundancy is used, the overall reliability is calculated as follows:

$$P_S = 1 - (1 - P_1)^n$$

where P_S = reliability of the system
P_1 = reliability of the individual elements in the redundancy
n = number of identical redundant elements

Example Suppose a unit has a reliability of 99.0 percent for a specified mission time. If two identical units are used in parallel redundancy, what overall reliability will be obtained?

$$R = 1 - (1 - 0.99)(1 - 0.99) = 0.999 \text{ or } 99.9 \text{ percent}$$

Table 10-3 **Reliability Prediction for a 1.38 Cubic Inch per Revolution Variable Displacement Pump**

(Only Parts with Significant Percent Failure Are Listed)	*First Analysis*			*Analysis after Changes*		
	Percent Failure at		*$/Tractor at*	*Percent Failure at*		*$/Tractor at*
Part Name	*500 Hours*	*1,500 Hours*	*500 Hours*	*500 Hours*	*1,500 Hours*	*500 Hours*
Pump drive coupling special screws	0.6	3.0	0.21	0.2	1.0	0.07
Pump drive coupling	0.07	0.8	0.01	0.07	0.8	0.01
Hydraulic pump shaft	0.02	0.06	0.01	0.02	0.06	0.01
Pump shaft oil seal	3.7	20.0	0.41	1.0	5.0	0.10
Pump shaft bushing	0.75	2.25	0.30	0.25	0.75	0.10
O-ring packings (11) (evaluated separately)	0.63	2.10	0.08	0.63	2.10	0.08
Stroke control valve	13.0	38.0	0.77	0.05	0.15	0.02
Assembly reliability (68 parts)						
At 500 hours		80%			97.3%	
At 1,500 hours		49%			92.0%	
Cost at 500 hours		$2.18			$0.77	

Redundancy is an old design technique invented long before the advent of reliability prediction techniques. However, the designer can now predict in *quantitative* terms the effect of redundancy on system reliability.

10-7 Reliability as a Function of Applied Stress and Strength

An individual component is satisfactory if its strength is greater than the stress applied to it. For the same design, strength will vary from component to component. The applied stress will also vary. The variation in each of the two parameters is depicted in Figure 10-6. Consider the *difference* between strength and applied stress in a given instance. The probability of successful performance ("reliability") is the probability that this difference is greater than 0. Figure 10-7 shows a distribution of the *difference* between strength and stress. Assuming independence of strength and stress:

$$\mu_{\text{difference}} = \mu_{\text{strength}} - \mu_{\text{stress}}$$

$$\sigma_{\text{difference}} = \sqrt{\sigma_{\text{strength}}^2 + \sigma_{\text{stress}}^2}$$

If normality is assumed, the probability of a difference greater than 0 can be estimated by finding the area under the curve.

Figure 10-6 Stress and strength distributions.

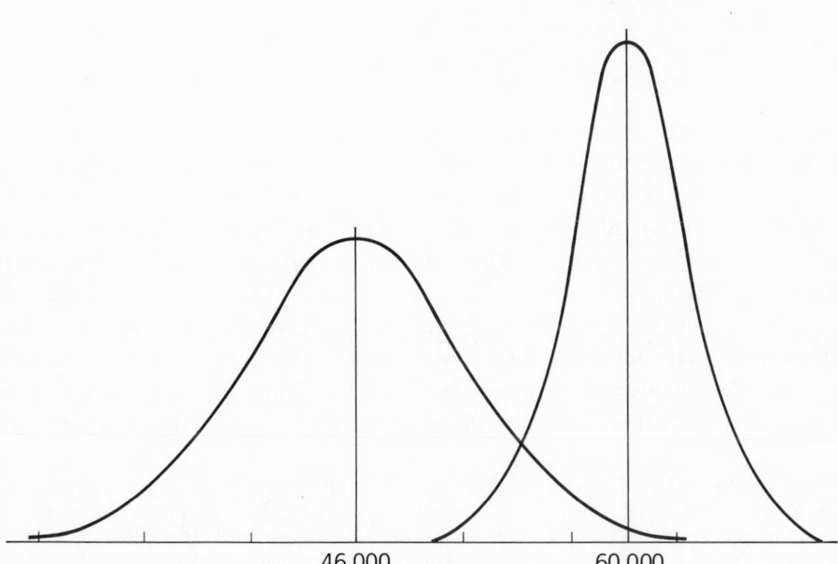

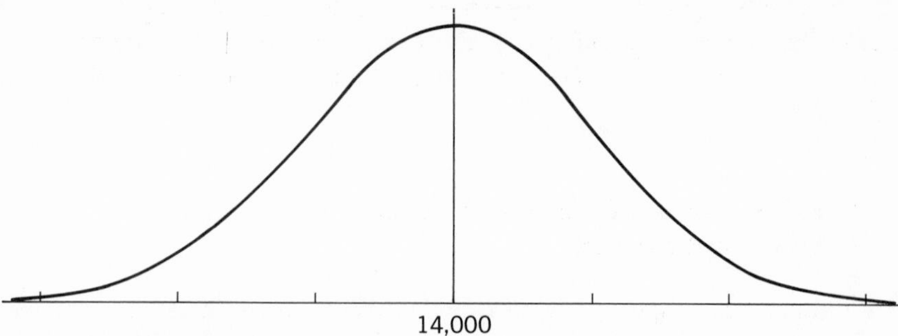

Figure 10-7 Distribution of (strength − stress).

Example Suppose the following estimates apply to a part:

	Strength	*Applied Stress*
Average	60,000 psi	46,000 psi
Standard deviation	3,000 psi	5,500 psi

Then average difference = 60,000 — 46,000 = 14,000

$$\sigma_{\text{difference}} = \sqrt{(3{,}000)^2 + (5{,}500)^2} = 6{,}260$$

$$K = \frac{0 - 14{,}000}{6{,}260} = -2.24$$

From Table A in the Appendix, the area greater than 0 is 0.9875. Thus, the reliability is 98.75 percent.

This discussion has been simplified in order to stress a basic concept. The key point is that *variation* in addition to average value must be considered in design. Designers have always recognized the existence of variation by using a "safety factor" in design. However, safety factor[9] is often defined as the ratio of average strength to the worst stress expected.

Note that in Figure 10-8, all the designs have the *same* safety factor. Also note that the reliability (probability of a part having a strength greater than the stress) varies considerably. Thus the uncertainty often associated with this definition of safety factor is in part due to its failure

[9] For a discussion of various definitions of safety factor and safety margin, see D. Kececioglu and D. Cormier, "Designing a Specified Reliability into a Component," *Proceedings of the Third Annual Aerospace Reliability and Maintainability Conference,* Washington, 1964, Society of Automotive Engineers, p. 546.

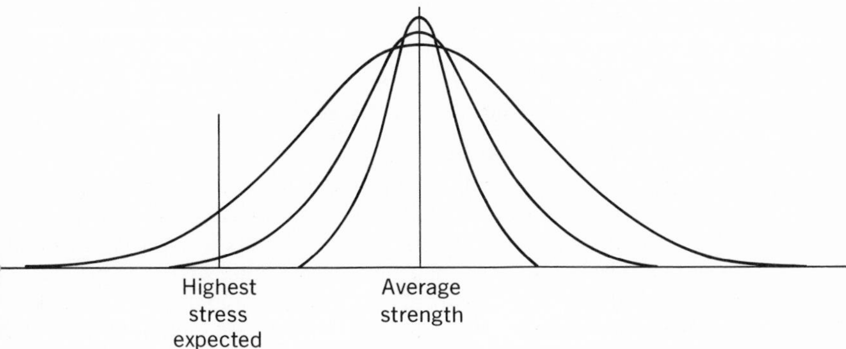

Figure 10-8 Variation and safety factor.

to reflect the *variation* in strength and *variation* in stress. Such variation is partially reflected in a "safety margin" defined as

$$\frac{\text{Worst stress} - \text{average strength}}{\text{Standard deviation of strength}}$$

This recognizes the variation in strength but is conservative because it does not recognize a variation in stress (see section 13-2).

The implications of recognizing variation are far-reaching. For example, consider the following basic formula in strength of materials:

$$S = \frac{Tr}{J}$$

The torque T, radius r, and polar moment of inertia J each have variability which contribute to the variability in shearing stress S. This variability is generally recognized in total by using a safety factor in design. The uncertainty in a safety factor can be narrowed by estimating standard deviations for T, r, and J and using statistical methods to estimate the variation in shearing stress in a manner similar to that used in calculating the standard deviation of the difference between strength and stress. The approach would also apply to many other design formulas.

10-8 Availability

"Availability" is sometimes a more appropriate measure than reliability. Whereas reliability is a measure of performing without failure, availability recognizes both reliability and maintainability:

1 Time availability is the percentage of operating time that an equipment is operational. This is usually computed as

$$P_A = \frac{\text{MTBF}}{\text{MTBF} + \text{MRT}}$$

where MTBF = mean time between failures
MTR = mean repair time

Chapter 8 illustrates this formula.

2 Equipment availability is the percentage of equipments which will be available for use after t hours of operation due to the combined effect of units which did not fail and failed units which were restored to service within a specified maximum downtime.

3 Mission availability is the percentage of missions of time t which will not have any failure which cannot be restored within a specified maximum downtime.

These definitions recognize that failures will occur but that a good maintainability program can improve the overall effectiveness of the product. (Calabro[10] discusses the formulas for these definitions of availability.)

These definitions of availability are stated here to emphasize that "mean time between failures" is not the only index of product effectiveness for complex products. A suitable index must be chosen or created to adequately reflect the needs of the user (and not the desires of the design engineer). Selecting such an index for each product has the advantage of forcing a meeting of the minds on what is meant by fitness for use quality.

PROBLEMS

1 A manufacturer claims that he can design and manufacture a radar set that will have a mean time between failures of 240 hours based on the exponential distribution. Suppose that a certain mission requires failure-free operation of the set for 24 hours. What is the chance that the set will complete a mission without failure?

2 A piece of ground support equipment for a missile has a specified mean time between failures of 100 hours. What is the reliability for

[10] S. J. Calabro, *Reliability Principles and Practices*, McGraw-Hill Book Company, New York, 1962, chap. 9.

a mission time of 1 hour? 10 hours? 50 hours? 100 hours? 200 hours? 300 hours? Graph these answers by plotting mission time versus reliability. Assume an exponential distribution.

3 Suppose a washing machine manufacturer desires that there be a 99 percent chance of his machine completing the cleaning of a load of clothes. The cleaning cycles take 0.5 hour. Assume an exponential distribution.

a What requirement should be set on the mean time between failures?

b Suppose this requirement were doubled. What would then be the reliability for a cleaning cycle?

4 The average life of subassembly A is 2,000 hours. Data indicate that this life characteristic is exponentially distributed.

a What percent of the subassemblies in the population will last at least 200 hours?

b The average life of subassembly B is 1,000 hours and the life is exponentially distributed. What percent of the subassemblies in the population will last at least 200 hours?

c These subassemblies are independently manufactured and then connected in series to form the total assembly. What percent of assemblies in the population will last at least 200 hours?

5 It is expected that the average time to repair a failure on a certain product will be 3 hours. Assume that repair time is exponentially distributed. What is the chance that a failure will be repaired in 4 hours or less?

6 It is expected that the average time to repair a failure on a certain product is 4 hours. Assume that repair time is exponentially distributed. What is the chance that the time for a repair will be between 3 and 5 hours?

7 The time between failures (TBF) for a certain assembly is exponentially distributed with a mean of 100 hours. Suppose the TBF had been normally distributed with a mean of 100 hours and standard deviation of 20 hours. What value of mission time (TBF) based on the exponential would have the same reliability as a 100-hour mission time based on normality?

8 Develop a functional success diagram for a product with which you are familiar. (Obtain prior approval of the product from the instructor.)

9 The following table summarizes basic failure rate data on components in an electronic subsystem:

Component	*Quantity*	*Failure Rate per Hour*
Silicon transistor	40	74.0×10^{-6}
Film resistor	100	3.0×10^{-6}
Paper capacitor	50	10.0×10^{-6}

Estimate the mean time between failures. (Assume an exponential distribution. All components are critical for subsystem success.)

10 A system consists of subsystems A, B, and C. The system is primarily used on a certain mission that lasts 8 hours. The following information has been collected:

Subsystem	*Required Operating Time during Mission*	*Type of Failure Distribution*	*Reliability Information*
A	8 hours	Exponential	50% of subsystems will last at least 14 hours
B	3 hours	Normal	Average life is 6 hours with a standard deviation of 1.5 hours
C	4 hours	Weibull with $\beta = 1.0$	Average life is 40 hours

Assuming independence of the subsystems, calculate the reliability for a mission.

11 A hydraulic subsystem consists of two subsystems in parallel, each having the following components and characteristics:

Components	*Failures/10⁶ Hours*	*Number of Components*
Pump	23.4	1
Quick disconnect	2.4	3
Check valve	6.1	2
Shutoff valve	7.9	1
Lines and fittings	3.13	7

The components within each subsystem are all necessary for subsystem success. The two parallel subsystems operate simultaneously and either can perform the mission. What is the mission reliability if the mission time is 300 hours? (Assume an exponential distribution.)

12 Reliability information has been collected[11] for a tractor transmission. The design has been divided into five groups of components. All components and groups may be considered in series because if any components fails, the complete transmission will fail. The failure rate information is stated in terms of a basic failure rate and modification factors K_A and K_{op} to reflect stress conditions and operational environment respectively. The actual failure rate for a component is equal to the basic failure rate multiplied by the two factors. A prediction of reliability for 5,000 hours is desired. The operating time for each group is listed after the name.

[11] *Product Reliability Planning and Control,* Allis-Chalmers Manufacturing Company, Milwaukee, December, 1968, p. 46.

Component	*Quantity*	*Basic Failure Rate*	K_A	K_{op}
Main shaft (5,000 hours)				
Bearing assemblies	2	0.0000015	2.0	1.17
Cap bearing	1	0.0000008	1.8	1.17
O-ring seal	1	0.00000068	1.4	1
Shims and snaps	1	0.0000005	1.1	1
Spacer	1	0.0000002	1.0	1
Main shaft	1	0.0000005	1.8	1
First gear (750 hours)				
Gear on main shaft	1	0.000038	2.1	1
Gear on pinion shaft	1	0.000017	1.5	1
Collar	1	0.0000014	1.1	1
Coupling	1	0.000002	1.1	1
Second gear (1,750 hours)				
Gear on main shaft	1	0.000008	1.9	1
Gear on pinion shaft	1	0.000007	1.8	1
Collar	1	0.0000062	1.4	1
Coupling	1	0.000005	1.3	1
Third gear (2,500 hours)				
Gear on main shaft	1	0.000007	1.7	1
Gear on pinion shaft	1	0.000008	1.5	1
Collar	1	0.0000062	1.4	1
Coupling	1	0.000005	1.4	1
Pinion shaft (5,000 hours)				
Bushings	4	0.0000003	1.3	1
Front bearing—pinion	1	0.000002	1.7	1.17
Rear bearing—pinion	1	0.000001	1.5	1.17
Snaps, washers, etc.	1	0.0000005	1.1	1
Pinion shaft	1	0.000007	3.1	1

a Calculate the reliability for each of the five groups of components.
b Calculate the overall transmission reliability.
c There are 27 components in the transmission. Which ones are the "vital few"?
d Suppose that an overall reliability of 80 percent is desired. Define several alternative approaches that should be investigated to achieve this level.

13 The following estimates, based on field experience, are available on three subsystems:

Subsystem	*Percent Failed at 1,000 Miles*	*Weibull Beta Value*
A	0.1	2.0
B	0.2	1.8
C	0.5	1.0

If these estimates are assumed to be applicable for similar subsystems which will be used in a new system, predict the reliability (in terms of percent successful) at the end of 3,000 miles, 5,000 miles, 8,000 miles, and 10,000 miles.

14 Breaking strength is a critical characteristic on a certain mechanical component. Past data indicate that 3,000 psi is a good estimate of the population standard deviation of strength of individual components. It is expected that the component will be subjected to an average stress of 30,000 psi with a standard deviation of 4,000 psi. What requirement should be set on average strength if a reliability of 98 percent is desired? Assume that strength and stress are normally distributed.

15 It is desired that a power plant be in operating condition 95 percent of the time. The average time required for repairing a failure is about 24 hours. What must be the mean time between failures in order for the power plant to meet the 95 percent objective?

16 A manufacturing process runs continuously 24 hours per day and 7 days a week (except for planned shutdowns). Past data indicate a 50 percent probability that the time between successive failures is 100 hours or more. The average repair time for failures is 6 hours. Failure times and repair times are both exponentially distributed. Calculate the availability of the process.

Statistical Aids for Planning and Analyzing Tests

11-1 Problems of Prototype and Pilot Run Tests

Here are some testing problems that require statistical techniques to achieve an optimum test program.

1 Comparison of test data between two alternative designs, or comparison of test data from one design with the specification values
2 Determination of the number of tests required to provide adequate data for evaluation
3 Determination of the reliability of a limited number of test results in estimating the true value of a product characteristic
4 Planning of experiments to assure that all questions can be answered with a minimum number of tests

Industrial situations often consist of a combination of these problems. This chapter presents the statistical methods for handling these problems.

11-2 Concept of Statistical Inference

Examine the following problems concerned with the evaluation of test data. For each, give a yes or no answer based on your intuitive analysis of the problem. (Write your answers on a piece of paper *now* and then check for the correct answers at the end of this chapter.) Some of the problems are solved in the chapter.

Examples of Engineering Problems That Can Be Solved Using the Concept of Significance Tests

1 A single-cavity molding process has been producing insulators with an average impact strength of 5.15 foot-pounds. A group of 12 insulators from a new lot shows an average of 4.952 foot-pounds. Is this enough evidence to conclude that the new lot is lower in average strength?

2 Past data show the average hardness of brass parts to be 49.95. A new design is submitted and claimed to have higher hardness. A sample of 61 parts of the new design shows an average of 54.62. Does the new design really have a different hardness?

3 Two types of spark plugs were tested for wear. A sample of 10 of design 1 showed an average wear of 0.0049. A sample of 8 of design 2 showed an average wear of 0.0064. Are these enough data to conclude that design 1 is better than design 2?

4 Only 11.7 percent of the 60 new-alloy blades on a turbine rotor failed on test in a gas turbine where 20 percent have shown failures in a series of similar tests in the past. Are the new blades better?

5 1,050 resistors supplied by one manufacturer were 3.71 percent defective. 1,690 similar resistors from another manufacturer were 1.95 percent defective. Can one reasonably assert that the product of one plant is inferior to that of the other?

You probably had some incorrect answers. If so, it was probably due to a failure to understand the concept of sampling variation. The "gossip" of a small sample size can be dangerous.

The concept of sampling variation will be introduced by an example. A battery is to be evaluated to assure that life requirements are met. An average life of 30 hours is desired. Preliminary data indicate that life follows a normal distribution and that the standard deviation is equal to 10 hours. A sample of four batteries will be selected at random from the process and tested. If the average of the four is close to 30 hours, it will be concluded that the battery meets the specification. Figure 11-1 plots the distribution of *individual* batteries from the population assuming that the true *average* of the population is exactly 30 hours.

If a sample of four is life tested, the following lifetimes might result: 34, 28, 38, and 24, giving an average of 31.0. However, this is a random sample selected from the many batteries in the population. Suppose another sample of four were taken. The second sample of four would likely be different from the first sample. Perhaps the results would be 40, 32, 18, and 29, giving an average of 29.8. If the process of drawing many samples (with four in each sample) were repeated over and over again, different results would be obtained in most samples.

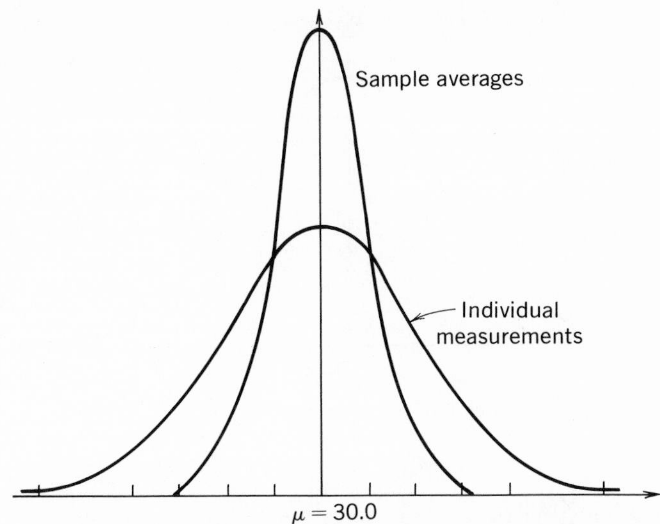

Figure 11-1 Distributions of individuals and sample averages.

This is significant because *all* the samples were drawn from the *same* process. This outcome of different sample results illustrates the concept of sampling variation.

Returning to the problem of evaluating the battery, a dilemma exists. In the actual evaluation, only one sample of four can be drawn (because of time and cost limitations). Yet the experiment of drawing many samples indicates that samples vary. The question is how reliable is the one sample of four that will be the basis of the decision. The final decision is a function of which sample result is obtained. The key point is that the existence of sampling variation means that any one sample cannot be relied upon to always give an adequate decision. The statistical approach analyzes the results of the sample, *taking into account the possible sampling variation that could occur.* Formulas have been developed defining the expected amount of sampling variation. Knowing this, a valid decision can be reached based on evaluating the one sample of data.

The problem, then, is to define how averages of samples vary. If sampling were continued and for each sample of four the average calculated, these averages could be compiled into a histogram. Figure 11-1 shows the resulting probability curve, superimposed on the curve for the population. The inner curve represents the distribution of life for the sample *averages* (where each average includes four individual batteries). The curve for averages is narrower than the curve for in-

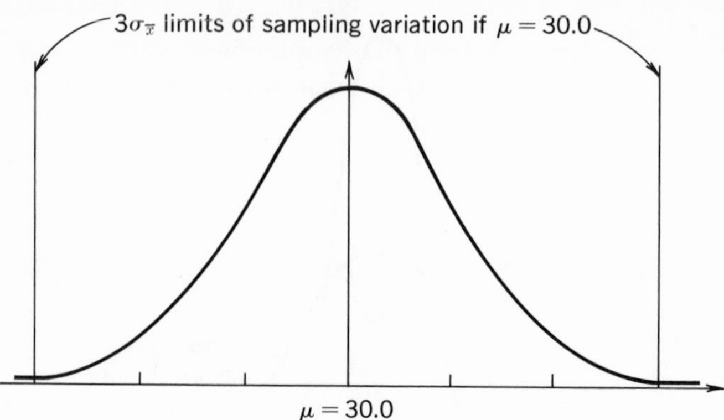

Figure 11-2 Distribution of sample averages.

dividuals because in calculating averages, extreme individual values are offset. The mathematical properties for the curve for averages have been studied and the following relationship developed:

$$\sigma_{\bar{x}} = \frac{\sigma}{\sqrt{n}}$$

where $\sigma_{\bar{x}}$ = standard deviation of averages of samples (sometimes called the standard error of the mean)
σ = standard deviation of individual items
n = number of items in *each* sample

This relationship is significant because if an estimate of the standard deviation of *individual* items can be obtained, then the standard deviation of sample averages can be calculated from the above relationship instead of running an experiment to generate sample averages.

The problems of evaluating the battery can now be portrayed graphically (Figure 11-2).

This shows the variation of sample averages if the true overall average is 30. If the average of four falls within the limits of sampling variation shown on the curve, then it will be concluded that the true average is 30 and that the deviation of the sample average from 30 was due to *sampling* variation. In other words, there is no evidence that the observed average differs *significantly* from 30.0. If the average exceeds the limits, it will be concluded that the sample could not have come from a population with an average of 30 because the variation of the sample average from 30 is larger than that expected for such a population; i.e., there is evidence that the observed average differs significantly from 30.0. In the actual test of hypothesis, the sample average is expressed as a deviation from the hypothesis average in units of standard deviations of sample averages. Thus, if the sample average were 36.0, the statistic calculated would be

$$U = \frac{36.0 - 30.0}{10/\sqrt{4}} = 1.2$$

That is, $\bar{X}$ is 1.2 standard deviations (of averages) from the hypothesis mean.

11-3 The Two Types of Sampling Error

In the battery problem, a limited number of units are to be selected from the process, evaluated, and a decision made. As the decision concerns the population, but will be based on the result of a sample of the population, it is possible to make a wrong decision. Two types of errors can be made. The problem will be structured by defining the specific statement ("hypothesis") to be statistically evaluated. The hypothesis will be that the true average of the population is 30.0 or H: $\mu_0 = 30.0$.

One error that could be made is to *reject* this hypothesis when it is true. This is called the type I error or the "level of significance" and is denoted by α. The error can be shown graphically in Figure 11-3. The area between the vertical lines represents the acceptance region for the test of hypothesis. If the sample result falls within the acceptance region, the hypothesis is accepted. Otherwise, it is rejected. Notice that there is a small portion of the curve which falls outside the acceptance region. This portion (α) represents the prob-

Figure 11-3 Type I or α error.

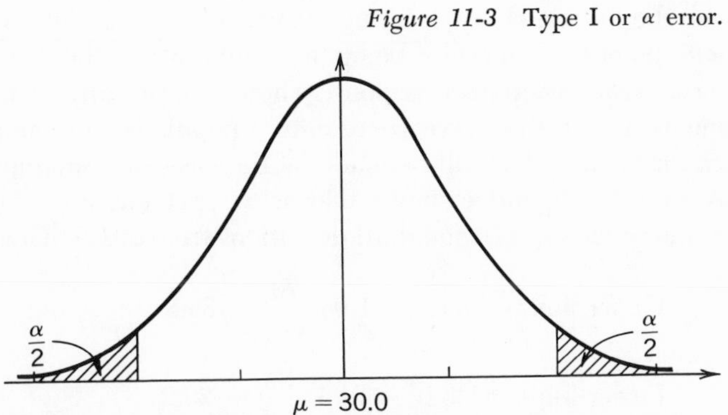

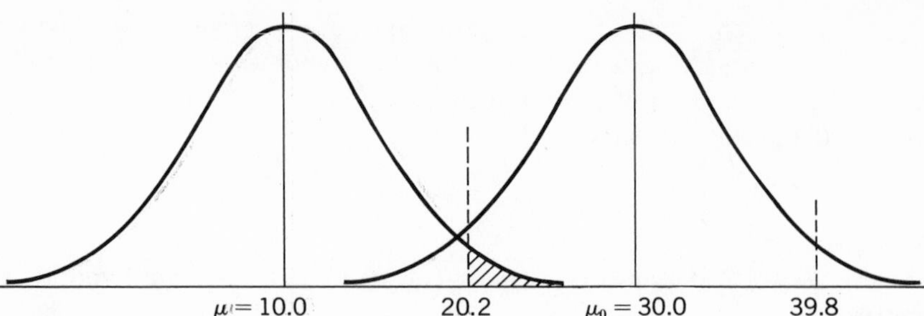

Figure 11-4 Type II or β error.

ability of obtaining a sample result outside the acceptance region, even though the hypothesis is correct.

The second type of error is called the type II error. This is the probability of *accepting* a hypothesis when it is *false.* This error is denoted by β. Figure 11-4 shows the type II error (shaded area). Notice it is possible to obtain a sample result within the acceptance region, even though the population has a true average which is *not* equal to the average stated in the hypothesis. *The numerical value of the type II error depends on the true value of the population average.* This leads to the concept of the operating characteristic curve.

The operating characteristic curve defines the probability of accepting a false hypothesis (the type II error). Every test of hypothesis has an operating characteristic curve. The literature contains these curves for most tests of hypotheses. The curve will be developed for the example in this section to illustrate the concept.

A hypothesis has been proposed that the average life in the population is 30.0 hours. Suppose it has been decided that the type I error must be 5 percent. This is the probability of rejecting the hypothesis when, in truth, the true average life is 30.0. The acceptance region can be obtained by locating values of average life which have only a 5 percent chance of being exceeded when the true average life is 30.0. The acceptance region is shown graphically in Figure 11-5. Remember that the curve represents a population of sample *averages* because the decision will be made on the basis of a sample average. Table A in the Appendix shows that a 3.5 percent area in each tail is at a limit which is 1.96 standard deviations from 30.0. Thus:

$$\text{Upper limit} = 30.0 + 1.96 \frac{10}{\sqrt{4}} = 39.8$$

$$\text{Lower limit} = 30.0 - 1.96 \frac{10}{\sqrt{4}} = 20.2$$

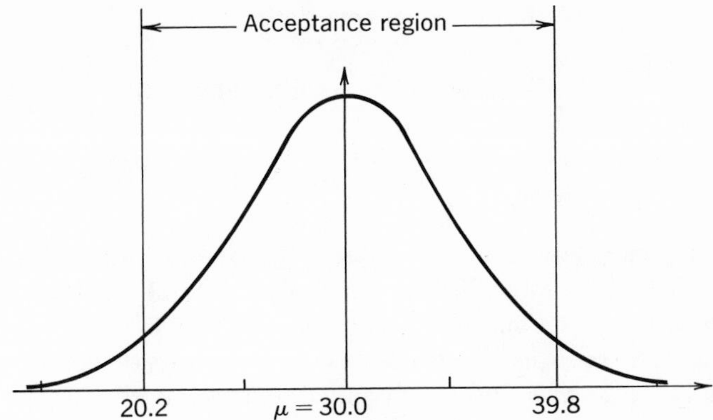

Figure 11-5 Acceptance region for H: $\mu_0 = 30.0$.

The acceptance region is thereby defined as 20.2 to 39.8. If the average of a random sample of four batteries is within this acceptance region, the hypothesis is accepted. If the average falls outside of the acceptance region, the hypothesis is rejected.

The problem now is to construct an operating characteristic curve to define the magnitude of the type II error. As the type II error is the probability of *accepting* the original hypothesis ($\mu_0 = 30.0$) when it is *false*, the probability of a sample average falling between 20.2 and 39.8 must be found when the true average of the population is equal to something other than 30.0.

Suppose the true average is equal to 35.0 Figure 11-6 shows the distribution of sample averages and the location of the acceptance region.

The type II error is the shaded area in Figure 11-6. This area

Figure 11-6 Acceptance region when $\mu = 35.0$.

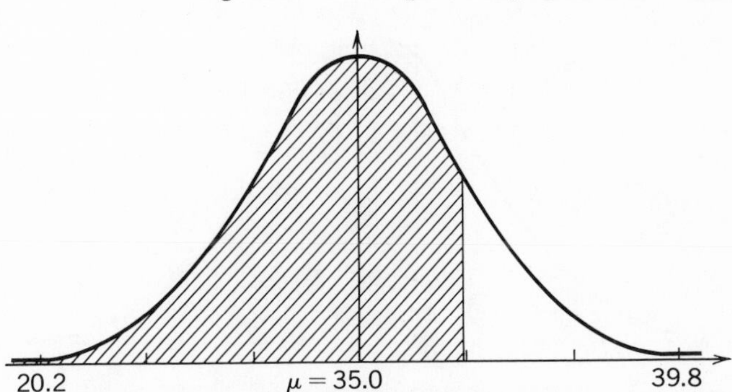

is found by the following calculations:

$$\frac{39.8 - 35.0}{10/\sqrt{4}} = +0.96 \qquad \text{area} > 39.8 = 0.1685$$

$$\frac{20.2 - 35.0}{10/\sqrt{4}} = -2.96 \qquad \text{area} < 20.2 = \frac{0.0015}{0.1700}$$

These calculations indicate that if the true average life in the population is 35.0, the probability of selecting a sample of four batteries having an average within the acceptance region is 0.830 (1–0.1700). This means there is an 83 percent chance of incorrectly accepting the hypothesis (that the average of the population is equal to 30.0).

Now suppose the true average is 22.0. Figure 11-7 shows the distribution of sample averages and the location of the acceptance region.

The shaded area again indicates the type II error, calculated as follows:

$$\frac{39.8 - 22.0}{10/\sqrt{4}} = +3.56 \qquad \text{area} > 39.8 = 0 \text{ (approx.)}$$

$$\frac{20.2 - 22.0}{10/\sqrt{4}} = 0.36 \qquad \text{area} < 20.2 = \frac{0.3594}{0.3594}$$

There is a 64 percent (100–35.94) chance that, if the true average of the population is 22.0, an average of four will fall between 20.2 and 39.8. Or, there is a 64 percent chance of incorrectly accepting the hypothesis that the true population average is 30.0.

Calculations as above are repeated for all possible values of the true average of population. The results form the operating characteristic (OC) curve shown in Figure 11-8. The OC curve is a plot of the

Figure 11-7 Acceptance region when $\mu = 22.0$.

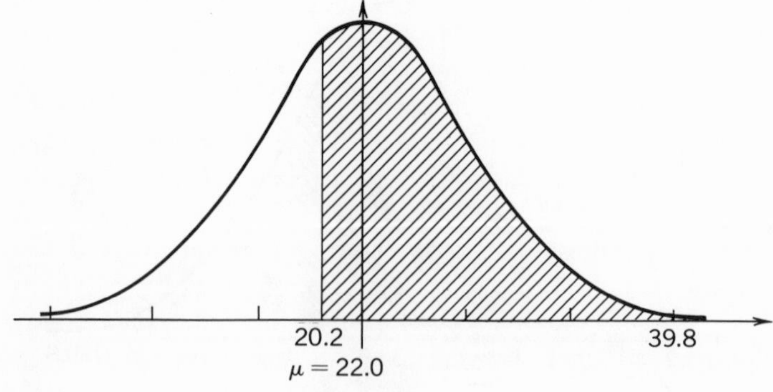

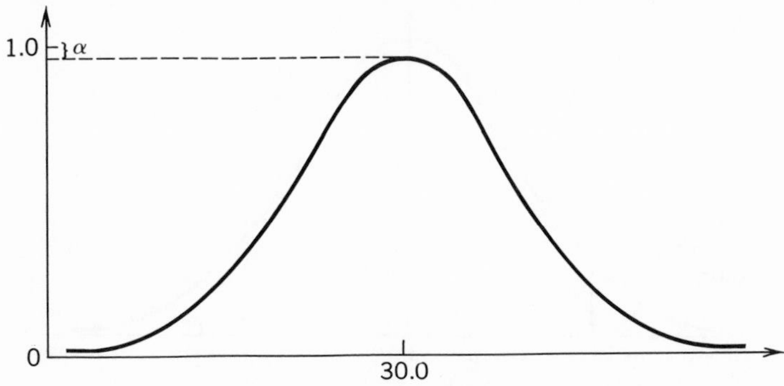

Figure 11-8 Operating characteristic curve.

probability of accepting the original hypothesis as a function of the true value of the population parameter. Note that for an average equal to the hypothesis (30.0), the probability of acceptance is $1 - \alpha$.

11-4 The Use of the Operating Characteristic Curve in Selecting an Acceptance Region

The acceptance region was determined by dividing the 5 percent allowable α error into two equal parts (see Figure 11-3). The entire 5 percent error could also be placed at either the left or the right tail of the distribution curve (Figure 11-9).

In Figure 11-9*a*, the acceptance region includes averages of four greater than 21.8. In Figure 11-9*b*, the acceptance region includes averages less than 38.2. Thus, there are three acceptance regions with the same error. A test of hypothesis having the first acceptance region is referred to as a two-tail test. A test having either of the other two acceptance regions is called a one-tail test. For a given sample size, it is desired to control the type I error to a specified value and to obtain a type II error as small as possible.

Operating characteristic curves for tests having the acceptance regions shown in Figure 11-9 can be developed following the approach used for the two-tail region. Although the α error is the same, the β error varies for the three tests.

In some problems, knowledge is available to indicate that if the true average of the population is *not* equal to the hypothesis value, then it is on one side of the hypothesis value. For example, a new material of supposedly higher average strength will have an average

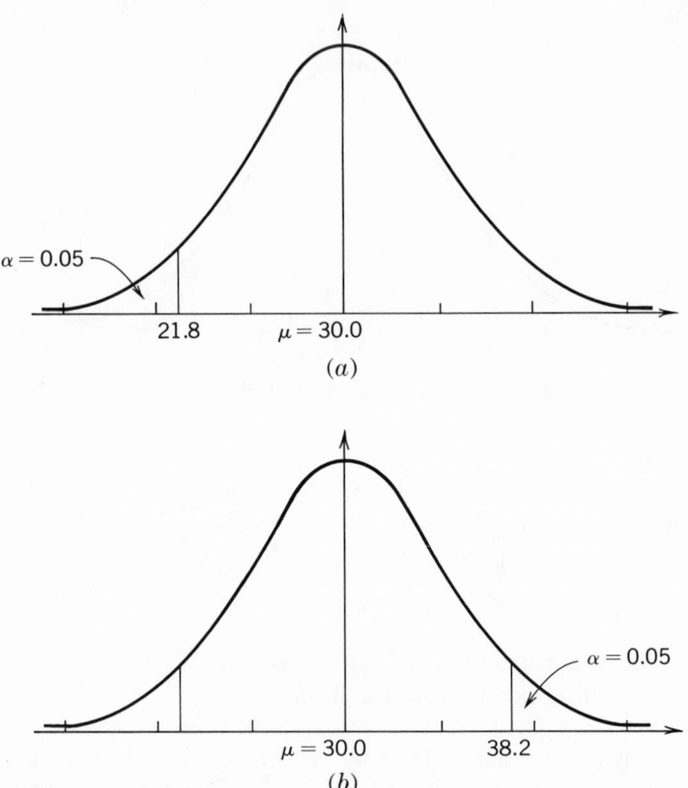

Figure 11-9a Entire α error on left tail. *b* Entire α error on right tail.

equal to or *greater than* that of the present material. Such information will help select a one-tail or two-tail test to make the error as small as possible. The following guidelines are based on the analysis of OC curves:

Use a one-tail test with the entire α risk on the right tail if (1) it is suspected that (if μ_0 is not true) the true mean is $> \mu_0$ or (2) values of the population mean $< \mu_0$ are acceptable and we are only interested in detecting a population mean $> \mu_0$.

Use a one-tail test with the entire α risk on the left tail if (1) it is suspected that (if μ_0 is not true) the true mean is $< \mu_0$ or (2) values of the population mean $> \mu_0$ are acceptable and we are only interested in detecting a population mean $< \mu_0$.

Use a two-tail test if (1) there is no prior knowledge on the location of the true population mean or (2) we are interested in detecting a true population mean $<$ or $>$ the μ_0 stated in the original hypothesis.

The selection of a one- or two-tail test will be illustrated in the examples in the next section.

11-5 Specific Tests of Hypotheses

The previous concepts are the basis for testing hypotheses concerning specific parameters of populations. A summary of some important tests is given in Table 11-1. (The t, chi-square, and F distributions have a parameter known as "degrees of freedom" as part of their definition. This is defined for each test.) Illustrative examples will now be presented to explain the specific tests. In these examples, a type I error of 0.05 will be assumed.

1 Test for difference between sample mean $\overline{X}$ and mean of population μ. (Standard deviation of the population is known.)

Example A single cavity molding press has been producing insulators with an average impact strength of 5.15 foot-pounds and with a standard deviation of 0.25 foot-pound. A new lot shows the following data from 12 specimens.

Specimen Number	*Impact Strength*
1	5.02
2	4.87
3	4.95
4	4.88
5	5.01
6	4.93
7	4.91
8	5.09
9	4.96
10	4.89
11	5.06
12	4.85
	$\overline{X} = 4.95$

Is the new lot from which the sample of 12 was taken significantly different in average impact strength from the past performance of the process?

Table 11-1 SUMMARY OF FORMULAS ON TESTS OF HYPOTHESES

Hypothesis	*Test Statistic and Distribution*
1 H: $\mu = \mu_0$ (the mean of a normal population is equal to a specified value μ_0; σ is known)	$U = \dfrac{\overline{X} - \mu_0}{\sigma/\sqrt{n}}$ Normal distribution
2 H: $\mu = \mu_0$ (the mean of a normal population is equal to a specified value μ_0; σ is estimated by s)	$t = \dfrac{\overline{X} - \mu_0}{s/\sqrt{n}}$ t distribution with $n - 1$ degrees of freedom (DF)
3 H: $\mu_1 = \mu_2$ (the mean of population 1 is equal to the mean of population 2; assume that $\sigma_1 = \sigma_2$ and that both populations are normal)	$t = \dfrac{\overline{X}_1 - \overline{X}_2}{\sqrt{1/n_1 + 1/n_2}\sqrt{[(n_1 - 1)s_1^2 + (n_2 - 1)s_2^2]/(n_1 + n_2 - 2)}}$ t distribution with DF $= n_1 + n_2 - 2$
4 H: $\sigma = \sigma_0$ (the standard deviation of a normal population is equal to a specified value σ_0)	$\chi^2 = \dfrac{(n - 1)s^2}{\sigma_0^2}$ chi-square distribution with DF $= n - 1$
5 H: $\sigma_1 = \sigma_2$ (the standard deviation of population 1 is equal to the standard deviation of population 2; assume that both populations are normal	$F = \dfrac{s_1^2}{s_2^2}$ F distribution with $\text{DF}_1 = n_1 - 1$ and $\text{DF}_2 = n_2 - 1$
6 H: $p = p_0$ (the fraction defective in a population is equal to a specified value p_0; assume that $np_0 \geq 5$)	$U = \dfrac{p - p_0}{\sqrt{p_0(1 - p_0)/n}}$ Normal distribution
7 H: $p_1 = p_2$ (the fraction defective in population 1 is equal to the fraction defective in population 2; assume that n_1p_1 and n_2p_2 are each ≥ 5)	$U = \dfrac{X_1/n_1 - X_2/n_2}{\sqrt{\hat{p}(1 - \hat{p})(1/n_1 + 1/n_2)}}$ $\quad \hat{p} = \dfrac{X_1 + X_2}{n_1 + n_2}$ Normal distribution

Solution: H: $\mu_0 = 5.15$ foot-pounds

(The average of the population from which the sample was taken is the same as the past process average.)

Test statistic

$$U = \frac{\overline{X} - \mu_0}{\sigma/\sqrt{n}}$$

Acceptance region Assuming no prior information and that a deviation on either side of the hypothesis average is important to detect, then a two-tail test (Figure 11-10) is applicable.

From Table A in the Appendix, the acceptance region is U between -1.96 and $+1.96$.

Analysis of sample data

$$U = \frac{4.95 - 5.15}{0.25/\sqrt{12}} = -2.75$$

Conclusion Since U is outside the acceptance region, the hypothesis is rejected. Therefore, sufficient evidence is present to conclude that the average impact strength of the new process is significantly different from the average of the past process. (The confidence limit concept presented in section 11-8 provides a means of estimating how *much* the difference is.) The answer to the question in section 11-2 is yes.

Figure 11-10 Distribution of U (two-tail test).

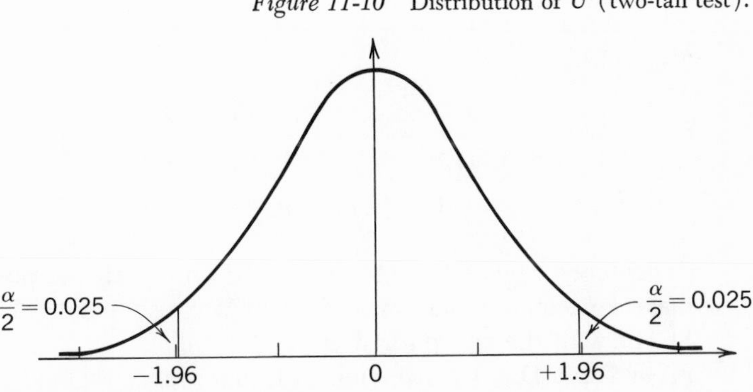

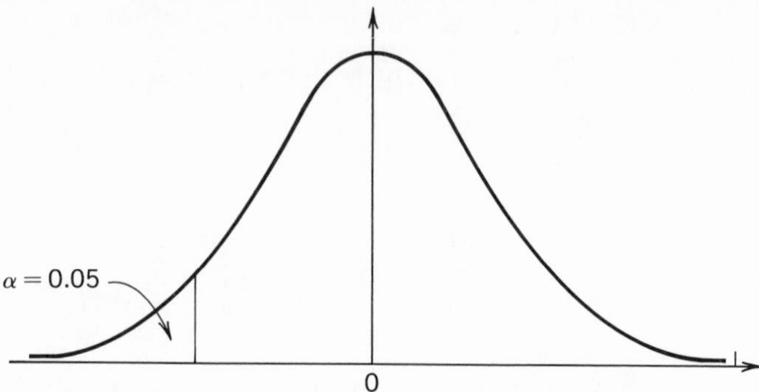

Figure 11-11 Distribution of t with α on left tail.

2 Test for difference between two sample means $\bar{X}_1$ and $\bar{X}_2$ when the standard deviation is unknown but believed to be the same for the two populations. (This assumption can be evaluated by test 5.)

Example Two types of spark plugs were operated in the alternate cylinders of an aircraft engine for 100 hours and the following data obtained:

	Make 1	*Make 2*
Number of spark plugs tested, n	10	8
Average wear per 100 hours $\bar{X}$, in.	0.0049	0.0064
Variability s, in.	0.0005	0.0004

Can it be said that make 1 wears less than make 2?

Solution: H: $\mu_1 = \mu_2$

Test statistic

$$t = \frac{\bar{X}_1 - \bar{X}_2}{\sqrt{1/n_1 + 1/n_2}\sqrt{[(n_1 - 1)s_1^2 + (n_2 - 1)s_2^2]/(n_1 + n_2 - 2)}}$$

with degrees of freedom $= n_1 + n_2 - 2$.

Acceptance region We are concerned only with the possibility that make 1 wears less than make 2; therefore, use a one-tail test (Figure 11-11) with the entire α risk in the left tail.
From Table D in the Appendix, the acceptance region is $t > -1.746$.

Analysis of sample data

$$t = \frac{0.0049 - 0.0064}{\sqrt{1/10 + 1/8}\sqrt{[(10-1)(0.0005)^2 + (8-1)(0.0004)^2]/(10+8-2)}} = -7.0$$

Conclusion Since t is outside the acceptance region, the hypothesis is rejected. Therefore, sufficient evidence is present to conclude that make 1 wears less than make 2. The answer to the question in section 11-2 is yes.

3 Test for difference between sample variability s and population variability σ.

Example For the insulator strengths tabulated in the first example, the sample standard deviation is 0.036 foot-pound. The previous variability, recorded over a period, has been established as a standard deviation of 0.25 foot-pound. Does the low value of 0.036 indicate that the new lot is significantly more uniform (i.e., lower standard deviation)?

Solution: H: $\sigma_0 = 0.25$ foot-pound

Test statistic

$$\chi^2 = \frac{(n-1)s^2}{{\sigma_0}^2}$$

with degrees of freedom $= n - 1$.

Acceptance region We believe that the standard deviation is smaller; therefore, we will use a one-tail test (Figure 11-12) with the entire risk on the left tail.
From Table E in the Appendix the acceptance region is $\chi^2 \geq 4.575$.

Figure 11-12 Distribution of χ^2 with α on left tail.

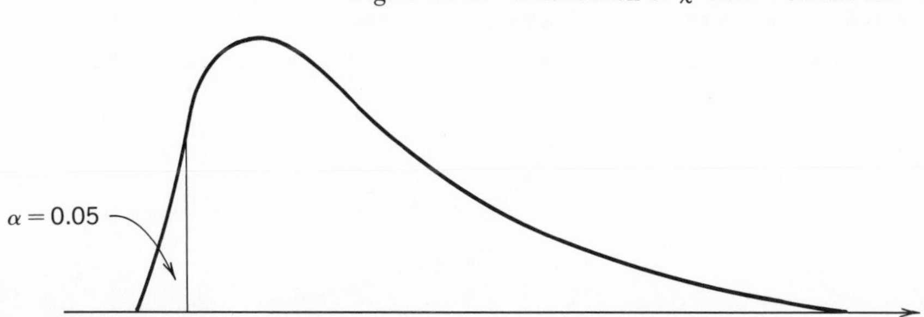

Analysis of sample data

$$\chi^2 = \frac{(12 - 1)(0.036)^2}{(0.25)^2} = 0.23$$

Conclusion Since χ^2 is outside the acceptance region, the hypothesis is rejected. Therefore, sufficient evidence is present to conclude that the new lot is more uniform.

4 Test for the difference in variability (s_1 versus s_2) in two samples.

Example A materials laboratory was studying the effect of aging on a metal alloy. They wanted to know if the parts were more consistent in strength after aging than before. The data obtained were:

	At Start (1)	*After* 1 *Year* (2)
No. of specimens n	9	7
Average strength $\overline{X}$, psi	41,350	40,920
Variability s, psi	934	659

Solution: H: $\sigma_1 = \sigma_2$

Test statistic

$$F = \frac{s_1^2}{s_2^2} \qquad \text{with } DF_1 = n_1 - 1,\ DF_2 = n_2 - 1$$

Acceptance region We are concerned with an improvement in variation; therefore, we will use a one-tail test (Figure 11-13) with the entire α risk in the right tail.

From Table F in the Appendix, the acceptance region is $F \leq 4.15$.

Figure 11-13 Distribution of F with α on right tail.

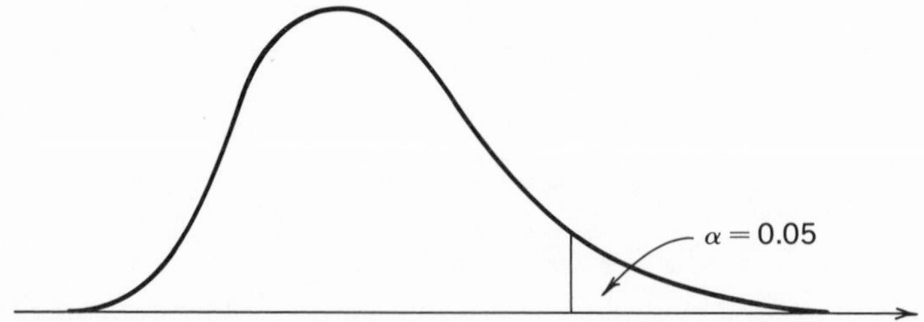

Analysis of sample data

$$F = \frac{(934)^2}{(659)^2} = 2.01$$

Conclusion Since F is inside the acceptance region, the hypothesis is accepted. Therefore, there is not sufficient evidence to conclude that the parts were more consistent in strength after aging.

11-6 What the Test of Hypothesis Does Not Tell

An "acceptance" of a hypothesis doeos *not prove* that it is correct. For each problem there existed a hypothesis that there was no difference between the two populations under comparison. In the language of change, it was assumed that the sample under study represented a population which had not changed from the original population. The data of the sample then were used to affirm or deny the hypothesis.

Under this method the sample is a piece of evidence used to affirm or deny an assumption.

But the fact that the sample gives no evidence of change is not proof that there has been no change. It is proof only that there has been no change of such size that chance alone cannot explain it. In the example of the aging of an alloy, the difference in standard deviations could reasonably have been caused by chance alone. If the difference had in fact been caused by aging, samples of nine and seven specimens were simply inadequate to recognize the existence of so small a change. It is then important to notice that, in testing for significance of differences under such a hypothesis, the sample does not detect change unless the change is greater than can be explained by chance alone. If a change has in fact occurred, but of an amount smaller than can be accounted for by chance, the test will not detect it. If it is essential to detect small changes, the remedy is to increase the sample size, thereby reducing the size of change that can be accounted for by chance alone.

If the change observed is large enough to show up as significant, the test itself does not say how much change occurred; it indicates only that a change of some size has occurred. The real amount of change is more likely to be the observed difference than to be any other single value (provided that the difference is significant). But the amount of change observed is subject to a band of error, the width of the band depending on the sample size. If the true amount of change must be determined from the observed change with great precision, it becomes necessary to use samples of large size (see the discussion on confidence limits, section 11-8).

11-7 Size of Sample for Tests of Hypotheses

Although not explicitly stated, an overall procedure was followed in the previous examples:

Design of the Test

1 State the hypothesis.
2 Define the desired statistical risks.
   α = level of significance = type I error = probability of rejecting the hypothesis when it is correct.
   β = type II error = probability of accepting the hypothesis when it is false.
3 Define the test statistic which will be used to test the hypothesis.
4 Determine the sample size n required to attain the risk levels defined above.
5 Determine the acceptance region for the test, i.e., the range of values of the test statistic which will result in a decision to accept the hypothesis.

Analysis of Test Results

6 Obtain a random sample of n items.
7 Compute the value of the test statistic for the sample.
8 Compare the value of the test statistic to the acceptance region and make a decision to accept or reject the hypothesis.

The sample size required will depend on (1) the sampling risks desired, (2) the size of the smallest true difference that is to be detected, and (3) the variation in the characteristic being measured. The sample size can be determined by using the "operating characteristic" curve for the test. Curves[1] are given (Figure 11-14) for a two-tail test of the hypothesis that the mean is equal to a specified value. Suppose it were important to detect the fact that the average of the battery cited previously was 35.0. Specifically, we want to be 80 percent sure of detecting this change (β = 0.2). Further, if the true average was 30.0 (as stated in the hypothesis), we want to have only a 5 percent risk

[1] A good source of OC curves is A. H. Bowker and Gerald J. Lieberman, *Handbook of Industrial Statistics*, Prentice-Hall, Inc., Englewood Cliffs, N.J., 1955.

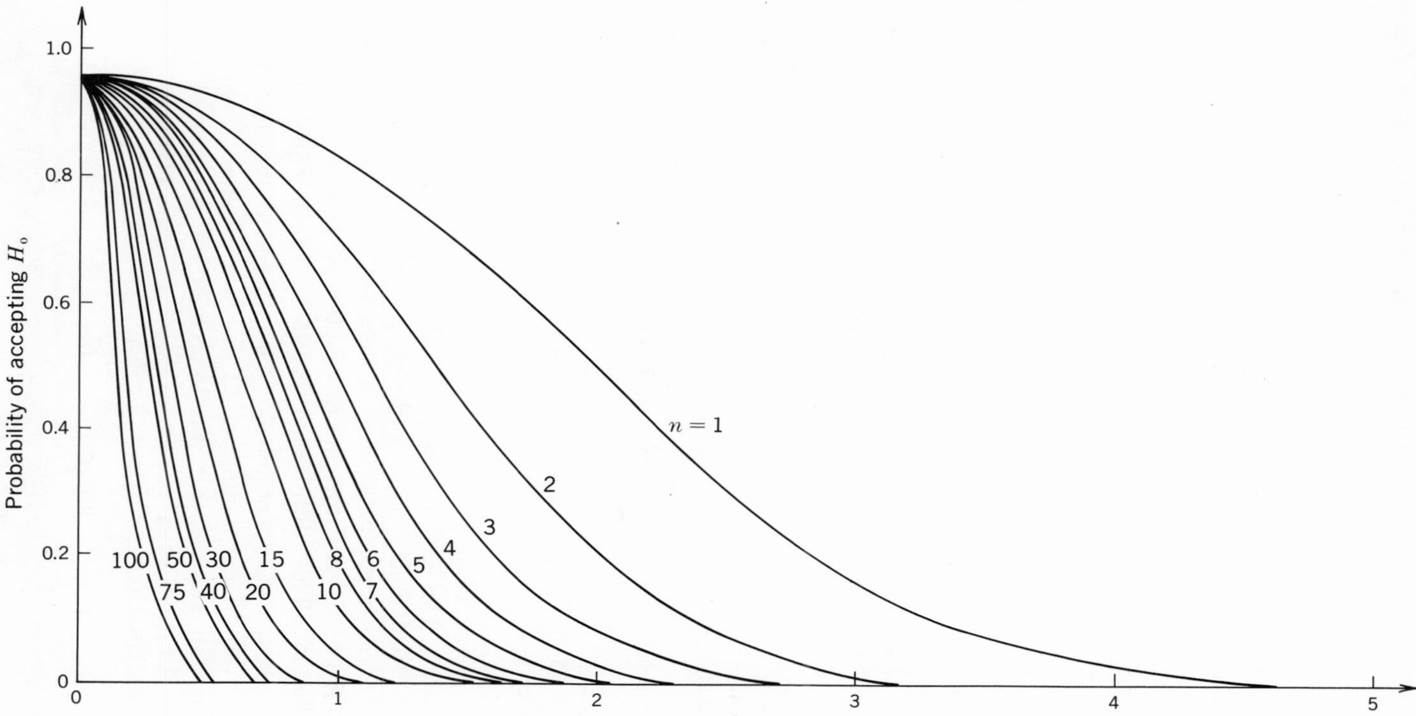

Figure 11-14 Operating characteristics of the two-sided normal test for a level of significance equal to 0.05. Source: Charles D. Ferris, Frank E. Grubbs, and Chalmers L. Weaver, "Operating Characteristics for the Common Statistical Tests of Significance," *Annals of Mathematical Statistics,* June, 1946.

of rejecting the hypothesis ($\alpha = 0.05$). In using Figure 11-14, d is defined as

$$d = \frac{\mu - \mu_0}{\sigma} = \frac{35.0 - 30.0}{10} = 0.5$$

Entering with $d = 0.5$ and $P_a = 0.2$ (the beta risk), the curves indicate that a sample size of about 30 is required.

11-8 Statistical Estimation: Confidence Limits

Estimation is the process of analyzing a sample result in order to estimate the corresponding value of the population parameter. For example, a sample of 12 insulators had an average impact strength of 4.952 foot-pounds. If this is a representative sample from the process, what estimate can be made of the true average impact strength of the entire population of insulators? Or, suppose a sample of 100 units is 5 percent defective. What can be said of the true percent defective of the population?

1 The "point estimate" is a single value used to estimate the population parameter. For example, 4.952 foot-pounds is the point estimate of the average strength of the population.
2 The "confidence interval" is a *range* of values which includes (with a preassigned probability called "confidence level") the true value of a population parameter. "Confidence limits" are the upper and lower boundaries of the confidence interval. Confidence level is the probability that an assertion about the value of a population parameter is correct.

Duncan[2] provides a thorough discussion of the concept of confidence limits. The explanation here will merely indicate the concept behind the calculations. Suppose the sample of 12 insulators mentioned above came from a population with a standard deviation σ of 0.25 foot-pound. Figure 11-15*a* shows the distribution of sample averages from the population. This distribution has an average equal to μ and $\sigma_{\bar{x}} = 0.25/\sqrt{12} = 0.072$ foot-pound. Now consider the two extreme possibilities for the one sample of 12 that was selected. Suppose this sample gave an average that was located at the extreme of the left tail of the distribution of sample averages (Figure 11-15*b*). Then μ could be estimated as

$$\bar{X} + 3\sigma_{\bar{x}} = 4.952 + 3(0.072) = 5.168$$

[2] Acheson J. Duncan, *Quality Control and Industrial Statistics*, 3d ed., Richard D. Irwin, Inc., Homewood, Ill., 1965, p. 456.

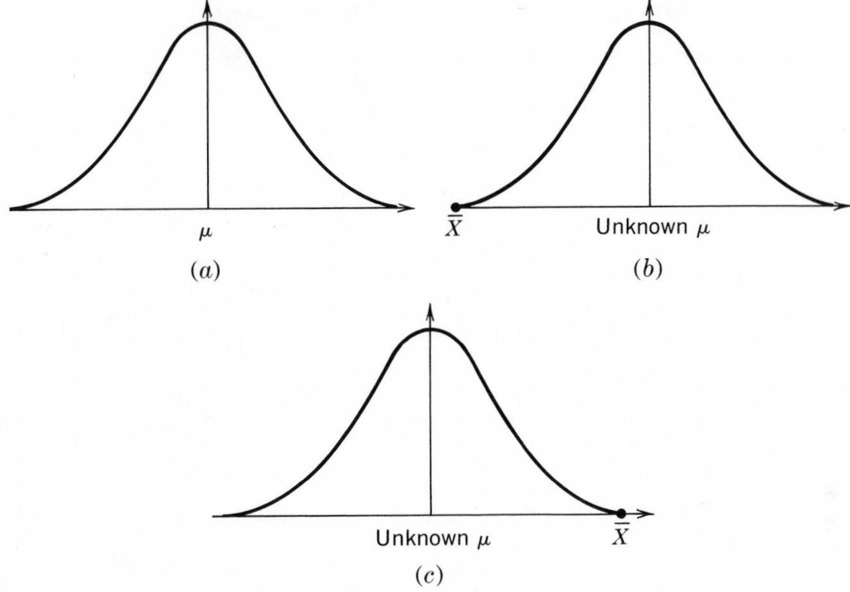

Figure 11-15a Distribution of sample averages. *b* Sample average at the left tail. *c* Sample average at the right tail.

On the other hand, suppose the sample average really came from the extreme right tail (Figure 11-15*c*). The μ could be estimated as

$$\bar{X} - 3\sigma_{\bar{x}} = 4.952 - 3(0.072) = 4.736$$

Thus two extreme possibilities have been assumed and an estimate of the true μ calculated for each assumption. It can then be concluded with 99.73 percent confidence that the true μ lies between 4.736 and 5.168. (More precisely, 99.73 percent of the sets of confidence limits so obtained will contain the true μ.) Table 11-2 summarizes confidence limit formulas for common parameters. The 99.73 percent is called the confidence level,[3] and the numbers 4.736 and 5.168 are called confidence limits. Confidence level is the degree of desired trust or assurance in a given result. A confidence level is always associated with an assertion based on actual measurements and measures the probability that the assertion is true. Confidence limits are limits which include the true value with a preassigned degree of confidence (the confidence level).

Example Sixty-one specimens of brass had a mean hardness of 54.62 and an estimated standard deviation of 5.34. Determine the 95 percent confidence limits.

[3] Confidence levels of 90, 95, or 99 percent are usually assumed in practice.

Table 11-2 SUMMARY OF CONFIDENCE LIMIT FORMULAS

Parameters	Formulas
1 Mean of a normal population (standard deviation known)	$\bar{X} \pm K_{\alpha/2} \frac{\sigma}{\sqrt{n}}$ where $\bar{X}$ = sample average K = normal distribution coefficient σ = standard deviation of population n = sample size
2 Mean of a normal population (standard deviation unknown)	$\bar{X} \pm t_{\alpha/2} \frac{s}{\sqrt{n}}$ where t = distribution coefficient (with $n - 1$ degrees of freedom) s = estimated σ
3 Standard deviation of a normal population	Upper confidence limit $= s\sqrt{\frac{n-1}{\chi^2_{1-(1-\alpha)/2}}}$ Lower confidence limit $= s\sqrt{\frac{n-1}{\chi^2_{(1-\alpha)/2}}}$ where χ^2 = chi-square distribution coefficient with $n - 1$ degrees of freedom $1 - \alpha$ = confidence level
4 Population fraction defective	See Table K in the Appendix
5 Difference between the means of two normal populations (standard deviations σ_1 and σ_2 known)	$(\bar{X}_1 - \bar{X}_2) \pm K_{\alpha/2}\sqrt{\frac{\sigma_1^2}{n_1} + \frac{\sigma_2^2}{n_2}}$
6 Difference between the means of two normal populations ($\sigma_1 = \sigma_2$ but unknown)	$(\bar{X}_1 - \bar{X}_2) \pm t_{\alpha/2}\sqrt{\frac{1}{n_1} + \frac{1}{n_2}} \times \sqrt{\frac{\Sigma(X - \bar{X}_1)^2 + \Sigma(X - \bar{X}_2)^2}{n_1 + n_2 - 2}}$
7 Mean time between failures based on an exponential population of time between failures	Upper confidence limit $= \frac{2rm}{\chi^2_{1-(1-\alpha)/2}}$ Lower confidence limit $= \frac{2rm}{\chi^2_{(1-\alpha)/2}}$ where r = number of occurrences in the sample (i.e., number of failures) m = sample mean time between failures $DF = 2r$

Solution:

$$\text{Confidence limits} = \bar{X} \pm t\frac{s}{\sqrt{n}}$$

$$= 54.62 \pm 2.00\frac{5.34}{\sqrt{61}}$$

$$= 53.25 \text{ and } 55.99$$

There is 95 percent confidence that the true mean hardness of the brass is between 53.25 and 55.99.

Example A radar system has been operated for 1,200 hours, during which time eight failures occurred. What are the 90 percent confidence limits on the mean time between failures for the system?

Solution:

$$\text{Estimated } m = \frac{1{,}200}{8} = 150 \text{ hours}$$

$$\text{Upper confidence limit} = \frac{2rm}{\chi^2_{1-(1-\alpha)/2}}$$

$$= \frac{(2)(8)(150)}{7.962} = 301.4$$

$$\text{Lower confidence limit} = \frac{2rm}{\chi^2_{(1-\alpha)/2}}$$

$$= \frac{(2)(8)(150)}{26.296} = 91.3$$

There is 90 percent confidence that the true mean time between failures is between 91.3 and 301.4 hours.

Some confusion has arisen on the application of the term "confidence level" to a reliability index such as mean time between failures. Suppose the numerical portion of a reliability requirement reads as follows:

The MTBF shall be at least 100 hours at the 90 percent confidence level. This means:

1 The minimum MTBF is 100 hours.
2 Actual tests shall be conducted on the complete product to demonstrate with 90 percent confidence that the 100-hour MTBF has been met.
3 The test data shall be analyzed by calculating the observed MTBF and the lower one-sided 90 percent confidence limit on MTBF.
4 The lower one-sided confidence limit must be $\geq$ 100 hours.

Thus the term "confidence level," from a statistical viewpoint, has great implications on a test program. Note that the observed MTBF must be greater than 100 if the lower confidence limit is to be $\geq$100. Confidence level means that sufficient tests must be conducted to demonstrate, with statistical validity, that a requirement has been met. Confidence level does *not* refer to the qualitative opinion about meeting a requirement. Finally, confidence level does *not* lower a requirement; i.e., a 100-hour MTBF at a 90 percent confidence level does not mean that 100 hours is desired but that (0.90) (100) or 90 hours is acceptable. These serious misunderstandings *have* occurred. When the term is used, a clear understanding should be verified and not assumed. Johnson and Leone[4] sum up estimation and tests of hypotheses:

"In estimation we are concerned with obtaining as good estimates of parameter value(s) as possible. Testing hypotheses comprises techniques for using data to check statements ('hypotheses') made before looking at the sample values about the population(s) from which the samples have been drawn. This division is evidently arbitrary in that there can be some overlapping."

11-9 Importance of Confidence Limits in Planning Test Programs

Additional tests will increase the precision of the estimates obtained from a test program. This obvious fact sometimes hides a more subtle point. The increase in precision usually does not vary linearly with the number of tests—doubling the number of tests usually does *not* double the precision (even approximately)! Examine the graph (Figure 11-16) of confidence interval for the average against sample size (a standard deviation of 50.0 was assumed): when the sample size is small, an increase has a great effect on the width of the confidence interval; after about 30 units, an increase has a much smaller effect. The inclusion of the cost parameter is vital here. The cost of additional tests must be evaluated against the value of the additional precision.

11-10 Determination of the Sample Size Required to Achieve a Specified Precision in an Estimate

Confidence limits can help to determine the size of test program required to estimate a product characteristic within a specified precision. It is desired to estimate the true average of the battery previously cited. The estimate must be within 2.0 hours of the true average if the estimate is to be of any value. A 95 percent confidence level is desired on

[4] Norman L. Johnson and Fred C. Leone, *Statistics and Experimental Design*, John Wiley & Sons, Inc., New York, 1964, vol. 1, p. 180.

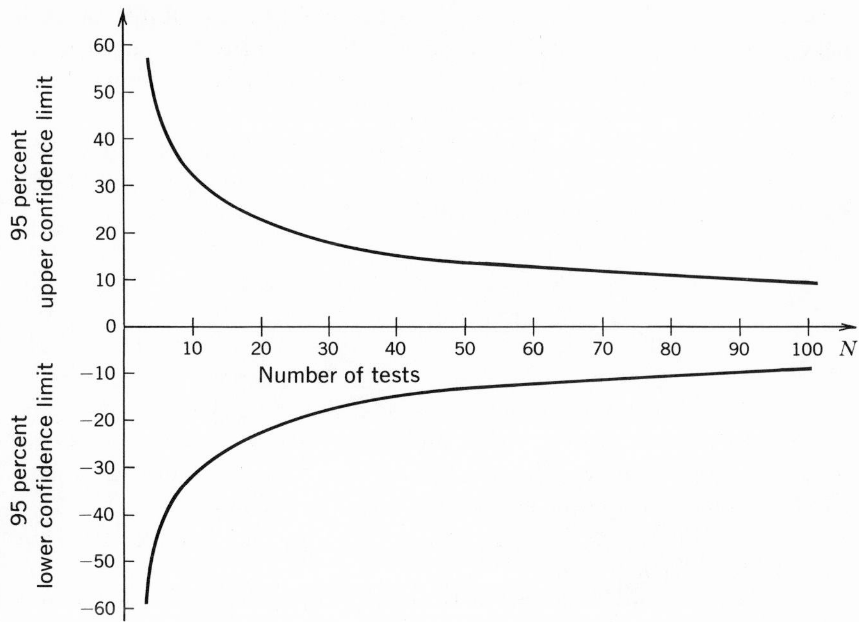

Figure 11-16 Width of confidence interval versus number of tests.

the confidence statement. The 2.0 hours is the desired confidence interval or

$$2.0 = \frac{(1.96)(10)}{\sqrt{n}} \qquad n = 96$$

A sample of 96 batteries will provide an average which is within 2.0 pounds of the true average (with 95 percent confidence). Notice the type of information required for estimating the average of a normal population: (1) desired width of the confidence interval (the precision desired in the estimate), (2) confidence level desired, and (3) variability of the characteristic under investigation. The number of tests required cannot be determined until the engineer furnishes these items of information. Past information may also have a major role in designing a test program (see section 11-11).

11-11 Use of Prior Information in Planning Tests

The previous sections have assumed that statements concerning product performance would be based strictly on actual tests conducted on the product. Recently, attention has been drawn to the possibility of using past ("prior") information to supplement the actual test data. The approach is based on a theorem originally proposed by Rev. Thomas Bayes in 1763. The theorem may be stated as follows:

If $\mu_1, \mu_2 \cdots \mu_n$ are mutually exclusive events (i.e., they cannot happen simultaneously) of which one must occur, and if these events can be associated with another event A, then

$$P(\mu_i/A) = \frac{P(\mu_i)P(A/\mu_i)}{\sum_{\text{all } i} [P(\mu_i)P(A/\mu_i)]}$$

An oversimplified example will illustrate the concept. Suppose it is desired to evaluate the average life of an engine. The requirement is 5,000 hours. The wealth of experience in the development department on similar engines has been summarized as follows:

We are fairly certain (80 percent) that the new engine will have a mean life of 5,000 hours.

There is a small chance (15 percent) that the new engine will have a mean life of only 2,500 hours.

There is a smaller chance (5 percent) that the new engine will be poor and only achieve a mean life of 1,000 hours.

The three values of mean life are the events μ_1, μ_2, and μ_3. The percentages are the values of $P(\mu_i)$, that is, the probability that the mean life of the population will be μ_i.

The following test program has been proposed: run eleven engines each for 1,000 hours; if four or fewer failures occur conclude that the design meets the 5,000-hour requirement. In this case, the event A is the passing of the above program (i.e., running eleven engines and having four or fewer failures). Assuming an exponential distribution, calculations for an operating characteristic curve can be made (see Chapter 17) to find the probability of passing the test given that the true mean is μ_i

These probabilities are $P(A/\mu_i)$ in Bayes' theorem. Summarizing:

μ_i	$P(\mu_i)$	$P(A/\mu_i)$
5,000	0.80	0.95
2,500	0.15	0.71
1,000	0.05	0.10

Suppose the test program is conducted and passed. The probability that the 5,000-hour mean life requirement has been met may be calculated as

$$P(\mu = 5{,}000/\text{test passed}) = \frac{0.8(0.95)}{0.8(0.95) + 0.15(0.71) + 0.05(0.1)} = 0.86$$

Thus, the test program would give reasonably high assurance (86 percent) that the design meets the requirement (if the test is passed). Simultaneously, if the design is extremely poor ($\mu = 1{,}000$), there is only a small chance (10 percent) that it will pass the test. (The adequacy of a 71 percent chance of passing a design with a mean of only 2,500 hours is a subject of discussion.)

The controversial question in all this is the determination of the values of $P(\mu_i)$; for example, what is the basis of saying that there is an 80 percent chance that the engine will achieve the 5,000-hour mean life? The writers believe that it *may* often be possible to synthesize such probabilities from past experience on similar products or parts of products.

This experience may be test data or proven scientific principles. As a simple example, suppose it is desired to predict the probability that an object will not break if it is dropped from a specified height. Six such objects are dropped and none breaks. Confidence limit calculations (based on the binomial distribution, not on Bayes' theorem) predict, with 90 percent confidence, that the probability of success on a seventh trial is at least 0.68. However, if scientific principles (e.g., the law of gravity and strength of material principles) tell us that the object should not break, then the prediction is conservative. Ideally, conclusions should be based only on tests of the current product. However, this is often impossible and engineering managers have used and will continue to use past experience in drawing conclusions on current products. They often do this in an informal and qualitative manner. The Bayes approach, when applicable, may provide an opportunity to evaluate this extremely valuable past experience in a more methodical manner. Judgment will be needed and danger will exist. Johnson and Leone[5] state: "It should be emphasized that the results obtained in this example are only applicable if the prior distribution is really *known* to be of the stated form (in practice, approximately so). If this is not remembered there is a danger of the use of Bayes' Theorem degenerating into a series of attractive mathematical exercises with little relevance to applied statistics."

There is controversy on the Bayes versus the conventional approach to analysis of data. The viewpoints as applied to evaluating a new

[5] *Ibid.*

Table 11-3 BAYES VERSUS CONVENTIONAL STATISTICAL APPROACH IN EVALUATING A NEW DESIGN

	Bayes	*Conventional*
Role of past experience	Past experience is valuable in helping to predict performance of a new design	Not applicable because the new design is different. Evaluation of new design must be based solely on test results of new design
Purpose of development test program on new design	Confirm or deny expected performance of new design as predicted from past experience	Supply the data for evaluating the new design
Validity of conclusions about the new design	Depends on ability to quantitatively relate past experience to new design	More conservative than Bayes because evaluation is based solely on test results of new design
Number of tests required on new design	Bayes approach requires less than conventional approach because it makes use of applicable past data	

design are summarized in Table 11-3. A further example of the use of prior information is presented in Chapter 17.

11-12 The Design of Experiments

The "design of experiments" is the term used to denote the *plan* of an experiment that recognizes statistical aspects.

The contrast between the classical method of experimentation (vary one factor at a time, hold everything else constant) and the modern approach is striking. Table 11-4 compares these two approaches for an experiment in which there are two factors (or variables) whose effects on a characteristic are being investigated. (The same conclusions hold for an experiment with more than two factors.)

Some key questions that arise are

1 How large a difference in the conditions being compared is considered significant from an engineering point of view? (How large a difference do we want the experiment to detect?)

Table 11-4 A COMPARISON OF THE CLASSICAL AND MODERN METHODS OF EXPERIMENTATION

Criteria	*Classical*	*Modern*
Basic procedure	Hold everything constant except the factor under investigation. Vary that factor and note the effect on the characteristic of concern. To investigate a second factor, conduct a separate experiment in the same manner	Plan the experiment to evaluate both factors in one main experiment. Include, in the design, measurements to evaluate the effect of varying both factors simultaneously
Experimental conditions	Care taken to have material, men, and machine constant throughout the entire experiment	Realizes difficulty of holding conditions reasonably constant throughout an entire experiment. Instead, experiment is divided into several groups or blocks of measurements. Within each block, conditions must be reasonably constant (except for deliberate variation to investigate a factor)
Experimental error	Recognized but not stated in quantitative terms	Stated in quantitative terms
Basis of evaluation	Effect due to a factor is evaluated with only a vague knowledge of the amount of experimental error	Effect due to a factor is evaluated by comparing variation due to that factor with the quantitative measure of experimental error
Possible bias due to sequence of measurements	Often assumed that sequence has no effect	Guarded against by "randomization"
Effect of varying both factors simultaneously ("interaction")	Not adequately planned into experiment. Frequently assumed that the effect of varying factor 1 (when factor 2 is held constant at some value) would be the same for any value of factor 2	Experiment can be planned to include an investigation for "interaction" between factors

Table 11-4 **A Comparison of the Classical and Modern Methods of Experimentation** (*Continued*)

Criteria	*Classical*	*Modern*
Validity of results	Misleading and erroneous if "interaction" exists and is not realized	Even if "interaction" exists, a valid evaluation of the main factors can be made
Number of measurements	For a given amount of useful and valid information, more measurements needed than in the modern approach	Fewer measurements needed for useful and valid information
Definition of problem	Objective of experiment frequently not defined as necessary	In order to design experiment, it is necessary to define the objective in detail (how large an effect do we want to determine, what numerical risks can be taken, etc.)
Application of conclusions	Sometimes disputed as only applicable to "the controlled conditions under which the experiment was conducted"	Broad conditions can be planned into the experiment, thereby making conclusions applicable to a wider range of actual conditions

2 How much variation has been experienced in the quality characteristics under investigation?

3 What risk do we want to take that the experiment incorrectly concludes that a significant difference exists when the correct conclusion is that no significant difference exists? (This is the type 1 error.)

4 What risk do we want to take that the experiment fails to detect the difference that really does exist? (This is the type II error.)

5 Do we have any knowledge about possible interactions of the factors? Do we wish to test for these interactions?

Many experimental problems can be handled with one of the standard experimental designs.[6]

11-13 Analysis of a Designed Experiment: The Analysis of Variance

The analysis is often performed by a technique known as "analysis of variance" (ANOVA). The statistician does not regard this as an ad-

[6] For a summary of five important types of designs, see J. M. Juran (ed.), *Quality Control Handbook* (*QCH*), 2d ed., McGraw-Hill Book Company, New York, 1962. Johnson and Leone, *op. cit.*, provide detailed plans and methods of analysis.

vanced technique. However, a knowledge of basic statistics *is* necessary to understand it. There are alternative techniques (e.g., control charts) which are approximate but simpler to explain and which warrant consideration when presentation is important.[7] ANOVA divides the total variation in an experiment into the variation due to (1) the factors under investigation, (2) interactions between factors, and (3) experimental error. The variation of the factor under investigation is compared to the experimental error and a conclusion is reached. It is assumed that the populations being compared are normally distributed with equal variances. An example will illustrate.

Consider an experiment for evaluating the ability of four different thermometers and three analysts to measure melting point (adapted from Wernimont[8]).

The coded data are given in Table 11-5.

Table 11-5 **Data on Thermometers and Analysts**

	Analyst				
Thermometer	*A*	*B*	*C*	*Totals*	$\bar{X}$
1	2.0	1.0	1.5		
	1.5	1.0	1.0	8.0	1.33
2	1.0	0	1.0		
	1.5	1.0	1.5	6.0	1.00
3	−0.5	−1.0	1.0		
	0.5	0	1.0	1.0	0.16
4	1.5	−1.0	0.5		
	1.5	0	1.0	3.5	0.58
Totals	9.0	1.0	8.5		

Two questions may be posed. First, is there a significant difference between thermometers in measuring melting point? Second, is there a significant difference between analysts in measuring melting point? The total variation may be divided into:

1 Variation due to thermometers (variable 1)
2 Variation due to analysts (variable 2)
3 Variation due to a possible interaction between thermometers and analysts
4 Variation due to a residual effect (assumed to be experimental error)

[7] The following note was attached by management to a report that used analysis of variance to evaluate the results of a "Latin square" design of experiment: "Good job of testing and reporting here (if only I could read Latin better!)."

[8] Grant Wernimont, "Quality Control in the Chemical Industry, Part II: Statistical Quality Control in the Chemical Laboratory," *Industrial Quality Control*, p. 8, May, 1947.

The formulas for this analysis are summarized below in terms of a variance definition formula, which is followed by an equivalent calculation formula for the numerator of the variance.

Total variation = variation due to variable 1 + variation due to variable 2 + variation due to interaction of variables 1 and 2 + variation due to residual (experimental error)

Total variation This is the numerator of the variance of all the individual values about the grand average ($\bar{\bar{X}}$) or

$$\frac{\Sigma(X - \bar{\bar{X}})^2}{rc - 1} \qquad \begin{aligned} r &= \text{no. of rows} \\ c &= \text{no. of columns} \end{aligned}$$

The calculation formula is

$$\text{Sum of squares} = \Sigma(X^2) - \frac{T^2}{rcn} \qquad \begin{aligned} T &= \text{grand total} \\ n &= \text{no. of measurements at each combination} \end{aligned}$$

Variation due to variable 1 (variation between rows) This is the numerator of the variance of the row averages about the grand average ($\bar{\bar{X}}$). The basic formula for the variance is

$$\frac{c\Sigma(\bar{X}_{\text{rows}} - \bar{\bar{X}})^2}{\text{no. of rows} - 1}$$

The calculation formula is

$$\text{Sum of squares} = \frac{R_1^2 + R_2^2 + \cdots + R_r^2}{cn} - \frac{T^2}{rcn} \qquad R = \text{row total}$$

Variation due to variable 2 (variation between columns) This is the numerator of the variance of the column averages about the grand average $\bar{\bar{X}}$. The variance is

$$\frac{r\Sigma(\bar{X}_{\text{col}} - \bar{\bar{X}})^2}{\text{no. of columns} - 1}$$

The calculation formula is

$$\text{Sum of squares} = \frac{C_1^2 + C_2^2 + \cdots + C_c^2}{rn} - \frac{T^2}{rcn} \qquad C = \text{column total}$$

Variation due to residual (experimental error) This is the numerator of the variance of individual readings taken under the same set of conditions (i.e., within one set of combination of variables 1 and 2) or

$$\frac{\Sigma(X - \overline{X}_{\text{comb}_1})^2 + \Sigma(X - \overline{X}_{\text{comb}_2})^2 + \cdots}{(n-1) + (n-1) + \cdots}$$

The calculation formula is

$$\Sigma(X^2) - \Sigma\,\frac{(w^2)}{n}$$

where w = total of the readings within a combination of variables 1 and 2

Variation due to interaction This is obtained by subtraction

Sum of squares = total SS − SS for residual − SS for columns − SS for rows

The resulting ANOVA table is

Source	*SS*	*DF*	*Mean Square (MS)*
Thermometer	5.021	$4 - 1 = 3$	1.67
Analyst	4.615	$3 - 1 = 2$	2.31
Interaction	2.729	$(4 - 1)(3 - 1) = 6$	0.45
Residual	2.625	$23 - (3 + 2 + 6) = 12$	0.22
Total	14.990	$24 - 1 = 23$	

Note that each variance (called mean square) is SS ÷ DF.

The first hypothesis to be tested concerns the interaction. It will be assumed that no interaction between thermometer and analysts exists (Figure 11-17). Assuming a type I error of 0.05, the acceptance region for testing this hypothesis is an F value equal to or greater than 3.00 (based on 6 and 12 degrees of freedom) from Table F in the Appendix.

It is desired to determine if the variation due to interaction is significantly greater than the variation due to the residual term. Therefore, the F test is

$$F = \frac{0.45}{0.22} = 2.04$$

As the calculated test value falls within the acceptance region, the hypothesis of no interaction is accepted.

At this point, a judgment must be made. If it appears that no interac-

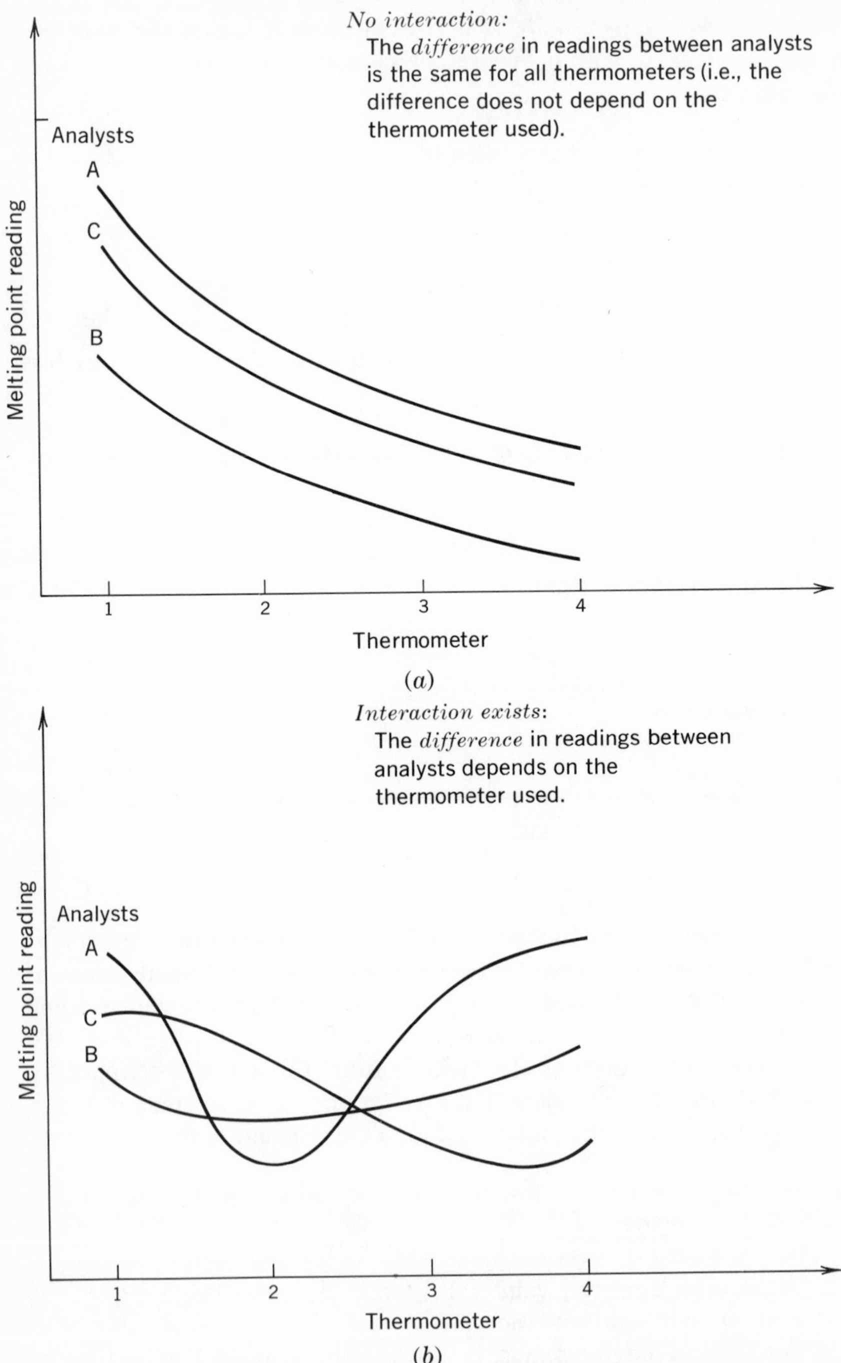

Figure 11-17a No indication of interaction. *b* Existence of interaction.

tion exists, then it is usually considered valid to assume that the calculated interaction is really a part of the residual variation and that the interaction term may be combined with the residual term and a new table constructed:

Source	*SS*	*DF*	*MS*
Thermometer	5.021	3	1.67
Analyst	4.615	2	2.31
Residual	5.354	18	0.30
Total	14.990	23	

Note that although the mean square for residual has increased slightly, the degrees of freedom for the residual have increased from 12 to 18. The increased degrees of freedom narrow the acceptance region, thereby increasing the likelihood that a true difference between thermometers or between analysts will be detected in the analysis.

The main analysis may now be completed. A hypothesis is stated that the thermometers yield the same average melting point reading; i.e.,

$$\text{H: } \mu_1 = \mu_2 = \mu_3 = \mu_4$$

The acceptance region for this hypothesis (assuming a type I error of 0.05) is an F value equal to or less than 3.16. The calculated F value compares the mean square for thermometers to the mean square for the combined residual term:

$$F = \frac{1.67}{0.30} = 5.56$$

As the calculated F value falls outside of the acceptance region, the hypothesis is rejected. The conclusion is that there *is* sufficient evidence to conclude that the thermometers yield different average melting point readings.

A second hypothesis states that the analysts are not different in their ability to read melting point; i.e.,

$$\text{H: } \gamma_A = \gamma_B = \gamma_C$$

The acceptance region for this test is an F value equal to or greater than 3.35.

The calculated F value compares the mean square for analysts to the residual mean square or

$$F = \frac{2.31}{0\ 30} = 7.70$$

Again, the hypothesis is rejected because the calculated F value falls outside of the acceptance region. It is concluded, then, that the analysts read melting point differently and therefore are one of the significant factors in the variation of all the melting point readings.

Additional tests[9] may be made to narrow the differences which have been detected to determine which thermometers and which analysts are the main causes of variation.

This example should now be reviewed in light of the advantages of the modern method of experimentation described in Table 11-4. Note that:

1 Two variables (thermometer and analyst) were evaluated in one experiment.
2 *All* readings were used in the evaluation of both thermometers and analysts. The example illustrates "balanced" designs—the same number of observations were taken by each analyst on each thermometer. In evaluating the differences among *thermometers* (rows), any possible effect due to one or more analysts is "balanced out" because the same number of analyst observations were taken with each thermometer. For example, a tendency for analyst A to read 2° high could be simulated by viewing 2° as a constant present in all the A observations. As this same constant would equally affect all thermometers, it would not affect an evaluation of the *difference* among thermometers. Similarly, in evaluating analysts, the effect of thermometers is balanced out. This simple but powerful concept is one reason why *all* the observations may be used to evaluate thermometers and analysts. The classical method of experimentation would require two separate experiments—one for evaluating thermometers (with the analyst held constant) and one for evaluating analysts (with the thermometer held constant).
3 Experimental error was quantified.
4 A possible interaction between thermometer and analyst was quantitatively evaluated (instead of being rationalized as nonexistent).
5 The conclusions concerning thermometers are applicable to three analysts. The conclusions concerning analysts are applicable to four thermometers.

The statistical design of experiments is relatively complicated. However, it is a tool that can help achieve the breakthrough in knowledge often required for solving difficult quality problems. Analysis of variance and regression analysis[10] (the study of the relationship between variables) are important tools for analysis.

[9] Duncan, *op. cit.*, p. 632.
[10] See *QCH*, 1962, sec. 15.

Answers to questions at beginning of chapter

1 No
2 Yes
3 Yes
4 No
5 Yes

PROBLEMS

Note: The specific questions have purposely been stated in nonstatistical language to provide the student with some practice in choosing techniques and making assumptions. When required, use a type I error of 0.05 and a confidence level of 95 percent. State any other assumptions needed.

1 A manufacturer of needles has a new method of controlling a diameter dimension. From many measurements of the present method, the average diameter is 0.030 inch with a standard deviation of 0.004 inch. A sample of 25 needles from the new process shows the average to be 0.028. If a smaller diameter is desirable, should the new method be adopted? (Assume that the standard deviation of the new method is the same as that for the present method.)

2 In the garment industry, the breaking strength of cloth is important. A heavy cotton cloth must have at least an average breaking strength of 200 psi. From one particular lot of this cloth, these five measurements of breaking strength (in psi) were obtained:

206
194
203
196
192

Does this lot of cloth meet the requirement of an average breaking strength of 200 psi?

3 In a drug firm, the variation in the weight of an antibiotic, from batch to batch, is important. With our present process, the standard deviation is 0.11 gram. The research department has developed a new

process that they believe will produce less variation. The following weight measurements (in grams) were obtained with the new process:

7.47
7.49
7.64
7.59
7.55

Does the new process have less variation?

4 A paper manufacturer has a new method of coating paper. The less variation in the weight of this coating, the more uniform and better the product. The following 10 sample coatings were obtained by the new method:

Coating Weights (in Weight/Unit Area × 100)	
223	234
215	229
220	223
238	235
230	227

If the standard deviation in the past was 9.3, is this proposed method any better? Should they switch to this method?

5 An essential property of product A is the viscosity stability. The product must undergo a viscosity degradation test. Two methods of mixing are now used. Ten samples taken from each method showed the following results:

Method 1	*Method 2*
6.8% degradation	5.7% degradation
6.1% degradation	6.3% degradation
6.2% degradation	5.3% degradation
5.8% degradation	5.8% degradation
6.0% degradation	5.9% degradation
6.9% degradation	6.1% degradation
5.9% degradation	5.7% degradation
6.2% degradation	6.4% degradation
6.1% degradation	5.3% degradation
5.7% degradation	5.6% degradation

a Is the variation of the two methods the same?
b Are both methods producing the same results?

6 A manufacturer of rubber products is trying to decide upon which "recipe" to use for a particular rubber compound. High tensile strength is desirable. Recipe 1 is cheaper to mix, but he is not sure if its strength is about the same as that of recipe 2. Five batches of rubber were made by each recipe and tested for tensile strength. These are the data collected (in psi):

Recipe 1	*Recipe 2*
3,067	3,200
2,730	2,777
2,840	2,623
2,913	3,044
2,789	2,834

Which "recipe" would you recommend that he use?

7 In the casting industry, the pouring temperature of metal is important. For an aluminum alloy, past experience shows a standard deviation of 15°. During a particular day, five temperature tests were made during the pouring time.

a If the average of these measurements was 1650°, make a statement about the average pouring temperature.
b If you had taken 25 measurements and obtained the same results, what effect would this have on your statement? Make such a revised statement.

8 At the casting firm mentioned in problem 7, a new aluminum alloy is being poured. During the first day of pouring, five pouring temperature tests were made with these results:

1705°
1725°
1685°
1690°
1715°

Make a statement about the average pouring temperature of this metal.

9 A manufacturer pressure-tests gaskets for leaks. The pressure at which this gasket leaked on nine trials was (in psi):

4,000	3,900	4,500
4,200	4,400	4,300
4,800	4,800	4,300

Make a statement about the average "leak" pressure of this gasket.

10 In a test of 500 electronic tubes, 427 were found to be acceptable. Make a statement concerning the true proportion that would be acceptable.

11 In a meat packing firm, out of 600 pieces of beef, 420 were found to be Grade A. Make a statement about the true proportion of Grade A beef.

12 Test runs with five models of an experimental engine showed that they operated, respectively, for 20, 18, 22, 17, and 18 minutes with 1 gallon of a certain kind of fuel. A proposed specification states that the engine must operate for an average of at least 22 minutes.

a What can we conclude about the ability of the engine to meet the specification?

b What is the probability that the difference between the sample average and the specification average was due to sampling variation?

c How low would the average operating minutes (of the engine population) have to be in order to have a 50 percent chance of concluding that the engine does not meet the specification?

13 A manufacturer claims that the average length in a large lot of parts is 2.680 inches. A large amount of past data indicates the standard deviation of individual lengths to be 0.002 inch. A sample of 25 parts shows an average of 2.678 inches. The manufacturer says that the result is still consistent with his claim because only a small sample was taken.

a State a hypothesis to evaluate his claim.

b Evaluate his claim using the standard hypothesis testing approach.

c Evaluate his claim using the confidence limit approach.

14 A specification requires that the average breaking strength of a certain material be at least 180 psi. Past data indicate the standard deviation of individual measurements to be 5 psi. How many tests are necessary to be 99 percent sure of detecting a lot that has an average strength of 170 psi?

15 Tests are to be run to estimate the average life of a product. Based on past data on similar products, it is assumed that the standard deviation of individual units is about 20 percent of the average life.

a How many units must be tested to be 90 percent sure that the sample estimate will be within 5 percent of the true average?

b Suppose funds were available to run only 25 tests. How sure would we be of obtaining an estimate within 5 percent?

16 Refer to the illustrative example in section 11-11. Suppose prior information had resulted in the following estimates:

μ_i	$P(\mu_i)$
5,000	0.60
2,500	0.30
1,000	0.10

If the same test program was conducted and was passed, what is the probability that the 5,000-hour mean life requirement has been met? Comment on how this result compares to the 0.86 determined in the illustrative problem.

17 An engineer wants to determine if the type of test oven or temperature has a significant effect on the average life of a component. He proposes the following design of experiment:

	Oven 1	*Oven 2*	*Oven 3*
550°	1	0	1
575°	0	1	1
600°	1	1	0

The numbers in the body of the table represent the number of measurements to be made in the experiment. What are two reasons why interaction cannot be adequately evaluated in this design?

18 Consider the 24 measurements listed in Table 11-5.

a Show the data to two different friends who do not have any knowledge of statistics. Ask each to comment on possible differences between thermometers or analysts. (Obtain their comments independently.) Did either of them mention the possibility of interaction?

b Devise one or more ways of graphing the original data.

12 Quality Specifications

12-1 Purpose and Content of Quality Specifications

A specification is a definition of a design. The design remains a concept in the mind of the designer until he defines it through verbal description, sample, drawing, writing, etc. The purpose of a specification then is *definition.*

Primitive forms of industrial society do not require specifications because the producer and consumer meet face to face in the marketplace with the product physically present for inspection and evaluation by both parties.

The spread of commerce gave rise to specification by sample and written specifications which can be traced back to the ancient Egyptians.[1]

In the complex industrial society of today, the quality specification takes on the characteristics of an industrial law. The modern factory is an industrial community complete with laws to specify what is right and wrong. One of these systems of industrial legislation is the quality specification—the quality law of the industrial community.

The subject matter of quality specifications may include: (1) materials (e.g., components, ingredients) and finished products, (2) processes, (3) method of test and criteria for acceptance and rejection, (4) method of use, and (5) complete programs. One specification may cover all these areas or separate specifications may be needed for one

[1] For a discussion of the history of quality specifications, see J. M. Juran (ed.), *Quality Control Handbook* (*QCH*), 2d ed., McGraw-Hill Book Company, New York, 1962, pp. 3-2 to 3-4.

or more areas. The following paragraphs will discuss the content of specifications for complete programs. Experience has evolved a fairly standardized table of contents[2] for the other principal forms of quality specifications.

Quality Program Specifications

Product complexity and additional product parameters such as reliability and maintainability have resulted in the increased use of specifications to define the *tasks* that should be conducted in a complete quality program. Thus, a quality program specification goes far beyond the process specification which specified the means of *manufacturing* an article.

An example of a military quality program specification is MIL-Q-9858A, *Quality Program Requirements.* This specification requires that a contractor establish a quality program. The major tasks required are

1 Definition of the organization of the quality program
2 Quality planning *activities*
3 Documented work instructions and control procedures
4 Formal control of drawings to assure currency with design changes
5 Maintenance of measuring and testing devices
6 Control of quality of purchased material
7 Inspection of the finished product
8 Documentation of quality costs
9 Provision for taking corrective action on quality problems

Chapter 28 discusses the problem of *evaluating* a program for compliance with such requirements.

In procuring highly reliable complex products, some customers have found it necessary to create specifications which define the elements of a reliability program that a contractor is expected to conduct during a contract. Typical elements in a reliability program specification for a major complex product are

1 Allocation of an overall reliability requirement to subsystems
2 Prediction of reliability during the design and development phase
3 Design review program
4 Formal program for reporting and following up on product failures
5 Types of component tests to be conducted
6 Reliability demonstration requirements
7 Formalization of the management plan for the reliability program

[2] See *QCH,* 1962, pp. 3-4 to 3-7.

8 System of reports to be issued to the customer describing reliability programs status

Various agencies of the government have prepared specifications[3] covering complete reliability programs and specific tasks such as prediction and testing.

12-2 The Scientific Approach to Tolerancing

An important element in many specifications is the tolerance limits. In theory, the cost of precision must be balanced against the value of precision (see Figure 4-3). In practice, the designer has great difficulty in achieving this balance because:

1 There are many tolerances to be established, and there is not enough time available to do a thorough job on each.
2 The designer does not have adequate facts about the costs of precision.
3 The designer does not have adequate facts about the value of precision.
4 The organization structure prevents the designer from working closely with the other functions that could help in setting tolerances.
5 The "cultural pattern" of the design department includes obsolete premises and habits.

Solutions for these problems are discussed below. The special problems of sensory qualities (qualities which must be measured by the senses of humans) are discussed in Chapter 13.

12-3 Functional and Nonfunctional Requirements

Formal methods for distinguishing between functional and nonfunctional requirements are important because they can help achieve both better product performance and lower manufacturing cost. The distinction is this: Functional requirements ensure performance for intended use, ensure long useful life, minimize accident hazards, protect lives or property, provide interchangeability in the field, and provide competitive sales advantage. Nonfunctional requirements inform the shop as to method of manufacture, reduce cost of manufacture, facilitate manufac-

[3] W. L. Hurd, Jr., "Interpretation of Reliability, Maintainability and Safety Requirements," *Annals of Reliability and Maintainability*, vol. 4, July, 1965, Spartan Books, Inc., New York, p. 41. Also see K. A. Frederiksen, "Incentives for Reliability," *Proceedings of the Tenth National Symposium on Reliability and Quality Control*, Institute of Electrical and Electronics Engineers, Inc., p. 274.

ure, provide interchangeability in the shop, and provide information o toolmakers.

Most drawings and specifications issued to the shop contain both ypes of requirements but the shop is usually *not* told which requirenents are functional and which are nonfunctional. Lack of this knowl-edge causes shop problems that increase costs. Two examples illustrate he significance:

Example A specification may call for a liquid to be of a certain specific gravity before being sent to a still for distillation. The only reason for the requirement may be to reduce the amount of heat required for distillation. Thus the requirement is a nonfunctional one. However, if it were not known to be nonfunctional, much valuable time could be wasted and production held up while one batch is brought to the specification called for.

Example One example worthy of note in this connection involved a water plug putty—a powder product which when mixed with water had to set hard in 30 seconds in order to quickly and effectively seal holes in masonry walls and foundations without appreciable shrinkage and cracking. A product batch met the functional requirements but fell outside the bulk density range specified. If the requirements had been classed properly, the inspector would have known what to do immediately. As a result, he had to hold up production to confer with his superior and the product designer to dispose of the problem. It should be understood, of course, that the variation in bulk density should be investigated and this matter cleared up; however, bulk density is another problem.

Section 12-5 presents further examples of how making a distinction points the way to reducing manufacturing costs. The distinction also helps to make clear who can waive the requirement. The design department must approve the waiver of functional requirements but design department approval is not essential for the waiver of nonfunctional requirements. In addition, the design department must have a major voice in deciding what is functional and what is not.

Several plans are in use for distinguishing between functional and other characteristics:

1 Classification of all characteristics as to importance, common classes being critical, major, minor. In this plan the vital functional defects are given the highest classification of importance (see section 12-4).
2 Identifying the functional requirements with a distinguishing letter

(such as E for engineering). When this is done, all other requirements are assumed to be nonfunctional.

3 Use of separate drawings or sheets for nonfunctional requirements. These are called by such names as "operation sheets" or "methods sheets."

4 Collection of all functional requirements into a test specification.

5 Compilation of manuals of shop practice.

The *mechanism* of making the distinction is of secondary importance. The key point is to make the distinction and inform all those affected.

12-4 Classification of Characteristics

In a classification of characteristics system, the designer indicates the degree of seriousness for each *characteristic* on a drawing.[4] (The levels are usually defined in terms of the effect on safety and function.) This differs from a classification of *defects* system, in which each possible defect is classified by the degree of seriousness.

A classification of characteristics system has several advantages:

1 Defines priorities for the inspection effort by establishing the relative importance of each characteristic. A review of current drawings may reveal areas of dramatic savings in inspection effort. Among the examples reported by Allen[5] is that of the machined part that had 18 dimensions (over 50 percent of the total) that the design function stated did not need to be controlled as long as the parts assembled properly. This meant a saving in inspection man-hours of 44 percent.

   If the relative importance is not defined by the design function, the decision will be made by default by the inspection function. If not provided with priorities, the inspection function has no choice but to establish priorities of inspection effort based on *their* opinion of the characteristics (perhaps flavored by the convenience of inspecting the various characteristics). For some products, the relative importance of characteristics is straightforward and does not require an extensive design background. However, for other products the designer is better qualified than anyone else to define the importance.

2 Impresses on the designer the need to carefully consider the importance of each characteristic in establishing tolerances. Thus, if the charac-

[4] For an example, see *QCH*, 1962, p. 3-10.

[5] Paul E. Allen, "Evaluating Inspection Costs," *Transactions of the National Convention of the American Society for Quality Control,* 1959, p. 585.

teristic is relatively unimportant, then an arbitrarily tight tolerance is not justified.

3 Reduces cost of rework and rejection. Where the classification results in changes of tolerances, some parts previously rejected will become acceptable, thereby eliminating the manufacturing and inspection labor costs and tooling costs to correct the previous rejections. Allen[6] cites an example of 95 pieces in a sequence of parts. If all dimensions had been inspected on all pieces, 10 rejection tags would have been written at a cost of about $150 and a rework cost of about $1,200 incurred. However, "exact conforming parts would be of no better quality than the ones already produced."

4 Increases inspection accuracy on important characteristics by concentrating effort on those characteristics. Allen[7] reports the results on submitting a lot with known defects to 17 inspectors. Each man inspected the lot to *all* the drawing characteristics and to the classified characteristics. "An average error rate of 44% on Major and Critical Characteristics was found when all drawing characteristics were checked, and an average error rate of 17% on Major and Critical Characteristics when only the classified characteristics were checked."

It is not unusual for a design function to strongly resist the classification of characteristics system. Discussions about the system often become less than completely objective. The designer tends to stress that the system will require too much time on his part. Further, he believes that he has *already* indicated priorities by assigning different tolerances to different characteristics. His view is that *every* tolerance is critical and must be met. The discussion becomes dramatic when he is informed that some of his tolerances are being ignored and the product still functions properly. (See section 12-3.)

12-5 Facts on the Costs of Precision

When a designer specifies a tolerance, he is, in effect, drawing a check on the company treasury. A tolerance defines the required precision for a product and both tooling and direct labor costs may be greatly influenced by the amount of precision desired.

Table 12-1 shows an example[8] of the cost for various levels of precision in machining holes.

[6] *Ibid.*

[7] *Ibid.*

[8] R. C. Johnson, "Value Engineering in the Bureau of Ships," *Journal of the American Society of Naval Engineers*, Fall, 1958, pp. 77–85.

Table 12-1 Example of Costs of Precision

Method	Lowest Tolerance Possible	Comparative Cost: Tool Cost (%)	Comparative Cost: Direct Labor Costs (%)
Punch (template)	+0.004 −0.002	100	100
Drill (jig)	+0.007 −0.002	175	300
Drill and ream (jig and bushings)	+0.0006 −0.0000	225	400
Bore (fixture) including punch, rough and finish bore on Borematic	+0.0004 −0.0000	540	700
Hone or lap (fixture including punch and bore)	+0.0002 −0.0000	730	1,100

The setup costs, based on 200-plate quantity (0.180-inch-thick brass plate 3.50 × 500 size), have been included in direct labor costs.

Note how the tolerance determines the manufacturing process. Data of the type in Table 12-1 show that significant cost reductions could be achieved by setting tolerances that strike a better balance between the cost of precision and the value of precision.[9] In general, the designer is *not* provided with usable information on the cost of precision. Such information could be generated if a company has a good manufacturing cost control system. However, the designer cannot be expected to compile this information. It should be generated (in the form needed by the designer) by an industrial engineering or quality control department.

12-6 Facts on the Value of Precision

For some products the established price scale recognizes differences in precision (purity of chemicals, precision of electrical and mechanical components; grades of finish; firsts, seconds, etc., in textiles).

However, in most situations there is no direct measure of value of precision. Nor do modern cost systems provide direct means to measure value of precision. Instead, indirect means must be used to quantify these values. Most important of these indirect means is the concept of tolerances based on the "market."

Many requirements for precision are specified by the customer or

[9] See *QCH*, 1962, p. 3-21, for examples.

by competitive practice. The problem of economics of tolerances then is one of comparing the costs of alternative ways of attaining the specified precision.

In other situations, while there is no quantified competitive practice, it is possible to determine market quality. Where such data have never been available before, a reduction in cost is possible (see Chapter 4). This is especially the case in establishing tolerances for sensory qualities (see section 13-6).

A further approach is the preparation of performance curves which show the relationship of the variable under consideration to product performance. The shape of the performance curve is usually conclusive as to the importance. The general concept is shown in Figure 12-1.

An example is seen in Figures 12-2 and 12-3. In each case, a product performance is determined in part by the adjustment of a component. In Figure 12-2, a doubling of the adjustment tolerance from A to B results in about a 3 percent drop in product performance. In Figure 12-3, a doubling of adjustment tolerance from A to B results in a 27 percent drop in performance.

There are also intangibles associated with value of precision—customer goodwill, sales department morale, etc. The importance of these matters is undeniable despite the absence of means for evaluating them. However, it is best to consider the intangibles only after the tangibles

Figure 12-1 Approach to functional tolerancing.

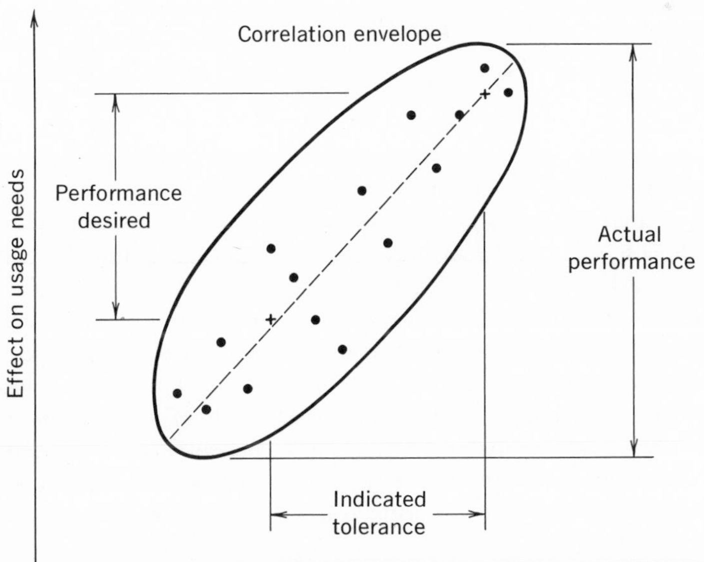

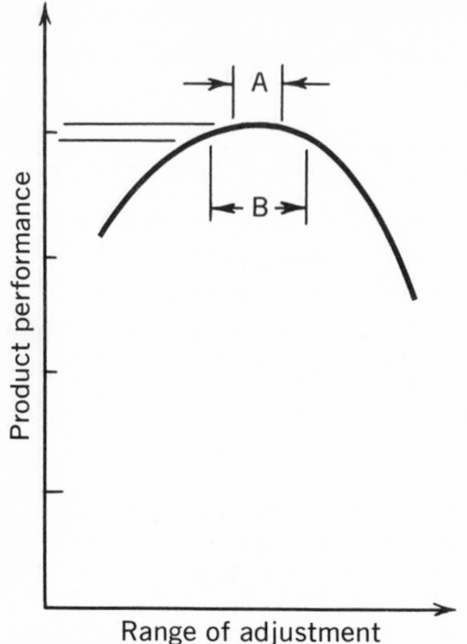

Figure 12-2 Small effect of component tolerance on product performance.

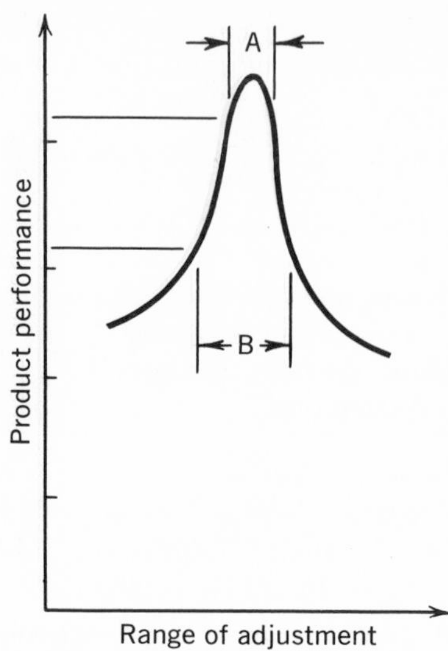

Figure 12-3 Large effect of component tolerance on product performance.

have been evaluated. Otherwise, emotional considerations may carry the day.

12-7 Tolerances for Interchangeability

Interchangeability is absolutely vital to the life of industrial civilization (see section 2-5).

From the design viewpoint, the problem is not merely to achieve interchangeability at all costs. Interchangeability is only one means of achieving a final result. Fitting in assembly or in service is another. In some instances fitting is a more economic means than is interchangeability (see Table 12-2).

In essence, interchangeability implies greater cost of components to achieve lowered cost of assembly and maintenance. The greater cost of interchangeable components is due to holding each piece to close limits so that, even when extremes come together, there will still be no misfit. A new philosophy makes misfits highly improbable rather than impossible. The effect of this difference on increasing assembly costs is negligible, but the effect on lowering cost of components can be substantial (see Chapter 13).

The economics of interchangeability in the shop differs from the economics of interchangeability in the field. In the shop, the assembly

Table 12-2 RELATIVE COSTS AND SAVINGS THROUGH INTERCHANGEABILITY

Interchangeability Requires Added Costs Because of	*Interchangeability Achieves Reduced Cost through*
Added engineering effort to design interchangeability	Lessening human skills to fit components
Added engineering effort to plan for manufacture	Lessening time required due to fitting in assembly
Greater precision in machines, tools, and measuring instruments	Lower cost of maintenance in the field when replacing parts
More extensive controls in fabricating components	

department often has facilities for taking care of instances which do not fit. In some designs, the misfits are merely thrown away, it being cheaper to use a replacement piece than to repair a piece. A stock of parts is normally at hand, and delays caused by misfits are negligible.

In the field, there seldom exists the means for fitting misfit parts, and there is seldom a stock of replacement parts. Spare parts must usually be ordered from the factory. The failure of a spare to assemble properly can give rise to substantial delay and irritation. The difference in application is of sufficient consequence that in many companies a special routine exists to ensure the conformity of spare parts shipped to the field.

12-8 Special Problems of Specification of Complex Products

The twentieth-century products have requirements for reliability, maintainability, availability, safety, compliance with laws and codes, and still other parameters over and above the already complex technological characteristics (frequency response curve, spectographic analysis, etc.). These new requirements pose serious contractual risks for both producer and user, due to present uncertainties in defining them and in measuring actual performance. These risks are well summarized by Johnson.[10]

The sheer number of specifications and characteristics has grown to a point that they cannot be grasped by the old bit-by-bit reading. They must now be grouped and arranged in a way which simplifies the job of seeing the interrelationships. One form[11] developed for this pur-

[10] R. H. Johnson, "Contracting for Reliability," *Proceedings of the Eighth National Symposium on Reliability and Quality Control,* 1962, Institute of Electrical and Electronics Engineers, Inc., p. 181.

[11] AFBM Exhibit 61-1, U.S. Air Force Systems Command, Minuteman Specification Program, Jan. 15, 1962.

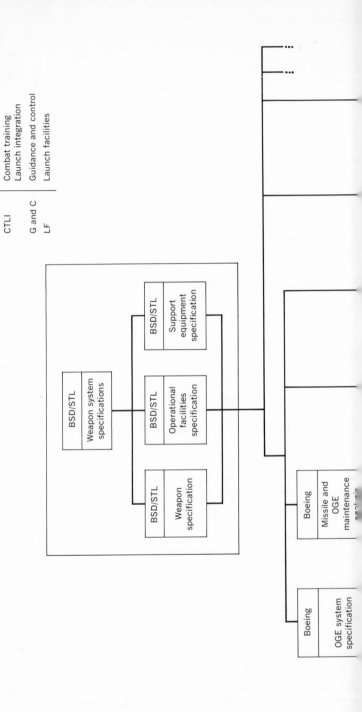

Symbol	Meaning
OGE	Operational ground equipment
CTLI	Combat training Launch integration
G and C	Guidance and control
LF	Launch facilities

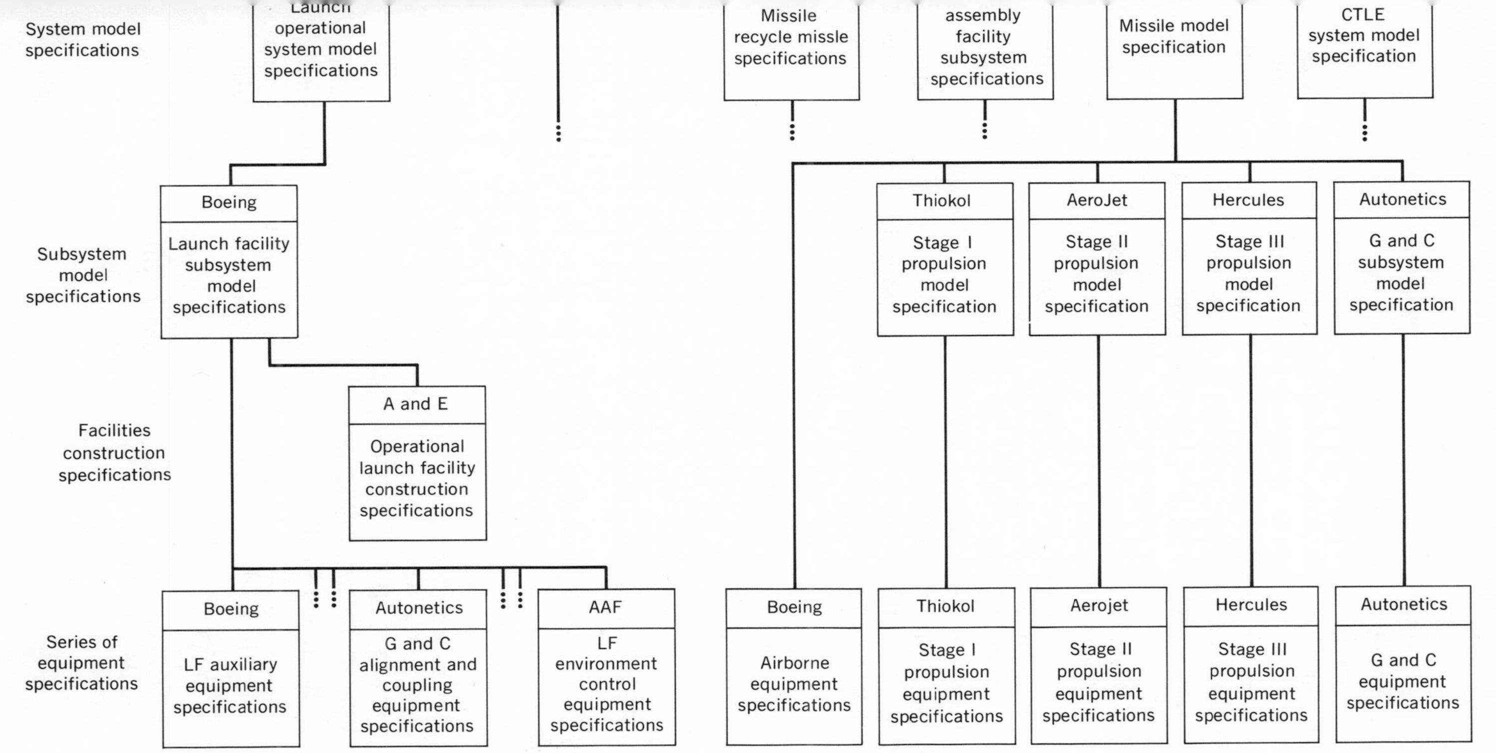

Figure 12-4 **Portion of a specification tree.**

pose is the "specification tree" (Figure 12-4). This set out all pertinent specifications, arranged so as to show their interactions.

In some cases the purchaser gives up on details and sets out only the end requirements. This form avoids the headache of detailed specification but can create a worse headache in operation and maintenance.

Specifications for complex products become closely interwoven with the contract between producer and user. (The old clear distinction between contract provisions and specifications provisions is now blurred.) This overlap has required new organization forms to coordinate the technological, commercial, and legal relationships within and between the companies involved.

The need for coordination is, in many companies, dismissed with the comment that "In this company anyone may talk with anyone else." However, providing the opportunity for coordination does *not* mean that the coordination will occur. The initiative and leadership may or may not spring up to make use of the opportunity. Upper management should not leave this to chance but should create specific machinery for coordination. Various forms of this machinery are discussed in Chapter 7.

At the least, these new products require a positive teamwork arrangement among the departments. In some companies, a separate coordinating department has been created to facilitate getting this teamwork done.

12-9 Standardization in Specification

As applied to quality specifications, standardization is the establishment of requirements which are adequate for a given purpose and which must be followed in new situations having the same purpose.

Standardization acts to reduce the number of types, sizes, and grades of products, provides a basis for interchangeability,[12] and provides predictable criteria for safety laws, building codes, purchase specifications, method of test, and many other activities.

An analysis of the number of different parts performing the same basic function is often revealing; e.g., in two airplanes of comparable weight, one was designed to use only 73 different kinds of bolts, and the other to use 601 kinds of bolts. These were some of the reasons for lack of standardization:

1 Personal preferences of the designer, coupled with the "Rube Goldberg" tendencies of some drafting-board men.
2 Personal preferences of the airplane user.

[12] See *QCH*, 1962, p. 3-16.

3 Lack of standardization in the screw industry or for merchandising reasons to pander to the prejudices of customers.
4 Stockkeeping and stock-identification errors within the services and industry.
5 Laxity in inspection and enforcement of specifications by inspectors of both the industry and the services.
6 An apathetic attitude of "what's the difference?"[13]

A simple and effective concept of "value engineering" is the questioning of requests for development of new components. A search sometimes reveals that a component already available can do the job. The payoff is not only savings on development costs but also the use of a component that is already proven to have adequate quality. This has properly resulted in an emphasis on using "standard, proven components." The key question is "Are the requirements of the new application within those previously established for the standard component?" This may be a difficult technical question (particularly when environmental conditions are involved), and so potential cost savings must be carefully evaluated against the failure cost of a component that does not meet current requirements.

Specifications are influenced by many organizations outside of the company. For example, specifications on product requirements and/or test methods may be issued by consumer organizations, engineering societies, or standardization organizations. The federal government is now involved in the evaluation of requirements for products beyond the original scope of food and drugs.

12-10 Jurisdiction in Product Specifications

All divisions have something to contribute to the decision of what should be the characteristics of the product. Sales presents the view of the customers. Design presents the technical limitations. Manufacture presents the cost and personnel limitations. The final design is properly a composite of these and other views.

However, once the decision is made, there must be one and only one source of authorized information for defining the product. If there is more than one source of information, confusion is inevitable. For this forceful reason, it is universal practice in industry that one and only one department of the company may have authority to issue official product specifications.

[13] From the report of the Joint Army-Navy-Industry Standardization and Simplification Conference, Philadelphia, May 8 and 9, 1947.

In the mechanical industries, the department which is given authority to issue official product specifications is often called the design department. In the chemical process industries, the research, technical, metallurgical, or such-named department has authority. In some companies, the name of the department is subject to local tradition; it may be called the products division (in a carpet company) or the art department (in a company making greeting cards). But the principle (that one and only one department shall have the right to issue official specifications) is thoroughly sound.

Some product design departments have come to misinterpret this grant of authority (to issue specifications). They have come to believe that since they have the sole right to issue official product specifications, they also have the sole voice in deciding on product designs. *This belief is erroneous.* The grant of authority is to avoid the confusion which results from a multiplicity of sources of product specification, and to prevent any one department from having a monopoly on product design decisions.

Ensuring that agreement is reached by all, while restricting issuance to one, requires that the specifications drafted by the design department be accepted by the other departments. A usual method is through an interdepartmental committee with representatives from all interested departments (see Chapter 7).

A specification, once officially issued, is to be regarded as an industrial law. No one in the company has, in theory, any right to deviate from the specification without securing a change authorization. To do otherwise is, in effect, rewriting the specificaton and thereby creating the situation of having more than one source of specification.

However, the practice of deviating from the specification is widespread, and it takes several forms:

1 Nonconforming product is tested, is found to be usable, and is shipped, but without getting changes in the specifications.
2 Nonconforming product is judged to be usable by a material review board (see Chapter 24) and is shipped, again without changing specifications.
3 Nonconforming product is shipped under the urgency of other standards (schedule, cost) without asking anyone.

As these nonconforming products are used, and turn out to perform adequately, there arises a growing awareness that the product meets the ultimate criterion of "fitness for use," and that therefore any move to bring product and specification into harmony should be to change the specification, not the product. However, the practice of deviations, both the authorized and (especially) the unauthorized, is so widespread, and of such long standing, that the backlog of needed specifications changes is simply enormous. In consequence, companies are unwilling to devote the engineering talent and money necesary to go through all these old and present specifications; they have too much work to do on new products.

How to move from this situation of unrealistic tolerances loosely enforced to a situation of realistic tolerances rigidly enforced is at present an unsolved problem. The solutions for the future involve several elements:

1 In any case, break the monopoly of the designers for setting tolerances by requiring a team review. (The designer still has the last word on structural integrity.)
2 For new products do the work needed to set realistic tolerances, and create the controls needed to enforce them strictly.
3 Since these new products involve important departures from present practice, distinguish the specifications clearly so that all concerned know which is which.

In this way, as new products continue to supersede the old, more and more of the practice will be of the new order, until the old becomes obsolete.

12-11 Cultural Patterns in the Design Department

Like any other society of human beings, the design department has a system of beliefs, habits, attitudes, and practices. This "culture" is a real factor in the behavior of the design department.

The elements of this culture include:

1 The concept of factor of safety
   Designers have been taught to allow for the unexpected or the unusual i.e., overloading of machines, use for an unintended purpose. The history of structural failures has amply justified this view of the designer.
   This concept of factor of safety was extended into the design of interchangeability. Not only were designs made so that misfits would in theory be impossible, but designs were also deliberately tightened as a factor of safety to anticipate failure of conformance by the shops. This was a logical extension of the concept of factor of safety, since the shops did in fact play loose with the tolerances.
   The evolution of the inspection department and its successors has had a profound effect on the extent to which the shops take the

tolerances seriously. The historical safety factors have become far less necessary. However, the accumulated designs, and the new designs which follow the precedent of the old, are still loaded with tolerances which are unnecessarily close because of fears which may be out of date.

2 The concept of a monopolistic right to decide tolerance questions
Tolerances involve many departments (see section 12-16). However, in many companies this has not been understood. In consequence, the design department has, by default, exercised a monopoly on this decision and it has come to be regarded as a vested right.

3 Lack of awareness of the quantitative effect of tolerance decisions on factory economics
As noted previously, the designer usually does not have information on this point, but even when he does, he usually lacks the tools needed to do something about it.

4 Prime interest in creating new designs rather than improving the old
This is not merely the result of the designer's personal preference; it is also the result of management's emphasis and budgeting. Inherently it is risky organization practice to give to a single department the responsibility for both construction and maintenance.

5 A vested and shockproof interest in making decisions based on past practice and bargaining rather than on experiment

Collectively, these and other elements of the cultural pattern of the design department constitute a formidable source of resistance to change, both good and bad. Those who are trying to introduce changes in design department practice must recognize the existence and meaning of this cultural pattern.

PROBLEMS

1 Study the specification for any of the following and report on their adequacy for their intended purpose: (*a*) a residential home, (*b*) a sampling of recipes from a cookbook, (*c*) the curriculum of a bachelor's degree, (*d*) the Ebers Papyrus, (*e*) a mechanical or electronic component, and (*f*) a complex system.

2 Prepare a product specification for any of the following and classify the quality characteristics as to seriousness: (*a*) a printed book, (*b*) a football, (*c*) a man's shirt, and (*d*) a desk.

3 Speak with *several* designers in the same organization (or college class) and find out their approach to setting tolerance limits on a characteristic. Speak with them independently, summarize your findings, and make recommendations.

13 Statistical Aids in Limits and Tolerances

13.1 The Problems

In quantifying the limits in specifications, the designer encounters a wide variety of problems, each involving much technology, and each susceptible of more precise solution through statistical aids. Specifically, the designer needs to:

1 Quantify the conditions of use (stresses, loads, environments)
2 Evaluate the effect of these conditions of use and quantify the capability of designs (strength)
3 Establish safety margins
4 Set tolerance limits for discrete components or piece parts
5 Set tolerance limits for coalesced products
6 Set tolerance limits for interacting dimensions and parameters
7 Set tolerances for sensory qualities

Not only must specifications be quantified; the numbers must show the limiting values essential to achieve fitness for use at optimum cost. The following criteria define these needs:

1 The limit must reflect the needs of the user.
All designers and nondesigners subscribe to this principle, and all have difficulty in carrying it out. There are difficulties in discovering the users' needs and these difficulties multiply when the designs are on the frontier of the present state of the art, e.g., environments in outer space. There are organizational difficulties in communicating the users' needs to the designer—there are so many "post offices" that the effort and delay drive the men involved to give up and guess at it.

Of course, the "right" way is to analyze the needs of the user, design with this knowledge, prove the design by test, and then specify. This is both costly and time-consuming, and as a practical matter, can at best be done only for the vital few quality characteristics. For the rest, the method used is "engineering judgment," a mixture of past practice, defensive overspecifying, and some shrewd quantifying of common sense.

2 The limit should be compatible with the capabilities of the available materials and manufacturing processes.

Generally, the designers are well informed as to the properties of materials, and their designs reflect this knowledge. However, the designers are not well informed as to the capability of manufacturing processes. As a result, many manufacturing problems develop, and much criticism is leveled at designers for lack of realism.

Actually, the designer prefers to be realistic. However, data on process capability are not often quantified, and even less often is this knowledge conveyed to the designer in a form suitable for ready use (like the handbook tables on properties of materials). As information on process capability is quantified and made available to the designer, he will learn how to use it to improve his realism. Until then, resort must be had to team review of limits by all organizations concerned.

3 The limit must balance the cost of quality with the value of quality.

In most cases the designer cannot as yet do this by himself—he lacks too much of the essential information. As a result, he tends to overspecify, since he is usually held directly responsible for design failure, but is not usually held directly responsible for high manufacturing costs.

See section 4-2 for elaboration on this.

4 The limit should be clear.

The vague specification can be a bigger troublemaker than the incorrect specification. The latter is easier to detect and to contest. The former is easy to overlook. Vagueness is also easy to sweep "under the rug." As in many other matters, it is more important to be clear than to be logical.

13-2 Quantifying Conditions of Use and Design Capability

The conditions of use and design capability can be reflected in specifications in the form of a minimum value of the *safety margin.*

Quantifying the conditions of use and capability requires data. The data are often difficult to obtain, but the resulting payoff (as will be shown) justifies the effort. For example,[1] operating temperature is a

[1] *Reliability for the Engineer, Book Five: Testing for Reliability,* Martin Marietta Corp., Orlando, Fla., 1966, pp. 29–31.

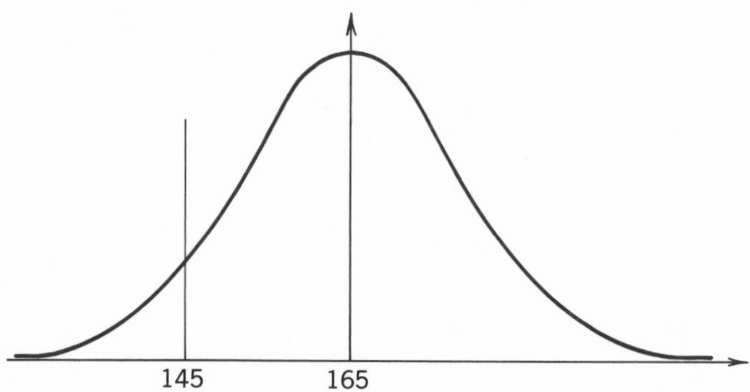

Figure 13-1 Distribution of strength.

critical parameter and the maximum expected temperature is 145°F. Further, the capability is indicated by a strength distribution having a mean of 165°F and a standard deviation of 13°F (see Figure 13-1). With knowledge of only the maximum temperatures, the safety margin is

$$\frac{145 - 165}{13} = -1.54$$

The safety margin says that the average strength is 1.54 standard deviations above the maximum expected temperature of 145°. Table A in the Appendix can be used to calculate a reliability of 0.938 (the area greater than 145°).

The reliability estimate of 0.938 assumes that the device will always be subjected to 145°F. Now suppose that data were available indicating that temperature was normally distributed with a mean of 85°F and a standard deviation of 20°F. The picture of stress and strength is then shown in Figure 13-2. Following the procedure outlined in Chapter 10, the reliability would then be evaluated in terms of a distribution of the *difference* between strength and stress.

$$\mu_{\text{difference}} = 165 - 85 = 80$$
$$\sigma_{\text{difference}} = \sqrt{(20)^2 + (13)^2} = 24$$

The distribution of the difference is shown in Figure 13-3. The safety margin is

$$\frac{0 - 80}{24} = -3.33$$

Table A in the Appendix would predict a reliability of 0.9996. This compares to 0.938 estimated without knowledge of the variation in stress.

In many instances, distribution information is not readily available on expected stress, and so the reliability boundary would have to be

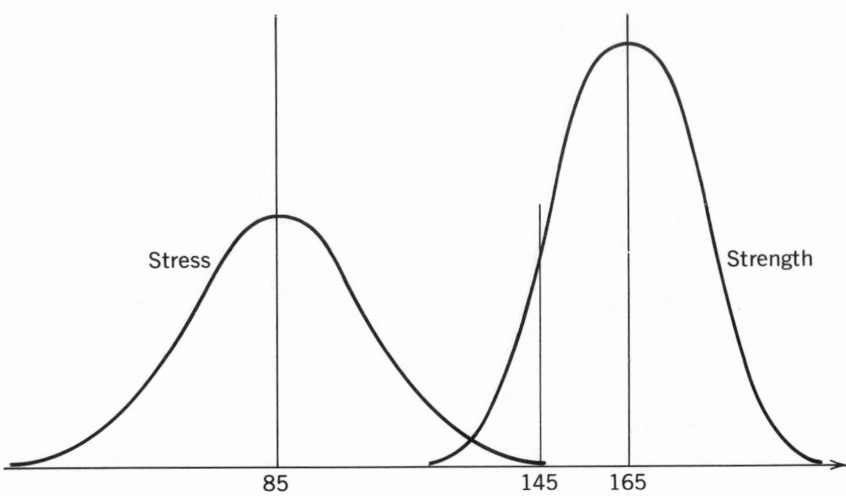

Figure 13-2 Distributions of stress and strength.

defined as as simple value based on engineering judgment instead of specifying the boundary in terms of a safety margin. However, the specification could require a strength safety margin of (say) 5.0. This would mean that tests would be required to show that the average strength is at least 5 standard deviations above the reliability boundary.

Lusser proposed the use of safety margins in specifications for critical products such as guided missiles (Figure 13-4).[2] He suggested safety

[2] R. Lusser, *Reliability through Safety Margins,* U.S. Army Ordnance Missile Command, Redstone Arsenal, Alabama, October, 1958.

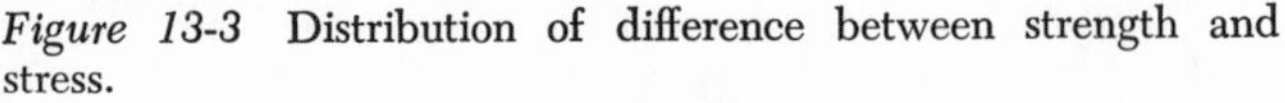

Figure 13-3 Distribution of difference between strength and stress.

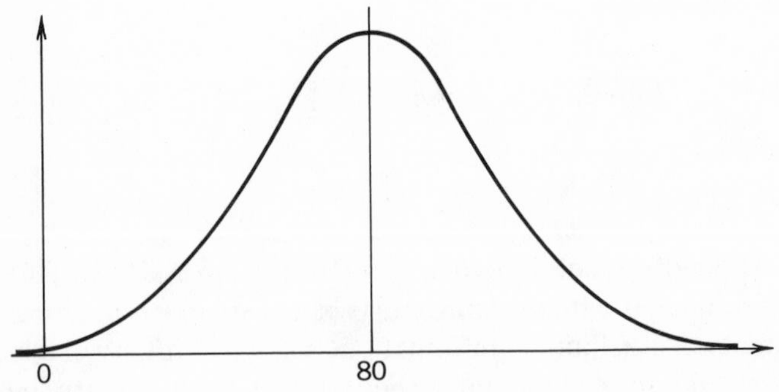

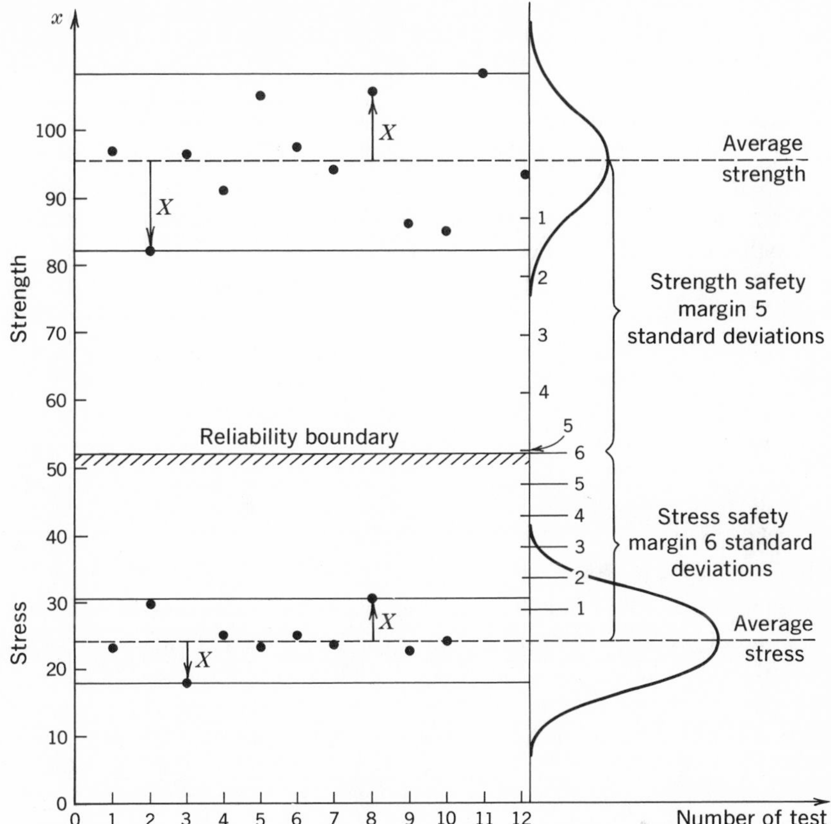

Figure 13-4 Illustrating how scatterbands of stresses and strengths shall be separated by a reliability boundary. Source: R. Lusser, *Reliability through Safety Margins,* U.S. Army Ordnance Missile Command, Redstone Arsenal, Alabama, October, 1959.

13-3 Analyzing Process Data to Set Limits on Discrete Components or Parts

margins for strength and stress (see Chapter 10 for the definitions of "safety margin" and "safety factor"). Specifically, the reliability boundary (maximum stress) was to be defined as 6 standard deviations (of stress) above the average stress. The average strength would have to be at least 5 standard deviations (of strength) above the reliability boundary. These minimum values of safety margin were suggested for products requiring extremely high reliability.

While designers think in terms of limits, processes do not. Intead, a process turns out product which exhibits (1) a central tendency whose location is determined by the controllable setting of the process, and (2) a scatter whose form and standard deviation are determined by the inherent precision or capability of the process.

The central tendency of the process can normally be regulated, and the designer is entitled to assume that the manufacturing people will do this regulation. However, the process variability is commonly inherent in the process and must therefore be accepted as a fact of life by the designer.

Generally, the designer will not be provided with information on process capability. His problem will be to obtain a sample of data from the process, calculate the limits that the process can meet, and compare these to the limits he was going to specify. (If he does not have any limits in mind, the capability limits calculated from process data provide him with a set of limits that are realistic from the viewpoint of producibility. These must then be evaluated against the functional needs of the product.)

Statistically, the problem is to predict the limits of variation of individual items in the total *population* based on a *sample* of data. For example, suppose a product characteristic is normally distributed with a population average of 5.000 inches and a population standard deviation of 0.001 inch. Limits can then be calculated to include any given percentage of the population. Figure 13-5 shows the location of the 99 percent limits. Table A in the Appendix indicates that plus or minus

Figure 13-5 Distribution with 99 percent limits.

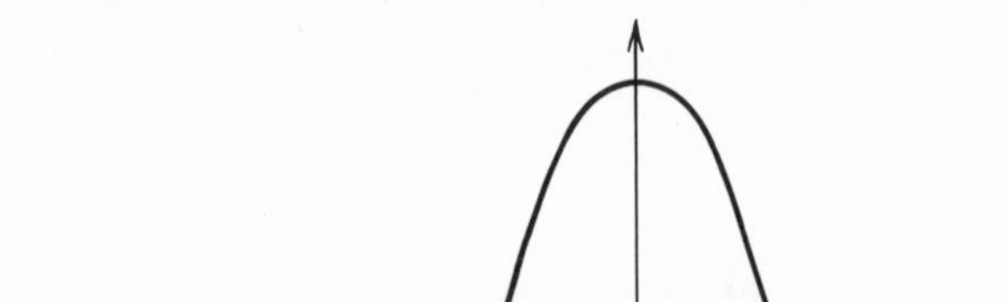

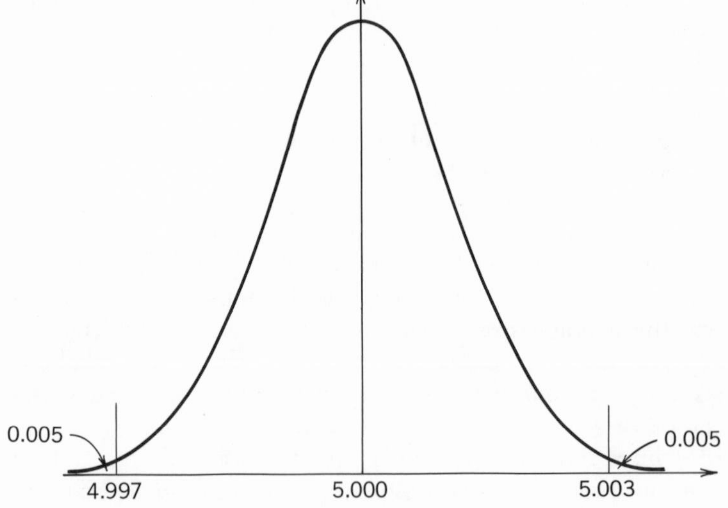

2.575 standard deviations will include 99 percent of the population. Thus, in this example, a realistic set of tolerance limits would be

$$5.000 \pm 2.575(0.001) = \begin{matrix} 5.003 \\ 4.997 \end{matrix}$$

Ninety-nine percent of the individual pieces in the population will have values within *these limits.*

In practice, the average and standard deviation of the population are *not* known but must be estimated from a sample of product from the process. As a first approximation, tolerance limits are sometimes set at

$$\bar{X} \pm 3s$$

Here the average $\bar{X}$ and standard deviation s of the sample are used directly as estimates of the corresponding population values. The multiple 3 in the relationship is rather arbitrary. If the true average and standard deviation of the population happen to be equal to those of the sample, and if the characteristic is normally distributed, then 99.73 percent of the pieces in the population will fall within the limits calculated above. These limits are frequently called "natural tolerance limits" (limits that recognize the actual variation of the process and therefore are realistic). This approximation ignores the possible error in both the average and standard deviation as estimated from the sample.

Methodology has been developed for setting tolerance limits in a more precise manner. For example, formulas and tables are available for determining tolerance limits based on a normally distributed population. (Natrella[3] presents tables for calculating statistical tolerance limits without assuming a normal distribution. This approach of "distribution-free" limits is appealing but generally requires high sample sizes.) For example, Table G in the Appendix provides factors for calculating tolerance limits that recognize the uncertainty in the sample mean and sample standard deviation. The tolerance limits are determined as

$$\bar{X} \pm Ks$$

The factor K is a function of the confidence level desired, the percent of the population to be included within the tolerance limits, and the amount of data in the sample.

For example, suppose a sample of 10 cylindrical rods from a process yielded an average and standard deviation of 5.0145 inches and 0.0493

[3] Mary Gibbons Natrella, *Experimental Statistics,* National Bureau of Standards Handbook 91, U.S. Department of Commerce, Aug. 1, 1963.

inch respectively. The tolerance limits are to include 99 percent of the rods in the population, and the tolerance statement is to have a confidence level of 95 percent. Referring to Table G in the Appendix, the value of K is 4.433, and tolerance limits are then calculated as

$$5.0145 \pm 4.433(0.0493) = \begin{matrix} 4.796 \\ 5.233 \end{matrix}$$

We are 95 percent confident that at least 99 percent of the rods in the population will have a diameter between 4.796 and 5.233 inches. This approach is more rigorous than the $\pm 3s$ natural tolerance limits, but the *two* percentages in the statement are a mystery to those without a statistical background.

All methods of setting tolerance limits based on process data assume that the sample of data is representative of a process that is sufficiently stabilized to be predictable. In practice, the assumption is often accepted without any formal evaluation. If sufficient data are available, the assumption can be checked with a statistical control chart. The concept of statistical control is discussed in Chapter 15. At this point, suffice it to say that statistical control means that the process is free from any assignable cause of variation which may result in excessively high variation in the final product.

13-4 Setting Tolerance Limits for Coalesced Products

For discrete products (e.g., bolts, pumps, toasters) the setting of tolerance limits on a quality characteristic includes several aspects that are clear:

1 The definition of a *unit* of products is a discrete piece (a bolt).
2 The product is tested and used on a unit basis. Therefore, tolerance limits apply to individual units of product and all units should meet the tolerances.

Coalesced products (gasoline, sheet steel, coal) are different:

1 The definition of a unit of product is a *mass* (e.g., a tank car of chemical).
2 The product is evaluated by testing *specimens* from the mass. The measurement on a single specimen is of little importance because the product is not used in the form of specimens. However, the measurement of a number of specimens is important in determining the characteristics of the mass.

For coalesced products, the specification must first define the mass in physical terms, set limits on the quality characteristics of the mass, and finally set limits on the measurement of specimens from the mass. The quality limits are usually defined in terms of average and standard deviation. For example, Rockwell C Scale hardness limits for a *mass* might be

Average = 58 ± 1; standard deviation = maximum of 1.5

The limits on the specimens from the mass must recognize the sampling variation within the mass. The formula for sampling variation of the average (see section 11-2) states that the variation depends on (1) the size of the sample and (2) the intrinsic variation of the mass itself. Thus, the test limits specified must be related to a defined number of measurements. (This is distinctly different from test limits for discrete products which apply to each unit of product.) The determination of the number of specimens to be tested and the corresponding limits is based on the concept of variables sampling plans (see Chapter 17).

The special problems of coalesced products can be a source of serious confusion. One company had a problem on the percent impurities in a chemical product. The specification stated that the percent impurities "shall not exceed 12%." Unfortunately, the manufacturing and quality people differed on the interpretation of the number 12 percent. The quality people interpreted the number to mean that any *single* measurement of impurity in the product must be equal to or less than 12 percent. The manufacturing people felt it meant that the *average* of the measurements taken over a 1-day period should be equal to or less than 12 percent. This lack of agreement had been going on for *months* because neither function fully realized the significance of the distinction between the mass and specimens.

13-5 Tolerance Limits for Interacting Dimensions

Interacting dimensions are those which mate or merge with other dimensions to create a final result. Consider the simple mechanical assembly shown in Figure 13-6. The lengths of components A, B, and C are interacting dimensions because they determine the overall assembly length.

Suppose the components were manufactured to the specifications indicated in Figure 13-6. A logical specification on the assembly length

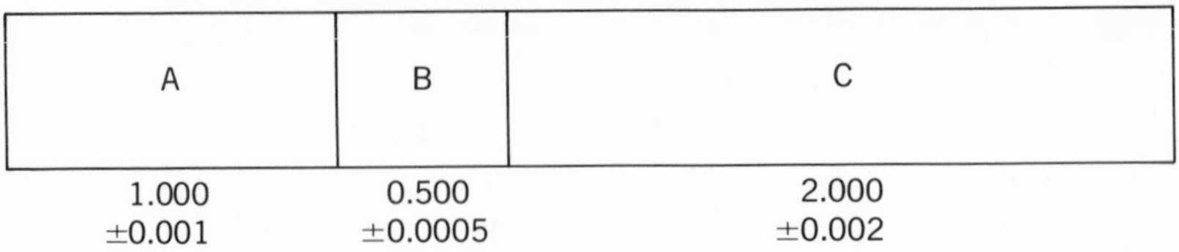

A	B	C
1.000 ±0.001	0.500 ±0.0005	2.000 ±0.002

Figure 13-6 Mechanical assembly.

would be 3.500 ± 0.0035, giving limits of 3.5035 and 3.4965. The logic of this may be verified from the two extreme assemblies:

Maximum	*Minimum*
1.001	0.999
0.5005	0.4995
2.002	1.998
3.5035	3.4965

The approach of adding component tolerances is mathematically correct but is often too conservative. Suppose that about 1 percent of the pieces of component A are expected to be below the lower tolerance limit for component A and suppose the same for components B and C. If a component A is selected at random, there is, on the average, 1 chance in 100 that it will be on the low side, and similarly for components B and C. The key point is this: If assemblies are made at random and if the components are manufactured independently, then the chance that an assembly will have all *three* components simultaneously below the lower tolerance limit is

$$\frac{1}{100} \times \frac{1}{100} \times \frac{1}{100} = \frac{1}{1{,}000{,}000}$$

There is only about one chance in a million that all three components will be too small, resulting in a small assembly. Thus, setting component and assembly tolerances based on the simple addition formula is conservative in that it fails to recognize the extremely low probability of an assembly containing all low (or all high) components.

The statistical approach is based on the relationship between the variances of a number of independent causes and the variance of the dependent or overall result. This may be written as

$$\sigma_{\text{result}} = \sqrt{\sigma^2_{\text{cause A}} + \sigma^2_{\text{cause B}} + \sigma^2_{\text{cause C}} + \cdots}$$

In terms of the assembly example, the formula is

$$\sigma_{\text{assembly}} = \sqrt{\sigma_A{}^2 + \sigma_B{}^2 + \sigma_C{}^2}$$

Now suppose that, for each component, the tolerance range is equal to ±3 standard deviations (or any constant multiple of the standard deviation). As σ is equal to T divided by 3, the variance relationship may be rewritten as

$$\frac{T}{3} = \sqrt{\left(\frac{T_A}{3}\right)^2 + \left(\frac{T_B}{3}\right)^2 + \left(\frac{T_C}{3}\right)^2}$$

or

$$T_{\text{assembly}} = \sqrt{T_A{}^2 + T_B{}^2 + T_C{}^2}$$

Thus, the *squares* of tolerances are added to determine the square of the tolerance for the overall result. This compares to the simple addition of tolerances commonly used.

The effect of the statistical approach is dramatic. Listed below are two possible sets of component tolerances which when used with the formula above will yield an assembly tolerance equal to ±0.0035.

Component	*Alternative 1*	*Alternative 2*
A	±0.002	±0.001
B	±0.002	±0.001
C	±0.002	±0.003

With alternative 1, the tolerance for component A has been doubled, the tolerance for component B has been quadrupled, and the tolerance for component C has been kept the same as the original component tolerance based on the simple addition approach. If alternative 2 is chosen, similar significant increases in the component tolerances may be achieved. This formula, then, may result in a larger component tolerance with *no* change in the manufacturing processes and *no* change in the assembly tolerance. Note that the *largest single* tolerance has the greatest effect on the overall result. If a component D with a tolerance of ±0.002 inch were added to either of the above tolerance sets, the overall assembly tolerance would increase only to ±0.0039. Now, if an overall tolerance is fixed but not being met, which component tolerances should be reduced? The formula can help determine which of the component tolerances have the greatest effect on the overall tolerance. This information, when coupled with economic information on achieving a smaller tolerance, can form the basis for a decision.

A risk is involved with this approach. It *is* possible that an assembly will result which falls outside of the assembly tolerance. However, the chance can be calculated and a judgment made on whether or not to accept the risk. To determine the probability that an assembly length will fall outside of its tolerance limits, a distribution of assembly lengths is determined (Figure 13-7). If the components are approximately normally distributed, then the assembly lengths will follow a normal distribution. The average assembly length is the addition of the nominal values of the components or 3.500. The standard deviation of assembly lengths is

$$\sigma_{\text{assembly}} = \sqrt{\sigma_A{}^2 + \sigma_B{}^2 + \sigma_C{}^2}$$

Suppose the component tolerances under alternative 2 are to be used and that process data indicate that these tolerances are about equal to 3 standard deviations. Then

$$\sigma_A = \frac{0.001}{3} = 0.00033 \qquad \sigma_B = \frac{0.001}{3} = 0.00033$$

$$\sigma_C = \frac{0.003}{3} = 0.001$$

The standard deviation of the assembly length then is

$$\sigma_{\text{assembly}} = \sqrt{(0.00033)^2 + (0.00033)^2 + (0.001)^2} = 0.0011$$

The probability that an assembly will exceed its lower tolerance limit can then be found by analyzing the area under the normal curve for assembly lengths. Thus:

$$K = \frac{3.4965 - 3.5000}{0.0011} = -3.2$$

Figure 13-7 Distribution of assembly length.

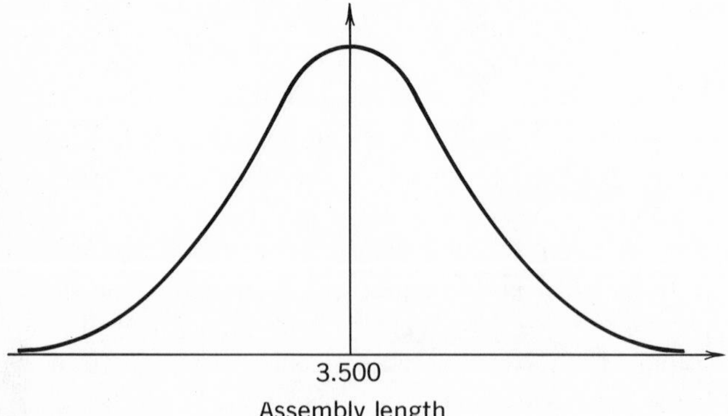

From Table A in the Appendix, the probability of an assembly being below 3.4965 is equal to 0.00069.

Similarly, the chance that an assembly will be above its upper tolerance limit is also 0.00069. This means that the chance of getting an out-of-specification assembly is 0.138 percent; i.e., about one-tenth of 1 percent of the assemblies will fail to meet the assembly specification.

The tolerance formula is not restricted to outside dimensions of assemblies. Generalizing, the left side of the equation contains the dependent variable or *physical result*, while the right side of the equation contains the independent variables of *physical causes*. Consider the problem of a shaft and a hole. Suppose it is desired that the *clearance* specification be 0.0007 ± 0.0005. With conventional theory, a difference of 0.0007 would be established between the nominal dimensions of the shaft and the hole, and the overall tolerance would be divided in half so that the shaft tolerance would equal the hole tolerance (or ±0.00025).

The statistical approach would yield the following relationship:

$$T_{\text{clearance}} = \sqrt{T^2_{\text{hole}} + T^2_{\text{shaft}}}$$

If the tolerance on the shaft is set equal to the tolerance on the hole,

$$T_{\text{clearance}} = \sqrt{2T^2}$$

where T = shaft tolerance = hole tolerance

$$0.0005 = \sqrt{2T^2}$$
$$T = \pm 0.00035$$

Thus the tolerance on the shaft and the hole may each be set at ±0.00035 instead of ±0.00025. The *physical resulting* dimension was reflected on the *left* side of the formula and the *physical causes* are reflected on the *right* side of the formula with *no* changes made in the signs underneath the square root. If the result is placed on the left and the causes on the right, then the formula always has *plus* signs under the square root—even if the result is an internal dimension (such as a clearance). The causes of variation are *additive* wherever the physical result happens to fall.

13-6 Assumptions of the Formula

The formula has these assumptions:

1 The component dimensions are independent and the components are assembled randomly. This assures that no assembly will have all components on either the high side or the low side. The formula

depends on a high value of a component being at least partially offset by a value below the nominal for one or more of the other components. This assumption is usually met in practice. A possible exception is where several of the components are identical. For example, two batteries selected for insertion into a flashlight might have been manufactured one right after another. In such a case, there is the chance that both batteries have a length which is similar, i.e., perhaps both on the high side. However, with identical components, the assumption of independence can be essentially achieved by mixing the identical components so that the two selected for one assembly would have been produced at different times and therefore would probably not both be extremes and in the same direction.

2 Each component dimension should be normally distributed. Some departure from this assumption is permissible.[4]

3 The actual average for each component is equal to the nominal value stated in the specification. For the original assembly example, the *actual* averages for components A, B, and C must be 1.000, 0.500, and 2.000 respectively. Otherwise, the nominal value of 3.500 will not be achieved for the assembly, and tolerance limits set about 3.500 will not be realistic. Thus, it is important to control the *average* value for interacting dimensions. A control chart for averages can be maintained on the process to assure that the average value meets the nominal specification. In addition, a variables sampling plan may also be used to check that the average value is close to the nominal specification. The need to control the average value for each component means that *variables measurement* is necessary. The less expensive (in the short run) attributes-type inspection is not appropriate for checking the average because attributes measurement classifies each piece as good or bad and does not provide an indication of the actual average.

Use caution if any of the assumptions are violated. Reasonable departures from the assumptions may still permit the *concept* of the formula to be applied. Notice that the formula resulted in the *doubling* of certain tolerances in the illustrative example. This much of an increase may not even be necessary from the viewpoint of process capability.

Further Applications of Statistical Tolerancing

The statistical method of tolerancing is *not* merely a change from an expression of tolerances in the form of limits on each component

[4] Irving W. Burr, "Tolerances for Mating Parts," *Industrial Quality Control,* vol. 15, no. 3, p. 18, September, 1958.

to a form of (1) upper and lower limits on the average $\bar{X}$ of the mass of components and (2) an upper limit to the scatter (σ) of the components.

The change is much more profound than mere form of the specification. It affects the entire cycle of manufacturing planning, production, inspection, quality control, service, etc. It is, in effect, *a new philosophy of manufacture.*

The first published example[5] of a *large-scale application* of statistical tolerancing appears to be that of the L-3 coaxial system (a broadband transmission system for multiple telephone or television channels).

The formula has been applied to a variety of mechanical and electronic products.[6] The concept may be applied to several interacting variables in an engineering relationship. The nature of the relationship need *not* be additive (assembly example) or subtractive (shaft and hole example). The tolerance formula can be adapted to predict the variation of results that are the product and/or the division of several variables.[7]

13-7 Setting Limits for Sensory Qualities

As with the specification of measurable qualities, setting limits on sensory qualities involves (1) discovering what consumers like, dislike, and prefer, (2) creating a product which will appeal to consumers, and (3) defining the product through specification.

In some industries, appearance of the product is the most important quality charactertistic (e.g., jewelry, silverware, textiles). In such industries inspectors at the "bottom" of the organization are often making decisions of the utmost importance to the life of the company. The inspectors have no choice, since the "top" has failed to provide standards. *In effect, the individual inspectors are setting the standards.*

With an approach that requires executive approval of standards, the way is paved not only for recognition of the standards, but also for audit to see that the standards are being met.

The organized approach to sensory standards includes:

1 Definition of various levels of sensory quality along the scale of "measurement." These levels are used as reference points throughout

[5] "The L-3 Coaxial System," *Bell System Technical Journal,* vol. 32, July, 1953. [For a summary, see J. M. Juran (ed.), *Quality Control Handbook* (*QCH*), 2d ed., McGraw-Hill Book Company, New York, 1962, pp. 3-45 to 3-46.]

[6] *QCH,* 1962, p. 3-45.

[7] G. Mouradian, "Tolerance Limits for Assemblies and Engineering Relationships," *Transactions of the Twentieth Annual Conference ofthe American Society for Quality Control,* 1966, pp. 598–606.

the organized procedure. (Sensory qualities have unique problems of defining a unit of measure and creating an instrument to measure in these units.)

2 Study of *yield* of the company's manufacturing process at each of these levels to discover the effect, on company cost, of the various potential decisions on sensory standards. (See Figure 13-8. Level 1 is highest quality.)
This determines the percentage of product being produced at the various defined levels. These percentages then become decisive as to the effect, on cost, of establishing the standard at any of the defined levels. For example, about 4 percent of production is at level 7 or worse. If the specification limit were set at level

Figure 13-8 Summary for executive decision on sensory standards.

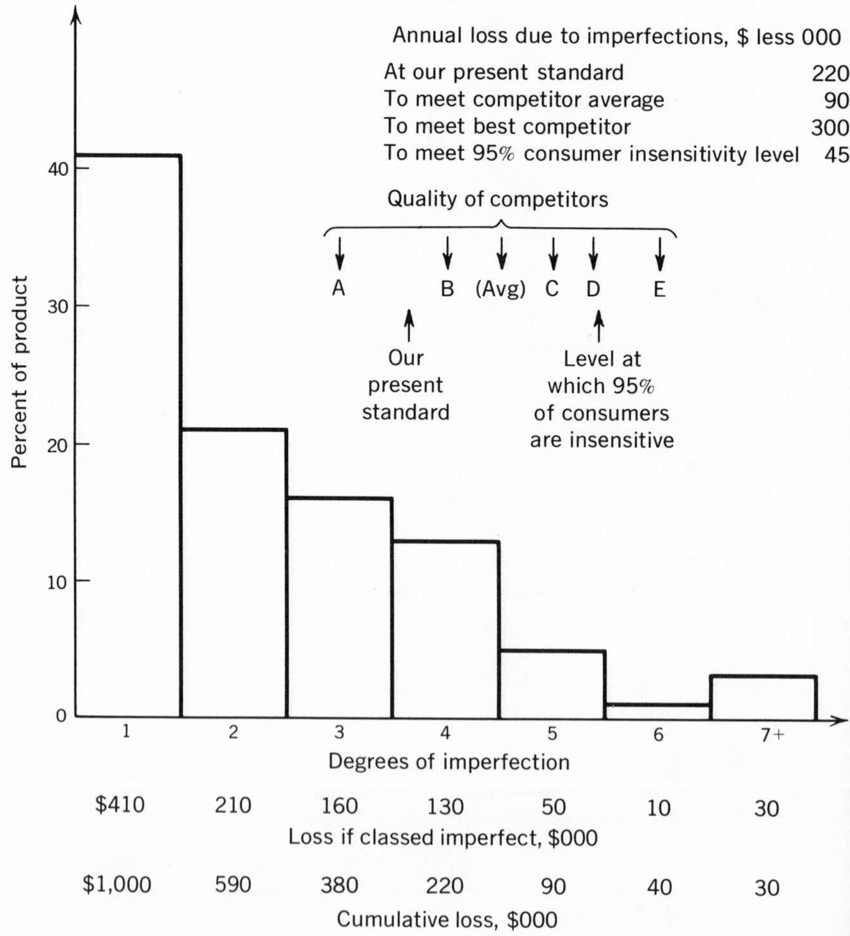

6, the 4 percent at level 7+ would mean a loss of $30,000.

3 Study of *consumer sensitivity* to discover the effect, on consumer acceptance, of a decision on sensory standards.

4 Study of *competitor quality* to discover what is "market quality" in relation to the various defined levels.

Competitor performance can be studied by buying adequate samples on the open market. Where this is burdensome, arrangements can usually be made to study the competitor products in the warehouse or showroom of a friendly dealer.

Competitor data are especially valuable to minimize any emotional considerations based on rumors or fragmentary information about competitor performance.

5 *Analysis and summary* of the foregoing for executive decision. The data on cost of setting standards at various levels, consumer sensitivity at various levels, and competitor performance can all be brought together for executive convenience (Figure 13-8). Note that the present standard is set at level 3, which is better than the industry average and second only to competitor. A. However, 95 percent of the consumers were insensitive to differences in levels through level 5.

While these data are helpful in clarifying the economics, they are not conclusive in making the decision. There are intangibles to be evaluated. There is also need for a spirit of participation in establishing company standards. Only through such participation will the various departments wholeheartedly implement the standards, i.e.. production meet them, quality control enforce them, sales support them, etc.

Therefore, the group (whether a committee or a single department) charged with recommending a sensory standard should provide for full and early participation of concerned departments. A recommendation on sensory standards prepared through such participation will seldom be upset by the higher supervision.

6 *Executive decision* on sensory standards.

The making of the executive decision should itself be formalized. The signature of the deciding executive(s) should be secured on the recommendations or on the samples themselves to clear the air on the fact of decision making, as well as to identify the standards unmistakably.

7 *Recording* the sensory standard.

For *stable* products the sensory standard can be selected, approved, and filed like any other master reference standard. Duplicates can be prepared for everyday use, and these can be checked against the master under a scheduled plan.

For *unstable* products there is a continuing problem of creating new masters at intervals which are shorter than the interval of deterioration. Sometimes special storage or preservatives can be employed to retard deterioration.

Another method of recording the standard, already used in visual inspection, is to standardize the conditions of inspection rather than the condition of the product. In this plan, a defect is anything objectionable which can be seen by the inspector (under the standard inspection conditions) at a specified distance from the product. In the graphic arts industries it is common for the manufacturer to work up samples for the approval of the customer. These samples, when approved, become the agreed-on specification for many sensory qualities. Samples are also commonly used as standards in the textile and paper products industries. Copies of these samples are filed with the pertinent written descriptions and orders. Photographs (and even stereophotographs) of defects are common forms of secondary standards.

8 Training the personnel in use of the standards.

9 Follow-up to ensure adequate use.

The foregoing approach is applicable to any sensory characteristic—not just those concerned with appearance. One example is the establishment of standards for sand and gas holes in piston rings.

PROBLEMS

1 A component is expected to encounter a maximum stress of 40,000 psi. Past data on similar components show that the standard deviation of *strength* is about 10 percent of the average strength. What specification should be set on average strength to achieve a safety margin of 3.0? 4.0? 5.0? (Assume that strength is distributed normally.)

2 Measurements were made on the bore dimension of an impeller. A sample of 20 from a pilot run production showed a mean value of 9.8576 inches and a standard deviation of 0.00015 inch. All the units functioned properly, and so it was decided to use the data to set specification limits for regular production.

a Suppose it was assumed that the sample estimates were exactly equal to the population mean and standard deviation. What specification limits should be set to include 99 percent of production?

b There is uncertainty that the sample and population values are equal. Based on the sample of 20, what limits should be set to be 95 percent sure of including 99 percent of production?

c Explain the meaning of the difference in (*a*) and (*b*).

d What assumptions were necessary to determine both sets of limits?

3 Evaluate the clarity of a numerical specification value of a coalesced product. Do this by making an inquiry at a local plant or distributor of the product. Possible examples are octane rating of gasoline and chemical characteristic of the local water supply. If the meaning is not clear, suggest a format (using symbols for numerical values) for a more definitive specification.

4 A circuit contains three resistors in series. Past data show this data on resistance:

	Mean, Ohms	*Standard Deviation, Ohms*
Resistor 1	125	3
2	200	4
3	600	12

a What percent of circuits would meet the specification on total resistance of 930 ± 30 ohms?

b Inquire from a local distributor if it is reasonable to assume that the resistance of a resistor is normally distributed.

5 Figure 13-9 shows a shaft (C), two rotating disks (B and D)

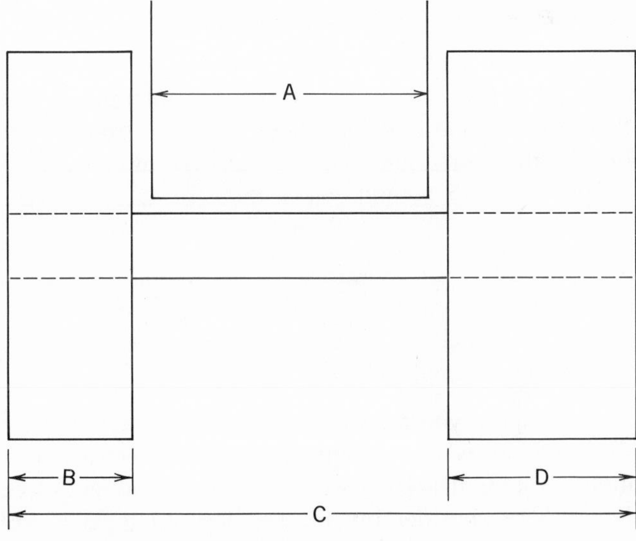

Figure 13-9 Shaft assembly.

keyed to the end of the shaft, and a frame (A) which must fit on the shaft between the disks.
The specifications are:

B: 1.000 inch $\pm$ 0.001 inch
C: 5.000 inches $\pm$ 0.004 inch
D: 1.500 inches $\pm$ 0.002 inch

Assume that all dimensions are normally distributed and the specifications are being met.

a What would be a reasonable specification to set on A?
b What are the maximum and minimum posssible dimensions for A?

6 A manufacturer of rotary lawn mowers received numerous complaints concerning the effort required to push its product. Studies soon found that the small clearance between the wheel bushing and shaft was the cause. Designers decided to increase the clearance to allow easy rotation of the wheel (or provide space for a heavy grease coating), yet be "tight enough" to prevent wobbling. Because of a large inventory of wheels and shafts, a decision was made to ream the bushings to a larger inside diameter (I.D.) and retain the shafts. The following specifications were proposed:

Shaft diameter = 0.800 inch $\pm$0.002 inch
New clearance = 0.008 inch $\pm$0.003 inch
Bushing I.D. = 0.808 inch $\pm$0.001 inch

A question was asked concerning the tolerances for the bushing I.D. Production people said that they could not economically hold such tolerances with their equipment. Design countered that they had created the product and should know their business. What comment would you make?

7 Research the literature on the planning of sensory tests for comparing several competing products (e.g., see *QCH*, 1962, p. 3-60). To gain an appreciation of the difficulties involved, plan and conduct a test for three brands of cola.

8 Suppose you are in charge of quality control for a small company (200 employees) manufacturing wound electrical coils on order for a number of industrial customers who produce electronic equipment. Orders vary considerably in size depending on whether the particular customer is, at the particular time, working on a development, a preproduction, or a production contract.

Coils are wound on a variety of cores, with and without "tuning plugs" (metallic cylinders whose position inside the core may be changed, permitting the user to adjust the electrical values), using wires of many sizes, materials, and numbers of turns. Your quality problem lies in the high rejection rate (12 percent) of finished coils at the 100 percent inspection and test operation. The principal defects accounting for 9 percent of the 12 percent rejections are out-of-tolerance electrical values (inductance, self-resonant frequency, Q, etc.) In addition, there are some customer returns, most of which have shown good but borderline electrical characteristics when retested on your equipment.

You have studied the variation patterns of at least 20 different high rejection items by making frequency histograms of lots just before final test. All these show either (1) wrong centering of the characteristic or (2) excessive variation compared with the tolerances.

Confronted with these data, the production foreman's reaction is that he follows engineering's specifications for core, wire size, number of turns, etc., and "this is how they come out." Engineering's reaction is along these lines:

> Customers often specify core, wire size, number of turns, etc., as well as the electrical values they want. Engineering then simply incorporates these into specifications for the plant. A small "sample shop" is maintained by engineering to make initial samples for customer approval before the order is processed. In this shop, a sample coil is wound per the specifications. If it checks OK, it is sent out for approval, after which the specification becomes firm. If it does not check OK, changes are made in the number of turns, etc., to get a good coil and the customer is so advised.

Quizzed about the time required for this procedure, the chief engineer admitted that it often takes a long time and that it is sometimes bypassed in order to meet delivery dates. Asked whether customers ever specify tolerances that are impossible to meet, he replied that if the shop would only make them all alike, he felt the tolerances could be met so long as the engineering sample was all right.

You have been asked by the owner-manager to recommend a program for reducing coil rejects. *Outline* the program, numbering and listing the steps you would recommend and, for each, the procedure, in brief, for accomplishing it.[8]

[8] Problem taken from course notes of Management of Quality Control Course, U.S. Air Force School of Logistics, 1959.

14 Manufacturing Planning for Quality

14-1 Quality Aspects of Planning for Manufacture

Planning for manufacture comprises the activities required to put the factory in readiness to meet its standards of quality, cost, and delivery date. The specific quality aspects of planning for manufacture include:

1 Choice of machines, processes, and tools capable of holding the tolerances
2 Choice of instruments of an accuracy adequate to control the processes
3 Planning the flow of manufacturing information and inspection criteria
4 Planning of process quality controls
5 Selection and training of production personnel
6 Planning the quality aspects of packing and shipping (Chapter 25)

Generally the planning of (1) through (3) is performed by a manufacturing planning department, (4) by a quality control department, (5) by the various production departments, and (6) in miscellaneous ways.

Table 14-1 shows the usual responsibilities for manufacturing planning.

The more complex the product, the greater is the need for formality in quality planning. This formality is exhibited in written procedures for production, inspection, packaging, etc., usually through process specifications, quality control manuals (see Chapter 29), written countdowns,

Table 14-1 TABLE OF RESPONSIBILITY FOR MANUFACTURING PLANNING

Activity	*Functions or Departments Available for Participation*					
	Manufacturing Planning	*Toolroom*	*Production*	*Inspection*	*Quality Control*	*Design*
Choice of machines and processes	XX		X		X	
Design, ordering, and tryout of tools	XX	X	X		X	
Design and ordering of measuring instruments for production	XX	X	X	X	X	
Tryout and approval of measuring instruments for production	XX	X	X	X	X	
Issuance of manufacturing information and inspection criteria	XX		X		X	X
Planning of process quality controls	X		X	X	XX	
Selection and training of production personnel			XX			

XX = prime responsibility
X = collateral responsibility

etc. In extreme cases the planning process itself must be formalized. An adaptation of the PERT network can be used for this.[1]

14-2 Basic Approaches to Quality Planning

There are three general categories of new (or changed) products, and each requires its own emphasis for quality planning.

1 Many "new" products are not very new. The change from the old is minor: a new size, a new color, a new configuration. The similar predecessor products have been in successful production, and the processes are proved, though with some known deficiencies. For such new products the planning is mainly concerned with (*a*) providing the new patterns, tools, etc., needed to deal with the new sizes and configurations and (*b*) avoiding any carryover of known deficiencies into the new products.

Once these needs have been met, production is allowed to proceed without the need for trial lots.

[1] Virgil Rehg, "QERT," *Transactions of the American Society for Quality Control*, 1966, pp. 107–114.

2 Where there is a substantial newness or change but not very large investments, the planning is done by analysis plus "debugging" during the first production lot. It is this category which puts the greatest premium on modern tools of analysis to get into production promptly while minimizing the risks of future quality troubles.
3 A third category is those new products which also represent a serious risk for the company because (*a*) they require new manufacturing processes, (*b*) they embody new functional features, and (*c*) the contemplated production will involve very large investments. In such cases it is usual to schedule a trial production lot for the purpose of debugging the process before going into full-scale production (see section 14-4).

14-3 Planning through Analysis

Planning consists mainly of utilizing prior knowledge to control future events. More specifically this requires that we:

1 Identify the "good" performances (high process yields, low costs, high customer satisfaction)
2 Discover, through analysis, what causes are decisive in obtaining these good performances
3 Utilize the resulting cause-and-effect relationships in new planning.

To a high degree, the good performances may be discovered from analysis of data already in the house—inspection and test reports, accounting reports, service reports. Sometimes the information is available from the factory floor.

For example, Weeks[2] documents a study of end thrust in rotors of a blower assembly. Measurements confirmed the shopmen's theory that the amount of thrust after assembly was less than the thrust before assembly. From these measurements and from process capability data, it was possible to change tolerances and eliminate 98 percent of the reworks previously necessary.

A fruitful source of prior knowledge is found in the studies conducted to eliminate chronic quality troubles. These studies require breakthroughs in knowledge before they yield breakthroughs in results. This new knowledge then becomes useful for future planning. For example, Haden[3] relates a problem in securing interchangeability be-

[2] Frank Weeks, Jr., "A Statistical Study of an Assembly Problem," *Industrial Quality Control (IQC)*, vol. 10, no. 6, pp. 76–78, May, 1954.
[3] Edward B. Haden, "Tolerance Corrections Applied to Plastics Fabrication," *Transactions of the National Convention of the American Society for Quality Control*, 1951, pp. 125–131.

tween fountain pen barrels and the mating sac sections. The study showed that:

1 The tolerances of the mating parts were based on "before polishing" rather than "after polishing" dimensions. Hence a change in design was needed.
2 The machines could reproduce pieces to a uniformity capable of holding the tolerances, but the operators had no adequate means for adjusting the average dimension up or down. Hence it was necessary to design and make a suitable adjusting device.

It may seem that it should be possible to plan by theoretical analysis alone. It does not work out that way. For example, it is possible to produce, in a laboratory, metallic "whiskers" whose tensile strength is an order of magnitude greater than that of the commercial forms of the same metals. However, until an economical process is developed, there can be no mass production or usage of this material.

The prior knowledge of "process capability" has widespread application, and is discussed in more detail in section 14-5.

14-4 Planning through Trial Lots

The trial lot is used to "clear the track" for full-scale production by:

1 Proving that the tools and processes can indeed turn this product out successfully, with economic yields
2 Proving, on test, that the product will possess the essential functional features
3 Proving, on use, that the product will achieve the intended field performance
4 Remedying the deficiencies in process or product before embarking on full-scale production

These proofs and remedies cannot be provided from the record of models made in the model shop. In the model shop the basic purpose is to prove engineering feasibility; in the production shop the purpose is to meet standards of quality, cost, and delivery. The model shop machinery, tools, personnel, supervision, motivation, etc., are all different from the corresponding situations in the production shop.

Despite their obvious desirability, trial lots are often avoided because of:

1 The *time,* which is so long that the urge is to gamble from model shop to mass production. This urge comes from the marketing

people or from top management itself, and may, in turn, be the result of competitive conditions.

2 The *cost*, which usually needs to be amortized over a very large production.

Where conditions permit a trial production run, there is much to be learned if all departments mobilize to participate in the trial and contribute to a better design, process, and product.

The need for a trial lot is related to the size of the risks taken. These risks in turn depend on:

1 The extent to which the product embodies new or untested quality features
2 The extent to which the fabrication contains new or untried tools and processes
3 The value of product which will be out in the field before there is conclusive evidence of the extent of process, product, and usage difficulties

An example of a special facility to aid in such trial lot quality planning is seen in the Quality Control Center reported by Ford Motor Company.[4] The facility is in the nature of an automobile assembly laboratory with a capacity of five cars per day. The center is intended to identify quality problems months before production-line assembly commences, and to permit training of key people on the operations involved.

14-5 Choice of Machines, Processes, and Tools

The planner should know in advance whether the process or machine he specifies is adequate to hold the design tolerance (assuming normal operation and maintenance of the process).

Several methods are in use for determining the adequacy of the machine:

1 Tryout of the product as made by the machine, either in subsequent operations or in actual use
2 Measurement of the machine itself, as, for example, the Schlesinger[5] method of testing machine tools
3 Measurement of the product (turned out by the machine) against the product tolerances

[4] "New Quality Control Center for Ford Motor Company," *IQC*, vol. 15, no. 7, p. 18, January, 1959.

[5] Georg Schlesinger, *Testing Machine Tools*, The Machinery Publishing Co., Ltd., London, and The Industrial Press, New York, 1945.

4 Measurement of the process capability of the machine in terms of 6σ of its product (the emerging practice)

The process capability is a measure of the inherent variability of the process. The planner must relate this capability to the tolerance. Usually he does this by an index known as the capability ratio:

$$\text{Capability ratio} = \frac{6\sigma \text{ variation}}{\text{total tolerance}}$$

One company uses as a rule of thumb for adequacy the following maxima for the capability ratio:

	Bilateral Tolerance, %	*Unilateral Tolerance,* %
Existing processes	75	88
New processes	67	83

Thus, a dimension with a bilateral tolerance should be assigned to an existing process having process variation of less than 75 percent of the tolerance range. Such rules of thumb stem from the histogram approach discussed in Chapter 20.

Most of the approaches for judging adequacy of processes apply to new machinery as well, but with some modifications. Some determinations can actually take place at the machine builder's plant.[6]

There is also emerging a new practice of specifying, on the purchase order for the machine, what is to be the maximum process variability. Some buyers require, as part of the approval procedure, that the histograms and capability analysis be submitted before final acceptance of the machine. Other provisions may require a demonstration of capability, either at the manufacturer's plant or at the buyer's plant.

In most companies the assignment for measuring process capability is given to the staff quality control department. This group needs these data for a variety of purposes, and possesses the statistical skills to collect and analyze the data correctly. In some companies, the quality control department publishes the summarized results of such studies. Each machine is listed, showing, for each type of operation (turn, face, bore, etc.), the 6σ value. Sometimes this is subdivided, depending on the materials processed.[7]

Collateral with the problem of choice of machinery is choice of

[6] See, for example, R. LeGrand, "How Ford Checks Machine Tools at Builder's Plant," *American Machinist*, vol. 95, no. 5, pp. 99–101, Mar. 5, 1951.

[7] For an example, see J. M. Juran (ed.), *Quality Control Handbook* (*QCH*), 2d ed., McGraw-Hill Book Company, New York, 1962, p. 25-28.

measuring instruments. Industry has, since World War II, undergone a rapid, drastic conversion from fixed-limit instruments to instruments which give variables measurements.

A rule for the planner is that the instrument should be sensitive enough to divide the tolerance preferably into ten parts, certainly no less than five parts. Unless such sensitivity is available, the setup men, operators, and inspectors will have difficulty in using inspection data to regulate the production processes.

The rule of one-tenth leads to a considerable pyramiding of accuracy. Using measure of length as an example, a product tolerance of 0.001 inch requires gages which can read to 0.0001 inch. In turn, the masters used to check the gage would require an accuracy to 10 millionths.

14-6 Planning the Flow of Information

"Essential" information includes:

1 The definition of things, i.e., the material, process, product, and test specifications
2 The definition of duties, i.e., the job specifications, table of responsibilities, decision criteria, training manuals, standard practice, etc.
3 The flow of data to permit control at all levels—operator, supervisor, executive

Figure 14-1 shows an example[8] of the flow of inspection and test information.

Industry has long accepted the need for clear definition of technological things. Hence the responsibilities for doing this are generally clear, and the procedures are well standardized. However, recognition of the need for definition of duties and for flow of data is more recent, the responsibilities are less clear, and the procedures are less well developed.

A usual pattern of responsibilities is shown in Table 14-2 on page 278.

14-7 Planning for Process Control

"Process control" or "correction" refers to the sequence of events by which a process is kept free of sporadic troubles, i.e., the means by which the status quo is maintained. This is in contrast to the problem of eliminating chronic conditions or changing the status quo (see section

[8] R. J. Smurthwaite and J. Kroehler, "The Quality Assurance Driving CAM-Corrective Action Motivation," *Transactions of the Twenty-first Annual Conference of the American Society for Quality Control*, 1967, pp. 275–285.

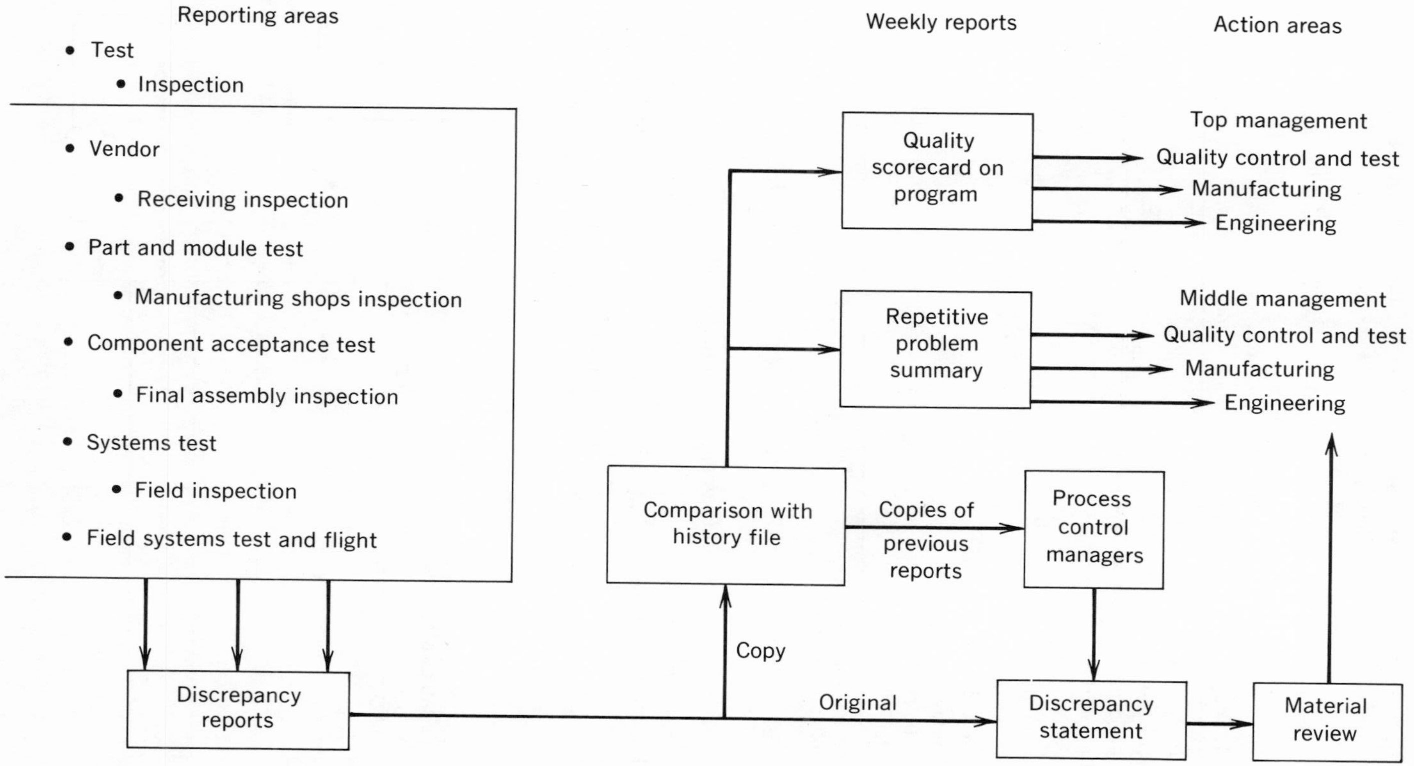

Reporting areas

- Test
  - Inspection
- Vendor
  - Receiving inspection
- Part and module test
  - Manufacturing shops inspection
- Component acceptance test
  - Final assembly inspection
- Systems test
  - Field inspection
- Field systems test and flight

Discrepancy reports

Copy

Original

Comparison with history file

Copies of previous reports

Process control managers

Discrepancy statement

Material review

Weekly reports

Quality scorecard on program

Repetitive problem summary

Action areas

Top management
Quality control and test
Manufacturing
Engineering

Middle management
Quality control and test
Manufacturing
Engineering

Figure 14-1 Corrective action information flow.

Table 14-2 **Table of Responsibilities for Flow of Data for Quality Control**

	Function or Department Participating					
Activity	*Quality Control*	*Manufacturing Planning*	*Production*	*Inspection*	*Accounting*	*Design*
Determination of data needed for control	XX	X	X	X	X	X
Design of forms	XX		X	X	X	
Recording of data			XX	XX		
Summary of data for management review	X				XX	
Summary of data for action on the factory floor				XX		
Issuance of reports to management	XX				X	

XX = prime responsibility
X = collateral responsibility

Figure 14-2 Servomechanism diagram for control generally.

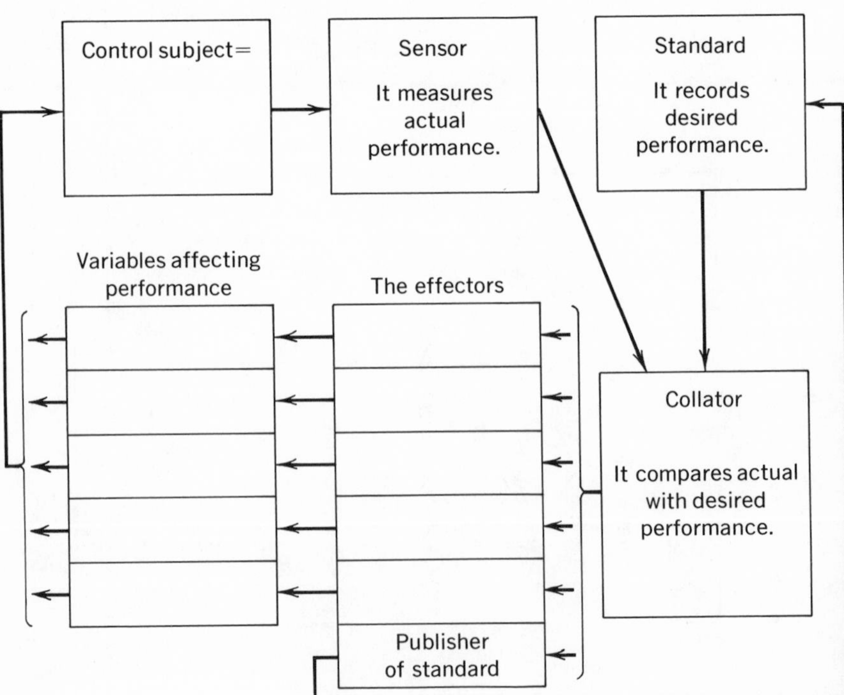

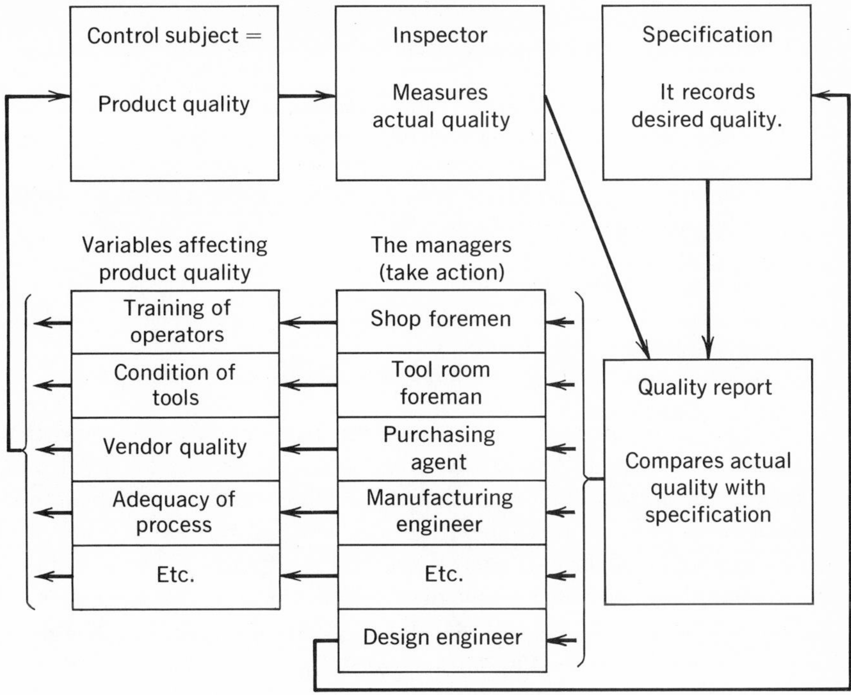

Figure 14-3 Servomechanism diagram for quality control.

1-10). The latter activity is often referred to as "defect prevention" or "quality improvement."

"Process control" also refers to current regulation on the factory floor, in contrast to "executive reporting" or "quality assurance" which deal with broader, summarized information, usually well after the fact.

The conceptual approach to process control follows the servomechanism cycle so common in engineering. Figure 14-2 shows this diagrammatically for the general case, and Figure 14-3 shows the same diagram as applied to process control.[9]

Achievement of process control requires that the personnel on the factory floor be in a state of self-control, i.e., possess all the elements of the servomechanism cycle:

1 A standard of what quality "should be"
2 Information on actual performance to compare with standard so as to determine any need for adjustments in process
3 Means for adjusting the process if necessary

[9] These diagrams are adapted from those first published in J. M. Juran, "Universals in Management Planning and Control," *The Management Review,* pp. 748–761, November, 1954.

The planner's approach to achieving self-control on a broad basis is through a well-known series of elements of process control planning:

1 Flow diagram of the process. For example,[10] see Figure 14-4.
2 Choice of control stations and of activities to be conducted at each control station. For example, see Figure 14-4.
3 Definition of characteristics to be controlled, and classification of relative seriousness (see Chapter 12).
4 Detailed inspection criteria.
5 Definition of responsibilities.
6 Data for self-control (see Chapter 23).
7 Special process control activities.
8 A surveillance plan for auditing the process control activity.

Planning for process control requires that quality characteristics be identified so that all hands are aware of just what is to be regulated. Because there are many characteristics and many stages of manufacture, it is common practice to concentrate the work of process control into a limited number of formal "control stations." At each of these stations, provision is made for process control of selected quality characteristics.

Choice of process control stations depends on specific job conditions. The most usual conditions suggesting control stations are discussed in section 16-2.

If there are numerous checks to be made at one station, the usual manner of organizing is to assign departmental employees to perform the checks. If there are few checks per station but numerous stations, resort may be had to a plan of process surveillance. Chapter 16 expands on this subject with an example of inspection planning for refrigerator cabinets.

14-8 Responsibility Patterns for Process Control

The two most fundamental decisions made on the factory floor are

1 Should the process run or stop (process acceptance)?
2 Does the product conform or not (product acceptance)?

Normally, the authority to make these two decisions is delegated to the production and inspection departments respectively.

This simplified division of responsibility is complicated by a wide assortment of variation, the most prevalent being shown in Table 14-3 on page 283.

[10] *Quality Control Manual,* Armour and Co., BLH Electronics Division.

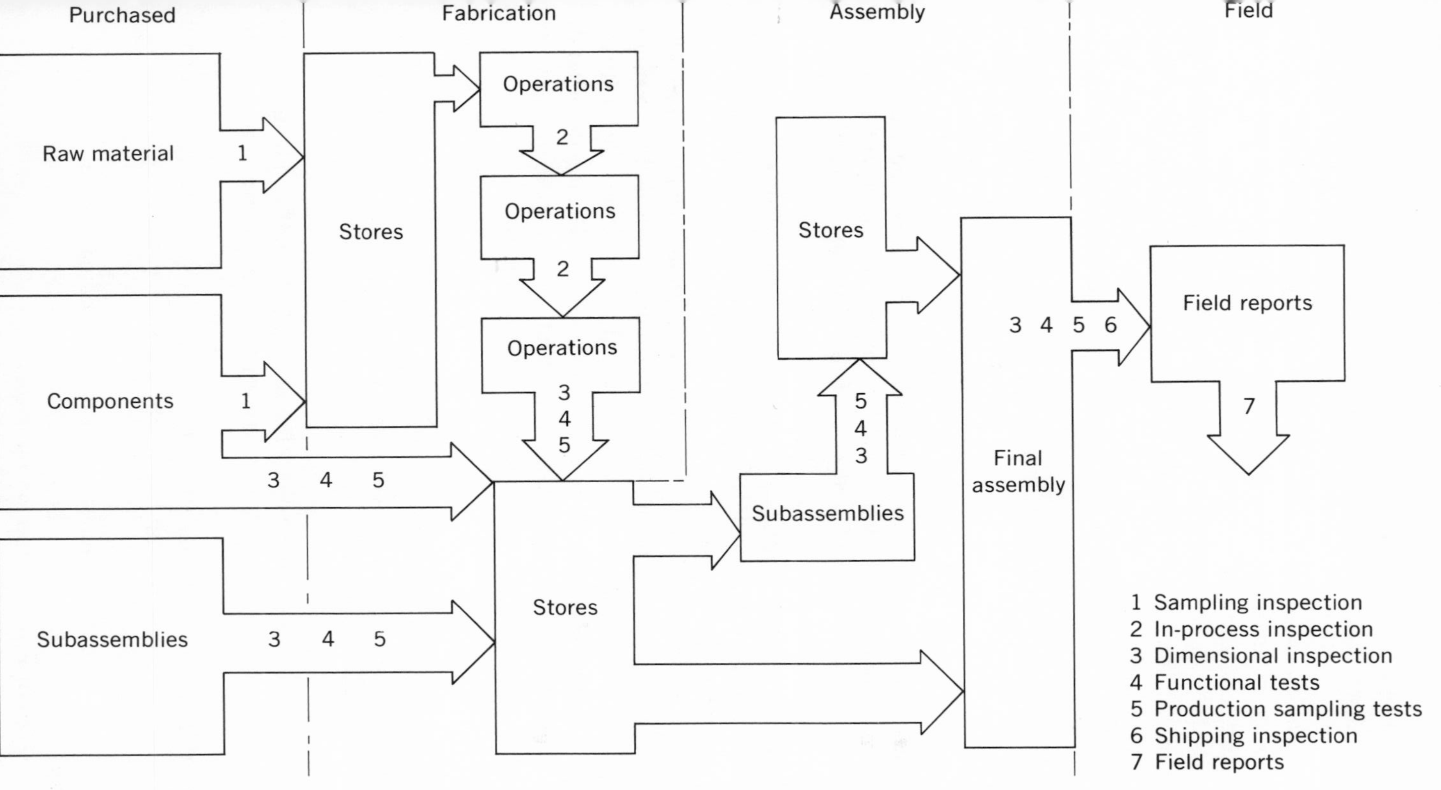

Figure 14-4 Flowchart showing quality and reliability check points. Source: *Quality Control Manual,* Armour and Co., BLH Electronics Division.

The pattern of activity surrounding process control decisions differs quite markedly from that surrounding product acceptance decisions.

Not only do these things become complicated; they become widely controversial as well. A widespread cause of controversy concerns the numerous cases in which the production department lets the process run despite evidence of nonconformance of some kind; i.e., product or process fails to conform to the respective specifications. (Normally, production has some plausibility for its case: the product fails to conform to its specification, but can be used; the process fails to conform to its specification but the product conforms to the product specification.)

Many inspection supervisors chafe at the idea of allowing the process to run while nonconformance exists, and some urge that they be given authority to shut down such processes. However, they already possess several weapons short of authority to stop the process:

1 Direct discussion with the production supervisor on the job.
2 Appeal up the line, with special emphasis on repeat instances.
3 Nonacceptance of the product. (This is usually sufficient.)

If the inspection supervisor is repeatedly ignored or overruled on all these counts, he is warranted in concluding that he is out of step with management policy.

On the constructive side, inspection can attack some of the root causes. Process specifications seldom distinguish clearly between advisory and mandatory tolerances. (In the absence of such distinctions, production tends to regard them all as advisory, while inspection tends to regard them all as mandatory.) Inspection can also question its own biases when the product is clearly fit for use, so that what needs changing is the specification.

14-9 Production Department Needs for Process Control

Once control stations have been chosen, the job of the planner is to provide each station with all the elements of the control cycle. The cycle may be fully automatic, in which case the feedback "loop" is closed by use of instruments and actuators (see section 14-12). Where the cycle is less than fully automatic, the loop must be closed by production department personnel.

The vast extent of process control effort demanded of production operators requires that there be complete understanding of the essentials of process control by production operators.

Table 14-3 Distinction between Tolerances

	Process Control	*Product Acceptance*
Purpose of the tolerances:	To provide a basis for making decisions on the process	To provide a basis for making decisions on the product
Tolerances published in:	Process specification	Product specification
Specification usually issued by:	Process engineering department	Product design department
Tolerances concern:	Process conditions	Product qualities
Instrumentation is usually:	An integral part of the process	Not an integral part of the process
Usual measurement to discover compliance is by:	Production department for advisory tolerances; inspection department for mandatory tolerances.*	Inspection department
Decisions on whether there is compliance made usually by:	Production department for advisory tolerances, inspection department for mandatory tolerances	Inspection department
Deviation from specification usually authorized by:	Production department for advisory tolerances, process engineering department for mandatory tolerances	Inspection department for nonfunctional tolerances, product design department for functional tolerances

* Advisory tolerances are supplied to operators for manufacturing convenience. Mandatory process tolerances must be met as a means of supplying essential product qualities, i.e., long life, safety, etc., when there is no immediate, economic test for the product quality.

The term "production" as used here refers to (1) the operators who peform work directly on the product (as distinguished from machine maintenance men, stockkeepers, and other service personnel), and (2) the hierarchy of supervision needed to select, train, and supervise the operators.

In order to control processes, production supervision must provide the operator with four basic needs:

1 Knowledge of what he is supposed to do, including
  a What decisions he is authorized to make
  b The meaning of the specification
  c The criteria to be used in making decisions
2 Knowledge of what he is in fact doing
3 Means for regulating or adjusting the process

4 A state of mind so that he will actually use his facilities and skills to meet the standard.

Of the foregoing, (1) through (3) must *all* be met before the operator can be held responsible for meeting standards. If any of these is missing, the resulting defects cannot be regarded as operator controllable.

Meeting these needs is no simple feat. Some entire industries provide no process control specifications, e.g., dimensional control in the metal cutting industries. In such cases, production personnel are forced to plan their own process control tolerances. Lacking scientific skills, their plans are often uneconomic.

Some planners have found it desirable to provide "narrow limits" to operators as a form of process control tolerances.[11] These can be of much aid to operators if the question of responsibility is cleared up; i.e., is the operator responsible for meeting the product specification tolerances or the process control tolerances?

The knowledge requirement also encounters serious contradictions. In former years, the official specifications were contradicted by the foreman's unofficial black book. More recently, the contradictions are seen in process control specifications versus product acceptance specifications (section 14-8) and in both of these versus the Shewhart control chart limits (section 15-3). The introduction of these control charts has affected the operator's sources of knowledge and of decision making, as is seen in Table 14-4.

Generally industry has not answered squarely the questions raised by production as a result of the introduction of control charts. First among these questions are those of the legitimacy or status of the control chart. For instance:

1 What is the official status of the control chart? Is it merely an informal suggestion by the quality control engineer, or is it a "legitimate" impersonal boss, like the specification, the collective bargaining agreement, etc.?
2 Is the control chart to be used in addition to, or instead of, the specification?
3 If "in addition to," what of the cases where there is conflict; i.e., control chart requires more frequent adjustment, process is not capable, etc.
4 If the control chart is to be used "instead of" the specification, why is this not legitimatized; i.e., why is there no formal order relieving the operator of the responsibility for meeting specifications?

[11] See section 23-4. For a broad discussion, see Ellis R. Ott and August B. Mundel, "Narrow Limit Gaging," *IQC*, vol. 10, no. 5, pp. 21–28, March, 1954.

Table 14-4 OPERATOR ACTION TABLE*

Type of Information or Decision Needed	*Sources of Operator's Information or Decision* — *Before Introduction of Control Charts; Operator Responsible for Meeting:* *Process Specification*	*Product Specification*	*After Introduction of Control Charts*
1 Information on what the process should be doing	Direct from process specification	Direct from product specification	From the control chart
2 Information on what the process actually is doing	From process instruments or inspectors	From measurements on the product	From the control chart
3 Decision on whether "is doing" differs from "should be doing" by an amount great enough to warrant process adjustment	From operator experience	From operator experience	From the control chart
4 Decision on extent of process adjustment needed	From operator experience	From operator experience	From operator experience

* From J. M. Juran, "Cultural Patterns and Quality Controls," *IQC*, vol. 14, no. 4, pp. 8–13, October, 1957.

Little has been done to answer these questions despite their evident importance to the operator, and, for that matter, to the entire production department.

Data needed for process control are usually provided by the instrumentation inherent in the process, or by measurements or observations made by the operator. In critical processes, these data are supplemented or even dominated by inspection and laboratory data.

The problem of state of mind, as well as other aspects of process control, are discussed in Chapters 21 and 22.

14-10 Special Process Control Activities

For certain products it may be desirable to consider special process control activities. Some of these are discussed below.

"Critical component control" is a term used to describe the special precautions taken (labeling, transporting, handling, and storage) for vital components. The activities are well defined and may include separate storage areas, locked transportation means, handling restricted to authorized personnel, etc.

"Configuration control" is a procedure for reviewing and incorporating design changes in complex assemblies in an orderly manner. This is discussed in Chapter 25.

"Traceability is the ability to trace the history, application, use, and location of an individual article or characteristic lot of articles, through use of recorded identification numbers."[12] The identification numbers may apply to individual items of hardware or to lots, or they may be date codes for continuously made materials, or a combination of these.

"Integrity control" is a procedure for assuring that no unauthorized work on a product takes place. This may include formal "quarantine" that requires written authority for work. It may also employ seals that can only be broken with the approval of Quality Control, coupled with a work authorization form.

"Certification of critical skills" is the procedure for formally testing the ability of operators and inspectors to perform critical operations. Recertification may also be required periodically. The procedure may include the issuance of a card which must be carried by the operator. Process surveillance (see below) may include a random check of operators and/or inspectors by requiring that the card be shown as evidence of current certification. (Chapter 16 includes a discussion of inspector training.)

14-11 Process Surveillance

Many features of process control are present at the operator's work station and are of such a nature that he can observe and regulate them. Other features may be outside of the operator's natural orbit of activity. Still others, though observable by the operator, are regulated by someone else. In addition, there are those operator controllable features of process control for which there is need to have added assurance that the operator is meeting his responsibility.

The proper regulation of the foregoing is sometimes left to the production operator. In other situations, it is left to the production supervisor to review process and operator performance periodically. In still other situations, a new post is created (quality auditor, patrol inspector, etc.) to conduct surveillance of the process.

The "specification" to guide such a patrol inspector consists of a

[12] Morris K. Dyer, "Product 'Traceability' for NASA Space Systems," *Transactions of the American Society for Quality Control,* 1966, pp. 202–208.

listing of the process characteristics he is to check, how often he is to check them, what results he is to record, and what criteria he is to observe for taking action. This "specification" is prepared by consultation among departments. When agreement is reached, a "patrol beat" is designed.[13] Record forms and report forms are designed for ease in recording and in feedback.

In addition to surveillance of characteristics, a program of "quality audit" for *any* activity affecting quality can also be conducted. Such audits are used to check certification of operators and inspectors, calibration status of measuring instruments, currency of drawings used by production operators, etc. This is further discussed in Chapter 28.

14-12 Automation and Quality Planning

"Automation" is essentially an acceleration of a long-range trend to turn over to machines the repetitive tasks formerly done by the backs, hands, and minds of human beings. Applied to quality problems, this trend occurs in:

1 Process control, where indicating, recording, computing, and controlling instruments are provided to aid human regulation of the process, or to make the process self-regulating (see section 23-7)
2 Product inspection, where automatic instruments sort the product and provide control data

One manufacturer[14] uses a computer system to:

1 Control numerically controlled drilling and assembly machines
2 Monitor and dispatch work in process from station to station along a matching line
3 Control in-process test equipment
4 Monitor instruments in the chemical processes
5 Arrange queues in optimum fashion before banks of similar machines
6 Monitor the performance of machines
7 Spell out machine maintenance requirements in "real time"
8 Provide immediate feedback after quality checks
9 Gather statistical data for analysis by production control, quality assurance, accounting, and other sections

While automation provides means for cost reduction and greater reliability in achieving quality, it also intensifies some old problems

[13] For an example, see *QCH*, 1962, p. 4-20.
[14] "A Step toward the 'Automatic Factory,'" *Production Magazine*, p. 75, July, 1965.

to an extent which requires new solutions. The new situations posed are:

1 The economics of automatic machinery presuppose essential freedom from jamming and delay due to feeding defective parts and materials to the machines. This requires new levels of perfection for the parts and materials.
2 On the production floor, the employee skill of feeding machines becomes obsolete and shifts to the skill of adjustment and maintenance of machinery and instruments, and of clearing trouble. The implications in selection and training of employees are profound.
3 In the inspection department, the man-hours devoted to measurement and visual inspection shrink drastically, leaving the main emphasis on interpretation of data, disposition of product, and reporting. Here again, there is a profound effect on selection and training.

The gaging problems presented by automated processes are remarkably different from those presented at a conventional inspection station. Gages used in automated processes may have to:

1 Probe for the presence of missing operations, broken taps, etc.
2 Prevent damage to machines from excessive loads or broken tools
3 Correlate several measurements to determine overall acceptability
4 Decide whether the parts should undergo transfer or not
5 Measure the product
6 Actuate devices which shut off the machines
7 Actuate instrument panels which show where the trouble lies
8 Actuate machine adjustment for self-correction

Additional new problems are presented. The gages may have to operate very rapidly (either during the machining cycle time or during the transfer time). Parts may be so hot immediately after machining that the error due to temperature must be compensated. Production may be so rapid that mechanical sampling is required. Gaging stations sometimes must be virtually as rugged as the machining stations.

All this means that gage planning for automation should be done at the same time as the original planning for manufacture.

PROBLEMS

1 For any manufacturing process acceptable to the instructor, prepare a plan for achieving and controlling quality during manufacture,

including flow diagram, selection of control stations, definition of work to be performed at each control station, and criteria to be used.

2 Visit a local manufacturing company and create a table similar to Table 14-1.

3 Visit a local manufacturing company and create a table similar to Table 14-2.

4 Read the Metal Containers, Inc., case in the Appendix and answer the following problem.

You are Lafferty, the quality control manager. During your 2-months' survey, you looked at how MC plans for quality. Your findings are summarized below:

Machines, tools, operation sequences These are specified by manufacturing engineering, except in Detroit, where these functions are handled by production supervision.

Gages The mechanics are provided with general use gages for dimensional control. (There are a few special gages for standard cans.) In addition, there are some special shears and microscopes to judge the adequacy of the all-important seam configurations. These gages are generally provided by manufacturing engineering.

Process control decisions The plan of decision making is as shown in the table below:

	Decision Made by		
Decision for	*Cleveland Milwaukee*	*Detroit*	*Chicago*
Setup acceptance	Mechanic	Mechanic	Mechanic plus QC
Running acceptance	Mechanic plus QC	Mechanic	Mechanic plus QC
Product acceptance	QC	QC	QC

There is a widespread belief in the company that can making is not yet a science, and that therefore much detail must be left to the discretion of the mechanic.

Control criteria Chicago has gone the farthest. For all control decisions (setup acceptance, running acceptance, and product acceptance) Chicago has formal criteria—how many to measure, the acceptable level, etc. Detroit has virtually none of this, leaving it all to the judgment of the mechanic and the inspector on the line. Cleveland and Milwaukee have formal criteria for product acceptance, but not for setup acceptance or running acceptance.

Records Except at Chicago, records are pretty sparse. At Chicago, the formal controls throw off a lot of records for so small a plant. These are analyzed and charted.

You have been struck by the fact that Detroit and Chicago operate at such extremes of informality or formality. Evidently the former owners of the Detroit plant had relied a good deal on the mechanics and had given them a lot of responsibility, which they had accepted. In contrast, the Chicago company had tended to use unskilled operators, and to rely on engineers, staff people, and consultants to supply instructions, controls, etc.

You are now faced with the question of what to recommend to the plant managers and to your boss as to planning for quality generally. The quality control supervisors in the plants want more participation in decision making and in the planning process. Generally, each plant manager feels that his plant is "different" and that what he has pretty well fits his special situation.

Prepare your recommendations on quality planning.

5 Read the Metal Containers, Inc., case in the Appendix and answer the following problem.

You are Kowalski, the quality control supervisor at MC's Cleveland plant. You are in a quandary over recent developments. Two months ago, the company hired a quality control manager from the outside—a man who had never been in the can business before. Your old boss Mr. Wallace did this. You have a lot of respect for Mr. Wallace, and so you are pretty sure there were some good reasons for setting up such a job and for bringing in an outsider for it. Not only that, the outsider seems to be all right—capable, patient, judicious. He ought to do OK. He has been raising questions about things which were swept under the rug years ago. That's good. Some of his questions have also made you reexamine some of your own views about responsibility. You have been giving this a lot of thought, and you find that it all boils down to two broad questions:

- ***a*** The fact that you will be having two bosses, in a way
- ***b*** The new look all of you are taking at responsibility on the factory floor

Mr. Wallace went out of his way to explain to you and your boss, Demos, the Cleveland plant manager, that Lafferty will have no responsibility for plant operations. Lafferty's job, as far as the plants are concerned, is to offer advice, make studies and recommendations, propose methods, etc. The plants have to decide whether they will go along with these proposals. Your boss is still the plant manager, but you are to work closely with Lafferty, supply him with information, help him try things out, etc. It all sounds good, but you wonder. They said about the same thing when a purchasing director was appointed. Now he is doing a lot more than advising. So far, Lafferty has only been going around and learning the ropes. But a fellow with that much on the ball isn't going to be satisfied to take it easy.

Now as to the factory floor. Lafferty has already been helpful in pointing out some vague spots in the definition of responsibility. He showed that while the responsibility for setup decisions is with the mechanic, there is no specification for what constitutes a good setup. Hence, the mechanic is really writing the specification, and thousands of cans are made before someone catches up with him. Chicago does have a setup specification, and the mechanic must follow it, make measurements, record them, and so on, as part of the setup procedure. This might be a good idea for Milwaukee, where so much depends on the setup being right.

While you have not been able to sell the idea of a quality control check on setups, you do have a responsibility for checking the lines while they are running. Your men make visual and gaging checks, and report their findings to the mechanics or floor supervisors. Many times nothing is done about it, and you have had plenty to say to the production supervision about their lack of interest in quality. Now along comes Lafferty and makes it look different. Lafferty pointed out that production and quality control were talking about two different things.

Quality control (says Lafferty) is concerned about whether the product is acceptable or not, whereas production is concerned about whether to run the process or not. Lafferty drew this diagram:

	Product Acceptable	*Product Not Acceptable*
Process is per process specification	RUN	
Process is not per process specification		STOP

What Lafferty was demonstrating was that quality control and production are together on only two of the four possible combinations. The other two combinations have not been talked out. On top of that, Lafferty thought that the company had at least two ideas about quality standards:

a The "excellence standard" (meaning that we should try to make MC a quality house). This would involve stopping machines which were outside the process specification even if the product were usable.

b The "usage standard" (meaning we should base decisions on customer needs). This would permit machines to run so long as the product was usable.

Lafferty's analysis was pretty helpful in explaining some of the seemingly strange actions of production. Some supervisors were willing to revise an off-standard process if they had to stop the machine for other reasons anyhow. But most of them saw no sense in changing a process if the product was acceptable.

You have been talking with Demos about all this. He too has been concerned about Lafferty's status. He is also interested in the questions Lafferty raised about responsibility on the factory floor. Demos has asked you to give these things some thought, and come back with your ideas.

What do you propose to Demos?

15 Statistical Tools in Manufacturing Planning

15-1 The Role of Statistical Tools in Manufacturing Planning

Two of the chief criteria for completeness of planning are that (1) the process will be able to hold the tolerances and (2) the process will be stable and free from unpleasant surprises.

Statistical tools have been developed to provide quantitative tests of processes for both of these criteria. To measure the ability of a process to hold tolerances, a quantitative study is made of "process capability." Process capability is the minimum variation that a process can achieve. To measure the stability of a process, statisticians have created a concept of a "state of statistical control" and have devised some elegant tools to detect the presence or absence of statistical control.

This chapter deals with the methods for making these tests of process capability and of state of statistical control.

15-2 Process Capability Analysis with Simple Charts

Process capability is determined in order to judge the ability of a process to meet a product specification. A process capability analysis consists of (1) measuring the process capability to discover whether the process is inherently capable of meeting the tolerances, and if necessary, (2) discovering why a process which is "capable" is nevertheless failing to meet tolerances. Before the statistical tools are considered, it will be well to present some examples of process data plotted against tolerances.

Often, the simple plotting of a chart of consecutive pieces drawn from the process is sufficient, without computations, to disclose all that needs to be known about a process which is failing to fulfill specifications.

Example An example[1] is shown in Figure 15-1. In this company the turning operations on the screw machines, making tiny watch parts, are of great importance. A study was conducted of the capability of one of the screw machines. The machine was studied for several hours, the personnel on the job having been asked to follow their usual practices. A diary, or "log," was kept of the actual happenings on the job. Meanwhile, the order of the pieces was carefully preserved so that the 500 pieces shown on the chart are in the actual manufacturing sequence. Then the pieces were carefully measured for each of the five critical dimensions, using a very precise gage. These 2,500 observations are shown on the chart together with the log of what transpired during the day. The disclosures of the chart are many:

1 Progressive changes in diameter were very slight. During the run of 500 pieces, the average diameter changed about 0.0001 inch, this being about a third of the narrowest tolerance range. This uniformity exceeded the expectations of most of the supervisors.
2 Changes caused by restocking the machine were less than expected, and of short duration. Note the effect of pieces no. 55 and no. 493. The effect at piece no. 271 is negligible.
3 The operators' gages were inadequate to "steer" the machines. Note at piece no. 140 how both short-end diameters dropped suddenly (these diameters are controlled by a single toolholder). The explanation was that the operator, from his gage, had concluded that the parts were becoming oversize and had adjusted the machine downward. At piece no. 197 he restored the adjustment following a check by the patrol inspector. At piece no. 392 the operator was again misled by his gage.
4 The machine capability for the pivot-shoulder distance was easily adequate. (The greater variability of the first 50 parts was due to the investigator's unfamiliarity with the precise gage.)

From these disclosures, the supervisors concluded that (1) the process was uniform enough for the job, (2) the gages were inadequate for the job, (3) it was possible to make more pieces with less frequent adjustment and with less frequent checking than had been

[1] The authors are indebted to G. P. Luckey and A. B. Sinkler, of Hamilton Watch Co., Lancaster, Pa., for this excellent example.

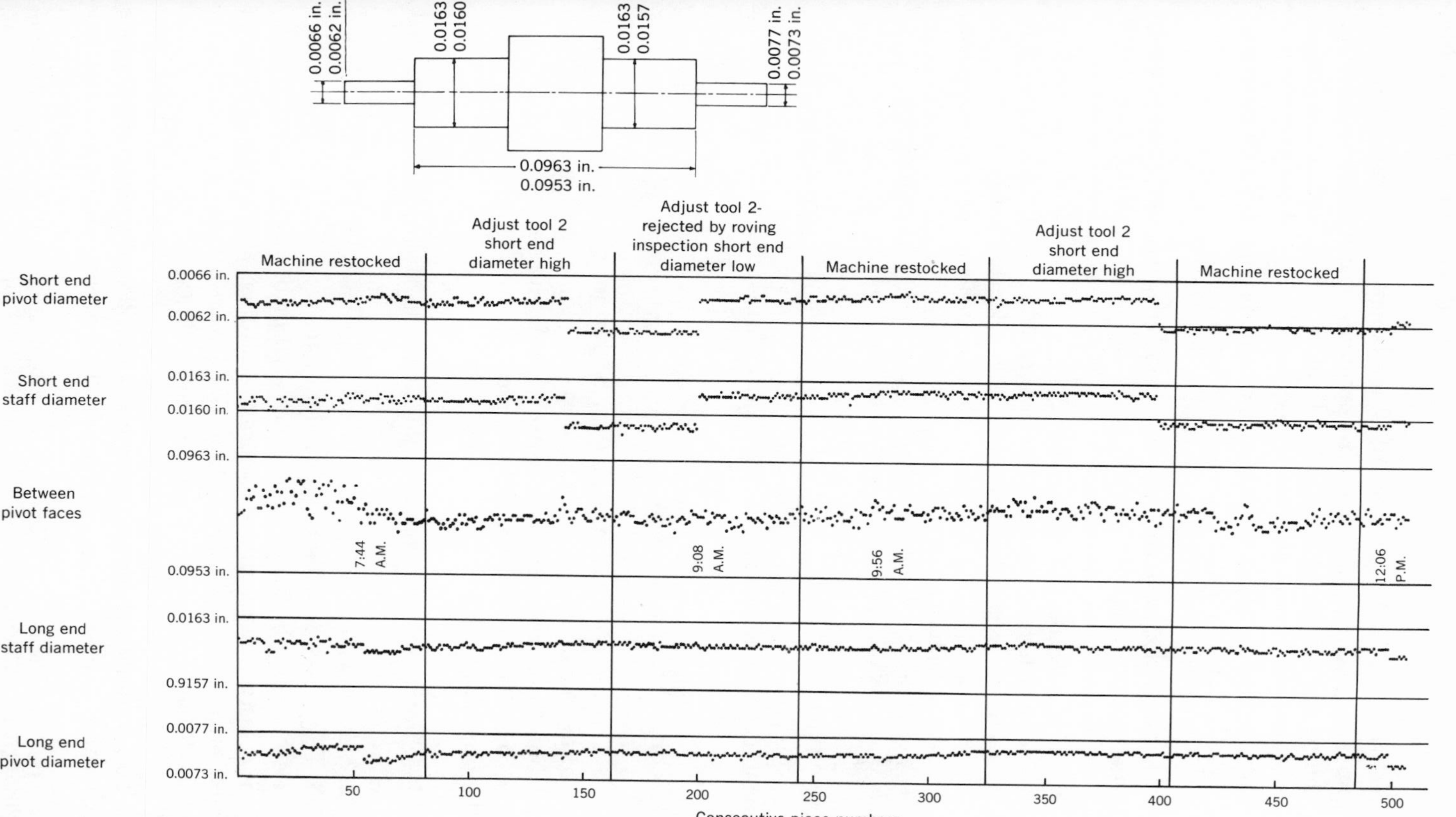

Figure 15-1 Process capability study of automatic screw machine.

realized, and (4) a number of other beliefs were unfounded. By providing better gages and by periodic measurement at the machines, the company reduced the level of defects in this department substantially, while also increasing production. An associated detailed gaging could thereupon be minimized. The consequence was a better product at substantially lower cost.

This classic example includes five basic elements of a process capability analysis:

1 The specification tolerance
2 Determination of whether the process is centered, i.e., whether the actual average is midway between the tolerance limits
3 Measurement of the natural or inherent variability of the process (i.e., the instantaneous degree of reproducibility)
4 Measurement of the actual variability over a period of time
5 Causes of the difference between (3) and (4)

These elements are further discussed in Chapter 30 in relation to improving an existing process. However, they also set the stage for a tool of manufacturing planning, i.e., the control chart.

15-3 Control Chart Concept

The data in the previous example can be further analyzed. For example, a histogram could be constructed to show the centering and variation in relation to tolerance limits. However, a histogram represents *performance* but does not necessarily depict the full *potential* capability of the process because:[2]

1 A histogram may contain values obtained from different machine settings.
2 A histogram may contain values from multiple station machines.
3 There may be within sample variation.
4 There may be sample to sample variation or time variation.
5 A histogram may have abnormal variation, which may or may not be attributable to the machine, but which comes from assignable conditions that cause the process to be unstable.

To manufacture product that meets specifications, it is often necessary that the process be "uniform." This means that the variability

[2] I. A. De Grote, "Machine Quality Capability Studies," *Transactions of the Sixteenth Midwest Quality Control Conference,* Oct. 19–20, 1961, pp. 155–169.

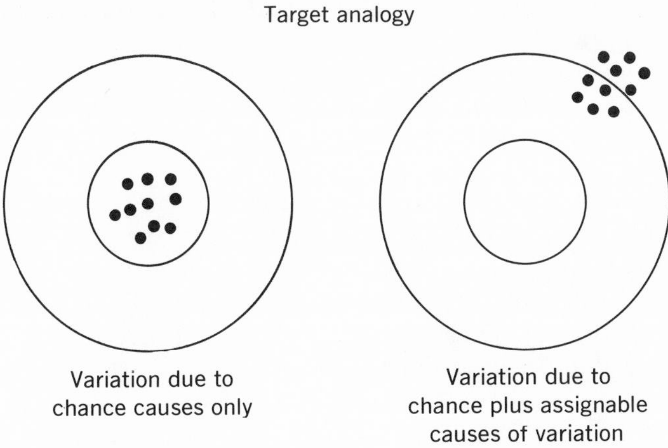

Figure 15-2 Distinction between chance and assignable causes of variation.

in the product should be due to numerous, small "chance" causes of variation rather than to a few large "assignable" causes of variation (see Figure 15-2). The objective is to restrict the causes of variation to the chance causes by detecting and eliminating the assignable causes. This is done by taking a series of samples from the process and testing these samples for significant differences from the aggregate of all the samples. For example, the formulas in Chapter 11 could test for a significant difference between the average of four units taken from a process at 10 A.M. and the average of all units made during the previous month. If the hypothesis of no significant difference was accepted, this would indicate that only chance causes of variation were present at the time the sample was taken.

While the hypothesis testing can be conducted by repeated calculation, Dr. W. A. Shewhart devised a graphic method for doing this testing continuously. The Shewhart control chart is thus a graphic continuous test of hypothesis. The control chart has been applied not only in process control work but also in many other areas such as analysis of experimental results, control of clerical errors, and analysis of time study data.

Control charts may be classified by the characteristic being tested:

1 The average of the measurements in the sample. This is known as an $\overline{X}$ chart. Averages are used because they are more sensitive to change than individual values.[3]

[3] See J. M. Juran (ed.), *Quality Control Handbook*, 2d ed., McGraw-Hill Book Company, New York, 1962, p. 13-46.

2 The range of the measurements in the sample. This is known as an R chart.
3 The percent defective in the sample. This is known as a p chart. (Shewhart's original chart, dated May 16, 1924, was a p chart.)
4 The number of defects in the sample. This is known as a c chart.

There are many variations of these four basic charts. In addition, it should be noted that a control chart can be constructed for any statistical measure, e.g., the median, correlation coefficient, etc. Examples of control charts and the economic implications of the concept are discussed in Chapter 23. The concept is introduced here because of the connection with process capability.

15-4 Establishing Trial Control Limits: $\bar{X}$ and R Charts

Trial control limits are usually set at 3 standard deviations above and below the central line or expected value. The control limits are an example of a principle in the control sequence defined in Chapter 3. These limits are a means of comparing actual performance with a standard. If all points fall within the control limits, then it is concluded that only chance variation is present. If any points fall outside the control limits, then assignable variation is assumed present in the process. Methods have been developed to simplify the calculations by eliminating the need to calculate standard deviations. Table 15-1 summarizes the formulas and constants needed.

Referring to Table 15-1, the shortcut formulas for the control limits on sample averages are

$$\text{Upper control limit} = \bar{\bar{X}} + A_2\bar{R}$$
$$\text{Lower control limit} = \bar{\bar{X}} - A_2\bar{R}$$

where $\bar{\bar{X}}$ = grand average = average of the sample averages
$\bar{R}$ = average of the sample ranges
A_2 = constant found from Table H in the Appendix

The shortcut consists of (1) computing, for each sample, the range (difference between largest and smallest) of the individuals, (2) averaging the ranges thus obtained, and (3) then multiplying the average range by a conversion factor to get the distance from the limit line to the expected average. The central line is merely the average of all the individual observations.

Table 15-1 CONTROL CHART LIMITS—ATTAINING A STATE OF CONTROL

Chart for	*Central Line*	*Lower Limit*	*Upper Limit*
Averages $\bar{X}$	$\bar{\bar{X}}$	$\bar{\bar{X}} - A_2\bar{R}$	$\bar{\bar{X}} + A_2\bar{R}$
Ranges R	$\bar{R}$	$D_3\bar{R}$	$D_4\bar{R}$
Percent defective p	$\bar{p}$	$\bar{p} - 3\sqrt{\frac{\bar{p}(1-\bar{p})}{n}}$	$\bar{p} + 3\sqrt{\frac{\bar{p}(1-\bar{p})}{n}}$
Number of defects c	$\bar{c}$	$\bar{c} - 3\sqrt{\bar{c}}$	$\bar{c} + 3\sqrt{\bar{c}}$

Values of A_2, D_3, and D_4, are given in Table H in the Appendix. A partial tabulation of these factors is reproduced here in Table 15-2 for the convenience of the reader in the following text.

Table 15-2 CONSTANTS FOR $\bar{X}$ AND R CHART

n	A_2	D_3	D_4	n
2	1.880	0	3.268	2
3	1.023	0	2.574	3
4	0.729	0	2.282	4
5	0.577	0	2.114	5
6	0.483	0	2.004	6
7	0.419	0.076	1.924	7
8	0.373	0.136	1.864	8
9	0.337	0.184	1.816	9
10	0.308	0.223	1.777	10

Referring to Table 15-1, the shortcut formulas for control limits on sample ranges are

$$\text{Upper control limit} = D_4\bar{R}$$
$$\text{Lower control limit} = D_3\bar{R}$$

where D_3 and D_4 are constants found from Table 15-2. The calculation of the control limits is illustrated in the next section.

15-5 Determination of Process Capability from a Control Chart Analysis

If a process is in statistical control, then it is operating with the minimum amount of variation possible (the variation due to chance causes). If, and only if, a process is in statistical control, the following relationship holds:

$$s = \frac{\bar{R}}{d_2}$$

Table I in the Appendix provides values of d_2. Knowing the standard deviation, process capability limits can be set at a number of multiples of the standard deviation to correspond to a desired percentage of the population to be within the capability limits. For example, process capability set at ± 3 standard deviations will include 99.73 percent of the individual items in the population *if* the process remains in statistical control. These "natural tolerance" limits can be compared to the specification tolerance limits. Thus, if information on process capability is known, in the *planning* of manufacturing, products can be assigned to those processes which will minimize the amount of defective work produced.

For example, suppose 25 samples of 5 items each were inspected from a process. A control chart analysis demonstrated statistical control on both the average and the range charts. The following data have been summarized from the analysis:

$$\bar{\bar{X}} = 14.34 \qquad \bar{R} = 0.39$$

The standard deviation can be calculated as

$$s = \frac{\bar{R}}{d_2} = \frac{0.39}{2.326} = 0.168$$

Suppose that the specification were 14.40 ± 0.45.

The natural tolerance limits are $\pm 3\ (0.168)$ or ± 0.51. As ± 0.51 is wider than the tolerance range of ± 0.45, it is concluded that this process is *not* inherently capable of meeting the tolerances. The natural tolerance limits of ± 0.51 represent the *best* the process can do because the process is operating under a state of statistical control (no assignable causes of variation).

It is worthwhile to diagram this situation (Figure 15-3).

The average dimension is a little lower than the desired nominal value. Therefore, the total percent defective would be reduced if the process were centered to an average value of 14.40. Even if this is done, defectives will still be produced because the variation in the process (± 3 standard deviations) is larger than the $\pm$ tolerance range. There are three alternative courses of action in such a situation:

1 Reduce the standard deviation. As the process is in a state of statistical control, the assignable causes of variation from the process have

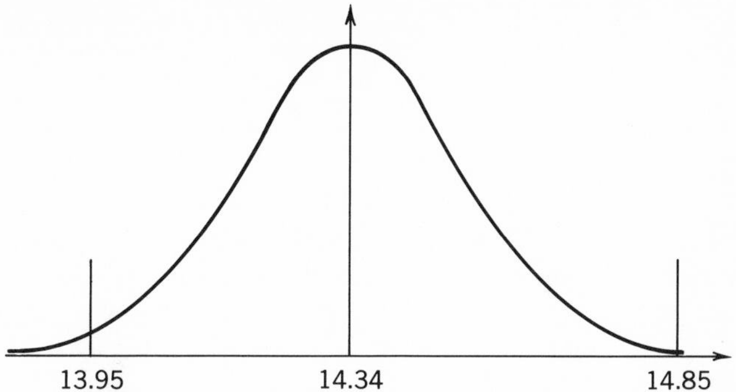

Figure 15-3 Natural tolerance limits with process average of 14.34.

been eliminated. Therefore, to reduce the standard deviation, a fundamental process change is required such as a complete overhaul of the machinery, a change in materials, etc.

2 Widen the product tolerances.

3 Live with the situation and be prepared to produce a number of defective units ("suffer and sort").

This situation illustrates a problem in breakthrough rather than control as defined in Chapter 3.

Another example[4] of control chart analysis for process capability consists of 20 samples of 5 readings each on a certain product characteristic with a specification of 37.0 ± 10.0 (coded values). The control chart for the data is shown in Figure 15-4. The original control limits were

$$\begin{aligned} \textit{Averages:}\quad \text{UCL} &= \bar{\bar{X}} + A_2\bar{R} = 33.6 + 0.577(6.2) = 37.2 \\ \text{LCL} &= \bar{\bar{X}} - A_2\bar{R} = 33.6 - 0.577(6.2) = 30.0 \\ \textit{Ranges:}\quad \text{UCL} &= D_4\bar{R} = 2.114(6.2) = 13.1 \\ \text{LCL} &= D_3\bar{R} = 0(6.2) = 0 \end{aligned}$$

Samples 9 and 13 are out of control. Assignable causes of variation were present in the process at the time those samples were taken. An investigation should be made to determine and eliminate these causes from the process. It is tempting to speculate on what would happen if these assignable causes were eliminated from the process. This can be made on paper by assuming that the causes were eliminated, thereby making it reasonable to eliminate the data from samples 9 and 13. If

[4] The original data are in Albert H. Bowker and Gerald J. Lieberman, *Engineering Statistics*, Prentice-Hall, Inc., Englewood Cliffs, N.J., 1959, p. 398.

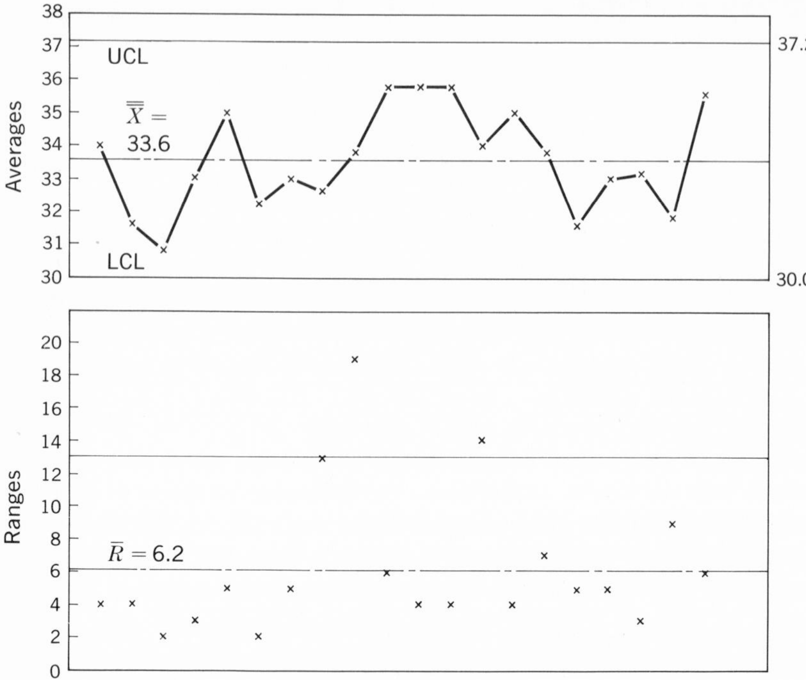

Figure 15-4 Average and range control chart.

this is done, the new $\bar{R}$ is 5.1 and the new $\overline{\overline{X}}$ is 33.6. A new set of control limits should now be calculated for the original data to see if any additional points fall out of control. Sample 8 then falls out of control. If these data are dropped and the limits are recalculated, sample 3 is out of control. If these data are dropped, the new limits are

$$\begin{aligned}
\textit{Averages:}\ \text{UCL} &= \overline{\overline{X}} + A_2\bar{R} = 33.7 + .0577(4.8) = 36.5 \\
\text{LCL} &= \overline{\overline{X}} - A_2\bar{R} = 33.7 - 0.577(4.8) = 30.9 \\
\textit{Ranges:}\ \text{UCL} &= D_4\bar{R} = 2.115(4.8) = 10.2 \\
\text{LCL} &= D_3\bar{R} = 0(4.8) = 0
\end{aligned}$$

With this final set of control limits, all the original points in the analysis fall within statistical control.

We have *no* justification for eliminating data from a process unless *actions* have been taken and *verified* that the causes of the out-of-control points have been eliminated. There is also a question of the effect of eliminating the assignable causes on the remaining data which were in control. Ideally, an entirely new study should be made. The rough approach of eliminating the data on paper sometimes helps to see if the process *would be* capable of meeting tolerances if all the assignable causes could somehow be eliminated. If the analysis indicates that even after the elimination of the assignable causes the process is still not capable of meeting the tolerances, then a large effort to eliminate assignable causes may *not* be the best approach because it will only reduce, not eliminate, the defective work. Such a conclusion may help to dramatize the need to review tolerances or make a major change in the entire process.

The standard deviation may now be calculated as

$$\frac{\bar{R}}{d_2} = \frac{4.8}{2.326} = 2.042$$

The natural tolerances, therefore, are $\pm 3(2.042)$ or ± 6.1. As these natural tolerances are smaller than the tolerance range of ± 10.0, it is concluded that if the assignable causes of variation are eliminated from the process, and if the process remains in statistical control, then it does have the inherent capability of meeting the proposed tolerance range. A sketch (Figure 15-5) can help to summarize this analysis. This diagram indicates that some centering of the process is necessary.

Calculating natural tolerance limits as $\pm 3s$ is common in practice but it is approximate. The calculated standard deviation is really an *estimate* of the true standard deviation of the population but is regarded as the true value for the population. This error can be taken into account with the formulas presented in Chapter 13. Here the natural tolerance limits are $X \pm Ks$ where K is a function of the amount of data in the original analysis, the confidence level desired, and the per-

Figure 15-5 Capability if assignable causes are eliminated.

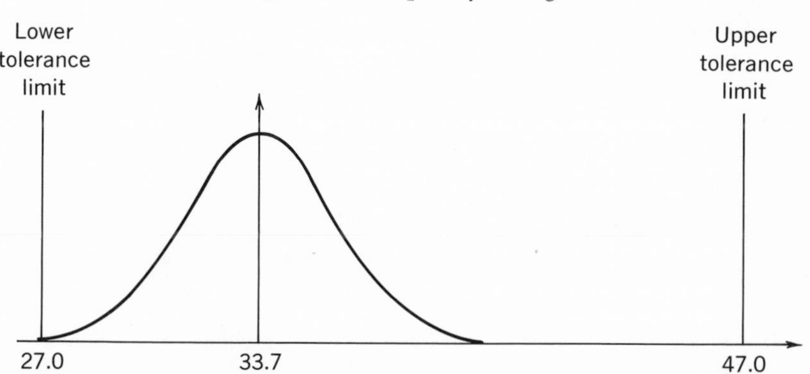

centage of the population which must be included within the tolerance limits. Bingham[5] compares the approximate and exact methods.

Another role of the control chart in quality planning is in controlling components which will go together in complex assemblies. Chapter 13 discusses the relationship between component tolerances and assembly tolerances. The formula assumed that each component was meeting its specified nominal dimension; i.e., the actual average was equal to the nominal value. This assumption can be continuously checked by an average chart with the nominal specification as the center line of the chart. If the chart remains in control, it is concluded that the process is maintaining an actual average equal to the nominal specification. If the process goes out of statistical control, it is concluded that the nominal specification is no longer being met and that action should be taken to correct the process. A significant application of this approach was made to the L-3 coaxial system (a broadband transmission system for multiple telephone or television channels).

15-6 The Assumption of Statistical Control and Its Effect on Process Capability

All statistical predictions assume a stable population. In a statistical sense, a stable population is one which is repeatable and therefore free of unusual causes of variation—that is, a population which is in a state of statistical control. The statistician rightfully insists that this be the case before predictions can be made. Simultaneously, the manufacturing engineer rightly insists that the process conditions (feeds, speeds, etc.) be fully defined.

In practice, a problem frequently arises. An original control chart analysis will often show the process to be out of statistical control. (It may or may not be meeting product specifications.) When an investigation is made of the out-of-control points, the reasons found are sometimes causes that *cannot* be economically eliminated from the process. Strictly speaking, the instability of the process means that a prediction of the process should not be made. However, the problem of estimating process capability still remains! Some comparison of capability to product tolerances must be made even though it must be viewed as a first approximation. It is better that an approximate analysis be made rather than delay the analysis because an assumption cannot be met. The danger in delaying the analysis is that the assignable causes may *never* be eliminated from the process and the indecision will simply revert back

[5] R. S. Bingham, Jr., "Tolerance Limits and Process Capability Studies," *Industrial Quality Control*, vol. 19, no. 1, pp. 36–40, July, 1962.

to the interdepartmental bickering on whether the "tolerance is too tight" or "manufacturing too careless."

A good way to start is by plotting individual measurement against *tolerance* limits (see section 15-2). This may show that the process can meet the product tolerances even with assignable causes present. If a process had assignable causes of variation, but is able to meet the product tolerances, then *no* problem exists. The statistician will properly point out that a process with assignable variation is unpredictable and therefore anything might happen. This point is well taken but in establishing *priority* of quality control efforts, processes that are meeting tolerances simply cannot be given high priority.

If a process is out of control and the causes cannot be economically eliminated, the standard deviation and natural tolerance limits can be computed (with the out-of-control points included). *These limits will be inflated because the process will not be operating at its best. In addition, the instability of the process means that the prediction is approximate.*

15-7 Relation of Control Limits to Product Tolerances

Control limits for a chart for averages represent 3 standard deviations of sample *averages* (not individual values). As tolerance limits usually apply to *individual* values, the control limits *cannot* be compared to tolerance limits, because averages inherently vary less than the individual measurements going into the averages (Figure 11-1). Therefore, tolerance limits should *not* be placed on a control chart for averages. The only valid comparison that can be made is to convert $\bar{R}$ to the standard deviation, calculate the natural tolerance limits, and compare to product tolerances. (Chapter 23 shows how special adaptations of the basic control chart permit exceptions to this statement.)

15-8 Illustration of the Use of Process Capability Data in Manufacturing Planning

Screw machines are subject to numerous small variables (e.g., clearances, end play) which combine to result in a relatively constant frequency distribution. For example, changes in depth or length of cut or variation between lots of the same material have only minor effect on the distribution and thus the measured capability is essentially a constant for a machine.

The measurement and publication of machine capability for different operations and different materials is of great importance in preventing quality problems. For example, the design engineer, acquainted with the attainable tolerances of the available equipment, has a more rational basis for the specification of tolerances. Many plants use these data as the basis for a complete review of all engineering tolerances.

The planning engineer, or dispatch clerk, can assign the jobs with the most rigid tolerances to the most capable machines and the less exacting work to the poorer machines. The tool designer is provided with a measure of the effectiveness of changes in his tooling; he can also spot the places where tooling improvements must be made. Capability information helps the foreman by highlighting individual machines that may require overhaul. The machine setup man learns which machines require the most attention to set up and which ones need only normal care. The purchasing agent has a means of comparing the actual performance of equipment with the manufacturer's claims. The patrol inspector and the machine operator have a better gage to show which machines need the closest watching in production.

Beyond the general uses above, there are problems arising from the relationship of machine capability to product tolerance. As applied to screw machine problems, the alternatives are

1 If the machine capability is inadequate to meet the tolerance:
a Try to shift the job to another machine with more adequate capability. If the order already is completed make provisions for proper assignment when the job occurs again.
b Try to improve the machine capability. This is particularly advisable if the value obtained differed markedly from that of similar machines. The machine may require overhauling or the tooling may need to be reviewed. For multispindle machines, the causes for spindle differences may need to be pinpointed. Control chart analysis may be helpful.
c Try to get a review of the tolerance. The availability of specific information showing what tolerance can be achieved may soften the engineer's attitude on liberalizing the tolerance.
d Sort the good product from the bad. The economics of this alternative should be brought to the attention of management.

2 If the machine capability is equal to the tolerance: This usually should be treated as in (1) above, since it means that tools must be set exactly at the nominal and gives no allowance for tool wear. However, if tool wear is negligible or where a small percentage of parts as in (3) below.
just outside the tolerance can be tolerated, this should be treated

3 If the machine capability is adequate to meet the tolerance:
a If the machine capability is of the order of two-thirds to three-

fourths of the tolerance or less, it is the acceptable situation. The machine should produce practically all good work over a long period of time if periodic samples are taken to check the setting of tools. (See section 14-5 for a specific guideline used by one company.)

b If the machine capability is less than one-half of the tolerance, consider reducing the tolerance. The ability to tighten the tolerance on one part in an assembly may permit loosening a difficult tolerance on another part. Closer guarantee of tolerances may also have competitive sales advantage.

c One hundred percent inspection of the product is not needed and a sampling procedure should be considered.

PROBLEMS

1 A large manufacturer of watches makes some of his own parts and buys some other parts from a vendor. The vendor submits lots of parts that meet the specifications of the horologist. The vendor thus wishes to keep a continuous check on his production of watch parts. One gear has been a special problem. A check of 25 samples of 5 pieces gave the following data on a key dimension:

$$\bar{\bar{X}} = 0.125 \text{ inch} \qquad \bar{R} = 0.002 \text{ inch}$$

What criterion should be set up to determine when the process is out of control? How should this criterion compare with the specifications? What are his alternatives if the criterion is not compatible with the specification?

2 A manufacturer of dustless chalk is concerned with the density of his product. Previous analysis has shown that his chalk has the required characteristics only if the density is between 4.4 gm/cc and 5.0 gm/cc. If a sample of 100 pieces gives an average of 4.8 gm/cc and a standard deviation of 0.2, is his process aimed at the proper density? If not, what should the aim be? Is the process capable of meeting the density requirements?

3 The specifications on the diameter of a wrist pin are 1.000 inch ±0.002 inch. Twenty samples of five pins show the average to be 1.001 inches and the average of the twenty ranges to be 0.002 inch. Are the specifications capable of being met by the process that makes the wrist pins? What assumption is necessary?

4 The head of an automobile engine must be machined so that both the surface that meets the engine block and the surface that meets the valve covers are flat. These surfaces must also be 4.875 inches ± 0.001 inch apart. Presuming that the valve cover side of the head is finished correctly, compare the capability of two processes for performing the finishing of the engine block side of the head. A broach set up to do the job gave an average thickness of 4.877 inches with an average range of 0.0005 inch for 25 samples of 4 each. A milling machine gave an average of 4.875 inches and average range of 0.001 inch for 20 samples of 4 each.

5 A critical dimension on a double-armed armature has been causing trouble and the designer has decided to change the specification from 0.033 inch ± 0.005 inch to 0.033 inch ±0.001 inch. To evaluate the proposed change, the manufacturing planning department has obtained the following coded data from the process:

Time	*Left Arm*			*Right Arm*			*Comments*
8:00	331	330	331	329	330	328	
8:30	332	331	329	327	331	329	
9:00	330	329	329	330	329	327	
9:30	332	330	331	331	332	328	
10:00	333	332	333	326	331	326	
10:30	332	331	332	329	330	331	
11:00	333	331	331	330	326	327	
11:30	332	332	333	327	326	329	
12:00	331	332	334	337	328	337	Adjustment
12:30	335	334	336	326	325	325	
1:00	333	332	332	329	332	330	
1:30	336	331	330	331	328	329	
2:00	332	334	329	332	330	329	
2:30	336	336	330	329	329	327	
3:00	329	335	338	333	330	331	Adjustment
3:30	341	333	330	329	331	332	

Comment on the proposal to change the specification.

6 Visit a local manufacturer and evaluate the procedures used to recognize process capability in assigning new products to machines.

16 Inspection

16-1 The Importance of Inspection

In theory, if all company departments do their job properly, the product could be shipped to the customer and would meet the criteria of fitness for use. However, the theory is not borne out by the realities. In consequence, companies follow an extensive practice of simulated use testing, with associated measurements and inspection. These are in the nature of a "countdown" designed to provide a proof of freedom from defects as well as ability to meet the criteria of use.

New forms of quality control methodology have not reduced the significance of these inspection and test countdowns. Here is an example[1] of some of the problems encountered in the aerospace industry.

> An early motor static firing was a complete failure because the hold-down bolts on the firing stand had not been tightened. We failed to spin in four flights because an insufficient quantity of booster pellets had been loaded into some spin rockets. Later we failed to ignite two second stages because some simple mechanical linkages in a safe arm device were sticky, and a third when the umbilical wouldn't retract because some other linkage was oversize and therefore hung-up. The list is endless, of flight and ground test failures whose causes are traced back to the lack of the simplest mechanical, electrical, or visual examination.

The point is that inspection has always been and always will be vital to a quality program.

[1] R. W. Smiley, "Government and the Inspector," *Industrial Quality Control* (*IQC*), vol. 20, no. 10, pp. 4–7, April, 1964.

16-2 Inspection Planning

"Inspection often involves a decision on product acceptance, but it has other purposes,[2] such as regulating manufacturing processes, rating overall product quality, and measuring inspection accuracy. However, the inspection act always includes (1) the interpretation of a specification, (2) measurement of the product, and (3) comparison of (1) and (2).

Inspection planning is the activity of designating the stations at which inspection should take place, and of providing inspectors with the facilities needed for knowing what to do and for doing it.

Inspection planning is done variously by inspectors, by inspection supervisors, or by special planners. The following are the more usual job conditions which are decisive in deciding who does the planning:

1 Complexity of measuring the quality characteristic. The inspection of simple or routine characteristics can be planned by the inspector himself. With added complexity, the planning will require the knowledge of a skilled inspection specialist or engineer.
2 Complexity of the product. For complex products having many interacting dimensions or components, the planning will require someone with a knowledge of the functioning of the product, the relative seriousness of different characteristics, and the complete manufacturing sequence. This will likely be a quality control engineer working in conjunction with test specialists.
3 Purpose of the inspection. Inspection for the purpose of product acceptance may require someone with knowledge of the functioning of the product (e.g., a quality control engineer), but inspection for process control purposes can often be planned by inspectors or inspection supervisors assigned to the process.
4 Size and organization of the inspection function. As the function grows, separating the planning from the doing becomes justified. This leads to inspection planners who develop and standardize the most efficient practices for inspectors to use.

In practice, the decision of who does the planning involves all the above criteria.

Selection of Inspection Stations

The most usual conditions suggesting inspection stations are

1 At receipt of goods from vendors, usually called "vendor inspection"
2 During setup of operations

[2] See J. M. Juran (ed.), *Quality Control Handbook (QCH)*, 2d ed., McGraw-Hill Book Company, New York, 1962, p. 1-33.

3 At delivery of goods from one department to another, usually called "inspection in process"
4 During the progress of high quality or high expense operations
5 Upon completion of all fabrication operations, usually called "finished goods inspection"
6 Before completing an irreversible, expensive operation (e.g., pouring a melt of steel)
7 At natural "peepholes" in the process

These general rules may need modifications in specific instances. A process operation may require a "station" in which the inspector patrols a large area while for other operations no inspection stations may be needed between departments. There may also be a need for a station at a vendor's plant or a station after packing, or at the customers' premises.

The fundamental planning tool for choosing process control product acceptance stations is the product and process flowsheet (see Chapter 14).

Information for Inspection

The inspector needs information on what the product is supposed to be. The sources of this information are

1 *The customer's order,* defining what products the customer asked for and any modifications he wants from the specification.
2 *The product specification,* defining the properties the product is supposed to have.
3 *The list of quality characteristics* which are to be checked at this station.
Preparation of defect lists can become complex when the specification is complex. In that event, there must be a breakdown by the planners to determine what characteristics should be checked at which station. In making this breakdown the planner is guided by the product and process flowsheet (see Chapter 14).
A common form of process inspection plan is a listing of the check areas, the quality characteristics to be checked, the standard, the frequency of check, the data to be collected, and the type of reporting to be done.
4 *The pertinent industry, shop, and other standards applicable.*
The foregoing concern definition of the product. However, there are other sources of information about the product:
5 *Results of prior measurements on the product under consideration.*
An obvious opportunity lies in the field of vendor relations, where

it is common practice for vendor and customer to perform duplicate measurement on the same product. This is explained in Chapter 22.

6 *Information on prior performance of the same source of manufacture* (i.e., vendor, operator, machine, process, etc.). This principle is recognized in sampling tables which provide for different levels of inspection for different historical process averages.

7 *General information* which can add to the inspector's perspective. This includes the manufacturing flowchart, usage information, and knowledge of prior complaints and troubles. Visits to the field are uncommonly helpful in improving the inspector's perspective.

8 *Detailed inspection or test procedure.* This procedure contains some or all of the following items:

a Tests to be conducted, with step-by-step procedure
b Measurements to be taken and equipment to be used
c Data to be recorded
d Sample size, for each test to be conducted
e Procedure for selecting sample units
f Required accuracy of test equipment
g Environmental conditions during tests (including tolerance limits on the environmental conditions)
h Criteria for "acceptance" or "rejection" decision on quality
(1) Maximum and minimum limits for each test
(2) Allowable number of failures for "acceptance" of lots
(3) Basis on which production is to be shut down or kept running
i Report to be prepared
j Action to be taken in case of a "rejection" decision
k Disposition of test specimens
l Requirement for "certification" of test results by independent agency

All this detailed information required by the inspector highlights the need to often separate the "planning" from the "doing" in the same way that this philosophy is followed for the production operator.

9 Responsibility of the inspector in accepting or rejecting the product, notifying the foreman of excessive defects, and requesting or requiring that the line be shut down.

An Example of Inspection Planning

The following is an example of inspection planning for refrigerator cabinets.[3]

[3] L. Sandholm, "Program to Reduce the Need for Servicing Domestic Appliances," *First Scandinavian Conference on Quality and Reliability,* Gothenburg, May 27–28, 1968.

The planning starts with an analysis of the product specification which contains the following information:

1 Applicable documents
2 Description of product
  a General features
  b Variants
3 Product provisions
  a Function: General data. Operating characteristics. Acceptable noise level. Reliability
  b Materials and workmanship
  c Grounding and insulation
  d Dimensions
  e Finish
  f Labeling
4 Manufacture
  a Special assembly requirements
5 Dispatch
  a Packing
  b Labeling and stockkeeping

An analysis is also made of the analogous component specifications.

On the basis of specification analyses, design reviews (with representatives from product development, manufacturing engineering, quality control, and service), and the results of pilot runs, the overall system is defined by flowcharts which show the locations and types of inspections to be made (Figure 16-1).

For every inspection station, an inspection instruction is prepared listing the inspection characteristics, inspection method, and sample size (Figure 16-2).

A quality rating program (see Chapter 28) is used to obtain information on outgoing quality as the customer will see it. A random sample of finished units is inspected every day in accordance with a quality rating instruction and the results are recorded on a quality rating record which are later summarized (see Chapter 28).

The complete quality control setup is described in a manual having the following information: (1) product specification, (2) flowchart, (3) receiving inspection, (4) inspection procedures in different manufacturing departments, (5) final inspection, (6) quality rating, and (7) quality information feedback.

Every product or product group has such a manual.

All this planning must be supplemented by a well-conceived program of selecting and training inspection personnel.[4]

[4] For specific procedures, see *QCH*, 1962, sec. 7.

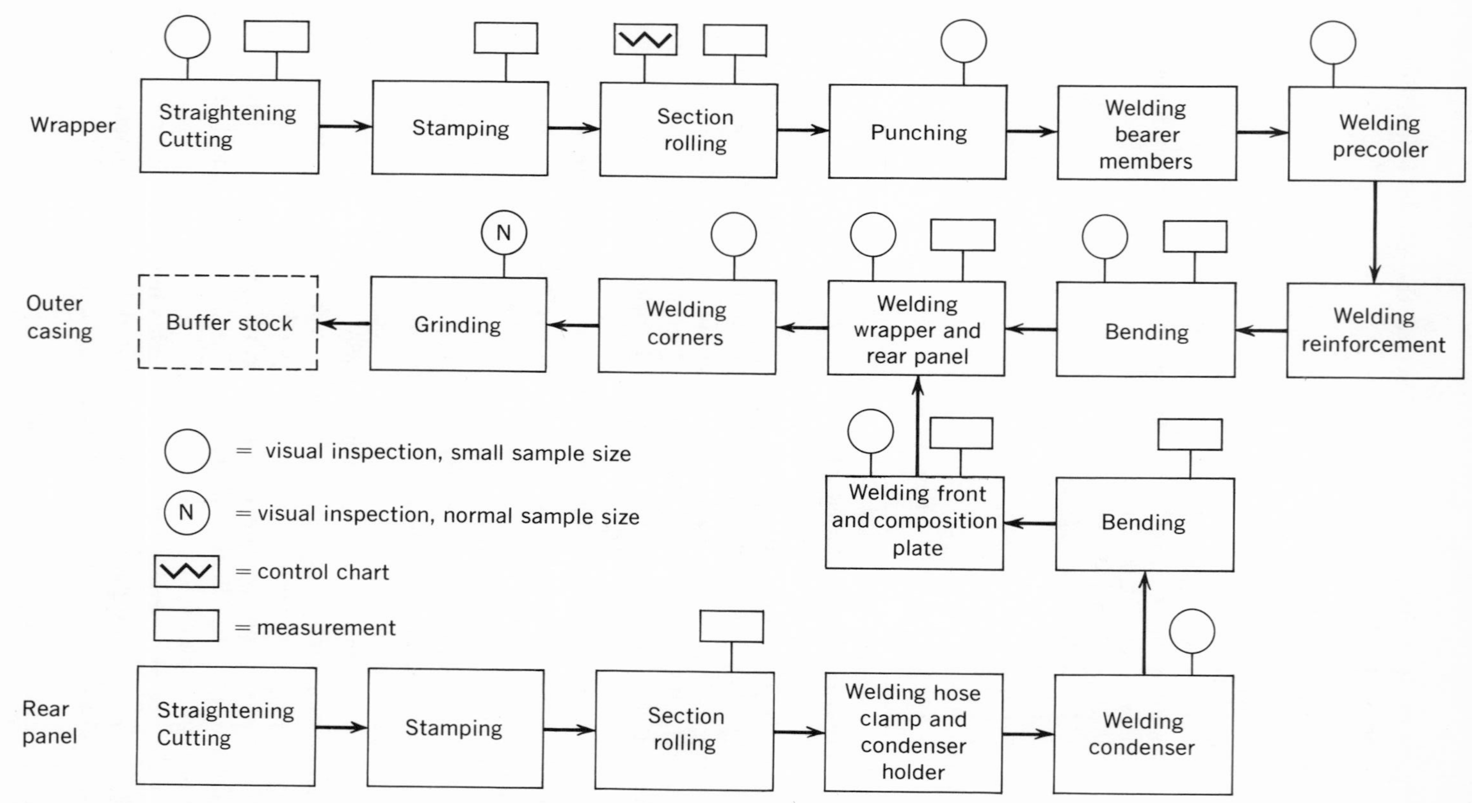

Figure 16-1 Flowchart showing inspection points in manufacturing of refrigerator cabinets. Source: L. Sandholm, "Program to Reduce the

Part name Casing		Part No. 200 10 68	Dwg. index 1	Rev. 1
Intended for Bending of outer casing		Insp. group alt. op. KAT	Date issued 10/10/70	Issued by RZ
Item No.	Characteristic	Inspection method		Sample size
1.1	Corner cut undamaged	Visual inspection		5 per day
1.2	Corner cut slit not exceeding 0.5 mm	Visual inspection		
1.3	Tube holders intact close to bending points	Visual inspection		
1.4	Height 1190 ± 1, 1490 ± 1, 1690 ± 1	Steel measuring tape		
1.5	Width 595 ± 0.5, 800 ± 0.5	Sliding caliper		
1.6	Lower hinge hole position 1106, 1406, 1606 ± 1	Steel measuring tape		
1.7	Intermediate bearer member position 800.5 ± 0.5 or 300.5 ± 0.5, 38.8 ± 0.5, 16.4	Fixture V-41071		
1.8	35 ± 0.5	Sliding caliper		
1.9	Tube flattening in bending minute 5.5	Sliding caliper		

Figure 16-2 Inspection instruction. Source: L. Sandholm, "Program to Reduce the Need for Servicing Domestic Appliances," *First Scandinavian Conference on Quality and Reliability,* Gothenburg, May 27–28, 1968.

16-3 Measuring Performance of Inspection

Highly repetitive inspection lends itself to measurement and the establishment of standards of a day's work by conventional time study methods. The approach is to divide the inspection work into small elements that occur in many inspection operations, e.g., place part in gage fixture, visual inspect, etc. Time measurements are then taken to obtain "standard" times for each element. The time required for a new inspection job is then synthesized by breaking it down into elements and adding up the standard times for the elements. If necessary, special allowances for fatigue or nonrepetitive operations (e.g., obtain a blueprint) may be incorporated in the final time allowance. One provision needs special emphasis. *The inspector must be given adequate time allowances to cover any added work arising from the presence of defectives in the product.* Writing up reports of defectives, preparing tags for identifying the defectives, taking second samples, etc., all take time. This must be recognized in a way that the level of product quality submitted for inspection has no effect on the output performance of the inspector.

> If inadequate time is allowed for handling defectives, the inspector has a personal stake in finding the product good. If too much time is allowed for handling defectives, the inspector has a personal stake in calling the product defective. (However, the latter is subject to immediate challenge by the operator.) Special attention should therefore be given to this provision by the time-study engineers so that the inspector will not be swayed in his quality judgment because he has an eye on his own output efficiency.[5]

This provision is extremely important.

An additional approach involves an overall evaluation of an entire inspection area. Saks[6] describes the evaluation of two inspection sections in a semiconductor plant by using work sampling concepts. In work sampling, observations are taken at random on a machine or an individual and its status noted and recorded into one of a set of predefined categories. With a sufficient number of observations, the ratio of the number of observations in a category to the total number of observations is a valid estimate of the *percentage of time* spent in the category. Three work samples were taken:

[5] J. M. Juran, *Management of Inspection and Quality Control,* Harper & Row, Publishers, Incorporated, New York, 1945, pp. 173–175.
[6] Gerald D. Saks, "An Operations Control Study of a Quality Control Department," *Journal of Industrial Engineering,* vol. 13, no. 6, pp. 460–464, Nov.–Dec., 1962.

1 A sample of 200 observations was taken on each of 33 pieces of equipment. The Pareto principle emerged; i.e., two of the equipments were used 75 to 85 percent of the time, a few equipments were used between 40 and 60 percent, and the rest were rarely used. The analysis concluded that a sufficient quantity of equipment was available and that the current problems of lack of equipment were due to poor scheduling and utilization and performance limitations on certain equipments. The low equipment utilization also revealed that personnel must be spending a great deal of time in activities other than testing. This led to the need for work sampling of the personnel.

2 A work sample was taken in each of the *two* inspection sections over a period of 1 month. About 400 observations were made in each section. The results are summarized in Table 16-1.

Table 16-1 WORK SAMPLE STUDY SUMMARIES BY ACTIVITY

	Reliability and Environmental Test Section (18 Inspectors)		*Final Quality Assurance Department (25 Inspectors)*	
	*Percent of Department Time**	*Total Percent*	*Percent of Department Time*	*Total Percent*
Inspection				
Electrical inspection	22		34	
Life or design test	14			
Visual inspection			7	
Total		36		41
Backup				
Preparation prior to test	8		6	
Paper work and analysis after completion	11		7	
Total		19		13
Other				
Walk with units	4		6	
Walk without units	8		8	
Talk to supervisor	4		3	
Talk to other	15		12	
Wait for equipment	1		2	
Idle	1		0	
Not located	12		15	
Total		45		46
Grand total		100		100

* Based on 8 hours per day (included in the results are two 10-minute coffee breaks).

The analysis concluded that much time was spent in test preparation and reporting and also in noninspection activities. Furthermore, personnel spent about 20 percent of their time in other physical areas and for another 12 to 15 percent of the time they could not be located at all.

The data made it possible to determine a better functional flow of the inspection activity, higher equipment utilization, and higher productivity.

16-4 Incentives for Inspectors

The functional role of an inspector as compared to an operator leads to some special problems in attracting and retaining qualified inspectors. The problem may be aggravated by the existence of a lower pay scale for inspectors as compared to operators. Three approaches have been used:

1 Provide incentive payment plans for inspection.
2 Make the inspection job a stepping-stone toward promotion, i.e., a career, not a job.
3 Provide nonfinancial incentives for the inspection job. These are in part inherent in the jobs. Operators generally deal with only their direct supervisors; inspectors have a wider range of contacts. The instructions for operators are usually highly standardized; inspectors often develop their own instructions. The range of judgment for operators is more limited than that for inspectors. The fact that the inspector judges the work of the operator puts him "above" the operator. All this gives the inspector a higher place in the social structure of the factory than the operator, whatever may be the job grades, the pay, or other formal measure used. Nonfinancial incentives are important, but they must usually be supplemented by other incentives such as incentive payment and/or promotion opportunities.

The success of piecework incentive plans for operators led to their consideration for inspectors. There were, however, important differences revealed.

1 The greater repetitiveness of production work more readily justified the industrial engineering studies needed to establish standards.
2 Sound means for measuring accuracy of inspection work were not widely understood.

3 There was a long-standing (and well-founded) prejudice against putting inspection work on a piecework payment basis.

Companies generally tried to meet these difficulties by the use of "measured daywork" schemes for inspection.

At Western Electric Co.'s Hawthorne plant, work was measured through standards derived from time study. The incentive was a "measured daywork" system, with rates of inspectors changing every 6 months, based on the performance for the preceding 6 months. Performance efficiency was equal to actual output divided by standard output. When the Juran-Melsheimer system (section 16-6) for measuring accuracy of inspectors was developed, the percentage of accuracy was multipled directly into the percentage of efficiency to compute the new rate.

In the case of patrol inspectors, one approach to quantity incentive has been to work up a historical relationship of patrol inspection time to production labor. Then the standard for patrol inspection is this ratio multiplied into the production labor hours for any given week.

16-5 Accuracy of Inspection

Inspector accuracy is the degree to which the inspector correctly makes decisions on product quality, i.e., accepts product which meets specifications and rejects product which fails to meet specifications. The inaccuracy is not only that bad product is accepted, but also that good product is rejected.

Inspection errors[7] are due to several types of causes:

1 Willful errors which include:
  a Criminal acts such as fraud or collusion
  b Falsification for the personal convenience of the inspector
2 Intermediate errors due to bias, rounding off, overzealousness, etc.
3 Involuntary or inadvertent errors due to blunder, fatigue, and other forms of human imperfection

Major quality errors are usually caused by some engineer, supervisor, or executive rather than an inspector. However, the collective errors of the inspectors may influence many managerial decisions and affect the quality reputation of the company.

For years it was believed that product which was 100 percent inspected would be completely clear of defects. This is not the case. Experience shows that inspectors make involuntary errors that are often

[7] See *QCH,* 1962, sec. 8, for discussion and examples of these errors.

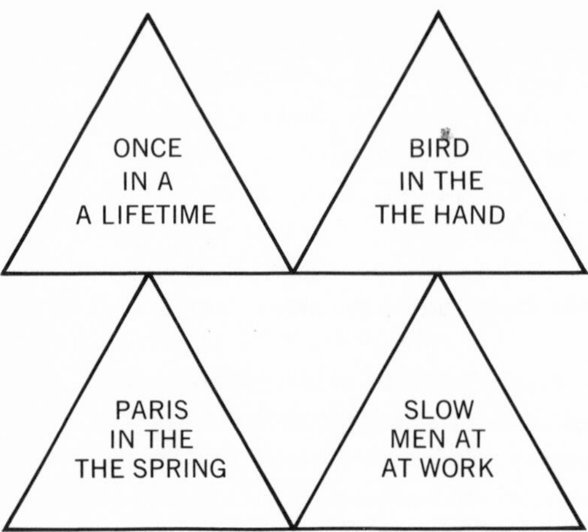

Figure 16-3 Effect of undue familiarity.

significant in size. Although the published evidence is limited,[8,9] it appears that when monotony and fatigue are present, the inspector will detect only about 80 percent of the defects.

Fatigue is one reason that human beings do not find all defects. This can be illustrated by the following sentence:

FEDERAL FUSES ARE THE RESULT OF YEARS
OF SCIENTIFIC STUDY COMBINED WITH
THE EXPERIENCE OF YEARS

The sentence is flashed before an audience for 30 seconds and each person is asked to count the number of times the letter "F" appears. It is usual to find that only about 80 to 90 percent of the "F's" to be found are actually found. Imagine the effect on the inspector who looks at the same product for 8 hours a day.

A similar effect occurs when there is undue familiarity present. Examine the diagrams in Figure 16-3.

There is a mistake in each diagram, but our familiarity with the sayings tends to make us overlook the mistakes.

[8] In two studies of on-wired electrical apparatus, inspectors averaged about 80 to 83 percent accurate on finding solder defects, with great variation between individual inspectors. Data on visual acuity tended to correlate with accuracy, leading to the decision to give prospective inspectors a visual acuity test before hiring. Reported by H. J. Jacobson, "A Study of Inspector Accuracy," *IQC*, vol. 9, no. 2, pp. 16–20, September, 1952.

[9] Alan D. Swain, "Technique for Human Error Rate Prediction," *Proceedings of the Symposium on the Quantification of Human Performance*, University of New Mexico, August, 1964.

Finally, involuntary inspector errors can also be due to the physical limitations of humans. An example of this is the difficulty in detecting high peaks on an oscilloscope during 1-minute intervals consisting of three 20-second sweeps. This and other examples are discussed in a later section of this chapter.

16-6 Evaluating the Accuracy of Inspectors

A variety of plans[10] have been developed to evaluate the accuracy of inspectors. These plans recheck the product after the inspector has finished his inspection. A disadvantage of most plans is that the inspector's accuracy depends on the quality of work submitted to the inspector—a factor which he cannot control. The Juran-Melsheimer plan, however, is independent of incoming quality.

In 1928, one of the authors in collaboration with C. A. Melsheimer designed a plan in which inspector's accuracy in finding defects is expressed by the formula

$$\text{Accuracy} = \text{percent of defects correctly identified} = \frac{d - k}{d - k + b}$$

where
d = number of defects reported by the inspector
k = number of good units rejected by the inspector
$d - k$ = true defects found by the inspector
b = defects missed by the inspector
$d - k + b$ = true defects originally in the product

Figure 16-4 illustrates how the percentage of accuracy is determined. If d is the number of cases reported by the inspector as defective (45, for example) and k is the number of cases reported by the inspector as defective but actually not defective (say 5), then $d - k$ will be the number of true defects identified by the inspector (40 in this instance). Now suppose b is the number of defects missed by the inspector but found by the "quality inspector" in his checking of the former's activity (10, in this example). Then $d - k + b$ will be the number of defects originally present, or the sum of those found by the inspector and those missed by him (50 in this case). Hence

$$\text{Percentage of accuracy} = \frac{d - k}{d - k + b} = \frac{45 - 5}{45 - 5 + 10} = 80$$[11]

In application of the plan, periodic check inspection is made of the inspector's work. Data on d, k, and b are accumulated over a

[10] See *QCH*, 1962, p. 8-30.
[11] Juran, *Management of Inspection and Quality Control*, pp. 53–56.

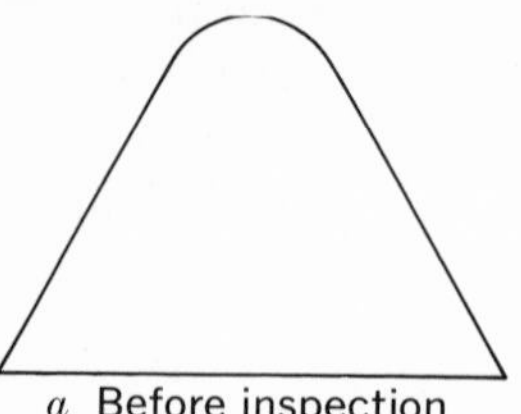

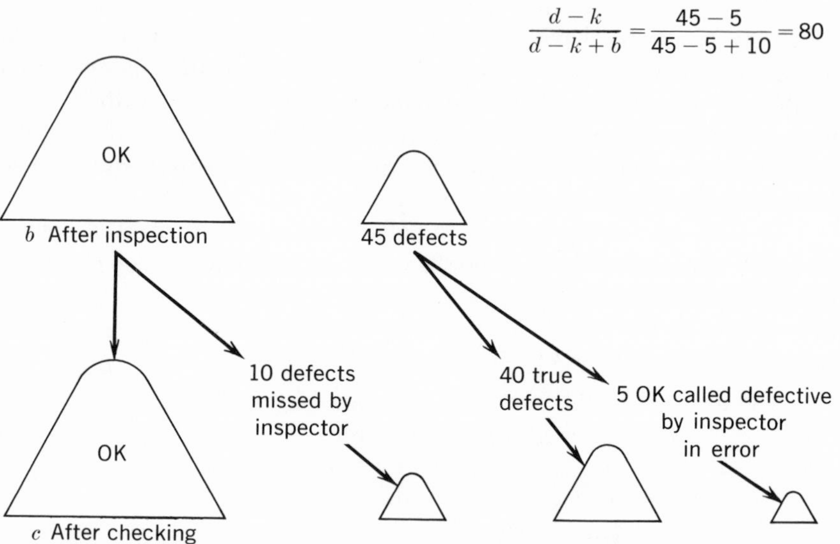

Figure 16-4 Process of arriving percentage of accuracy of inspectors.

period of months to summarize the inspector's accuracy. Some compromise with theory is made to avoid undue emphasis on any one lot checked.

Consider a series of inspections and check inspection as follows:

Job Number	*Total Pieces*	*d*	*b*	*k*
3	1,000	10	0	0
19	50	3	1	0
42	150	5	1	0
48	5,000	10	4	0
.	.	.	.	.
.	.	.	.	.
.	.	.	.	.
Totals		200	30	0

A numerical total would give for percent accuracy:

$$\frac{d}{d+b} = \frac{200}{230} = 87 \text{ percent}$$

In practice, special situations sometimes occur. For example, the various inspection jobs may vary greatly in the inspection time required or an occasional lot may be extremely defective for one characteristic and the inspector may miss it on every piece. Such situations can easily be handled[12] by simple adaptations of the basic formula.

Another approach to defining accuracy considers only defective units missed by the inspector. "Escape probability" is then defined as the probability of a discrepant item passing through established checkpoints with an ensuing system failure as a result. Kidwell et al.[13] describe calculations for escape probability in both the preproduction and production phases of the product life cycle.

The discussion in this chapter concerns the accuracy of the *inspector* and the approaches in this section are evaluations of the inspector. Chapter 28 includes a discussion of sampling plans designed to verify the validity of the results of the entire inspection system, of which the inspector is only a part.

16-7 Improvement of Inspector Accuracy

Any program of improvement of inspector accuracy can make use of the concept of self-control (section 1-12). Chapter 21 discusses in detail the principles of self-control as they apply to the production operator. Much of the detail discussed in Chapter 21 also applies to the inspector. As applied to the inspector, the principles can be stated as follows:

1 The inspector must know what he is supposed to do.
2 The inspector must know how he is doing.
3 The inspector must have the means of regulating his performance.

Further, the inspector must have the state of mind which makes him so desirous of making correct inspection decisions that he will make use of the knowledge of performance and the means regulating the inspection act.

For the inspector to know "what he is supposed to do," he must

[12] See *QCH*, 1962, p. 8-34.

[13] John L. Kidwell, Nicholas L. Squeglia, and H. J. Lavender, "Escape Probability as a System Design Consideration," *IQC*, vol. 23, no. 4, pp. 166–171, October, 1966.

have a specification that is completely clear. Ludwig[14] traces the history of inspection specifications on certain molecular electronic circuits. The first specification was one page in length and consisted of three or four criteria which were not descriptive. It was difficult to reject anything even if "a device looked terrible." Then the pendulum swung finally to a 10-page document consisting of 54 descriptive items. However, the more definitive specification still contained some subjective requirements and required a training period of 8 to 12 weeks. There was so much for the inspector to remember that inspection times increased and efficiency and morale dropped. Worst of all, the escape rate (percent rejects remaining after 100 percent inspection) was uncontrollable. The specification was changed four more times, but merely condensed the 54 items to 39. The program to lower the escape rate consisted of (1) a "lot acceptance" program in which 100 percent inspection was transferred to manufacturing and a lot "buy-off" made by a quality and reliability department, (2) stricter process controls to prevent defects, and (3) a program of "real" defect identification. Ludwig concludes: "It is an absolute necessity to replace human judgment with more precise evaluation tools such as thermal, electrical, and mechanical stress techniques. For the most reliable product, visual inspection must be limited to the vital few and can abandon the trivial many requirements."

For characteristics that cannot be expressed in quantitative form, it may be necessary to use photographs,[15] stereophotographs, or actual physical examples to clearly define the limits of acceptability. The inspector must also have clear instructions on how the inspection is to be performed. As small plants become larger and operations become more complex and varied, the need for formal "inspection instruction sheets" can become critical. These sheets are often regarded as unnecessary paper work and the resistance to a system of formal inspection instructions can be strong. A recent development is the use of videosonic systems (tape recordings combined with slides) to provide instructions to operators and inspectors.

For the inspector to know how he is doing, he must be provided with information that describes his ability to make correct inspection decisions. This may mean some type of check inspection plan as illustrated by the Juran-Melsheimer approach. A periodic "recertification" (by a formal test consisting of actual inspections) of his inspection skills is another method. Whatever the program, it must include con-

[14] H. D. Ludwig, "Microelectronics Visual Inspection—Fact or Fiction," *Transactions of the Twenty-first Annual Conference of the American Society for Quality Control*, 1967, pp. 629–635.

[15] For an example, see *QCH*, 1962, p. 3-65.

structive detailed information to highlight any areas that need improvement.

The problem of regulating the inspection process may require extensive management action. Consider the effect, discussed earlier, of monotony and fatigue on inspector accuracy. To minimize this problem, it may be necessary to purposely vary the kind of work performed by the inspector (job rotation) or to enlarge the scope of his work ("job enlargement") to reduce the monotony and fatigue effect associated with a highly specialized and short-time span task. Shainin[16] described how a sampling procedure "virtually eliminates the fatigue or boredom factor." Further, "the procedure represents considerably more than 80% improvement over the 100% checking in its ability to indicate the presence of individual defective pieces." These are strong statements, particularly when the company involved is making a critical product (aircraft propellers).

Another problem area in some inspection processes is the visual identification of the defect. It may be necessary to magnify the size of defects so that the human senses (visual or other) can consistently detect the defect. Kidwell et al.[17] described a 100 percent inspection acceptance test in which the inspector had to observe peaks on an oscilloscope during 1-minute intervals consisting of three 20-second sweeps. Each sweep contained an average of 20 peaks at irregular intervals. The human eye simply was not capable of following and recording movements on the oscilloscope. The solution was to photograph the oscilloscope patterns and have the inspector review the peaks. (Herrnstein[18] describes some experiments in which pigeons became far more effective in finding defects, in diodes and other products, than human beings.)

Sometimes the inspection equipment is capable of making the inspection but is so inconvenient that it is inevitable that any inspector would be tempted to falsify results. Consider a procedure[19] for inspecting tin cans. Each can must be cut into 20 pieces with a pair of hand-held tin snippers. The pieces are then submerged in chloroform to remove enamel, and finally thickness measurements are taken with a micrometer. Not only is the cutting by hand tedious but the chloroform has a biting effect on hands. The process is capable but would inspection errors be inspector controllable or management controllable?

The motivation of inspectors involves aspects similar to operator

[16] D. Shainin, "The Hamilton Standard Lot Plot Method of Acceptance Sampling by Variables," *IQC*, vol. 7, no. 1, pp. 33–34, July, 1950.
[17] Kidwell, Squeglia, and Lavender, *op. cit.*
[18] Richard J. Herrnstein, "In Defense of Bird Brains," *Atlantic Monthly*, pp. 101–104, September, 1965.
[19] From a term paper.

motivation, i.e., emphasis on job importance, measurement of performance, incentives, etc. These are discussed in Chapter 22.

It is important to know whether problems of inspector accuracy are mainly management controllable or operator controllable because the road to improvement differs markedly. Each company must determine this for itself by evaluating its adherence to the principles of self-control. Management controllable errors will require improvements in inspection plans, equipment, and methods. Operator controllable errors will require the tools of the behavioral sciences.

Another worthwhile distinction to make is the one discussed in Chapter 22 for three types of operator errors: (1) willful errors, (2) errors due to lack of skill, and (3) inadvertent errors. These three types correspond to the three causes of inspector error defined in the first section of this chapter. The distinction may also be helpful in developing remedies for *inspector* errors. Thus, in a willful inspector error, the inspector is deliberately committing the error and the remedy must determine and change the inspector's attitude. Errors due to lack of skill will require measurement to determine who the consistently good and bad inspectors are, diagnosis to find the unique practices of the good inspectors, and retraining of the poor inspectors (and/or change in inspection procedures) to standardize on the most effective practices. Inadvertent errors will require a remedy that makes an error impossible, i.e., foolproofing, redundancy, or some other means.

16-8 Budgeting for Inspection

A budget is a standard for expenditures but, more important, it is an end result of planning. When the planning of deeds (programs, improvements, etc.) is completed, the accounting process "prices" the deeds to arrive at a "budget," which states the money equivalent of these deeds.

The planning phase of the budget provides the main opportunity for increased efficiency. Past practices are reexamined, ideas for change are considered, and opportunities for breakthrough are seized. This is the *cost reduction* phase of budgeting.

Coupled with a system of cost accounting to report actual expenditures, the budget gives each manager a knowledge of what his costs should be, what his costs are, and the responsibility for doing something about the difference. This is the *cost control* phase of budgeting.

For a budget to be effective:

1 The basis of the budget must be clearly understandable by the lowest level of supervision that will be held accountable for performance.

2 There should be at least general agreement with the allowances permitted by the budget. This can best be accomplished by strong supervisory participation in the budget preparation and final decision.
3 Supervisors should be held responsible only for expenditures clearly controllable by them.

Neglect of any of these items defeats effective use of a budget. The budgeting concept with its emphasis on cost reduction and cost control is basic to *all* company activities. Accountants and industrial engineers are heavily experienced in budgeting. The inspection function should obtain the early participation of these specialists not only to avoid mistakes in budget preparation but also to enlist support to have the budget approved by management.

Most inspection budgets are established on the basis of past history, an engineering study, or the market (other companies). The most prevalent is the historical standard; i.e., we will be guided by past experience. For example,[20] past cost data can be analyzed to establish the dollars of inspection labor spent per $100 of direct shop labor. This index can then be applied to the direct labor estimate for new jobs to establish the variable inspection cost for the new job. Fixed costs (e.g., supervision) are added to arrive at a total inspection budget.

Hutter[21] combined a historical approach with 11 factors used to quantitatively evaluate inspection needs for 1 product.
Data from 8 actual projects was gathered in terms of the operator-to-inspector ratio and the total points from the inspection factor evaluation for each project. The results yielded a smooth curve as shown by the solid line in Figure 16-5. The dashed curve represents a 5 percent improvement from historical performance.

On a new project, the inspection factors are evaluated and the total points calculated. The dashed curve in Figure 16-5 provides an estimate of the operator-to-inspector ratio which is then applied to the budgeted number of operators to yield the inspection manpower.

A second basis for standards is the engineered standard; i.e., we will be guided by engineering studies in determining what ought to be spent.

Conventional industrial engineering study methods are applicable to highly repetitive inspection and have in fact been applied to such work. Engineered standards are also common for test work of a more intricate nature, though the variability is, of course, greater.

[20] See *QCH*, 1962, pp. 8A-5 to 8A-6.
[12] R. G. Hutter, "Inspection Manpower Planning," *IQC*, vol. 22, no. 10, pp. 521–523, April, 1966.

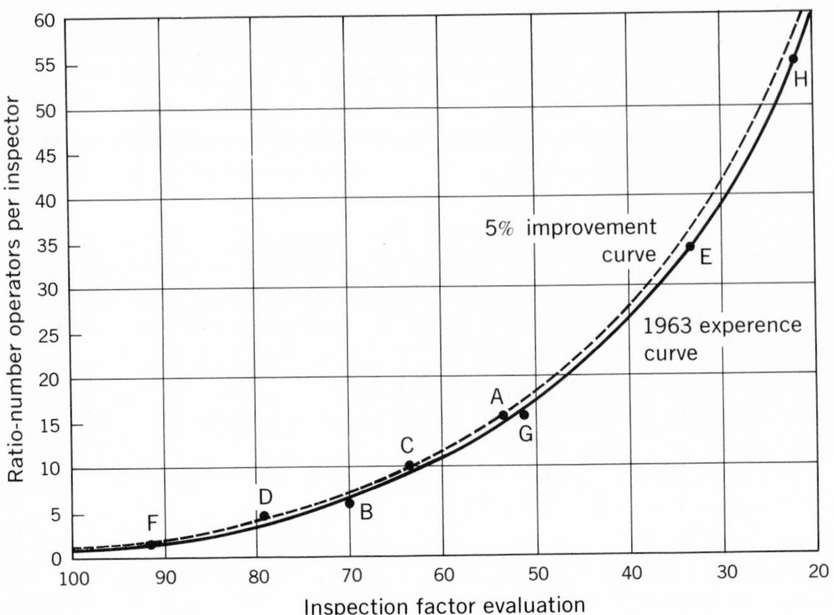

Figure 16-5 Project inspection experience and forecasting curve. Source: R. G. Hutter, "Inspection Manpower Planning," *Industrial Quality Control,* vol. 22, no. 10, pp. 521–523, April, 1966.

A third form of standard is based on the market.[22]

Here the "standard" is what other companies do. The theory is that other companies are faced with similar operating problems. Some of these companies are backward; some are progressive. Collectively they arrive at an assortment of levels of performance which reflects all practice, from the best to the worst. What could be more practical and realistic than to have an awareness of this market performance? Not only would there be some useful numbers against which to make comparison; the numbers would identify where lies the burden of proof. If our specialist department looks good compared to the market, the burden of proof is on those who challenge the specialist. If our specialist department looks poor compared to the market, the specialist can no longer hide behind a jungle foliage of jargon; the burden of proof is on him to justify why he is different.

In the quality function, some pioneering studies have already come over the horizon. A committee of the Aircraft Industries Associa-

[22] J. M. Juran, "What Is Par?" *IQC,* vol. 23, no. 4, pp. 196–197, October, 1966.

tion studied various management practices in "Quality Control" departments, including personnel ratios to production and other activities. Volume II of the Proceedings of the 8th European Organization for Quality Control Conference (1964) consists of a survey of management practices of 43 European companies, including some questions on ratios of inspection personnel to production personnel.

However, a look at the detailed findings of such studies soon reaffirms that anticipation is better than realization. It urns out there are shocking variations among companies, even in the same industry.

Thus, although the concept of market-based standards is sound, more research is needed to provide the data.

16-9 Approaches to Cost Reduction

nspection costs can be reduced by:

Planning of inspection. This was discussed earlier in this chapter.

Defect prevention. One hundred percent inspection caused by poor quality can be eliminated with a program for defect prevention. The potential savings can help sell a prevention program.

Industrial engineering improvement studies. The tools of "work simplification" can be applied to inspection work.

Barry[23] described a plan for simplifying inspection work as follows:

1 Choose one of the more troublesome inspection operations for study.
2 Consider eliminating the operation entirely.
3 Failing this, break the job down by:
   a Listing the work elements.
   b Listing the equipment used.
   c Preparing flow diagrams and workplace sketches.
   d Discussing the job with the inspector to get his ideas.
   e Questioning every detail (a checklist of 33 questions is provided).
4 Prepare a new method as the result of the study.
5 Install the new method, and follow up.

The substitution of sampling plans for 100 percent inspection may be a significant area for cost reduction and may sometimes result in

[23] Elmer N. Barry, "Work Simplification Applied to Inspection," *IQC*, vol. 15, no. 11, pp. 56, 58, May, 1959; also no. 12, pp. 19–20.

a better outgoing quality level. Soth[24] describes this in the inspec tion of paper sheets for defects. One hundred thirty-two girls wer employed to 100 percent inspect the sheets. This inspection re duced the percent defective from 3.5 to 0.04 percent. The primar objective of the original study was to improve quality but analysi revealed it would be impossible to do any better using 100 percen inspection. (One of the several reasons was monotony and fatigue In one day a girl looked at 230,000 edges of 57,500 sheets "whil standing under glaring lights in a huge room full of distractions." A sampling plan was instituted in which 100 percent inspectio was performed only when sampling results indicated the need. A improvement in quality was achieved because the effect of fatigu was reduced and less actual handling of paper resulted in less dam age. This was achieved with a reduction of over 15 percent i the inspection force (a saving of over $100,000 a year).

Census of inspection activity. A broad approach to the improvemen of inspection cost can be made through a census of inspection ac tivity. This is *not* a census of people called inspectors, but o work done to inspect or test product. An example[25] is the worl sampling study in section 16-7.

Choice of Cost Centers and Activity Indexes

For budgeting purposes, the inspection activity should be subdi vided into logical pieces ("cost centers") so as to fix responsibility mor closely and to provide some detail for control. Further, budgets mus recognize that most inspection costs vary with the amount of plant activ ity. This requires the use of an activity index. For example, the inde for receiving inspection might be "number of lots."[26]

The establishment of cost centers, activity indices, and a budge provides a measure of what costs *should* be. To this must be addec provision for measuring actual costs and feeding back the informatio to operating personnel for possible action.

Reports on costs versus budgets are prepared and distributed by the budget office on a regular schedule, usually weekly or monthly Supervisory review is concentrated on significant departure from stan dard. This review is often conducted as a scheduled *group review* of performance against budget, since there is much interrelation betweer causes in one department and effects in another. At these review ses sions, alternate courses of action are proposed and considered.

[24] Roger M. Soth, "A Practical Approach to Statistical Quality Control in Pape Finishing Operations," *Journal of Industrial Engineering,* vol. 13, no. 6, November-December, 1962, pp. 491–495.

[25] See also *QCH,* 1962, p. 8A-3.

[26] For elaboration, see *QCH,* 1962, pp. 8A-4 to 8A-5.

The very act of systematically reviewing performance and questioning variances is a main factor in holding costs to standard. If variances are sure to be questioned, supervisors do their utmost to prevent variances in the first place. If there is laxity in the review, performance will soon become lax as well.

PROBLEMS

1 For any product acceptable to the instructor, prepare an inspection plan, including choice of inspection stations, definition of work to be performed at each station, criteria to be used, facilities to be supplied, and the records to be kept.

2 Discuss the inspection of new homes with the appropriate municipal department. Determine the purpose of their inspection, the specification used, and how the inspection is conducted. Comment on this inspection from the viewpoint of the purchaser of a home.

3 Discuss the inspection of new roads with the appropriate municipal department. Determine the purpose of their inspection, the specification used, and how the inspection is conducted.

4 Your company has an inspector accuracy check on the final inspectors, as well as an independent check inspection. Both of these have been in force for 3 months, and the monthly reports on both are carefully reviewed by management.

In the fourth month, the results of the inspector accuracy check remained at about the same level while check inspection showed a shift to worse quality. (See Figure 16-6.) Arguments have arisen as to the validity of these results. (The data and computations for both charts have already been examined and have been found to be free from arithmetic error, etc.). The main question raised was: "Which of these two charts is correct?"

In order to analyze the situation, what factors should be investigated to get at the roots of the problem? Is it possible for both charts to be correct?[27]

5 Read the Metal Containers, Inc., case in the Appendix and answer the following problem.

[27] Taken from course notes of Management of Quality Control Course, U.S. Air Force School of Logistics, 1959.

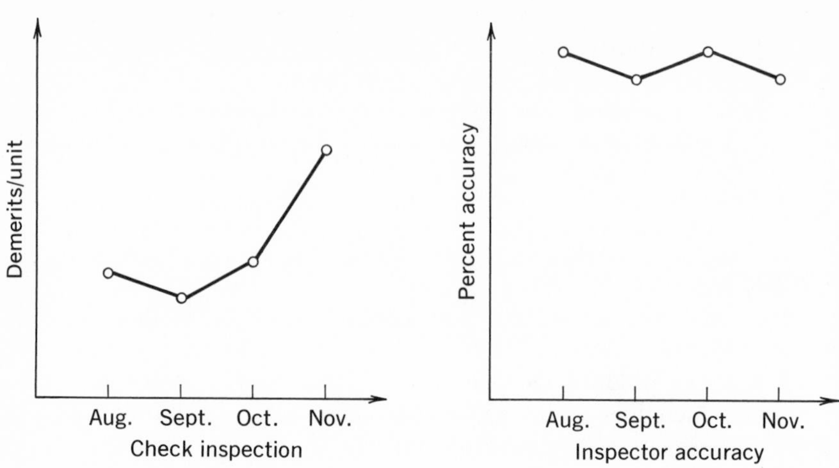

Figure 16-6 Results of check inspection and inspector accuracy. Source: Course notes of Management of Quality Control Course, U.S. Air Force School of Logistics, 1959.

You are Lafferty, the quality control manager. When Wallace hired you, one of his comments was as follows:

> One thing I'd like to know is how many inspectors do we really need, and what should they be doing. Whenever we have a blow-up of some sort, the quality control supervisors have a long list of causes, and invariably, they include in that list "not enough inspectors."
>
> It's amazing to me how the plants manage on such different plans. At Cleveland, 1 man in 20 is in the quality control department. At Milwaukee we have 400 people and 20 are in the quality control department—again 1 in 20. At Detroit, it's only 1 in 25. At Chicago it's 1 in 11. Are these differences justified? If we could operate them all like Detroit we could reduce the payroll by 15 people. It isn't only how many. What are we getting? Are we really getting control of quality? Are these people really working on what is important to the company? You are supposed to be the expert, so after you've had a look around, you give me some answers.

Now you have had your look around. What you have seen certainly doesn't square with your own experience. You have had three previous jobs in 16 years, as follows:

a Test engineer in the motor division of a huge electrical company. There everything was spelled out in great detail—job descriptions, test procedures, records, and the rest.

b Chief of test for a medium-sized independent making electrical motors—a competitor of your previous company. (You had published a technical paper which attracted attention.) In your new job you found good use for all that procedural know-how. They had little of it, and they needed it. You supplied it, and all agreed you had done them a lot of good.

c Chief of test for a company making air-conditioners. (They were customers of your previous company, and you had gotten acquainted while on official business.) By now you were spending some of your time in management of the quality function. You attended some QC conventions and seminars. At one of these you met Wallace.

From all this experience, you had come to feel that managing the inspection job should be done by:

a Making a flowsheet of the process, and identifying the proper control stations. In MC the key people know the flow fairly well. The inspectors know their areas fairly well. Control stations, while not spelled out, are generally identified.

b Preparing inspection manuals, instruction sheets, job descriptions, patrol beats, and other forms of detailed lists of duties and responsibilities. Except at Chicago, MC has virtually none of this. Mostly people have these things in their heads. You are uncomfortable about this—you doubt that people are really doing the things they say they are doing.

c Determining, by estimate or time study, how many man-hours are needed to carry out these duties, and justifying the manpower that way. MC has not done this, and there is a wide variation in quality control manpower, whether related to sales, production, or production manpower.

d Keeping records for feedback of quality data. Except at Chicago, there is little of this.

e Auditing to see that the schedules of inspection are adhered to, records kept, etc. This concept is new to MC.

f Analysis of records, plus supplemental studies, for quality improvement. In MC there are few records to analyze, and the work on quality improvement has been done by the production supervision and the manufacturing engineers, with little help from quality control.

Several other things bother you. The plants are managing to operate without any incoming inspection of tinplate other than the incidental visual look taken by the lithographers and shear operators. There is no quality control at shipping. On the other hand, all plants have a vague number of sorters on the production payroll. Some of these sorters are regulars on the lacquer line. Other sorting is done as part of the means of getting out of a mess. Generally, the production payroll is not subject to as serious challenge as the nonproduction payroll.

The Chicago plant is a puzzler. If all plants were to operate on such a formalized basis, it would be necessary to add over 40 people to the payroll. You are sure no one would buy that. You have grave doubt that the Chicago charts are doing the company any good.

The Milwaukee plant manager did exhibit some interest in inspection planning. He would like to see the entire process go further down the road to a science rather than being left to the judgment of the mechanics. He points out that whereas 2 bad cans out of 1,000 used to be tolerated by the canners and brewers, their automation has proceeded to a point that they now think in terms of 25 bad cans per million as a tolerable level.

Finally, you consider your personal responsibilities. You are now top quality dog in a company, for the first time. But you do not have line authority over the quality control supervisors—not yet anyway. You are a staff man, and you are to help the plants do a better, more economic quality job. You are also to help Wallace and the other people at headquarters to understand better what the company should do with respect to quality. You are aware that you have never been in this business before, that these plants are small compared to those in which you have operated, and that there are other differences which may trap you into doing something stupid.

What do you propose be done with respect to managing the quality control departments?

17 Acceptance Sampling

17-1 The Advantages of Acceptance Sampling

Acceptance sampling is the process of evaluating a portion of the product in a lot for the purpose of accepting or rejecting the entire lot as either conforming or not conforming to a quality specification.

Acceptance sampling can be performed by using either attributes measurements or variables measurements. This chapter presents the background and examples of plans for each of the two types.

The main advantage of sampling is economy. Despite some added costs to design and administer the sampling plans, the lower costs of inspecting only part of the lot result in an overall cost reduction. In addition to this main advantage there are others:

1 The smaller inspection staff is less complex and less costly to administer.
2 There is less damage to the product; i.e., handling incidental to inspection is itself a source of defects.
3 The lot is disposed of in shorter (calendar) time so that scheduling and delivery are improved.
4 The problem of monotony and inspector error induced by 100 percent inspection is minimized.
5 Rejection (rather than sorting) of nonconforming lots tends to dramatize the quality deficiencies and to urge the organization to look for preventive measures.
6 Proper design of the sampling plan commonly requires study of the actual level of quality required by the user. The resulting knowledge is a useful input to the overall quality planning.

Sampling involves risks but the advantages are significant.

17-2 The Concept of Sampling

The measurements in a sample provide two types of information:

1 Information about the product in the sample. This permits decisions about the product in the sample itself.
2 Information about the variables present in the manufacturing process. This leads to conclusions about the *process* at the time the sample was taken.

The knowledge of the process, in turn, gives information about the uninspected product. Acceptance sampling is valid because the uninspected product came from the same process which produced the sample of inspected product. Of course, there are risks.

Suppose that we wish to accept all lots which are not more than 3 percent defective, and that we wish to reject all lots which are over 3 percent defective. Along comes a lot of 1,000 pieces, of which 4 percent are actually defective. An inspector takes a sample of 20 pieces and finds none defective, and so he accepts the lot (the sample shows 0 percent defective). Clearly, the sampling plan has led to a wrong decision, since the lot is actually 4 percent defective.

Actually, the sample of 20 pieces could, by the luck of probability, have contained 1, 2, etc., even 20 defectives. This *sampling variation* is critical in the design of sampling plans. Through bad luck in sampling, there can be two errors in decision making:

1 A good lot may be rejected.
2 A bad lot may be accepted.

Modern design of sampling plans quantifies these risks and adjusts the sampling criteria to balance these risks in the light of the economic factors involved.

17-3 Economics of Sampling versus 100 Percent Inspection

An economic evaluation of sampling versus 100 percent inspection requires a comparison of *total* costs under each of the two procedures. Suppose it is assumed that no inspection errors occur and the cost to replace a defective found in inspection is borne by the producer or is small compared to the damage done by a defective in use. Then if

N = number of items in lot
n = number of items in lot sample
p = proportion defective in lot
A = damage cost incurred if a defective slips through inspection
I = inspection cost per item

total cost with 100 percent inspection = NI and total cost with sampling = $nI + (N - n)\,pA$. If these two costs are equated, a break-even point may be found in terms of proportion defective:

$$NI = nI + (N - n)pA$$

$$p_b = \frac{(N - n)I}{(N - n)A} = \frac{I}{A}$$

If the proportion defective in the lot is less that p_b, the total cost will be lowest with sampling. If p is greater than p_b, then using 100 percent inspection will result in a lowest total cost.

For example, consider the inspection of pieces of mica film used in condensers. Pinholes or conducting inclusions in the mica cause a short circuit in the finished condenser. It costs 0.1 cent to inspect one piece. If a bad piece of mica is allowed to go into a condenser, a damage cost of 10 cents is incurred, because the total cost of manufacturing the condenser is lost when it is ruined by the defective piece of mica. Therefore,

$$p_b = \frac{0.1}{10} = 0.01 = 1.0 \text{ percent}$$

If it is expected that the percent defective will be greater than 1.0 percent, then 100 percent inspection should be used. Otherwise, sampling should be instituted.

The variability in quality from lot to lot is important. If past history shows that the quality level is much better than the break-even point and is stable from lot to lot, little if any inspection may be needed. If the level is much worse than the break-even point, and consistently so, it will usually be cheaper to use 100 percent inspection rather than sample. If the quality is at neither of these extremes, then sampling will pay for itself.

17-4 Sampling Risks: The Operating Characteristic Curve

There is no way known, whether by sampling or by detailed (human) inspection, to be 100 percent sure that the material accepted is entirely

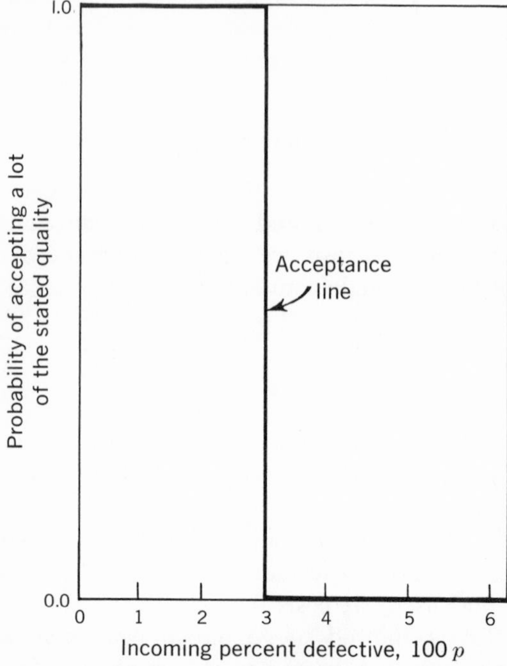

Figure 17-1 An ideal sampling plan performance.

free of defects. Sampling involves the risk that the sample will not reflect conditions in the lot. Detailed inspection, aside from its expense, involves the risk that inspectors will not find all defects (see Chapter 16).

Two types of errors can be made with sampling plans:

1 Good lots can be rejected (the producer's risk). This risk corresponds to the α risk discussed in Chapter 11.
2 Bad lots can be accepted (the consumer's risk). This risk corresponds to the β risk discussed in Chapter 11.

The operating characteristic (OC) curve for a sampling plan quantifies these risks. The OC curve for an attributes plan is a graph of the fraction defective in a lot versus the probability that the sampling plan will accept a lot which has specified fraction defective. Figure 17-1 shows an "ideal" OC curve where it is desired to accept all lots 3 percent defective or less and reject all lots having a quality level greater than 3 percent defective. All lots less than 3 percent defective have a probability of acceptance of 1.0 (certainty); all lots greater than

3 percent defective have a probability of acceptance of 0. Actually, however, no sampling plan exists that can discriminate perfectly; there always remains some risk that a "good" lot will be rejected or that a "bad" lot will be accepted. The best that can be achieved is to make the acceptance of good lots more likely than the acceptance of bad lots.

Figure 17-2 shows the OC curve if an inspector were instructed to take from a large lot a sample of 150 pieces and to accept the lot if he found no more than 4 defective pieces.

A 3 percent defective lot has one chance in two of being accepted. However, a lot 3.5 percent defective, though technically a "bad" lot, has 39 chances in 100 of being accepted.[1]
In like manner, a lot 2.5 percent defective, though technically a good lot, has 34 chances in 100 of not being accepted.

[1] The probability of 0.39 only applies when *and if* a quality level of 3.5 percent is submitted. As this is "bad" quality, relatively few such lots should be submitted by the producer. Therefore, only 39 percent of the few bad lots submitted will be accepted.

Figure 17-2 An actual sampling plan performance.

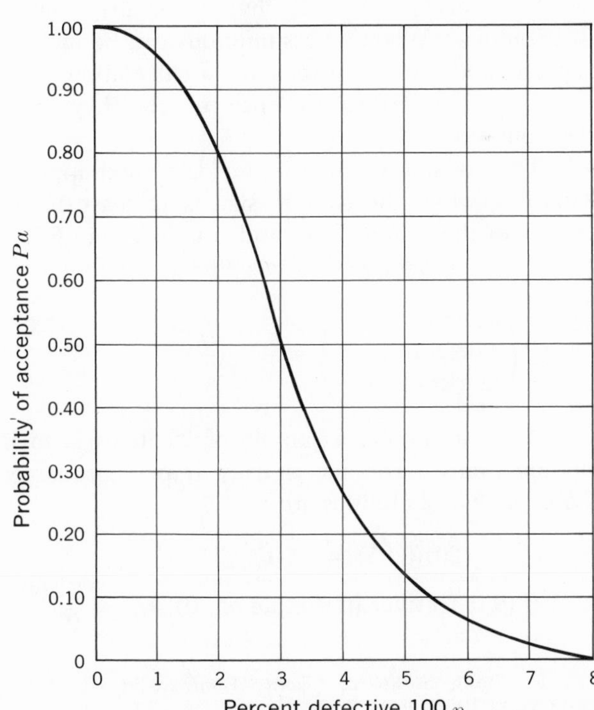

The lack of a sharp distinction may not be too serious. Economically, it is a matter of indifference whether a lot 3 percent defective is accepted or sorted. For lots 2.5 or 3.5 percent it makes a difference, but the difference is not great in relation to the cost of obtaining greater precision. Finally, there are imponderables which are still present in computing break-even points.

It should be noted that the curve does *not* predict the quality of the lots submitted for inspection. For example, it is *wrong* to say that there is a 39 percent chance that the lot quality is 3.5 percent defective. The curve merely tells the chance of accepting lots which are at any given fraction defective before inspection.

17-5 Constructing the Operating Characteristic Curve

An OC curve can be developed by determining the probability of acceptance for several values of incoming quality, p. The probability of acceptance is the probability that the number of defectives in the sample is equal to or less than the acceptance number for the sampling plan. There are three distributions that can be used to find the probability of acceptance: the hypergeometric, binomial, and the Poisson distribution. When its assumptions can be met, the Poisson distribution is preferable because of the ease of calculation.

Grant[2] describes the use of the hypergeometric and binomial distributions.

The Poisson distribution yields a good approximation for acceptance sampling when the sample size is at least 16, the lot size is at least ten times the sample size, and p is less than 0.1. The Poisson distribution function as applied to acceptance sampling is

$$P\begin{pmatrix}\text{exactly} \\ r \text{ defectives} \\ \text{in sample of } n\end{pmatrix} = \frac{e^{-np}(np)^r}{r!}$$

As an example, a sample of 20 items is to be selected from a lot of 400 items. The lot is truly 5 percent defective or p is equal to 0.05. The probabilities are

$$np = (20)(0.05) = 1.00$$

$$P\text{ (0 defectives in sample of 20)} = \frac{e^{-1.00}(1.00)^0}{0!} = 0.368$$

[2] E. L. Grant, *Statistical Quality Control,* 3d ed., McGraw-Hill Book Company, New York, 1964, chap. 9.

$$P\text{ (1 defective in sample of 20)} = \frac{e^{-1.00}(1.00)^1}{1!} = 0.368$$

$$P\text{ (2 defectives in sample of 20)} = \frac{e^{-1.00}(1.00)^2}{2!} = 0.184$$

Suppose the lot is to be accepted if 2 or fewer defectives are found in the sample. The probability of acceptance is then equal to $0.368 + 0.368 + 0.184 = 0.920$.

Table J in the Appendix provides values of the probability of r or fewer occurrences in a sample of n taken from a lot having quality p. In the above example, $np = 1.00$. From Table J, the probability of 2 or fewer defectives is read directly as 0.92.

17-6 Analysis of Some Rule-of-thumb Sampling Plans

No sampling plan should be adopted without first seeing its OC curve. "Rule-of-thumb" sampling plans are frequently found to be inadequate when evaluated by an OC curve. The "10 percent sampling rule" is an example. For any lot size, a sample equal to 10 percent of the lot is selected. If there are no defectives found in the sample, the lot is accepted. If any defectives are found in the sample, the lot is rejected. Figure 17-3 shows the operating characteristic curve for this rule-of-thumb plan based on four different lot sizes. Notice that the sample is 10 percent of the lot for each of the plans. It is usually presumed that with the sample always a constant percentage (10 percent) of the lot, the sampling risks will be constant. Figure 17-3 shows that this is not so. Suppose a 4 percent defective lot is submitted to such a plan. If the lot size is 50, the probability that the lot will be accepted by the plan is 80 percent. However, if a lot of 200 is submitted (and is 4 percent defective), then the probability of acceptance drops to about 44 percent. Similarly, a 4 percent lot of 1,000 pieces would have only about a 3 percent chance of the lot passing the acceptance plan. Thus, the 10 percent rule does *not* provide the same risks (not even approximately).

Another rule-of-thumb plan is an extreme one that might be proposed for a destructive test. The plan is this: Test one unit, and if it is acceptable, pass the entire lot. If it is defective, test a second unit. If the second unit is acceptable, pass the lot. If the second unit is also defective, reject the lot. Figure 17-4 shows the operating characteristic curve. Destructive testing necessitates small sample sizes but the operating characteristic curve shows that the sampling risks are so high for this plan that there is little protection against accepting bad

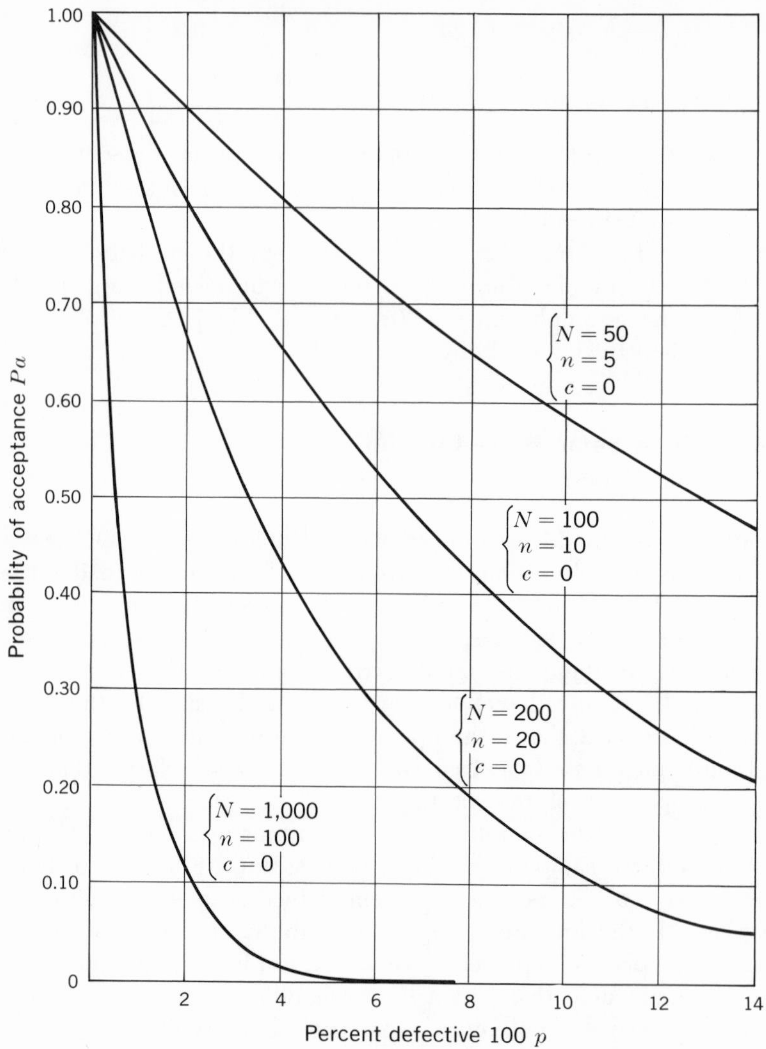

Figure 17-3 Comparison of operating characteristics curves for four sampling plans involving 10 percent samples. Source: E. L. Grant, *Statistical Quality Control,* 3d ed., McGraw-Hill Book Company, New York, 1964, p. 334.

quality. For example, if the lots submitted were about 70 percent defective, the probability of accepting such lots is 50 percent.

The point in showing the inadequacy of these rule-of-thumb plans is to emphasize that the intuitive approach can be so grossly in error as to seriously jeopardize the evaluation of a product. By deriving the operating characteristic curve the sampling risks become known. Even when these sampling risks are known, other considerations may dictate that a sampling plan be used which is inadequate from a statistical point of view. However, it is important that the size of the risk be known before a final judgment is made.

17-7 Evaluation of Parameters Affecting Acceptance Sampling Plans

Sampling risks are affected by lot size, sample size, and the acceptance number (see Figure 17-5). In Figure 17-5*a* the lot size is changed but the sample size and acceptance number are held constant. Notice that the lot size has little effect on the probability of acceptance. If the lot size is moderately large (at least ten times the sample size) it has a small effect on the probability of acceptance. It is tempting to conclude that lot size may be ignored in deriving a sampling plan.

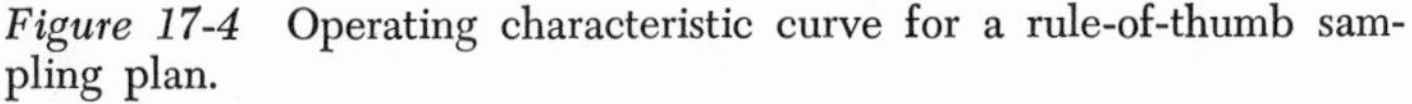

Figure 17-4 Operating characteristic curve for a rule-of-thumb sampling plan.

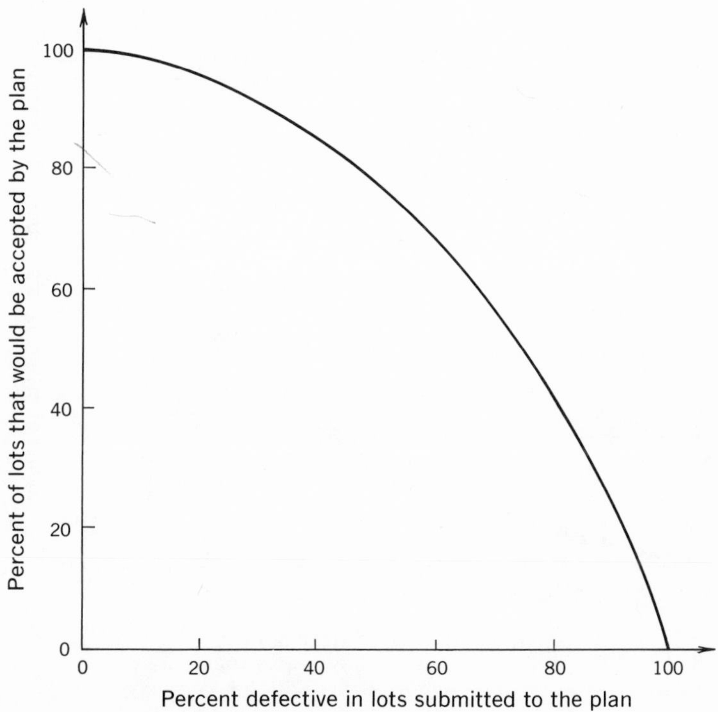

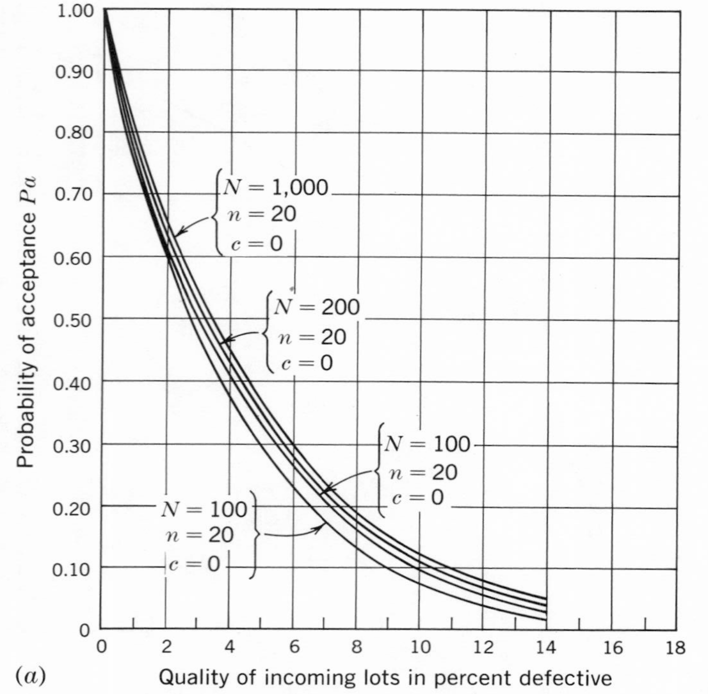

(*a*)

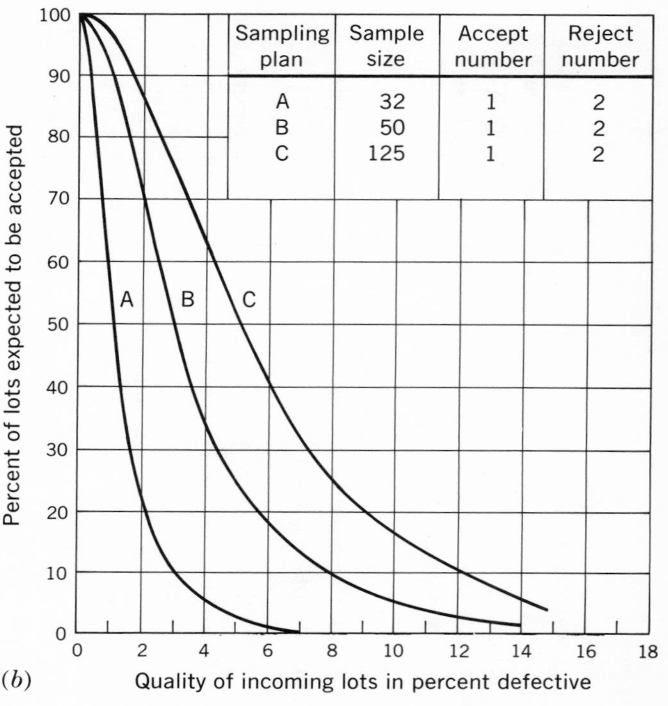

Sampling plan	Sample size	Accept number	Reject number
A	32	1	2
B	50	1	2
C	125	1	2

(*b*)

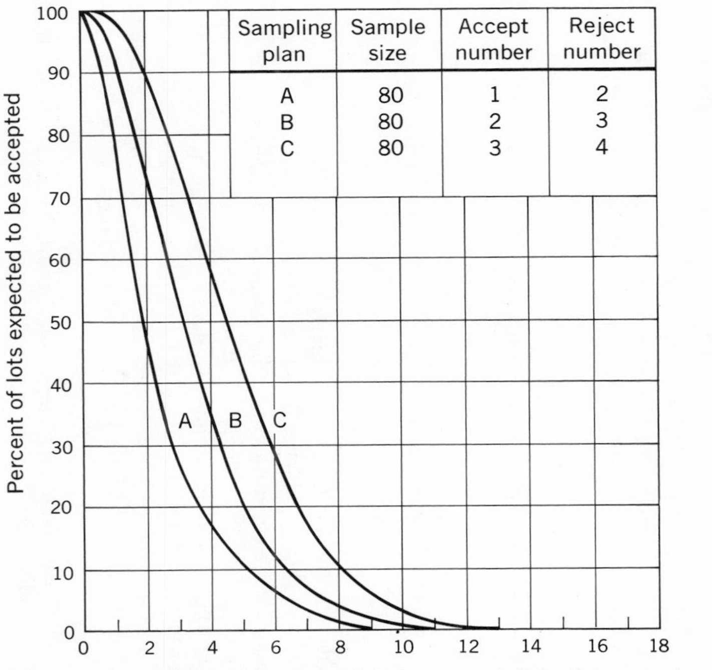

(c)

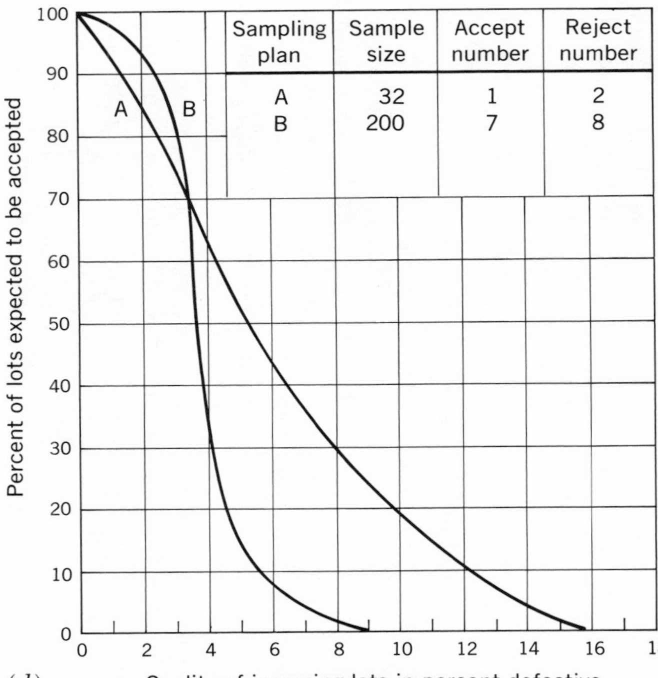

(d)

Figure 17-5a Effect of changing lot size on OC curve. Source: E. L. Grant, *Statistical Quality Control,* 3d ed., McGraw-Hill Book Company, New York, 1965, p. 335. *b* Effect of changing sample size on OC curve. Source: *Guide for Sampling Inspection,* Handbook H53, Government Printing Office, 1965. *c* Effect of changing acceptance number on OC curve. Source: *Guide for Sampling Inspection,* Handbook H53, Government Printing Office, 1965. *d* Effect of simultaneous change of sample size and acceptance number on OC curve. Source: *Guide for Sampling Inspection,* Handbook H53, Government Printing Office, 1965.

However, it is usually desired that larger lot sizes have a better operating characteristic curve in order to reduce the risk of error for large amounts of product. Therefore, most sampling tables do show lot size as a parameter. (As the lot size increases, the absolute sample size increases but the ratio of sample size to lot size decreases.) The fact remains, however, that if everything else is constant, the lot size has essentially no effect on the probability of acceptance. Figure 17-5*b* shows the effect of changing sample size with all other things constant. As the sample size gets larger and larger, the operating characteristic curve becomes steeper; that is, the probability of acceptance becomes smaller. This simply means that the larger the number of items in the sample, the more possibilities of finding enough defectives to exceed the acceptance number. Figure 17-5*c* shows the effect of changing only the acceptance number. As the acceptance number increases, the operating characteristic curve gets flatter; that is, the probability of acceptance increases. The larger the number of allowable defectives, the more the chance that an acceptance decision will be made. Figure 17-5*d* shows the effect of varying both the sample size and the acceptance number. Notice how the curve starts to approach the ideal OC curve when the acceptance number is changed from zero. This is generally the case. However, when the acceptance number is made larger than zero, then the sample size must be increased to achieve probability levels that provide adequate protection.

These aspects have been investigated and the conclusions incorporated in the published sampling tables discussed in later sections.

17-8 Quality Indices for Acceptance Sampling Plans

Many of the published plans can be categorized in terms of one of several quality indices:

1 *Acceptable quality level(AQL).* This is usually defined as the worst quality level that is still considered *satisfactory.* The units of quality level can be selected to meet the particular needs of a product. Thus, MIL-STD-105D[3] defines AQL as "the maximum percent defective (or the maximum number of defects per hundred units) that, for purposes of sampling inspection, can be considered satisfactory as a process average." If a unit of product can have a number of different defects of varying seriousness, then demerits can be

[3] *Sampling Procedures and Tables for Inspection by Attributes,* MIL-STD-105D, Government Printing Office, 1963.

assigned to each type of defect and product quality measured in terms of demerits. As an AQL is an *acceptable* level, the probability of acceptance for an AQL lot should be high (see Figure 17-6).

2 *Rejectable quality level (RQL)*. This is a definition of *unsatisfactory* quality. Different titles are sometimes used to denote an RQL; for example, in the Dodge-Romig plans, the term "lot tolerance percent defective (LTPD)" is used. As an RQL is an *unacceptable* level, the probability of acceptance for an RQL lot should be low (see Figure 17-6). In some tables, this probability is known as the consumer's risk, is designated as P_c, and has been standardized at 0.1.

The consumer's risk is not the probability that the consumer will actually receive product at the RQL. The consumer will in fact *not* receive 1 lot in 10 at RQL fraction defective. What the consumer actually gets depends on actual quality in the lots *before* inspection, and on the probability of acceptance.

3 *Indifference quality level (IQL)*. This is a quality level somewhere between the AQL and RQL. It is frequently defined as the quality level having a probability of acceptance of 0.50 for a given sampling plan (see Figure 17-6).

Figure 17-6 Quality indices for sampling plans.

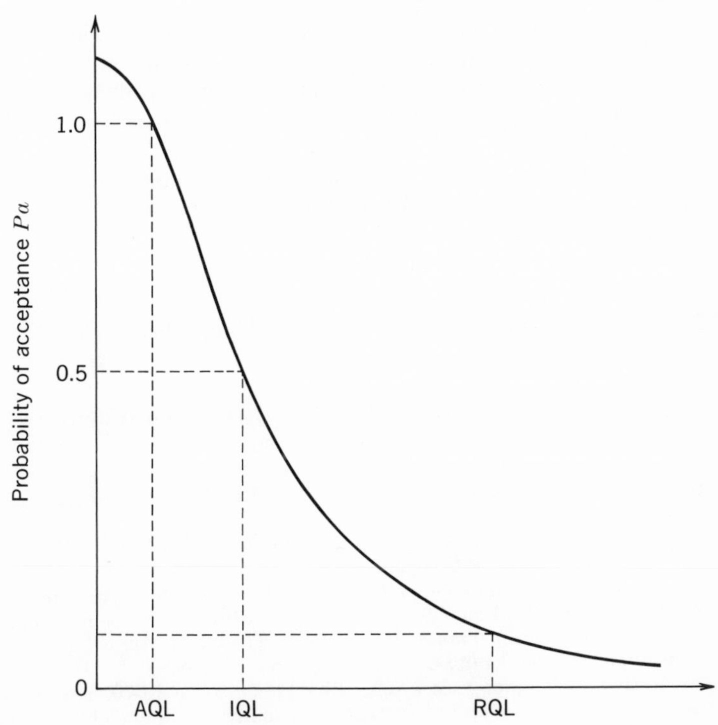

4 *Average outgoing quality limit.* A relationship exists between the fraction of defectives in the material before inspection (incoming quality p) and the fraction of defectives remaining after inspection (outgoing quality AOQ): AOQ $= pP_a$. Obviously, when incoming quality is perfect, outgoing quality must likewise be perfect. However, when incoming quality is very bad, outgoing quality will also be perfect,[4] because the sampling plan will cause all lots to be rejected and detail inspected. Thus at either extreme—incoming quality very good or very bad—the outgoing quality will tend to be very good. Between these extremes is the point at which the percent of defectives in the outgoing material will reach its maximum. This point is known as the average outgoing quality limit (AOQL).[5]

17-9 Types of Sampling Plans

Sampling plans are of two[6] types:

1 Attributes plans. A sample is taken from the lot and each unit classified as good or bad. The number bad is then compared with the allowable number stated in the plan, and a decision is made to accept or reject the lot. This chapter will illustrate attributes plans based on AQL, RQL, and AOQL.

2 Variables plans. A sample is taken and a *measurement* of a specified quality characteristic is made on each unit. These measurements are then summarized into a simple statistic (e.g., sample average) and the observed value compared with an allowable value defined in the plan. A decision is then made to accept or reject the lot. This chapter will describe an AQL variables plan.

In both cases, it is assumed that the sample is drawn in a random manner from the lot. This essentially means that each possible sample has an equal chance of being drawn. The lot is assumed to consist of product of homogeneous quality.[7]

17-10 Comparison of Attributes and Variables Plans

A comparison of attributes and variables sampling is given in Table 17-1.

[4] It is assumed that inspection is perfect, i.e, that the inspector finds all the defects which may exist in the pieces he inspects.

[5] For a sample calculation, see J. M. Juran (ed.), *Quality Control Handbook (QCH)*, 2d ed., McGraw-Hill Book Company, New York, 1962, p. 13-83.

[6] It is possible to combine an attributes and a variables plan into one. See, for example, John H. K. Kao, "Simple—Sample Attri-Vari Plans for Item—Variability in Percent Defective." *Transactions of the Twentieth Annual Conference of the American Society for Quality Control*, 1966, pp. 743–758.

[7] For further elaboration, see *QCH*, 1962, pp. 13-78, 13-79, and 24-16 to 24-18. Also see *Guide for Sampling Inspection*, Quality and Reliability Assurance Handbook H53, Government Printing Office, 1965.

Table 17-1 COMPARISON OF ATTRIBUTES AND VARIABLES SAMPLING PLANS

	Attributes	*Variables*
Type of inspection required for each item	Each item classified as defective or acceptable. Go-not-go type of gages may be used	Measurement must be taken on each item. Higher skill level of inspection required
Size of sample		Saving of at least 30%* in sample size (if only one characteristic must be measured on each item and if single sampling is used)
Assumption of underlying distribution	None	Some distribution must be assumed (usually normal)
Number of characteristics that can be reviewed in one sample	Any number	A separate sampling plan is required for each characteristic to be reviewed
Type of information provided for use in correcting process	Number of defectives (if go-not-go gages are used)	Valuable information on the process average and variation is available to indicate type of process correction required

* See A. H. Bowker and H. P. Goode, *Sampling Inspection by Variables*, McGraw-Hill Book Company, New York, 1952, pp. 32–33.

The key advantage of a variables sampling plan is the additional information provided in each sample which in turn results in smaller sample sizes as compared with an attributes plan having the same risks. However, if a product has several important quality characteristics, each must be evaluated against a separate variables acceptance criterion (e.g., obtain numerical values and calculate the average and standard deviation for each characteristic). In a corresponding attributes plan, the sample size required may be higher but the several characteristics could be treated as a group and evaluated against one set of acceptance criteria.

17-11 Single Sampling, Double Sampling, and Multiple Sampling

In single-sampling plans, the decision to accept or reject a lot is based on the results of inspection of a single group of specimens drawn from the lot (Figure 17-7). In double-sampling plans (Figure 17-8), a smaller initial sample is usually drawn, and a decision to accept or reject is reached on the basis of this smaller first sample, if the number of defectives is either quite large or quite small. A second sample is taken if the results of the first are not decisive. Since it is only necessary to draw and inspect the second sample in borderline cases, the average number of pieces inspected per lot is generally smaller with double sampling. In multiple-sampling plans, one or two or several still smaller samples are taken, usually continuing as needed, until a decision to accept or reject is obtained. Thus, double- and multiple-sampling plans *may* mean less inspection but are more complicated to administer.

In general, it is possible to derive single-, double-, or multiple-sampling schemes with essentially identical operating characteristic curves.

Figure 17-7 Schematic operation of single sampling.

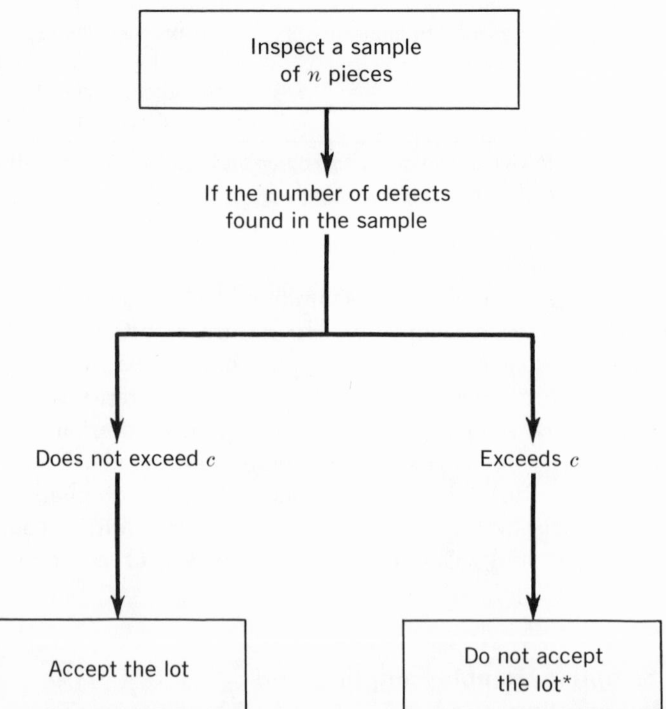

* In practice the lot may be repaired, junked, etc. Sampling tables usually assume that the lot is detail inspected and that the defective pieces are all repaired or replaced by good pieces.

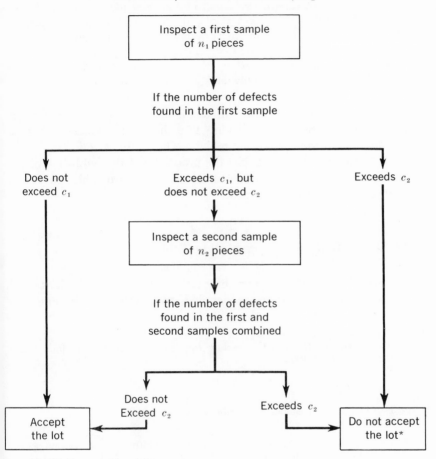

* Inspect the remainder of the pieces, replacing or repairing defective pieces.
Figure 17-8 Schematic operation of double sampling.

17-12 Characteristics of a Good Acceptance Plan

An acceptance sampling plan should have these characteristics:

1 The index (AQL, AOQL, etc.) used to define "quality" should reflect the needs of the consumer and producer and not be chosen primarily for statistical convenience.
2 The sampling risks should be known in quantitative terms (the OC curve). The producer should have adequate protection against the rejection of good lots; the consumer should be protected against the acceptance of bad lots.

3 The plan should minimize the *total* cost of inspection of all products. This requires careful evaluation of the pros and cons of attributes and variables plans, and single, double, and multiple sampling.
4 The plan should have built-in flexibility to reflect changes in lot sizes, quality of product submitted, and any other pertinent factors.
5 The measurements required by the plan should provide information useful in estimating individual lot quality and long-run quality.

Fortunately, published tables are available which meet these characteristics for many applications. Figure 17-9 shows the sequence of steps in choosing a plan from published tables. The double arrows emphasize the need to balance desires against the properties of the published plans. Three basic plans will be discussed here. There are other published plans developed for specific purposes.[8]

17-13 Use of Dodge-Romig Sampling Tables[9]

Dodge and Romig provide four different sets of attributes plans. All lots rejected are assumed to be detailed, and both the sampling and

[8] See *QCH*, 1962, pp. 13-84 to 13-117.
[9] H. F. Dodge and H. G. Romig, *Sampling Inspection Tables*, 2d ed., John Wiley & Sons, Inc., New York, 1959.

Figure 17-9 Choosing a sampling plan from published tables.

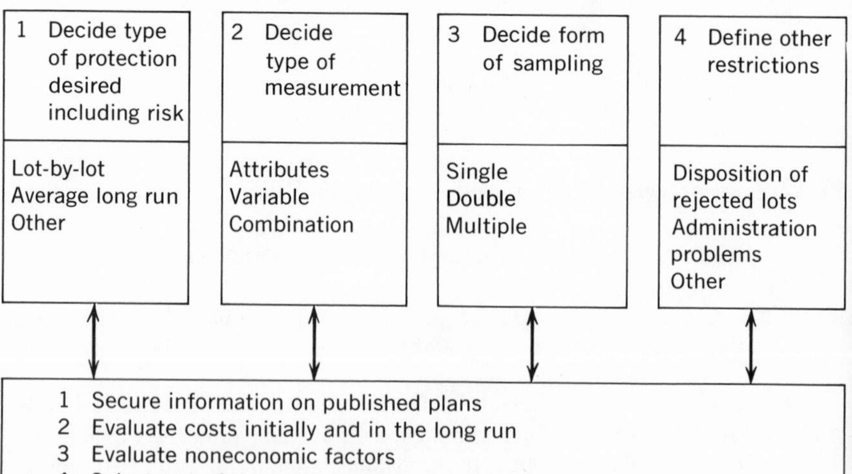

the 100 percent inspection were considered in finding the plan which would give minimum inspection per lot. This is appropriate for a manufacturer's inspection of his own product. The same theory applies where the sampling is done by the purchasing company and the detailing of rejected material is done by the supplying company, since in the long run all the supplier's costs to provide material of a specified quality are reflected in his price.

The Dodge-Romig plans provide emphasis on either lot-by-lot quality or long-run quality. The first and second sets of tables are classified according to lot tolerance percent defective p_t at a constant consumer's risk of 0.10. Available lot tolerance plans range from 0.5 to 10.0 percent defective. The third and fourth sets of tables are classified according to the average outgoing quality limit (AOQL) which they assure. Available AOQL values range from 0.1 to 10.0 percent. Lot tolerance plans emphasize a constant low consumer's risk (with varying AOQLs) and thus assure that individual lots of poor material will seldom be accepted. The AOQL plans emphasize the limit on poor quality in the long run but do not attempt to offer uniform protection against individual lots of low quality.

Table 17-2 shows a Dodge-Romig table for single sampling on the lot tolerance basis. All the plans listed in this table have the same risk (0.10) of accepting submitted lots that contain exactly 5 percent of defective articles. For example, if the estimated process average percent defective is between 2.01 and 2.50 percent, the last column at the right gives the plans that will provide the minimum inspection per lot. However, the probability that a lot of quality p_t will be rejected is the same for all columns, so that an initial incorrect estimate of the process average would have little effect except to increase somewhat the total number of pieces inspected per lot. The selection of a plan thus requires only two items of information: the size of lot to be sampled and the prevailing average quality of the supplier for the product in question.

Process average is determined from past records, modified by any supplemental knowledge useful for predicting the expected quality level.

A double-sampling plan for the same lot tolerance (provided in another table) will show that for comparable situations, fewer articles will be inspected under double sampling, provided that a decision can be reached on the basis of the first sample. If a borderline quality requires drawing a second sample, the total number of pieces inspected will be greater for double sampling than for single sampling.

Table 17-3 shows a typical table of AOQL plans using double sampling. In contrast to the lot tolerance tables, this table gives plans which differ considerably as to lot tolerance but which have the same

Table 17-2 Single Sampling Table for Lot Tolerance Percent Defective (LTPD) = 5.0%*

Lot Size	Process Average 0 *to* 0.05%			Process Average 0.06 *to* 0.50%			Process Average 0.51 *to* 1.00%			Process Average 1.01 *to* 1.50%			Process Average 1.51 *to* 2.00%			Process Average 2.01 *to* 2.50%		
	n	*c*	AOQL %	*n*	*c*	AOQL %	*n*	*c*	AOQL %	*n*	*c*	AOQL %	*n*	*c*	AOQL %	*n*	*c*	AOQL %
1–30	All	0	0	All	0	0	All	0	0	All	0	0	All	0	0	All	0	0
31–50	30	0	0.49	30	0	0.49	30	0	0.49	30	0	0.49	30	0	0.49	30	0	0.49
51–100	37	0	0.63	37	0	0.63	37	0	0.63	37	0	0.63	37	0	0.63	37	0	0.63
101–200	40	0	0.74	40	0	0.74	40	0	0.74	40	0	0.74	40	0	0.74	40	0	0.74
201–300	43	0	0.74	43	0	0.74	70	1	0.92	70	1	0.92	95	2	0.99	95	2	0.99
301–400	44	0	0.74	44	0	0.74	70	1	0.99	100	2	1.0	120	3	1.1	145	4	1.1
401–500	45	0	0.75	75	1	0.95	100	2	1.1	100	2	1.1	125	3	1.2	150	4	1.2
501–600	45	0	0.76	75	1	0.98	100	2	1.1	125	3	1.2	150	4	1.3	175	5	1.3
601–800	45	0	0.77	75	1	1.0	100	2	1.2	130	3	1.2	175	5	1.4	200	6	1.4
801–1,000	45	0	0.78	75	1	1.0	105	2	1.2	155	4	1.4	180	5	1.4	225	7	1.5
1,001–2,000	45	0	0.80	75	1	1.0	130	3	1.4	180	5	1.6	230	7	1.7	280	9	1.8
2,001–3,000	75	1	1.1	105	2	1.3	135	3	1.4	210	6	1.7	280	9	1.9	370	13	2.1
3,001–4,000	75	1	1.1	105	2	1.3	160	4	1.5	210	6	1.7	305	10	2.0	420	15	2.2
4,001–5,000	75	1	1.1	105	2	1.3	160	4	1.5	235	7	1.8	330	11	2.0	440	16	2.2
5,001–7,000	75	1	1.1	105	2	1.3	185	5	1.7	260	8	1.9	350	12	2.2	490	18	2.4
7,001–10,000	75	1	1.1	105	2	1.3	185	5	1.7	260	8	1.9	380	13	2.2	535	20	2.5
10,001–20,000	75	1	1.1	135	3	1.4	210	6	1.8	285	9	2.0	425	15	2.3	610	23	2.6
20,001–50,000	75	1	1.1	135	3	1.4	235	7	1.9	305	10	2.1	470	17	2.4	700	27	2.7
50,001–100,000	75	1	1.1	160	4	1.6	235	7	1.9	355	12	2.2	515	19	2.5	770	30	2.8

n = sample size; c = acceptance number
"All" indicates that each piece in the lot is to be inspected.
AOQL = average outgoing quality limit
* From H. F. Dodge and H. G. Romig, *Sampling Inspection Tables*, 2d ed., John Wiley & Sons, Inc., New York, 1959.

AOQL (1 percent). (The corresponding lot tolerances are, however, given.)

AOQL plans are appropriate only when all rejected lots are 100 percent inspected. The averaging of the perfect quality of the detailed lots with the poor quality of unsatisfactory lots occasionally accepted (owing to the unavoidable consumer's risk) determines the average outgoing quality and makes a limit possible. AOQL schemes are open to question where rejected lots are returned to an outside supplier, since there is no assurance that they will be 100 percent inspected and returned.[10]

Sampling is uneconomical if the average quality submitted is not considerably better than the AOQL specified. For this reason the Dodge-Romig AOQL tables do not give any plans for process averages which exceed the AOQL. Similarly, lot tolerance plans are not given for process averages greater than one-half of the specified lot tolerance. Actually 100 percent inspection is often less expensive than sampling if 40 percent or more of submitted lots are rejected, since the expenses of administration of the sampling plan and of double handling of rejected lots are eliminated.

17-14 Minimum Inspection per Lot

All Dodge-Romig plans were constructed to give minimum total inspection per lot for product of a given process average. This is an important feature of the Dodge-Romig plans and deserves further explanation.

Assume that a consumer establishes acceptance criteria as follows:

Lot tolerance fraction defective $p_t = 0.05$
Consumer's risk $P_c = 0.1$

A great many sampling plans meet these criteria; i.e., there are many combinations of sample size and acceptance number that would have an OC curve going through this point.

The total number of articles inspected is made up of two components: (1) the sample which is inspected for each lot and (2) the remaining parts which must be inspected in those lots which fail to pass the sampling inspection.

For small acceptance numbers the total inspection is high because many lots need to be detailed. For large acceptance numbers the total is again high, this time because of the large size of samples. The mini-

[10] This is one of the prime reasons for the preference for AQL schemes by the procurement agencies of the armed forces.

Table 17-3 **Double Sampling Table for Average Outgoing Quality Limit (AOQL) = 1.0%***

Lot Size	Process Average 0 to 0.02%						Process Average 0.03 to 0.20%						Process Average 0.21 to 0.40%					
	Trial 1		Trial 2				Trial 1		Trial 2				Trial 1		Trial 2			
	n_1	c_1	n_2	n_1+n_2	c_2	p_t %	n_1	c_1	n_2	n_1+n_2	c_2	p_t %	n_1	c_1	n_2	n_1+n_2	c_2	p_t %
1–25	All	0	—	—	—	—	All	0	—	—	—	—	All	0				
26–50	22	0	—	—	—	7.7	22	0	—	—	—	7.7	22	0	—	—	—	7.7
51–100	33	0	17	50	1	6.9	33	0	17	50	1	6.9	33	0	17	50	1	6.9
101–200	43	0	22	65	1	5.8	43	0	22	65	1	5.8	43	0	22	65	1	5.8
201–300	47	0	28	75	1	5.5	47	0	28	75	1	5.5	47	0	28	75	1	5.5
301–400	49	0	31	80	1	5.4	49	0	31	80	1	5.4	55	0	60	115	2	4.8
401–500	50	0	30	80	1	5.4	50	0	30	80	1	5.4	55	0	65	120	2	4.7
501–600	50	0	30	80	1	5.4	50	0	30	80	1	5.4	60	0	65	125	2	4.6
601–800	50	0	35	85	1	5.3	60	0	70	130	2	4.5	60	0	70	130	2	4.5
801–1,000	55	0	30	85	1	5.2	60	0	75	135	2	4.4	60	0	75	135	2	4.4
1,001–2,000	55	0	35	90	1	5.1	65	0	75	140	2	4.3	75	0	120	195	3	3.8
2,001–3,000	65	0	80	145	2	4.2	65	0	80	145	2	4.2	75	0	125	200	3	3.7
3,001–4,000	70	0	80	150	2	4.1	70	0	80	150	2	4.1	80	0	175	255	4	3.5
4,001–5,000	70	0	80	150	2	4.1	70	0	80	150	2	4.1	80	0	180	260	4	3.4
5,001–7,000	70	0	80	150	2	4.1	75	0	125	200	3	3.7	80	0	180	260	4	3.4
7,001–10,000	70	0	80	150	2	4.1	80	0	125	205	3	3.6	85	0	180	265	4	3.3
10,001–20,000	70	0	80	150	2	4.1	80	0	130	210	3	3.6	90	0	230	320	5	3.2
20,001–50,000	75	0	80	155	2	4.0	80	0	135	215	3	3.6	95	0	300	395	6	2.9
50,001–100,000	75	0	80	155	2	4.0	85	0	180	265	4	3.3	170	1	380	550	8	2.6

1–25	All	0	—	—	—	—	All	0	—	—	—	—	All	0				
26–50	22	0	—	—	—	7.7	22	0	—	—	—	7.7	22	0	—	—	—	7.7
51–100	33	0	17	50	1	6.9	33	0	17	50	1	6.9	33	0	17	50	1	6.9
101–200	43	0	22	65	1	5.8	43	0	22	65	1	5.8	47	0	43	90	2	5.4
201–300	55	0	50	105	2	4.9	55	0	50	105	2	4.9	55	0	50	105	2	4.9
301–400	55	0	60	115	2	4.8	55	0	60	115	2	4.8	60	0	80	140	3	4.5
401–500	55	0	65	120	2	4.7	60	0	95	155	3	4.3	60	0	95	155	3	4.3
501–600	60	0	65	125	2	4.6	65	0	100	165	3	4.2	65	0	100	165	3	4.2
601–800	65	0	105	170	3	4.1	65	0	105	170	3	4.1	70	0	140	210	4	3.9
801–1,000	65	0	110	175	3	4.0	70	0	150	220	4	3.8	125	1	180	305	6	3.5
1,001–2,000	80	0	165	245	4	3.7	135	1	200	335	6	3.3	140	1	245	385	7	3.2
2,001–3,000	80	0	170	250	4	3.6	150	1	265	415	7	3.0	215	2	355	570	10	2.8
3,001–4,000	85	0	220	305	5	3.3	160	1	330	490	8	2.8	225	2	455	680	12	2.7
4,001–5,000	145	1	225	370	6	3.1	225	2	375	600	10	2.7	240	2	595	835	14	2.5
5,001–7,000	155	1	285	440	7	2.9	235	2	440	675	11	2.6	310	3	665	975	16	2.4
7,001–10,000	165	1	355	520	8	2.7	250	2	585	835	13	2.4	385	4	785	1,170	19	2.3
10,001–20,000	175	1	415	590	9	2.6	325	3	655	980	15	2.3	520	6	980	1,500	24	2.2
20,001–50,000	250	2	490	740	11	2.4	340	3	910	1,250	19	2.2	610	7	1,410	2,020	32	2.1
50,001–100,000	275	2	700	975	14	2.2	420	4	1,050	1,470	22	2.1	770	9	1,850	2,620	41	2.0

Trial 1: n_1 = first sample size; c_1 = acceptance number for first sample.
"All" indicates that each piece in the lot is to be inspected.
Trial 2: n_2 = second sample size; c_2 = acceptance number for first and second samples combined.
p_t = lot tolerance percent defective with a consumer's risk P_C of 0.10.
* From H. F. Dodge and H. G. Romig, *Sampling Inspection Tables*, 2d ed., John Wiley & Sons, Inc., New York, 1959.

mum sum occurs at a point between these extremes. All the plans in the Dodge-Romig tables meet this criterion of minimum total inspection.[11]

17-15 MIL-STD-105D Tables

The Dodge-Romig tables provide plans that emphasize the protection of the *consumer* against *accepting bad* lots. The plans incorporate the concept of minimum total inspection (assuming rejected lots are 100 percent inspected). The MIL-STD-105D tables provide plans that emphasize the protection of the producer against *rejecting good* lots. The concept of minimum total inspection is *not* incorporated and therefore it is not assumed that rejected lots are 100 percent inspected.

The quality index in MIL-STD-105D[12] is the acceptable quality level (AQL). The AQL is the maximum percent defective (or number of defects per 100 units) that is satisfactory as a process average. The probability of accepting material of AQL quality is always high, but not exactly the same for all plans. For lot quality just equal to the AQL, the "percent of lots expected to be accepted" ranges from about 88 to 99 percent. Defects are classified as "critical," "major," or "minor." The government may, at its option, specify separate AQLs for each class or it may use a further subdivision and specify an AQL for each individual kind of defect which a product may show. The choice may be made from 26 available AQL values ranging from 0.010 to 1000.0. (Values of AQL of 10.0 or less may be interpreted as percent defective or defects per hundred units.)

The tables specify the relative amount of inspection to be used as "inspection level" I, II, or III; level II is regarded as normal. The concept of inspection level permits the user to balance the cost of inspection against the amount of protection required. The three levels involve inspection in amounts roughly in the ratio 0.4 to 1.0 to 1.6. (Four additional inspection levels are provided for situations requiring "small-sample inspection.")

A plan is chosen from the tables as follows:

1 The following information must be known:
   a Acceptable quality level (AQL)
   b Lot size
   c Type of sampling (single, double, or multiple)
   d Inspection level (usually level II)
2 Knowing the lot size and inspection level, obtain a code letter from Table 17-4.
3 Knowing the code letter, AQL, and type of sampling, read the sampling

[11] For a sample derivation, see *QCH*, 1962, pp. 13-80 to 13-81.
[12] *Sampling Procedures and Tables for Inspection by Attributes, op. cit.*

plan from Table 17-5 (Table 17-5 is for single sampling; the standard also provides tables for double and multiple sampling).

For example, suppose that a purchasing agency has contracted for a 1 percent AQL. Suppose also that the parts are brought in lots of 1,500 pieces. From the table of sample size code letters (Table 17-4) it is found that letter K plans are required for inspection level

Table 17-4 **SAMPLE SIZE CODE LETTERS—MIL-STD-105D (ABC STANDARD)**

	Special Inspection Levels				*General Inspection Levels*		
Lot or Batch Size	*S*-1	*S*-2	*S*-3	*S*-4	I	II	III
2–8	A	A	A	A	A	A	B
9–15	A	A	A	A	A	B	C
16–25	A	A	B	B	B	C	D
26–50	A	B	B	C	C	D	E
51–90	B	B	C	C	C	E	F
91–150	B	B	C	D	D	F	G
151–280	B	C	D	E	E	G	H
281–500	B	C	D	E	F	H	J
501–1,200	C	C	E	F	G	J	K
1,201–3,200	C	D	E	G	H	K	L
3,201–10,000	C	D	F	G	J	L	M
10,001–35,000	C	D	F	H	K	M	N
35,000–150,000	D	E	G	J	L	N	P
150,001–500,000	D	E	G	J	M	P	Q
500,001 and over	D	E	H	K	N	Q	R

II. Table 17-5 states the sample size is 125. For AQL = 1.0, the acceptance number is given as 3 and the rejection number as 4. This means that the entire lot of 1,500 articles may be accepted if three or fewer defective articles are found, but must be rejected if four or more are found. Where an AQL is expressed in terms of "defects per hundred units," this term may be substituted for "defective articles" throughout.

MIL-STD-105D includes provision for tightened inspection if quality deteriorates. If two out of five consecutive lots are rejected on original inspection, a tightened inspection plan (in a separate table) is imposed. The sample size is usually the same as normal, but the acceptance number is reduced. (The tightened plans do require larger sample sizes if the probability of acceptance for an AQL lot is less than 0.75.)

Table 17-5 Master Table for Normal Inspection (Single Sampling)—MIL-STD-105D (ABC Standard)

Acceptable Quality Levels (Normal Inspection)

Sample Size Code Letter	Sample Size	0.010 Ac Re	0.015 Ac Re	0.025 Ac Re	0.040 Ac Re	0.065 Ac Re	0.10 Ac Re	0.15 Ac Re	0.25 Ac Re	0.40 Ac Re	0.65 Ac Re	1.0 Ac Re	1.5 Ac Re	2.5 Ac Re	4.0 Ac Re	6.5 Ac Re	10 Ac Re	15 Ac Re	25 Ac Re	40 Ac Re	65 Ac Re	100 Ac Re	150 Ac Re	250 Ac Re	400 Ac Re	650 Ac Re	1,000 Ac Re
A	2	↓	↓	↓	↓	↓	↓	↓	↓	↓	↓	↓	↓	↓	↓	0 1	↓	↓	1 2	2 3	3 4	5 6	7 8	10 11	14 15	21 22	30 31
B	3	↓	↓	↓	↓	↓	↓	↓	↓	↓	↓	↓	↓	↓	0 1	↑	↓	1 2	2 3	3 4	5 6	7 8	10 11	14 15	21 22	30 31	44 45
C	5	↓	↓	↓	↓	↓	↓	↓	↓	↓	↓	↓	↓	0 1	↑	↓	1 2	2 3	3 4	5 6	7 8	10 11	14 15	21 22	30 31	44 45	↑
D	8	↓	↓	↓	↓	↓	↓	↓	↓	↓	↓	↓	0 1	↑	↓	1 2	2 3	3 4	5 6	7 8	10 11	14 15	21 22	30 31	44 45	↑	↑
E	13	↓	↓	↓	↓	↓	↓	↓	↓	↓	↓	0 1	↑	↓	1 2	2 3	3 4	5 6	7 8	10 11	14 15	21 22	30 31	44 45	↑	↑	↑
F	20	↓	↓	↓	↓	↓	↓	↓	↓	↓	0 1	↑	↓	1 2	2 3	3 4	5 6	7 8	10 11	14 15	21 22	↑	↑	↑	↑	↑	↑
G	32	↓	↓	↓	↓	↓	↓	↓	↓	0 1	↑	↓	1 2	2 3	3 4	5 6	7 8	10 11	14 15	21 22	↑	↑	↑	↑	↑	↑	↑
H	50	↓	↓	↓	↓	↓	↓	↓	0 1	↑	↓	1 2	2 3	3 4	5 6	7 8	10 11	14 15	21 22	↑	↑	↑	↑	↑	↑	↑	↑
J	80	↓	↓	↓	↓	↓	↓	0 1	↑	↓	1 2	2 3	3 4	5 6	7 8	10 11	14 15	21 22	↑	↑	↑	↑	↑	↑	↑	↑	↑
K	125	↓	↓	↓	↓	↓	0 1	↑	↓	1 2	2 3	3 4	5 6	7 8	10 11	14 15	21 22	↑	↑	↑	↑	↑	↑	↑	↑	↑	↑
L	200	↓	↓	↓	↓	0 1	↑	↓	1 2	2 3	3 4	5 6	7 8	10 11	14 15	21 22	↑	↑	↑	↑	↑	↑	↑	↑	↑	↑	↑
M	315	↓	↓	↓	0 1	↑	↓	1 2	2 3	3 4	5 6	7 8	10 11	14 15	21 22	↑	↑	↑	↑	↑	↑	↑	↑	↑	↑	↑	↑
N	500	↓	↓	0 1	↑	↓	1 2	2 3	3 4	5 6	7 8	10 11	14 15	21 22	↑	↑	↑	↑	↑	↑	↑	↑	↑	↑	↑	↑	↑
P	800	↓	0 1	↑	↓	1 2	2 3	3 4	5 6	7 8	10 11	14 15	21 22	↑	↑	↑	↑	↑	↑	↑	↑	↑	↑	↑	↑	↑	↑
Q	1 250	0 1	↑	↓	1 2	2 3	3 4	5 6	7 8	10 11	14 15	21 22	↑	↑	↑	↑	↑	↑	↑	↑	↑	↑	↑	↑	↑	↑	↑
R	2 000	↑	↑	1 2	2 3	3 4	5 6	7 8	10 11	14 15	21 22	↑	↑	↑	↑	↑	↑	↑	↑	↑	↑	↑	↑	↑	↑	↑	↑

↓ = use first sampling plan below arrow. If sample size equals, or exceeds, lot or batch size, do 100% inspection.
↑ = use first sampling plan above arrow.
Ac = acceptance number.
Rc = rejection number.

MIL-STD-105D also provides for reduced inspection where the supplier's record has been good. When the process average is better than the AQL by 2σ or more, a reduced inspection plan may be used. A table of lower limits for the process average and rules governing normal and reduced inspection are provided in the standard. Under reduced sampling, the sample size is usually 40 percent of the normal sample size.

The standard not only provides OC curves for most of the plans, but also average sample size curves for double and multiple sampling. The latter curves show the average sample sizes expected as a function of the product quality submitted. Although the OC curves are roughly the same for single, double, and multiple sampling, the average sample size curves vary considerably because of the inherent differences among the three types of sampling. The standard also states the AOQL that would result if all rejected lots were screened for defectives.

The sampling plans in MIL-STD-105D are sufficiently varied in type (single, double, multiple), amount of inspection, etc., to be useful in a great number of situations. The reader is also referred to an excellent supplementary publication,[13] which discusses and provides recommendations on many of the practical problems that arise during the day-to-day application of the standard. Dodge[14] has developed a procedure for sampling inspection based on the AQL concept which, among other features, results in a simpler set of tables than MIL-STD-105D.

17-16 Selection of a Numerical Value of the Quality Index

The problem of selecting a value of the quality index (for example, AQL, AOQL, or lot tolerance percent defective) is one of balancing the cost of finding and correcting a defective against the loss incurred if a defective slips through an inspection procedure.

Enell[15] has suggested that the break-even point (section 17-3) be used in the selection of an AQL. The break-even point for inspection was defined as the cost to inspect one piece divided by the damage

[13] *Guide for Sampling Inspection, op. cit.* For additional comments on H53, see Earl K. Yost, "DOD Handbook H53 Guide for Sampling Inspection," *Transactions of the Twentieth Annual Conference of the American Society for Quality Control,* 1966, pp. 103–106.

[14] Harold F. Dodge, "A General Procedure for Sampling Inspection by Attributes—Based on the AQL Concept," *Transactions of the Seventeenth Annual Conference of the American Society for Quality Control,* 1963, pp. 7–19.

[15] J. W. Enell, "What Sampling Plan Shall I Choose?," *Industrial Quality Control* (*IQC*), vol. 10, no. 6, pp. 96–100, May, 1954.

done by one defective. For the example cited in section 17-3 the break-even point was 1 percent defective.

As a 1 percent defective quality level is the break-even point between sorting and sampling, the appropriate sampling plan should provide for a lot to have a 50 percent probability of being sorted or sampled; i.e., the probability of acceptance for the plan should be 0.50 at a 1 percent defective quality level. The operating characteristic curves in a set of sampling tables such as MIL-STD-105D can now be examined to determine an AQL. For example, suppose the mica is expected in lots of 3,000 pieces. The operating characteristic curves for this case (code letter K) are shown in MIL-STD-105D. The plan closest to having a P_a of 0.50 for a 1 percent level is the plan for an AQL of 0.40 percent. Therefore, this is the plan to adopt.

Hansen[16] has summarized the common bases used to set quality level standards:

1 Historical Data—Past data are analyzed to arrive at an historical estimate of the process quality average. The standard is set equal to or at a large fraction of the historical average.
2 Empirical Judgment—Standard is set at a level approximating a proven satisfactory level for a similar item.
3 Engineering Judgment—Standard is based upon engineering estimates of the quality requirements for function, life, interchangeability, assembly, safety, and so forth.
4 Experimental—Tentative standard is set and adjusted as indicated by quality performance.
5 Minimum Total Cost—Standard is based upon an analysis of the costs of obtaining quality versus the costs of not having quality.
6 Consistent—Each category has a set standard which does not change.

Some plans (for example, MIL-STD-105D) include a classification of defects to help determine the numerical value of the AQL. Defects are first classified as critical, major, or minor according to definitions provided in the standard. Different AQLs may be designated for groups of defects considered collectively, or for individual defects. Major defects might be assigned a low AQL, say 1 percent, and minor defects a higher AQL, say 2.5 percent.

Additional research[17] has been conducted on the problem of selecting sampling plans.

[16] B. L. Hansen, *Quality Control: Theory and Applications*, Prentice-Hall, Inc., Englewood Cliffs, N.J., 1963, p. 280.
[17] See *QCH*, 1962, p. 13-96.

17-17 Acceptance Sampling Plans by Variables

In *attributes* sampling plans, each item inspected is classified as either defective or acceptable. The total number of defectives in the sample is then compared with the acceptance number and a decision made on the lot. In variables sampling plans, a *measurement* is taken and recorded for each item in the sample. An index (such as an average) is calculated from these measurements, compared with an "allowable" value, and a decision is made on the lot. The sample size and allowable value are a function of the desired sampling risks.[18]

This chapter explains one published plan. (Chapter 20 explains another variables plan known as the "lot plot" method.)

An example of a variables plan is *MIL-STD-414 Sampling Procedures and Tables for Inspection by Variables for Per Cent Defective.*[19] The format and terminology are similar to those of MIL-STD-105D; for example, the concepts of AQL, code letters, inspection levels, reduced and tightened inspection, and OC curves are all included.

MIL-STD-414 assumes a normal distribution and that information on variability is available or will be obtained in the sample. To provide flexibility in application, a number of alternative procedures are included. Only one of the procedures will be described here. A plan is selected as follows:

1 Select an acceptable quality level (AQL). Levels range from 0.04 to 15 percent.
2 Select a sample size code letter based on the lot size and inspection level. Five inspection levels are provided. Level IV is considered normal and is used unless another level is specified. Table 17-6 is used to determine the sample size code letter.
3 Select the sampling plan from a master table in section B, C, or D. Sections B and C contain plans for the case when the variability is unknown and is measured by the standard deviation or range, respectively. Section D provides the plans when the variability is known (in terms of standard deviation). A plan from section B will be selected for the following problem:

Example[20] The maximum temperature of operation for a device is specified as 209°F. A lot of 40 is submitted for inspection. Inspection level IV with an AQL of 1 percent is to be used. The standard deviation is unknown. Assuming that operating temperature follows a normal distribution, what variables sampling plan should be used to inspect the lot?

[18] See *QCH*, 1962, pp. 13-100 to 13-104.
[19] Government Printing Office, 1957.
[20] *Ibid.*

Table 17-6 SAMPLE SIZE CODE LETTERS*

Lot Size	*Inspection Levels* I	II	III	IV	V
3–8	B	B	B	B	C
9–15	B	B	B	B	D
16–25	B	B	B	C	E
26–40	B	B	B	D	F
41–65	B	B	C	E	G
66–110	B	B	D	F	H
111–180	B	C	E	G	I
181–300	B	D	F	H	J
301–500	C	E	G	I	K
501–800	D	F	H	J	L
801–1,300	E	G	I	K	L
1,301–3,200	F	H	J	L	M
3,201–8,000	G	I	L	M	N
8,001–22,000	H	J	M	N	O
22,001–110,000	I	K	N	O	P
110,001–550,000	I	K	O	P	Q
550,001 and over	I	K	P	Q	Q

* Sample size code letters given in body of table are applicable when the indicated inspection levels are to be used.

One type of plan in section B requires that n measurements be taken, the average and standard deviation calculated, and an evaluation made of the number of standard deviations between the sample average and the specification limit. More specifically,

1 Compute the sample average $\overline{X}$ and the estimate of the lot standard deviation s. Also compute $(U - \overline{X})/s$ for an upper specification limit U, or $(\overline{X} - L)/s$ for a lower specification limit L.

2 If the fraction computed in step 1 is equal to or greater than k, accept the lot; otherwise, reject the lot.

Table 17-6 provides the code letter as D and Table 17-7 (Master Table B-1 in MIL-STD-414) provides the values of n and k as 5 and 1.53 respectively.

Now suppose the measurements were 197°, 188°, 184°, 205°, and 201°. This yields an X of 195° and an s of 8.81°. Then $(U - X)/s = (209 - 195)/8.81 = 1.59$. As 1.59 is greater than 1.53, the lot is accepted. The OC curve for the plan is included in the standard.

The reader is referred to the standard itself for other procedures and tables.[21]

[21] A summary is provided in *QCH*, 1962, pp. 13-104 to 13-110.

Table 17-7 **Master Table for Normal and Tightened Inspection for Means Based on Variability Unknown, Standard Deviation Method**

(Single Specification Limit, Form 1)

		Acceptable Quality Levels (Normal Inspection)													
Sample Size Code Letter	*Sample Size*	0.04 *k*	0.065 *k*	0.10 *k*	0.15 *k*	0.25 *k*	0.40 *k*	0.65 *k*	1.00 *k*	1.50 *k*	2.50 *k*	4.00 *k*	6.50 *k*	10.00 *k*	15.00 *k*
B	3	↓	↓	↓	↓	↓	↓	↓	↓	↓	1.12	0.958	0.765	0.566	0.341
C	4	↓	↓	↓	↓	↓	↓	↓	1.45	1.34	1.17	1.01	0.814	0.617	0.393
D	5	↓	↓	↓	↓	↓	↓	1.65	1.53	1.40	1.24	1.07	0.874	0.675	0.455
E	7	↓	↓	↓	↓	2.00	1.88	1.75	1.62	1.50	1.33	1.15	0.955	0.755	0.536
F	10	↓	↓	↓	2.24	2.11	1.98	1.84	1.72	1.58	1.41	1.23	1.03	0.828	0.611
G	15	2.64	2.53	2.42	2.32	2.20	2.06	1.91	1.79	1.65	1.47	1.30	1.09	0.886	0.664
H	20	2.69	2.58	2.47	2.36	2.24	2.11	1.96	1.82	1.69	1.51	1.33	1.12	0.917	0.695
I	25	2.72	2.61	2.50	2.40	2.26	2.14	1.98	1.85	1.72	1.53	1.35	1.14	0.936	0.712
J	30	2.73	2.61	2.51	2.41	2.28	2.15	2.00	1.86	1.73	1.55	1.36	1.15	0.946	0.723
K	35	2.77	2.65	2.54	2.45	2.31	2.18	2.03	1.89	1.76	1.57	1.39	1.18	0.969	0.745
L	40	2.77	2.66	2.55	2.44	2.31	2.18	2.03	1.89	1.76	1.58	1.39	1.18	0.971	0.746
M	50	2.83	2.71	2.60	2.50	2.35	2.22	2.08	1.93	1.80	1.61	1.42	1.21	1.00	0.774
N	75	2.90	2.77	2.66	2.55	2.41	2.27	2.12	1.98	1.84	1.65	1.46	1.24	1.03	0.804
O	100	2.92	2.80	2.69	2.58	2.43	2.29	2.14	2.00	1.86	1.67	1.48	1.26	1.05	0.819
P	150	2.96	2.84	2.73	2.61	2.47	2.33	2.18	2.03	1.89	1.70	1.51	1.29	1.07	0.841
Q	200	2.97	2.85	2.73	2.62	2.47	2.33	2.18	2.04	1.89	1.70	1.51	1.29	1.07	0.845
		0.065	0.10	0.15	0.25	0.40	0.65	1.00	1.50	2.50	4.00	6.50	10.00	15.00	
		Acceptable Quality Levels (Tightened Inspection)													

All AQL values are in percent defective.

↓ Use first sampling plan below arrow, that is, both sample size as well as *k* value. When sample size equals or exceeds lot size, every item in the lot must be inspected.

17-18 Sampling Procedures Based on Prior Information

In this chapter no assumption was made about the quality of incoming lots. By implication, it was assumed that all possible quality levels had an equal chance of occurring. In practice, the probability of occurrence is *not* the same for all possible quality levels. (However, it is difficult to estimate the likelihood of each quality level.) The use of prior information to establish these probabilities of occurrence is called the Bayesian approach. In the Bayesian approach (see Chapter 11) the derivation of a sampling plan requires an assumption of a probability distribution for incoming quality levels. If desired, the expected dollar loss due to accepting bad lots and rejecting good lots can be taken into account in deriving the sampling plan. (The incorporation of prior probabilities and losses into the conventional analysis is called "statistical decision theory.")

The approach will be illustrated in terms of an example.[22] Consider the following three sampling plans:

Sample Size	*Allowable Number of Defectives for Lot Acceptance*
50	2
50	3
50	4

Suppose 5 percent defective is the dividing line between good and bad quality. Ideally, all lots less than 5 percent defective should be accepted; all lots greater than 5 percent should be rejected. The above three plans are submitted as reasonable attempts to approach the ideal. There are two errors that the plan can make: (1) rejecting lots having less than 5 percent defective, and (2) accepting lots having more than 5 percent defective. The likelihood of these errors can be read from the OC curves. For example, a 1 percent defective lot for the first plan has a P_a of about 0.91 or a probability of 0.09 of incorrectly being rejected by the plan. A summary of the probability of an incorrect decision for various values of incoming quality is given in Table 17-8. The table also provides estimates of the loss incurred due to an incorrect decision (a good lot rejected or a bad lot accepted) and the probability

[22] Adapted from L. G. Locke, "Bayesian Statistics," *IQC*, vol. 20, no. 10, pp. 18–22, April, 1964.

Table 17-8 SAMPLE CALCULATION OF EXPECTED LOSS

True Fraction Defective p	Probability That p Will Occur	Probability of Incorrect Decision on Lot $n = 50$ $c = 2$	$n = 50$ $c = 3$	$n = 50$ $c = 4$	Loss Incurred If Incorrect Decision Is Made	Expected Loss $n = 50$ $c = 2$	$n = 50$ $c = 3$	$n = 50$ $c = 4$
0	0.05	0	0	0	$25	0.00	0.00	0.00
0.01	0.05	0.0894	0.0138	0.0016	20	0.09	0.01	0.00
0.02	0.10	0.2642	0.0784	0.0178	15	0.40	0.12	0.03
0.03	0.15	0.4447	0.1892	0.0628	10	0.67	0.28	0.09
0.04	0.20	0.5995	0.3233	0.1391	5	0.60	0.32	0.14
0.05	0.20	0.7206	0.4595	0.2396	0	0.00	0.00	0.00
0.06	0.10	0.1900	0.4162	0.6473	5	0.10	0.21	0.32
0.07	0.05	0.1265	0.3108	0.5327	10	0.06	0.16	0.27
0.08	0.05	0.0827	0.2260	0.4253	15	0.06	0.17	0.32
0.09	0.05	0.0532	0.1605	0.3303	20	0.05	0.16	0.33
						2.03	1.43	1.50

that a lot having a true fraction defective p will be submitted for evaluation. The expected loss for each plan is calculated as

$$\sum_{p=0}^{p=1.0}\left[\begin{pmatrix}\text{probability that}\\ \text{quality level}\\ \text{will occur}\end{pmatrix}\begin{pmatrix}\text{probability that}\\ \text{incorrect decision}\\ \text{will be made}\end{pmatrix}\begin{pmatrix}\text{dollar loss}\\ \text{due to incorrect}\\ \text{decision}\end{pmatrix}\right]$$

Thus, the concept first considers the probability that a specific quality will be submitted and then, assuming that it is submitted, it considers the probability of making an incorrect decision. The product of these two probabilities is the chance that a specific quality will be submitted and an incorrect decision made. This is the only case that involves a dollar loss. Multiplying this probability by the loss gives the "expected loss."[23]

Table 17-8 indicates that the plan which allows three or fewer defectives will minimize the expected loss. (The optimum sample size must recognize both the expected loss and the cost of sampling. Smith[24] discusses procedures for determining the optimum sample size using the Bayesian approach of incorporating past information and the economic consequences into one overall model.)

[23] The expected loss is a special case of the expected value concept. The expected value is the average value which will occur in the long run. Thus, the expected value in a toss of two dice is $2(1/36) + 3(2/36) + \ldots + 11(2/36) + 12(1/36) =$ 7.0.

[24] Barnard E. Smith, "The Economics of Sampling Inspection," *IQC*, vol. 21, no. 9. pp. 453–458 March, 1965.

17-19 Product Acceptance for a Statistically Controlled Process

Chapter 15 mentioned that conventional control limits cannot be compared to tolerances, because control limits refer to sample *averages* and tolerances to *individual* items. It is possible for a process to be within limits on an average and range chart but have a high percent of individual items of product outside tolerances.

Adaptations of the basic control chart (e.g., narrow-limit gaging) have control limits based on the tolerance limits. These are meant to detect only process changes which are large enough to cause defectives. It is tempting to use such charts as a means of formally accepting product. If this is done, the control limits must reflect the two types of risk i.e., rejecting good product and accepting bad product.[25] Also, careful review of the method of selecting samples should be made. When a control chart is used for detecting process changes, it is usually best that the individual items within each sample be consecutively produced over a relatively short time period. This tends to yield statistical control limits that are more sensitive to detecting process changes than if the items in the sample were selected at random over a longer time period.[26] However, such a sample is not representative of the entire production and, therefore, is not valid for acceptance purposes. A judgment will be necessary for each case to decide if the chart is to serve as a control tool only or as both a control and an acceptance tool. However, the decision need not be "black or white." For example, in a given case, it *may* be best to use the chart as both a control and an acceptance tool, but to supplement it with a formal acceptance sampling plan using a small sample size.

A process in statistical control does have another advantage from the viewpoint of acceptance sampling. As lot size alone does not affect sample size, lots should be as large as possible to take advantage of sampling economies. However, it is important that any lot submitted for acceptance sampling be *homogeneous* in quality. This is critical because the sample must be representative of the entire lot. Simon[27] suggests that the statistical control chart be the criterion for homogeneity. The entire production from a process in statistical control can be considered homogeneous in quality and, therefore, can be viewed as one large lot for sampling purposes.

[25] See Richard A. Freund, "Acceptance Control Charts," *IQC*, vol. 14, no. 4, pp. 13–23, October, 1957.
[26] For an excellent discussion of "rational subgroups," see Grant, *op. cit.*, chap. 7.
[27] L. E. Simon, "The Industrial Lot and Its Sampling Implication," *Journal of the Franklin Institute*, vol. 237, pp. 359–370, May, 1944.

17-20 Knowledge beyond Sampling

Sampling data are only one of the sources of information on which to base decisions concerning the process. Other sources include:

1 Data on process capability. If we have learned that some aspects of the process are predictable (e.g., the standard deviation, the time-to-time drift in the average), then this knowledge can be used to supplement the sampling data in making decisions on the process and on the product. The Shewhart average and range control chart is an example. The sample sizes are small, but by utilizing data on prior knowledge of process capability, the effectiveness of these small samples is greatly magnified.

2 Knowledge of a scientific or engineering character pertinent to the process. For example, the inherent nature of a press operation is such that if the first piece and the last piece in a lot show perforated holes, it is inevitable that all the pieces in that lot will show the same perforated holes. In such a situation, a sample size of two pieces is adequate to make the decision provided that (*a*) the tools and the operation are of an "all or none" type, and (*b*) provision is made to preserve the order of manufacture.

3 Sampling data from other sources, e.g., vendors' test data, operators' measurements, automatic recording charts. Through the concept of audit of decisions (see section 20-5) it is possible to establish confidence in these added sources of measurement, and then to use the data as an added input to decision making.

PROBLEMS

1 A small manufacturer of television equipment has been inspecting all the raw materials he receives. His operations are expanding and there is now some question concerning the economics of the present inspection system. It costs 3 cents to inspect one unit of raw material, and if a unit of defective raw material is allowed to get into production it will cost him 25 cents to repair the damage done to the equipment it is used in. What level of quality makes the present inspection system economical?

2 A manufacturer of industrial truck batteries spends 73 cents to inspect each battery. If a defective battery is allowed to be shipped it costs the company $32.12 to replace it. What must the production quality level be to make a sampling plan economically acceptable?

3 A large gray iron foundry casts the base for precision grinders. These bases are produced at the rate of 18 per day and are 100 percent inspected at the foundry for flaws in the metal. The castings are then stored and subsequently shipped in lots of 300 to the grinder manufacturer. The grinder manufacturer has found these lots to be 10 percent defective. Upon receiving a lot the manufacturer inspects 12 of them and rejects the lot if 2 or more defectives are found. What is the chance that he will reject a given lot?

4 A manufacturer of pipe fittings produces continuously with an average of 1 defective for every 1,000 fittings. If a sample of 500 pieces is inspected, what is the likelihood that this sample will be equal to or less than 0.2 percent defective?

5 Verify the operating characteristic curve for a lot size of 200 in Figure 17-3. Calculate at least three points by the Poisson method.

6 Using only the addition and multiplication rules for probability, verify the operating characteristic curve in Figure 17-4. Calculate at least three points.

7 The following double sampling plan has been proposed for evaluating a lot of 50 pieces:

Sample	*Sample Size*	*Acceptance Number*	*Rejection Number*
1	3	0	3
2	3	2	3

Using only the addition and multiplication rules for probability, calculate the probability of accepting a lot which is 10 percent defective.

8 A single destructive test on a certain product involves a great deal of time and cost. For this reason, lots of 100 are to be evaluated by the following sampling plan: Test two units. If both function, accept the lot. If both fail, reject the lot. Otherwise, test one additional unit. If it functions, accept the lot; otherwise, reject the lot.

Prepare an operating characteristic curve by calculating at least three points.

9 Table 17-2 provides an LTPD plan for a lot size of 800 and process average of 0.50 percent. The plan calls for a sample size of 75 and an acceptance number of 1. Prepare an operating characteristic curve by calculating at least three points.

10 A manufacturer wishes to sample a purchased component used in his assembly. He wishes to reject lots which are 5 percent defective and receives the components in lots of 1,000. From the past he knows the supplier usually submits lots 2 percent defective or less. The vendor has agreed to 100 percent inspect all rejected lots. Find a sampling plan to meet these conditions.

11 A men's clothing manufacturer supplies a chain of discount stores with men's sport shirts. The stores inspect these shirts and wish to use sampling techniques. The shirts are received in lots of 5,000 and have in the past been 0.5 percent defective. The stores want to be sure that long-run quality is 1 percent defective or better.

a Propose a plan.

b State all assumptions.

c Make a statement about the possibility of accepting an individual lot which has poor quality.

12 Refer to MIL-STD-105D with the following conditions:

Lot size = 10,000
Inspection level II
Acceptable quality level = 4 percent

a Find a single sampling plan for normal inspection.

b Suppose a lot is sampled and accepted. Someone makes the statement: "This means the lot has 4 percent or less defective." Comment on this statement. (Assume the sample was randomly selected and no inspection errors were made.)

c Calculate the probability of accepting a 4 percent defective lot under normal inspection.

13 A manufacturer sells his product in large lots to a customer who uses a sampling plan at incoming inspection. The plan calls for a sample of 200 units and an acceptance number of 2. Rejected lots are returned to the manufacturer. If a lot is rejected and returned, the manufacturer has decided to gamble and send it right back to the customer without screening it (and without telling the customer it was a rejected lot). He hopes that another random sample will lead to an acceptance of the lot. What is the probability that a 2 percent defective lot will be accepted on either one or two submissions to the customer?

14 A customer is furnishing resistors to the government under MIL-STD-105D. Inspection level II has been specified with an AQL of 1.0 percent. Lot sizes vary from 900 to 1,200.

a What single sampling plan will be used?

b Calculate the quality (in terms of percent defective) that has an equal chance of being accepted or rejected.

c What is the chance that a 1 percent defective lot will be accepted?

15 Refer to problem 12. What sample size would be required by MIL-STD-414? (Assume inspection level IV, normal inspection, variability measured by the standard deviation, single specification limit, form 1.)

16 Refer to problem 14. What sample size would be required by MIL-STD-414? (Assume inspection level IV and Table 17-7 applies.)

18 Measurement

18-1 Growing Importance of the Measurement Process

Precision of measurement is fundamental to our technological civilization. Without such precision we cannot discover the properties of materials, test our theories, improve our designs, or meet a thousand other needs of technology.

Measurement is as old as history, but precise measurement is hardly a century old. Since the first practical micrometer screw (1867), measurement of length has improved by several orders of magnitude, as has measurement of mass, temperature, force, time, and many, many other natural phenomena. These advances in *metrology* (science of measurement) have made possible the mass production of modern ultraprecise apparatus, yet there is still action at the frontier. For example, a 1958 Aerospace Industries Association survey disclosed "variations totaling over 80 microinches on an identical inside diameter of a sample part, as measured by six of the most competent laboratories in the country, or 80 times the accuracy of the best gage block measurements of the period, with the best laboratories of the government and the gage manufacturers unable to agree to better than 70 microinches."[1]

A later A.I.A. survey[2] polled industry on measurement problems in the "vital 10" measurement categories: dimensional; optical; tempera-

[1] J. K. Emery, "Dimensional Metrology Standardization," *Mechanical Engineering*, pp. 40–45, February, 1965.
[2] *Calibration and Standards Questionnaire,* Aerospace Industries Association, January 9, 1959.

ture and humidity; shock, vibration, and force; microwave; radio frequency; electrical; pressure; vacuum and fluid flow; infrared and radiological. (Note how the new technological areas have extended the measurement problem far beyond dimensional measurement and even included areas of nondestructive testing.) The following examples of problem areas were reported in the survey:

44 percent of companies replying to the category reported insufficient measurement accuracies on internal diameters below 0.250 inch.
67 percent reported insufficient measurement accuracies on continuous wave voltage on unbalanced lines.
21 percent reported insufficient accuracies in the "liquid flow" subcategory of pressure.

There are still many industries in which a crude form of measurement is adequate, but some of the industries that now require precise measurement were first confronted with this need within the past decade.

18-2 Economics of Measurement

The economics of measurement has been greatly affected by an increased variety of purposes of measurement. In the past many engineers viewed instruments primarily as a means of *sorting* good and bad product. The emphasis on *prevention* concepts in the quality function has broadened the purposes of measurement. Figure 18-1 stresses that measurement provides information or decisions not only on individual units of product (the sorting operation) but also on lots, processes, and the measuring instruments themselves.

These new purposes require that measuring systems be selected not on initial cost alone but also by evaluating the total cost to obtain

Figure 18-1 Uses of measurement.

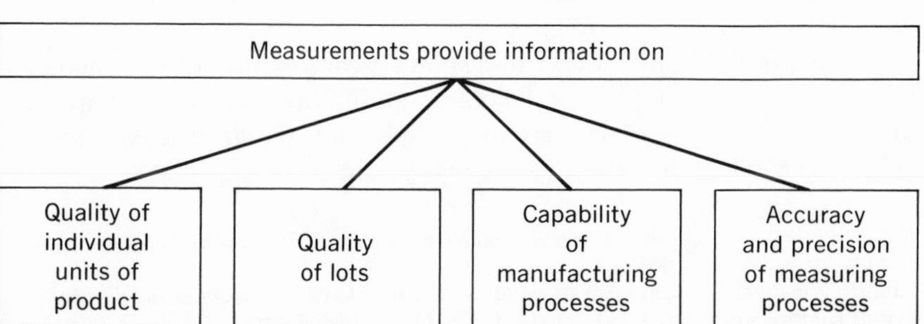

the kind of data needed for prevention purposes. This affects the comparison of attributes and variables measurement systems (see Table 18-1).

The high initial cost of the variables systems pays off in terms of more information value per observation. One study[3] compared three systems of evaluating a lot of 5,000 pieces. Sorting with attributes gages

Table 18-1 COMPARISON OF ATTRIBUTES AND VARIABLES MEASUREMENT SYSTEMS

Characteristic	*Attributes System*	*Variables System*
Cost of measuring instrument	Low	High
Grade of operator	Unskilled	Skilled
Speed of use	Fast	Slow
Recording of data	Simple	Complex
Overall cost per observation	Low	High
Information value per observation	Low	High
Number of observations to get good data	Many	Few

Table 18-2 COMPARISON OF MEASUREMENT METHODS

Factor	*Method A*	*Method B*
Man-time/test	7 minutes	25 minutes
$/man-hour	$5.00	$5.00
Sample/test	0.5 pound	0.5 pound
$/pound for sample	$0.20	$0.20
Test frequency	3/day	3/day
Base price	$2,000	$1,200
Installation	$400	$400
Man-time/year	$639	$2,281
Sample/year	$109	$876
	$3,148	$4,757

cost $34.25; sampling with attributes gages cost 65 cents; and sampling with variables gages cost 37 cents.

As another example, Table 18-2 shows a comparison[4] of two methods of measurement in a plastics laboratory. In this case, the equipment costing $800 more initially has an initial- and 1-year use cost which is about $1,600 less than the cheaper equipment.

[3] J. M. Juran (ed.), *Quality Control Handbook* (*QCH*), 2d ed., McGraw-Hill Book Company, New York, 1962, p. 9-3.

[4] Albert Woodward, "Administration of a Plastics Control Laboratory," *Industrial Quality Control* (*IQC*), vol. 23, no. 2, pp. 68–71, August, 1966.

18-3 Accuracy and Precision

Even when correctly used, a measuring instrument may not give a true reading of a characteristic. The difference between the true value and the measured value can be due to problems of:

1 Accuracy. The accuracy of an instrument is the extent to which the average of a long series of repeat measurements made by the instrument on a single unit of product differs from the true value of that product. This difference is usually due to a systematic error in the measurement process. In this case, the instrument is said to be "out of calibration."

2 Precision. The precision of an instrument is the extent to which the instrument repeats its results when making repeat measurements on the same unit of product.[5] The scatter of these measurements may be designated as σ_E, meaning standard deviation of measurement. The lower the value of σ_E the more "precise" is the instrument (see Figure 18-2).

There is much confusion on terminology. Some literature uses the terms "accuracy" and "precision" interchangeably. The term "bias" is sometimes used to indicate a lack of accuracy. The confusion extends

Table 18-3 **Comparison of Catalog Statements**

	Manufacturer		
Instrument	*A*	*B*	*C*
Outside micrometer (0–1 inch)	Accuracy 0.0001 inch	Accuracy is maintained at 0.00005 inch	No statement
Dial caliper	Accuracy to 0.001 inch per 6 inches	Accuracy guaranteed within 0.001 inch	No statement
Electronic comparator	Repeatable accuracies to 0.000004 inch	It repeats to within 0.000002 inch	Total error is less than 1½% of full-scale reading

to instrument catalogs. Table 18-3 shows the statements on error of measurement as listed in three catalogs. The confusion is compounded because none of these catalogs defines the word "accuracy."

[5] For an excellent discussion of "accuracy" and "precision," see C. Eisenhart, "The Reliability of Measured Value: Part I Fundamental Concepts," *Photogrammetric Engineering*, vol. 18, no. 3, pp. 542–561, June, 1952.

Youden[6] raises a pertinent question: Is a procedure with a small systematic error to be preferred to one with no systematic error but with much poorer precision? These alternatives are illustrated in Figure 18-3. With measurement procedure A, a systematic error of minus 10 units is present and the standard deviation of repeat measurements among themselves is 10 units. With measurement procedure B, there is no systematic error (the average value of all measurements equals the true value), but the standard deviation of the measurements among themselves is equal to 20 units. Thus, procedure B is only half as precise as procedure A. To compare these two systems, one could calculate the percent of results which would be within given limits of the

[6] W. J. Youden, "How to Evaluate Accuracy," *Materials, Research and Standards,* vol. 1, no. 4, pp. 268–271, April, 1961.

Figure 18-2 Distinction between accuracy and precision.

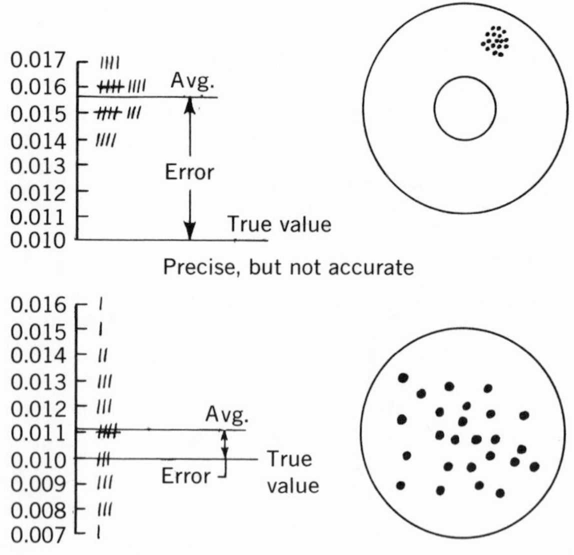

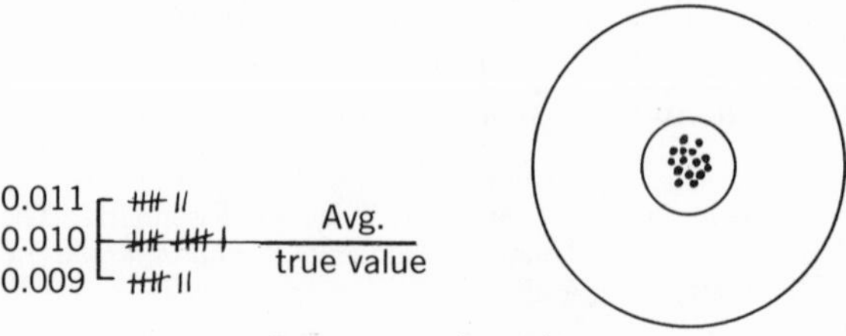

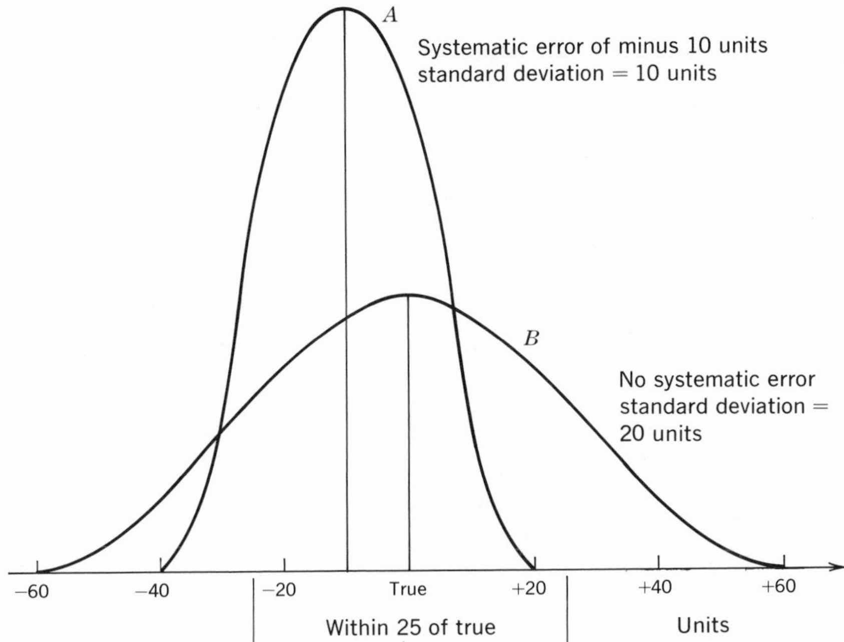

Figure 18-3 Which is the preferable measurement process? Source: W. J. Youden, "How to Evaluate Accuracy," *Materials, Research and Standards,* vol. I, no. 4, pp. 268–271, April, 1961.

true value. For example, under procedure A, 93.3 percent of the results will be within plus or minus 25 units of the true value compared to 78.9 percent with procedure B, suggesting that procedure A is preferable. However, this question illustrates how statistical analysis must be integrated with other facets of a problem. For example, if profit is related to the *average* content of an ingredient in the product, procedure B might be preferred because in the long run the average value will equal the true value, whereas in procedure A the average value will be 10 units less than the true value. In any event, both accuracy and precision are important in evaluating test procedures.

18-4 Effect of Measurement Error on Acceptance Decisions

The error of measurement is the difference between a true value and a measured value. This error can affect quality decisions on individual units of product and also decisions on lots submitted to sampling plans. These effects will now be discussed.

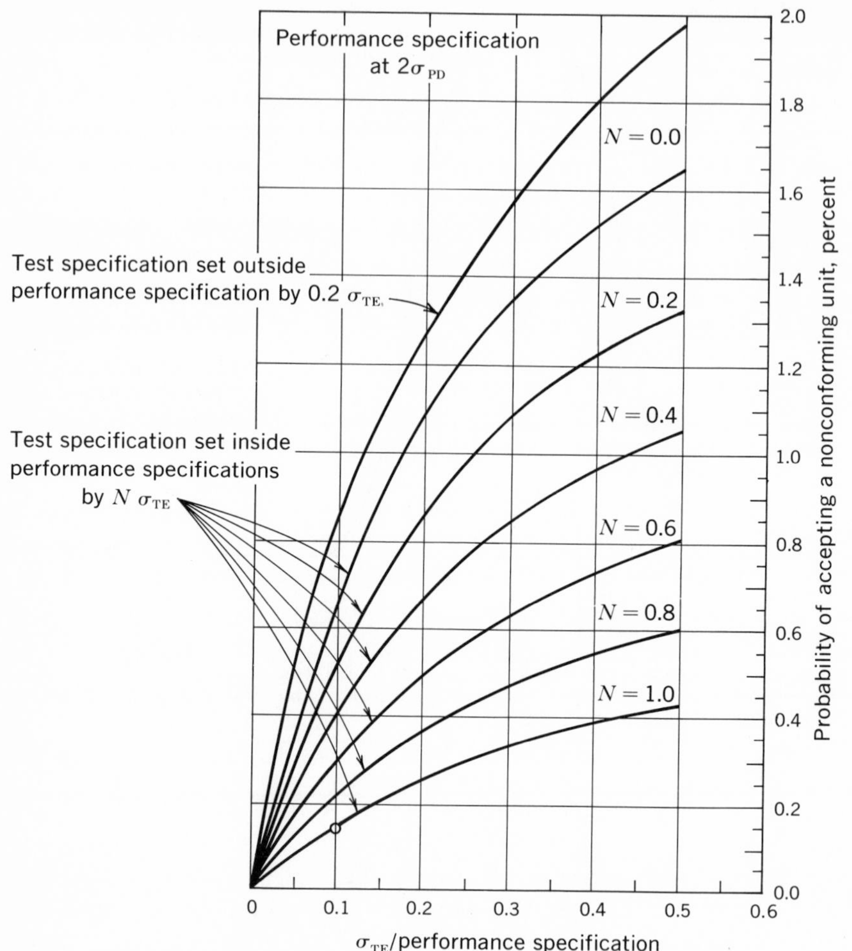

Figure 18-4 Probability of accepting a nonconforming unit. Source: A. R. Eagle, "A Method for Handling Errors in Testing and Measuring," *Industrial Quality Control*, pp. 10–14, March, 1954.

Measurement accuracy can affect the probability of two types of errors occurring in the classification of a product: (1) accepting a bad unit, and (2) rejecting a good unit. Figure 18-4 shows[7] the probability of accepting a nonconforming unit as a function of the measurement error (called $\sigma_{T.E.}$ by Eagle).

The horizontal axis of Figure 18-4 expresses the measurement error

[7] A. R. Eagle, "A Method for Handling Errors in Testing and Measuring," *IQC*, pp. 10–14, March, 1954.

as the standard deviation of the measurement error divided by the plus or minus value of the specification range (assumed equal to 2 standard deviations of the product).

For example, if the measurement error is one-half of the tolerance range, then the probability is about 1.65 percent that a nonconforming unit will be read as conforming (due to the measurement error) and, therefore, will erroneously be accepted.

Figure 18-5 shows the percent of *conforming* units which will be *rejected* as a function of the measurement error.

For example, if the measurement error is one-half of the plus or minus tolerance range, about 14 percent of the units which are really

Figure 18-5 Probability of rejecting an acceptable unit. Source: A. R. Eagle, "A Method for Handling Errors in Testing and Measuring," *Industrial Quality Control,* pp. 10–14, March, 1954.

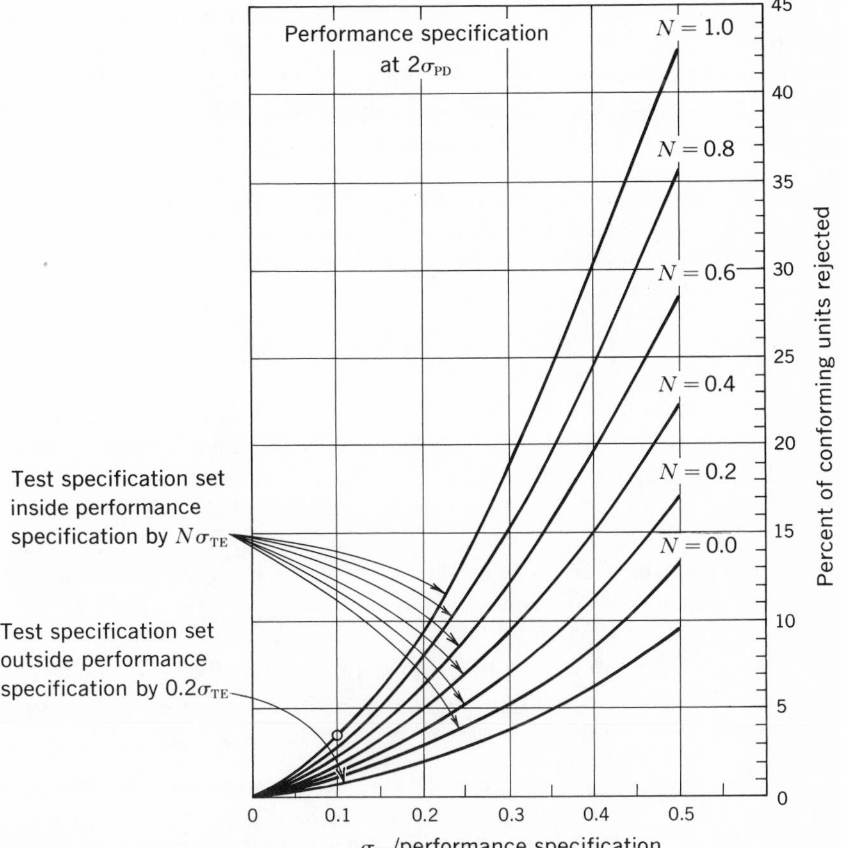

within specifications will be erroneously rejected because the measurement error will show these conforming units as being outside of specification.

As a rule of thumb, if the ratio of 3 standard deviations of measurement error to product tolerance is less than about 25 percent, then the effect of measurement error on decisions can usually be ignored.

The error of measurement can be allowed for adjusting test specification limits. Methods[8] have been developed reflecting different viewpoints on the relative importance of the producer's risk (of having a good unit rejected) and the consumer's risk (of having a bad unit accepted).

18-5 Statistical Aspects of Calibration

Because measuring instruments do not repeat their readings identically, any single reading is really a sample of the numerous readings possible. Suppose[9] a meter has a tolerance of ±1 percent and the calibration check requires an input of 1.000 volt. To be within calibration, then, the meter reading must be between 0.990 and 1.010. One reading will be taken and one of three conclusions drawn:

1 Meter is satisfactory and should remain in service.
2 Meter is unsatisfactory but should be (and can be) adjusted.
3 Meter is unsatisfactory but cannot be adjusted and must therefore be repaired.

Suppose a reading of 1.007 is obtained. Any single reading must be recognized as merely one value from a *population* of measurements having a certain mean and standard deviation. Thus, a reading of 1.007 might come from any one of many possible populations (a *few* of which are shown in Figure 18-6).

If a decision is made to adjust the meter by 0.007, there is a high likelihood that this will not only be wrong but that it will also perhaps make the total measurement error worse than it presently is. Note that only for the second distribution shown would an adjustment of 0.007 be correct.

Table 18-4 summarizes the comparison of possible decisions made from the results of a calibration check against the actual condition of the instrument. Making decisions on the basis of one measurement

[8] See *QCH*, 1962, p. 9-10.
[9] J. L. Berkowitz, "The Fallacy of Single Point Estimates in the Calibration Process," *Transactions of the Twentieth Annual Conference of the American Society for Quality Control*, 1966, pp. 962–972.

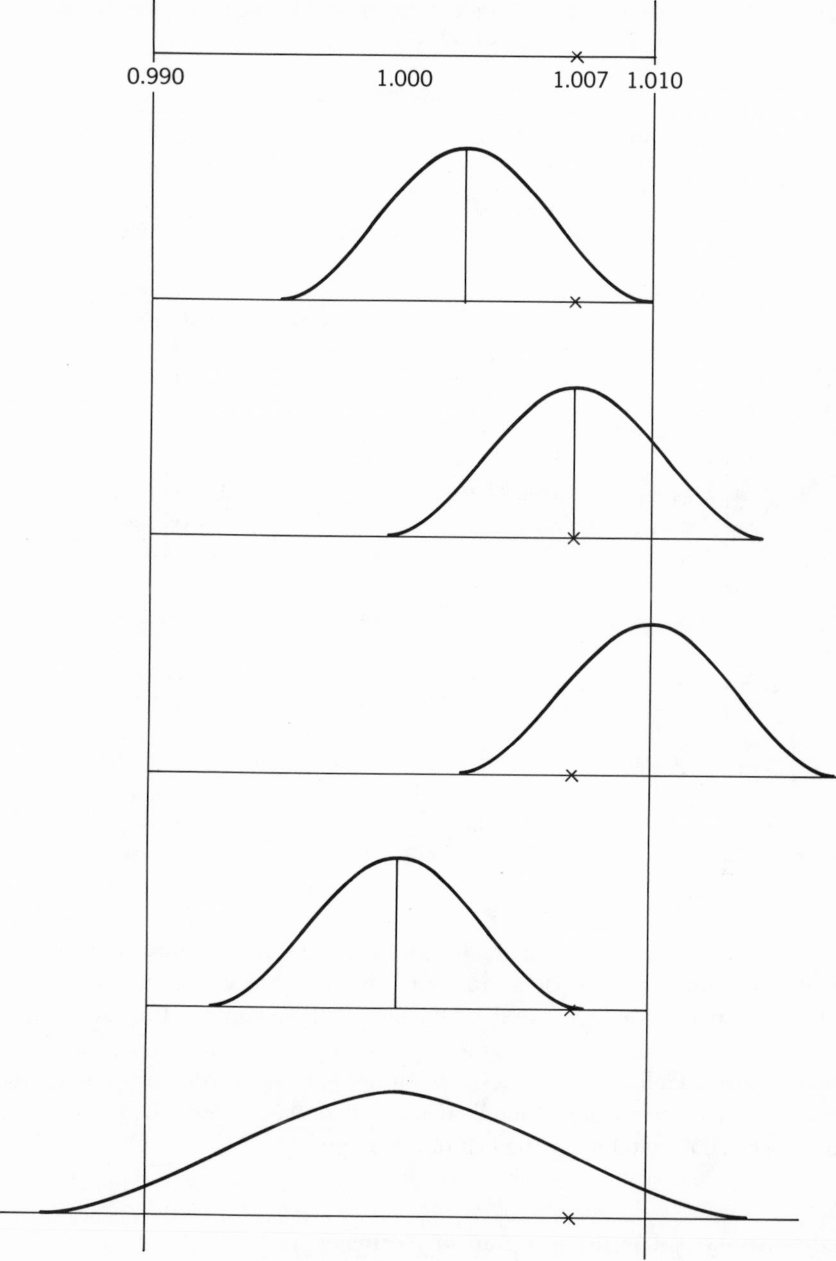

Figure 18-6 Some populations which could yield a reading of 1.007.

Table 18-4 COMPARISON OF POSSIBLE DECISION VERSUS ACTUAL INSTRUMENT CONDITIONS*

Actual Instrument Condition	Technician Decision: *Satisfactory (Continue in Operation)*	*Requires Adjustment*	*Requires Repair*
Satisfactory	Case 1 Proper decision	Case 2 Unnecessary adjustment expense, could cause bias, poor decision	Case 3 Unlikely since there is little cause to indicate repair is necessary
Unsatisfactory but Adjustable	Case 4 Acceptance of biased instrument, poor decision	Case 5 Proper decision, technician must now estimate the bias	Case 6 Unnecessary repair expense, poor decision
Unsatisfactory, Not Adjustable Requires Repair	Case 7 Acceptance of incapable instrument, poor decision	Case 8 Unnecessary adjustment could lead to instrument remaining in service, poor decision	Case 9 Proper decision

* J. L. Berkowitz, "The Fallacy of Single Point Estimates in the Calibration Process," *Transactions of the Twentieth Annual Conference of the American Society for Quality Control*, 1966, pp. 962–972.

will often result in the errors listed in the table. Figure 18-6 shows that by taking repeat measurements important knowledge about the distribution can be generated. Of course, the *value* of the knowledge from a distribution of measurements must be evaluated against the additional *cost*. Calibration checks using repeat measurements could not economically be justified for all measuring processes but the "vital few" processes could well meet the test of economics.

18-6 Analytical Determination of Accuracy and Precision

The accuracy of measurement can be quantified by determining the difference between the true value of a characteristic and the average value of multiple measurements made on the characteristic. The question of "how many measurements" can be answered by confidence limit concepts (Chapter 11). For example, suppose σ_E is estimated as 2.0 and it is desired to estimate accuracy within 0.5 unit at the 95 percent confidence level. Then 62 repeat[10] measurements should be taken on the same unit of product and the average of these calculated. The difference between the true value and the average of these repeat measurements is a measure of accuracy (or "calibration error" or "systematic error").

The basic procedure for determining the *precision* of a measuring process is to calculate the measurement variation that results if repeat measurements are made on one unit of product. The precision is calculated as the standard deviation of these repeat measurements *about their own average.*

It is often *assumed* that 6 of these standard deviations is less than 10 percent of the total tolerance range of the product being measured by the gage and thus "measurement error" is negligible. One company[11] decided to verify this by conducting an experiment. Repeat measurements (on the same part) were made by one inspector using one dial indicator to measure a 2-inch outside diameter. The precision in terms of 6 standard deviations was 30 percent (instead of the 10 percent assumed) of the tolerance range on the diameter.

18-7 Resolving Observed Values into Components of Variation

In drawing conclusions about measurement error, it is worthwhile to study the causes of variation of observed values. The relationship is

$$\sigma_{\text{observed}} = \sqrt{\sigma^2_{\text{cause A}} + \sigma^2_{\text{cause B}} + \cdots + \sigma^2_{\text{cause N}}}$$

The formula assumes that the causes act independently.

It is significant to find the numerical values of the components of observed variation, because the knowledge may suggest where effort

[10] For some products, particularly chemical products, there may be a choice between (1) taking a number of repeat measurements on the *same* item of product and (2) spreading the repeat measurements over a *number of samples* of the same product. The exact plan to be followed requires an analysis of the economics of obtaining a number of samples versus the cost of running an individual test. Other considerations are also involved. For more information, see Owen L. Davies (ed.), *Design and Analysis of Industrial Experiments,* Hafner Publishing Co., New York, 1954, chap. 4.

[11] Private communication to one of the authors.

should be concentrated to reduce the variation in the product. An analysis of product variation into the true product variation plus the other causes of variation may indicate important causes other than the manufacturing process. Thus, if the *measurement* error is found to be a large percentage of the total variation, then some fundamental questions should be posed on the direction of the quality improvement program. If the measurement error is large, then finding the detailed components (e.g., instrument, operator) of this error may help to reduce the measurement error which in turn may completely eliminate a quality problem.

For example, the observations from an instrument used to measure a series of different units of product can be viewed as a composite of (1) the variation due to the measuring method and (2) the variation in the product itself. This can be expressed as

$$\sigma_0 = \sqrt{\sigma_P^2 + \sigma_E^2}$$

where $\sigma_0 = \sigma$ of the observed data
$\sigma_P = \sigma$ of the product
$\sigma_E = \sigma$ of the measuring method

Now solving for σ_P

$$\sigma_P = \sqrt{\sigma_0^2 - \sigma_E^2}$$

If σ_E is less than one-tenth σ_0, then the effect upon σ_P will be less than 1 percent. A common rule of thumb requires that the instrument should be able to divide the tolerance into about 10 parts. As the effect is less than 1 percent, the rule of thumb seems uneconomically conservative for most applications.

For example, in the use of a certain new design of indicator gage, it was found that the σ_0 was 0.0011 inch. The shop questioned the validity of the gage. An experiment was conducted by having the gage check the same unit of product over and over again. The σ_E of these repeat readings was 0.0002. This includes the variation due to the instrument and the operator. Then,

$$\sigma_P = \sqrt{(0.0011)^2 + (0.0002)^2} = 0.00108$$

This was convincing proof that the errors in the indicator did not materially exaggerate the variation in the product.

In another instance, involving the efficiency of an air-cooling mechanism, σ_0 was 23 and σ_E was 16. Then

$$\sigma_P = \sqrt{(23)^2 - (16)^2} = 16.5$$

The measurement variation was almost as great as the product variation. Further study disclosed that the measurement variation was caused by several variables:

Variable	σ	σ^2
A	14	196
B	5	25
All other	7.2	52
Total		273

It became clear that real progress could be made only by improving variable A, and the engineers took steps accordingly.

If the product is changed or destroyed by test, repeat tests are not feasible. In that event, it is sometimes possible to conduct, on the same unit of product, two or more different methods of test simultaneously.

Simon[12] relates the results of three simultaneous measures of the burning time of fuses for projectiles. The three observers A, B, and C all had electric clocks which were started automatically by the firing of the gun. Each observer stopped his clock at the instant of seeing the flash of the bursting projectile. The analysis revealed the relative importance of the measurement error and the relative effectiveness of the three observers.

Sometimes a carefully designed experiment is necessary to develop quantitative estimates of the component causes of variation. Table 18-5 shows the plan and results of an interlaboratory study[13] of the Eberstadt method for the determination of acetyl in cellulose acetate. A test consisted of two individual measurements made by the same analyst on the same day in the same laboratory.

In comparing the measurements of several different laboratories, it was felt that the variation could be due to the following sources:

1 Differences between laboratory procedures and conditions
2 Differences between analysts within each laboratory
3 Differences from day to day in measurements made by the same analyst within one laboratory
4 Differences in replicate measurements made on the same day by the same analyst within the same laboratory

In designing an experiment to evaluate measurement variation, it is *essential* that those who are experienced with laboratory procedures

[12] For a summary, see *QCH*, 1962, pp. 9-5 to 9-7.

[13] This example is based on "Design and Interpretation of Interlaboratory Studies of Test Methods," by Grant Wernimont, *Analytical Chemistry*, vol. 23, no. 11, pp. 1572–1576, November, 1951.

Table 18-5 **Summary of Results for Interlaboratory Study of Eberstadt Method for Determination of Acetyl in Cellulose Acetate***

Laboratory	Analyst	Day	Individuals		Within Days R_1	Within Days $\bar{X}_1$	Within Analysts R_2	Within Analysts $\bar{X}_2$	Within Laboratories R_3	Within Laboratories $\bar{X}_3$	Among Laboratories R_4	Among Laboratories $\bar{X}_4$
1	1	1	39.01	39.22	0.21	39.12	0.19	39.01	0.08	39.05	0.75	39.04
		2	38.97	38.98	0.01	38.98						
		3	38.86	39.00	0.14	38.93						
	2	1	39.16	39.11	0.05	39.14	0.09	39.09				
		2	39.13	39.06	0.07	39.10						
		3	39.05	39.05	0.00	39.05						
2	1	1	39.20	39.27	0.07	39.24	0.20	39.12	0.14	39.19		
		2	39.16	39.01	0.15	39.08						
		3	39.11	38.96	0.15	39.04						
	2	1	39.31	39.36	0.05	39.34	0.22	39.26				
		2	39.39	39.28	0.11	39.34						
		3	39.13	39.11	0.02	39.12						
3	1	1	39.26	39.39	0.13	39.32	0.20	39.24	0.05	39.22		
		2	39.34	39.24	0.10	39.29						
		3	39.13	39.11	0.02	39.12						
	2	1	39.15	39.27	0.12	39.21	0.08	39.19				
		2	39.24	39.21	0.03	39.22						
		3	39.07	39.20	0.13	39.14						
4	1	1	39.32	39.38	0.06	39.35	0.12	39.40	0.08	39.36		
		2	39.37	39.42	0.05	39.40						
		3	39.50	39.44	0.06	39.47						
	2	1	39.39	39.31	0.08	39.35	0.05	39.32				
		2	39.26	39.33	0.07	39.30						
		3	39.29	39.34	0.05	39.32						
5	1	1	38.94	39.08	0.14	39.01	0.04	39.02	0.05	39.00		
		2	39.01	39.01	0.00	39.01						
		3	39.07	39.03	0.04	39.05						
	2	1	38.89	39.01	0.12	38.95	0.08	38.97				
		2	38.92	38.97	0.05	38.94						
		3	39.03	39.02	0.01	39.02						
6	1	1	39.02	39.07	0.05	39.04	0.18	39.15	0.03	39.13		
		2	39.22	39.16	0.06	39.19						
		3	39.27	39.16	0.11	39.22						
	2	1	39.05	39.03	0.02	39.04	0.14	39.12				
		2	39.09	39.27	0.18	39.18						
		3	39.16	39.11	0.05	39.14						
7	1	1	38.77	38.84	0.07	38.80	0.19	38.72	0.01	38.72		
		2	38.79	38.69	0.10	38.74						
		3	38.69	38.53	0.16	38.61						
	2	1	38.69	38.67	0.02	38.68	0.14	38.73				
		2	38.84	38.80	0.04	38.82						
		3	38.69	38.69	0.00	38.69						
8	1	1	38.85	38.74	0.11	38.80	0.28	38.65	0.08	38.61		
		2	38.61	38.66	0.05	38.64						
		3	38.50	38.54	0.04	38.52						
	2	1	38.66	38.59	0.07	38.62	0.11	38.57				
		2	38.51	38.66	0.15	38.58						
		3	38.40	38.62	0.22	38.51						
		Av.			0.0790	39.04	0.1444	39.04	0.0650	39.04		

* From Grant Wernimont, "Design and Interpretation of Interlaboratory Studies of Test Methods," *Analytical Chemistry*, vol. 23, no. 11, pp. 1572–1576, November, 1951.

and those who will *use* the results help decide what causes of variation will be investigated in the experiment. In this experiment, the tests were made on one batch of the material and an experiment was designed to investigate the causes of variation listed above.

The control chart method[14] provides a simple graphical analysis of the data in Table 18-5. This is an alternative to the more complicated analysis of variance technique. An average and range control chart was used to evaluate the data listed under "within days" in Table 18-5. Each point on the average chart represented the average of the two measurements taken on a single day by a single analyst in one of the laboratories. Some of the averages were "out of control." A point on the range chart was the difference between the two measurements comprising the average plotted on the average chart. Thus, the range chart provides the comparison of the variation *between* days to the variation *within* days. The range chart indicated that the variation between duplicate measurements on the same day is essentially the same from day to day for each analyst and also for each laboratory.

Another control chart for averages and ranges based on the data labeled "within analysts" in Table 18-5 was prepared. Each point on the average chart was the average of the three daily averages compiled by one analyst in one laboratory. The range chart indicated that the day-to-day variation within the 16 analysts was fairly stable, although somewhat greater than the variation of duplicate tests made on the same day. The control limits for the average chart were based on the $\bar{R}$ within analysts. Again, some averages were out of control, indicating a significant variation between the laboratories, i.e., variation not simply due to the random day-to-day variation of analysts.

The approach mentioned in Chapter 15 for estimating a standard deviation from the average range can now be used to set up the following equations:[15]

$$\sigma_1 = \sigma_{\text{tests within days}} = \frac{\bar{R}_1}{d_2} = \frac{0.0790}{1.128} = 0.0700$$

$$\sigma_2 = \sigma_{\text{days within analysts}} = \frac{\bar{R}_2}{d_2} = \frac{0.1444}{1.693} = 0.0853$$

$$\sigma_3 = \sigma_{\text{analysts within labs}} = \frac{\bar{R}_3}{d_2} = \frac{0.0650}{1.128} = 0.0576$$

$$\sigma_4 = \sigma_{\text{labs}} = \frac{\bar{R}_4}{d_2} = \frac{0.75}{2.847} = 0.26$$

[14] The construction of the two control charts in this example is listed as a problem assignment at the end of this chapter.

[15] These equations have been approximated to simplify the discussion. See the original reference for a more exact approach.

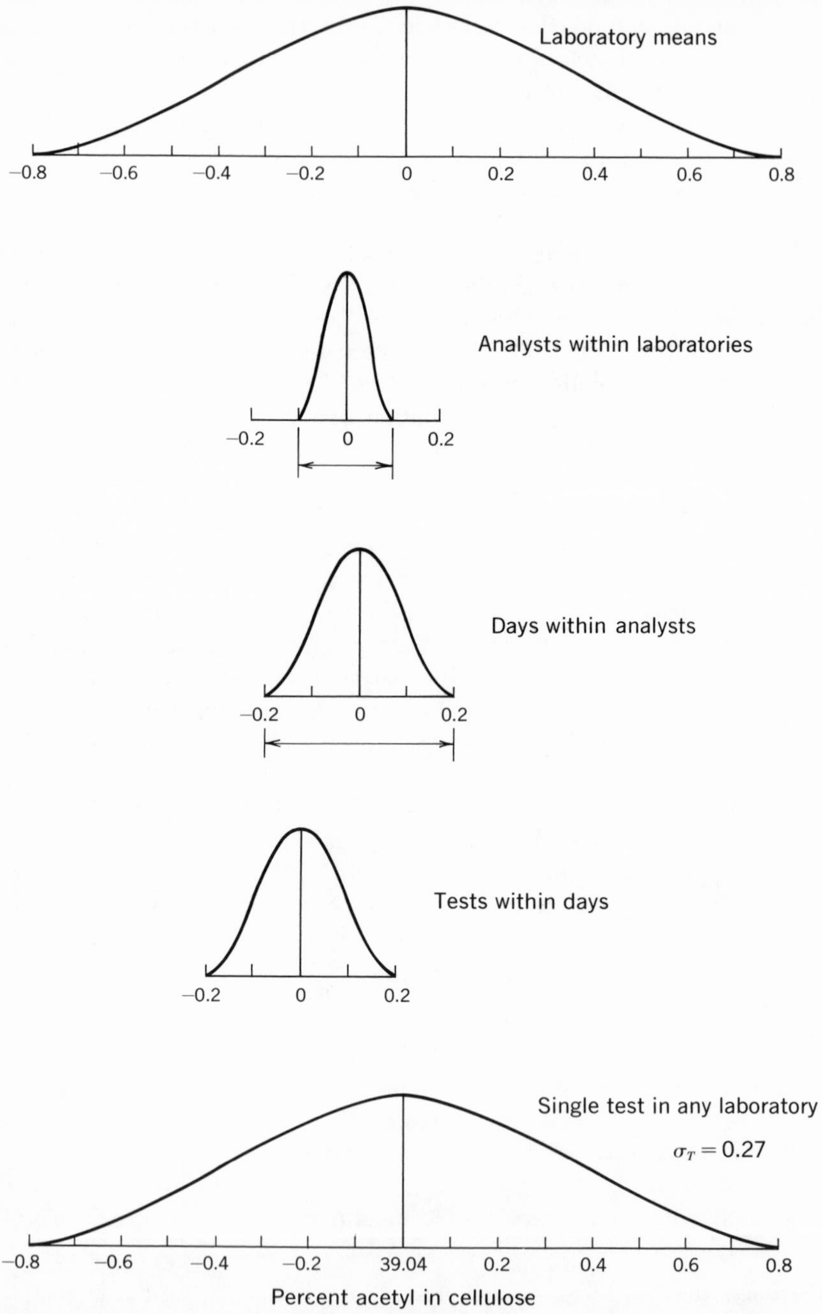

Laboratory means
−0.8
−0.6
−0.4
−0.2
0
0.2
0.4
0.6
0.8
Analysts within laboratories
−0.2
0
0.2
Days within analysts
−0.2
0
0.2
Tests within days
−0.2
0
0.2
Single test in any laboratory
σ_T = 0.27
−0.8
−0.6
−0.4
−0.2
39.04
0.2
0.4
0.6
0.8
Percent acetyl in cellulose

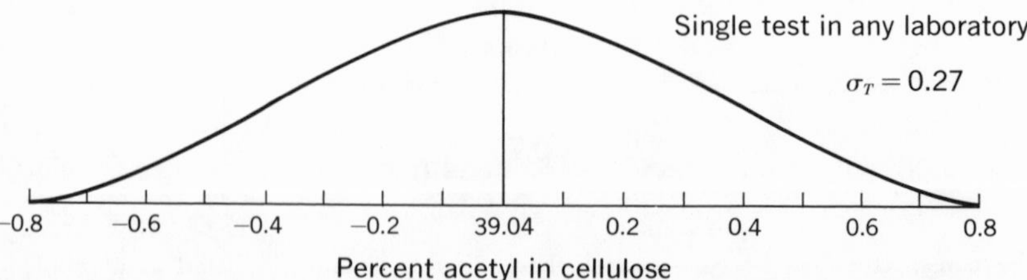

Figure 18-7 Distribution curves corresponding to components of variance. Source: Grant Wernimont, "Design and Interpretation of Interlaboratory Studies of Test Methods," *Analysis Chemistry*, vol. 23, no. 11, pp. 1572–1576, November, 1951.

The expressions provide the standard deviations of the component causes of variation in the overall measurement. The standard deviation of a single determination made in a given laboratory can now be computed as

$$\sigma_{\text{observed}} = \sqrt{(0.0700)^2 + (0.0853)^2 + (0.0576)^2 + (0.26)^2}$$
$$\sigma_{\text{observed}} = 0.29$$

(Note how the standard deviation[16] of observed values *mainly* depends on the largest single standard deviation of causes and *not* on the number of causes.) The relationship between the four components of variation and the total variation is shown graphically in Figure 18-7. The means of the curves were deliberately not lined up to show that the mean of each component's curve moved randomly between limits of the curve above it in the hierarchy. Thus, the means for analysts move (within limits) about their own laboratory mean. The variation between laboratories (standard deviation of 0.26) is much larger than any of the other component causes of variation and is, therefore, the *significant* cause. Once again, a statistical tool has helped identify the "vital element from the trivial many." This example also illustrates that diagnostic studies requiring special data are sometimes needed to obtain the breakthrough in knowledge for complex quality problems.[17]

18-8 Rounding Off of Measurements

Elimination of unneeded accuracy is a widespread practice. One aspect of this is rounding off, i.e., reading instruments to the nearest major calibration line. Such rounding off not only affects the precision of the measuring process but sometimes causes confusion in determining the probability distribution that the characteristic follows. Histograms of rounded-off measurements sometimes have so many peaks and valleys that the underlying shape of the distribution curve is lost in the detail of the peaks and valleys.

Figure 18-8 shows in detail how the tendency to round off can arise. Figure 18-9 shows the results of rounding off. The alternate high and low bars are the result of scale markings at the even-numbered values but not at the odd-numbered values. The inspector also flinched at the tolerance limit of 30 and rounded off values slightly above 30 to call them equal to 30 and therefore within the limit.

[16] If the more exact analysis of variance technique is used instead of ranges, then $\sigma_{\text{observed}} = 0.27$

[17] An excellent reference on interlaboratory studies is *ASTM Manual for Conducting an Interlaboratory Study of a Test Method*, ASTM Special Technical Publication, no. 335, 1963.

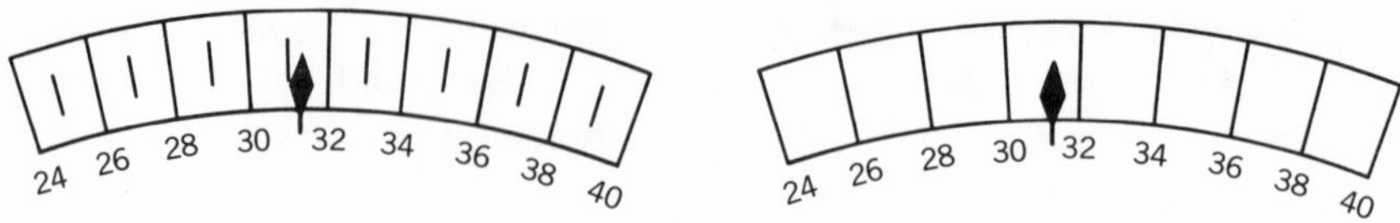

Figure 18-8 Why inspectors round off numbers at the gage markings.

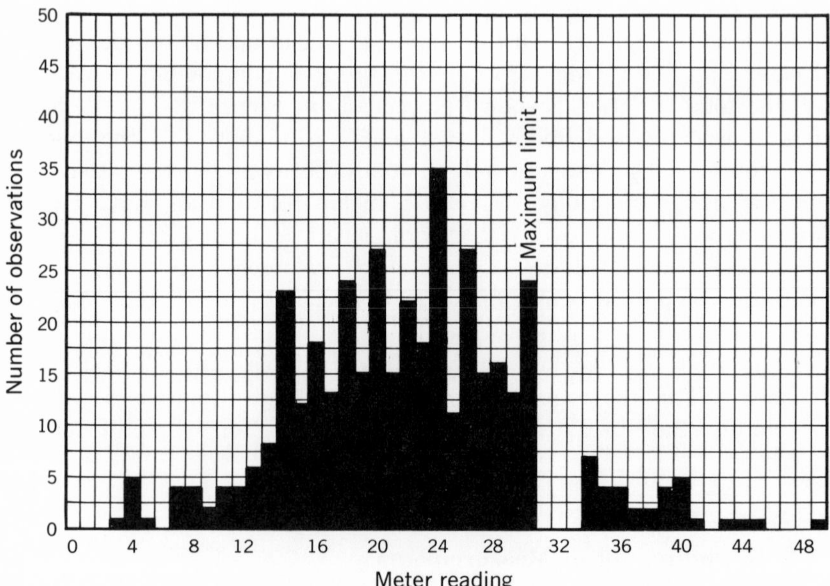

Figure 18-9 Results of rounding off numbers and flinching at design maximum.

Rounding off is easy to detect by analysis of the inspector's data. A good analyst can, from the data alone, reconstruct the pattern of scale markings of an instrument, without ever having seen the scale itself.

The prevalence of rounding off depends on many factors: the design of the instrument scale and pointer, the relation of instrumental precision to the product (or process) tolerances, the pressure of time, the intensity of supervision, the atmosphere of quality-mindedness.

18-9 Reducing and Controlling Errors

Accuracy and precision of measurement can be reduced either through:

1 Discovery of the causes of variation and remedy of these causes. This is the fundamental approach, and is at its best in discovering

inadequate training, perishable reagents, and other such soluble problems. This fundamental approach also points to other causes for which the remedy is unknown or uneconomic, i.e., basic redesign of the test procedure. In such cases resort is had to:

2 Multiple measurements and statistical methodology to control the error of measurement. The use of multiple measurements is based on the following relationship (see Chapter 11):

$$\sigma_{\bar{X}} = \frac{\sigma}{\sqrt{n}}$$

The formula states that to halve the error of measurement requires quadrupling (not doubling) the number of measurements.

After a certain number of tests have been conducted a significant reduction in the error of measurement can only be achieved by taking a *large* number of additional tests. This raises a question concerning the cost[18] of the additional tests versus the value of the slight improvement in overall accuracy. The alternative of reducing the causes of variation (by control charts or other techniques) must also be considered. An example of the use of a control chart will be presented next.

To check the measuring process on a viscosity test in the manufacture of cellulose esters, two tests were run each day on a known check batch.[19] A control chart for averages and ranges was set up and the average and range plotted for each day. This is shown in Figure 18-10. The averages for days no. 314 and 322 were beyond control limits, indicating a significant cause of variation in the measuring process. A general high trend also appeared during this period of time. A detailed investigation revealed that a change in the source of the solvent used in the test was made on day no. 308. It was concluded that the solvent caused the higher readings, and a return to the original source of the solvent was made on day no. 324. The chart then came back within control limits at the lower level.

Kroll[20] has reported extensive use of the control chart in a chemical laboratory. Over a 2-year period, the quality control program was able to place 75 to 80 percent of the routine inspection work load in statistical control. A reduction of about 50 percent in the standard deviation of measurement was usually experienced; for some tests, the error of

[18] See O. L. Davies, "Some Statistical Aspects of the Economics of Analytical Testing," *Technometrics*, vol. 1, no. 1, pp. 49–61, February, 1959.

[19] This example was taken from James A. Mitchell, "Control of the Accuracy and Precision of Industrial Tests and Analyses," *Analytical Chemistry*, vol. 19, no. 12, pp. 961–967, December, 1947.

[20] Frank W. Kroll, Jr., "Effective Quality Control Program for the Industrial Control Laboratory," *Transactions of the National Convention of the American Society for Quality Control*, 1958.

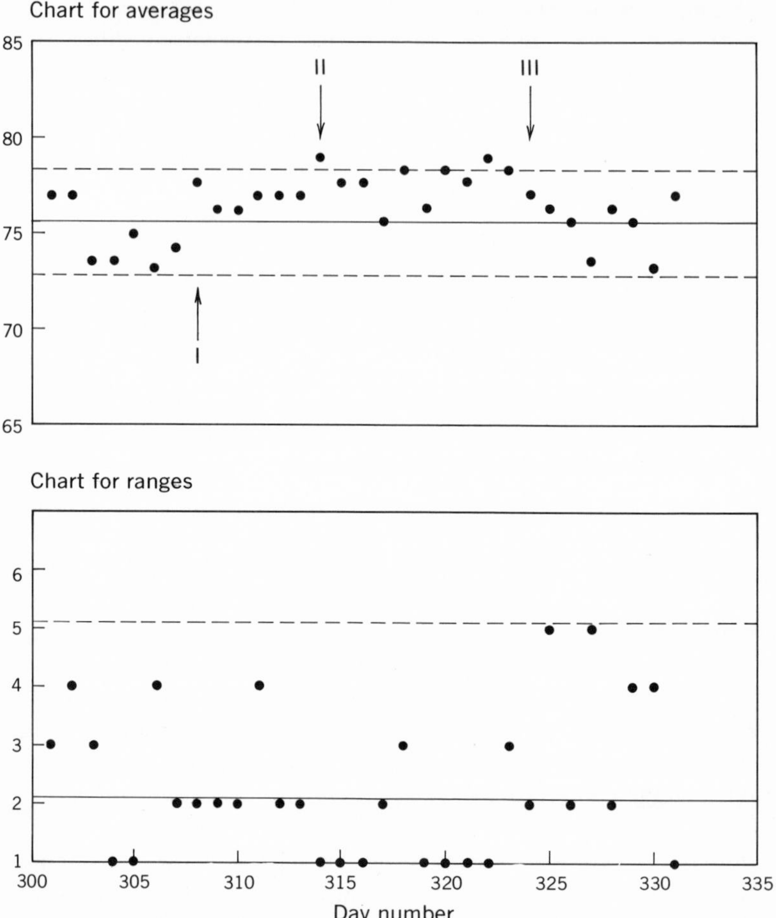

Figure 18-10 Control chart for laboratory test.

measurement was reduced to one-fifth of the error prior to use of control charts. When a test procedure showed no improvement after a control chart had been in use for several months, it was concluded that a major change in the procedure was required. Such a change was investigated from an economic point of view because an alternative to a new test procedure is to take repeat measurements using the old procedure and report average values instead of individual values. Kroll presents an example (Table 18-6) to show the dollars saved by a control chart (as the σ of the old procedure was 3 times as large, multitesting would have required 9 tests for each 1 under the new procedure).

Table 18-6 SAMPLE CALCULATION OF DOLLAR SAVINGS DUE TO REDUCTION OF STANDARD DEVIATIONS

Test Method	Vapor pressure
Unit Cost	X dollars*
Average no. of tests run per day	13
Standard deviation before control chart	1.0
Standard deviation after control chart	0.33
	Dollars per Day
Present cost to handle work load	13X
Cost of control chart analyses	X
Total cost of present system	14X
Cost of multitesting [(9)(13)(X)]	117X
Dollars per day saved	103X

* Unit costs vary from 2 to 20 dollars.

The control chart approach (1) provided an inexpensive method of improving testing variability for many tests at one time, and (2) helped to decide where concentrated analytical development effort should be applied. Only after these two facets had been investigated was duplicate testing considered as a means of reducing the error of measurement.

18-10 Programs for Maintaining Measuring Equipment

From the moment an instrument is put into use it begins to deteriorate in accuracy. To a degree, this takes place even when the instrument is not being used. If accuracy is to be maintained, there must be provision for checking and recalibrating the instruments.

Where there is no checking system in effect, a useful test for the need of a system is to check a sample of the instruments in use to see what is the actual condition prevailing. The facts from such a sample can form the basis of selling a broad gage control program. Thus, the breakthrough in attitude is achieved by facts instead of theoretical logic. It is not unusual to find that at least 25 percent of the measuring equipment is giving false readings.[21]

The procedure for maintaining accuracy of instruments involves many details. These details can be summarized in the form of a policy.

[21] See *QCH,* 1962, p. 9-16.

An example of this is the following,[22] which has been prepared for the "Mythical Corporation."

CONTROL OF MEASURING EQUIPMENT

Accuracy shall be controlled on all equipment used to measure:

1 Quality characteristics of material, parts, assemblies, and finished products.
2 Process conditions on which control of quality depends.

Authority
The order of authority of measuring standards is: National Bureau of Standards or an approved calibration agency; Mythical Central Standards to the extent available; Standards Laboratories within Divisions; Working Standards; Equipment directly used to measure quality characteristics and process conditions.

For most basic measurements, the results at the International Bureau of Weights and Measures are viewed as the ultimate standard. The U.S. National Bureau of Standards (NBS) derives its standards from the International Bureau. The hierarchy can then extend downward to a company standards laboratory that has "primary" standards traceable to NBS. The primary standards are used only to check "secondary" standards which in turn are used to check the equipment which performs the actual measurement on the product. This, of course, raises a question about the accumulation of errors through the hierarchy. It has been stated that each echelon should be ten times as accurate as the next one. Crow[23] has investigated this and concluded that the optimum ratio (based on minimizing total cost) should only be in the range of about 1.3 to 5.

Calibration

1 Each unit of measuring equipment shall be calibrated periodically against appropriate equipment above it on the authority list.
2 Schedules for calibration shall be established for each unit of measuring equipment. Among the factors considered in establishing such schedules shall be: severity of environment, severity of use, frequency of use, delicacy of the measuring equipment, accuracy of

[22] *Quality Management*, American Society for Quality Control, Aircraft and Missiles Division, sec. 5, 1965.
[23] Edwin L. Crow, "Optimum Allocation of Calibration Errors," *IQC*, vol. 23, no. 5, pp. 215–219, November, 1966.

measurement required, calibration history, importance of the characteristic or condition measured.
3 Calibration shall be performed to: published standard practices, manufacturer's written instructions, Mythical written instructions, other Mythical approved instructions.
4 Posted schedule: shall represent the maximum calibration interval; can be based on interim maintenance when
a Interim maintenance is scheduled.
b Interim maintenance is noted in the instructions; shall reference the document that "authorized and detailed the basis for" the calibration intervals.

The frequency of calibration has been surveyed.[24] Table 18-7 summarizes the results reported.

Table 18-7 SUMMARY OF CALIBRATION SURVEY

Area of Measurement	*Frequency of Calibration*
Dimensional; optical; temperature and humidity; shock, vibration, force; microwave electrical; pressure, vacuum, fluid flow	Six months or longer (according to 75 to 80 percent of the companies—depending on the specific area)
Radio frequency	47 percent of the companies calibrated every 1 to 3 months; 45 percent calibrated 6 months or longer
Infrared	All companies calibrated every 12 months or longer
Radiological	86 percent of the companies calibrated every 12 months or longer

The checking interval is determined by agreement among the interested supervisors, considering such factors as the kind and frequency of usage and the rate of deterioration. Sometimes the checking interval is in terms of time—every week, every hour, every month; sometimes the checking interval is based on amount of use—every 5,000 parts, every lot. For electrical test equipment, the amount of use is related to the length of time of "power on." There is an emerging practice of using an electrical elapsed time indicator working on the coulometer principle to record cumulatively the "power on" time. Outwardly, the device looks like a small calibrated thermometer tube.[25]

[24] *Calibration and Standards Questionnaire*, Aerospace Industries Association, Jan. 9, 1959.
[25] Dr. C. Beusman, "How Much Time on The Equipment?," *Electronic Products*, November, 1966.

Adherence to the checking schedule makes or breaks the system. In a time-interval system, each instrument is numbered, a gage card established, and a "flag" placed at the appropriate place on the calendar printed at the top of the card. If these cards are sorted in sequence of date, the cards that "mature" on a given day identify which instruments are to be called in for check. (If the information is analyzed by data processing equipment, a computer printout provides a list of equipment to be checked each day.) If there is a central measurement laboratory the instruments are called in. Some companies have found it possible to achieve substantial savings by introducing either (1) traveling gage carts or (2) "branch" gage-checking stations.

The branch checking station is useful for servicing a large shop department. Often it is set up in conjunction with the tool crib.

Any decentralization of the gage-checking facilities involves some added investment in "masters" and in other equipment. However, it usually reduces investment in spare gages by reducing the time the gage is out of service while being checked.

In some instances it is feasible to provide the operator with means of checking his own instrument. This is done through the use of masters. For example, the operator of a sorting machine may be given a go and a not-go master to try the machine for calibration. The machine must pass the go master a specified number of times in a row and similarly must reject the not-go master in order to be considered in calibration. The masters are used, and so checked, infrequently.

Returning to Mythical Corp. statements from *Quality Management:*[26]

Identification

1 Positive identification of calibration status of all measuring equipment shall be provided by means such as: color codes, tags, labels, stamps.

2 Calibration status identification shall, when practical, include: date of calibration, identification of person performing calibration, due date for next calibration.

3 When calibration status identification is not practical an audit system will be provided that will: monitor recall records, provide record of monitoring operation.

Production tooling used as media of inspection.

1 When production jigs, fixtures, tool masters, and other such devices are used as media of inspection, they shall be initially inspected or, by other suitable means, proved for accuracy prior to release for production use.

[26] *Op. cit.*

2 These devices shall be reinspected or proved at established intervals.
3 Inspection and/or proving shall conform to requirements of item 3 under "Calibration."

Environmental controls during calibration.

1 Measuring and test equipment and measuring standards shall be calibrated in an environment controlled to the extent necessary to assure continued measurements of the required accuracy.
2 Measuring and test equipment and measuring standards shall be utilized in an environment controlled to the extent necessary to assure continued measurements of the required accuracy.
3 Controlled environment includes, as a minimum: temperature, humidity, vibration, cleanliness, storage, handling.
4 When applicable, compensating corrections shall be applied to calibration results obtained in an environment which departs from standard conditions.

Records shall be maintained that provide:

1 Calibration source per MIL-C-45662A para 3.2.5.
2 Individual record of:
a Items noted under "Calibration."
b Items noted under "Production tooling used as media for inspection."

Individual records shall provide for:

1 Nomenclature (description) (include source when applicable).
2 Identification number.
3 Calibration interval (if not provided elsewhere in system).
3 Reference to calibration Instruction (if not provided elsewhere in system).
5 Identification of person performing the calibration.
6 Date of calibration.
7 Due date for next calibration.
8 Actual measurements obtained and/or other entries specified by Instruction.
9 Location of item or name of person to whom item is assigned.

Subcontractor calibration:

1 Subcontractors are required to maintain a calibration system that essentially meets these requirements.

2 When MIL-C-45662 is specified by contract, it shall be contractually extended to subcontractor.

The foregoing procedure for Mythical Corporation is a rather strict, rigorous system which is responsive to the exacting demands of aircraft and missile production. Less complex products require less complex systems. For a very practical approach, see Meckley.[27]

18-11 Conflict in Measurement

Much industrial controversy arises because of differences in measurement. Production departments disagree with inspection results. Test laboratories obtain conflicting results. Vendor and buyer measurements

[27] D. G. Meckley, III, "How to Set Up a Gaging Policy and Procedure," *American Machinist,* vol. 99, no. 6, pp. 133–144, Mar. 14, 1955. (This is summarized in *QCH,* 1962, pp. 9-19 to 9-20.)

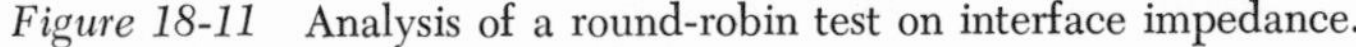

Figure 18-11 Analysis of a round-robin test on interface impedance.

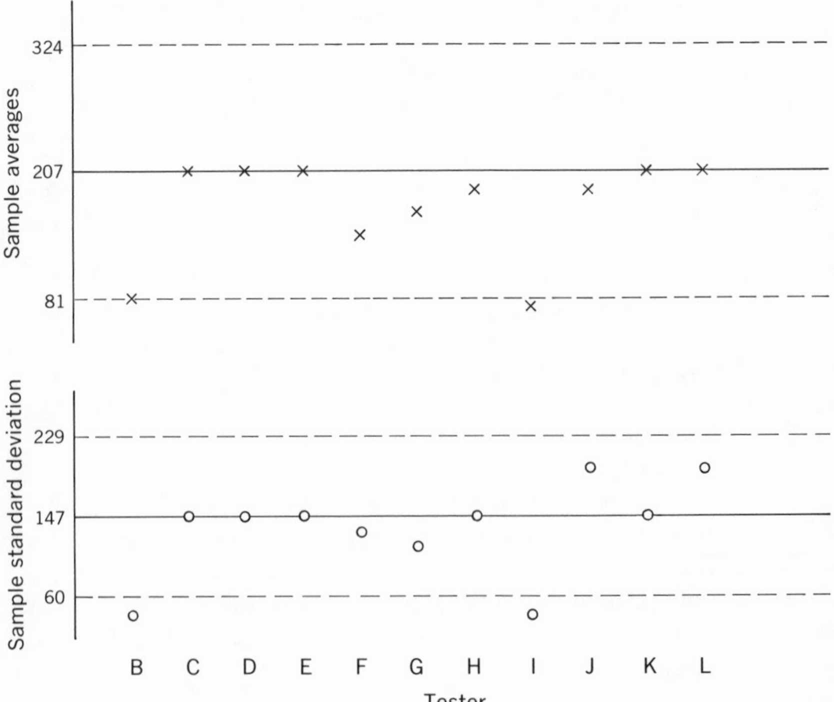

disagree due to a difference in measuring instruments or procedures. To resolve these cases, pairs of readings should be made by the two instruments on identical units of product and an analysis made for differences in both accuracy and precision.

When pairs of values cannot be obtained (such as in destructive tests) the order of production and measurement should be recorded so that samples are as alike as the process permits. Averages and standard deviations should be computed and tested for significance.

In some instances it may be necessary to develop a new system of measurement through the cooperative action of many companies in an industry. A series of "round-robin" tests may have to be organized to secure adequate data and to disclose to each company the relative position of its measuring instruments to the aggregate.

Such a round-robin test requires much advance planning and collaboration. The best means for achieving so broad-scale a result is through a committee established by some recognized standards committee or standards association.

Figure 18-11 shows a control chart analysis of a round-robin test on interface impedance of a certain tube. The same 12 tubes were sent to each of 11 test laboratories. Note that one laboratory read significantly low on the average but two laboratories were significantly *low* in variation. A nonstatistical review of the data incorrectly concluded that there was no significant difference between the laboratories.

Other graphical procedures[28] have also been applied in comparing laboratories.

PROBLEMS

1 From literature or discussions with appropriate people, compare the claims on measurement error of several brands of one of the following: (*a*) a measuring instrument acceptable to the instructor, (*b*) a tire pressure gage, (*c*) a household scale for weighing adults, and (*d*) a body temperature thermometer.

2 Define the objectives and plan an experiment to learn about the measurement error of one of the devices in problem 1.

3 An instrument has been used to measure the length of a part. The result was 2.638 inches. An error of measurement study was made on the instrument with the following results:

[28] See *QCH*, 1962, p. 9-9.

Accuracy: +0.001 inch (on the average, the instrument reads 0.001 inch high)

Precision: 0.0004 inch (1 standard deviation)

Make a statement concerning the true value of the part just measured. State all assumptions needed.

4 The inside diameter of 20 rings was measured at the same location by each of four inspectors using two types of bore gages. The *total* tolerance on the dimension was 0.0007 or 7.0 in units of ten-thousandths of an inch. Variability due to various causes was calculated in terms of 6 standard deviations and is summarized below:

Source	*Variability*
Product	4.6
Air gage	1.1
Mechanical gage	4.4
Inspector A	4.3
B	3.7
C	5.1
D	9.3

a Calculate the observed variability with the *best* combination of product, gage, and inspector. What percent of rings will be rejected in the long run?

b Calculate the observed variability with the worst combination of product, gage, and inspector. What percent of rings will be rejected in the long run?[29]

5 In section 18-6, it was stated that 62 measurements were required to achieve a specified accuracy. Show the calculations leading to this sample size.

6 The precision of a certain mechanical gage is indicated by a standard deviation (of individual repeat measurements) of 0.001 inch. Investigate the effect on precision of making multiple measurements. Consider 2, 3, 4, 5, 10, 20, and 30 as multiples. Graph the results.

7 A control chart analysis was made on a test method that "had been in use for a number of years and was thought to be very reliable." "Blind and random" submissions of a standard sample showed the following results for samples of four measurements of the percent metal in product 1301:

[29] Data taken from Roy A. Wylie, "How Accurate Are Your Measurements?," *Transactions of the National Convention of the American Society for Quality Control, 1963,* p. 92.

Date	*Average*	*Range*	*Date*	*Average*	*Range*
3/29	10.27	0.05	7/2	10.24	0.18
4/6	10.25	0.11	7/9	10.30	0.04
4/13	10.26	0.03	7/12	10.29	0.12
4/19	10.29	0.02	7/15	10.27	0.12
5/3	10.27	0.28	7/27	10.29	0.10
5/7	10.26	0.24	8/5	10.30	0.18
5/13	10.31	0.12	8/11	10.30	0.15
5/24	10.19	0.17	8/17	10.24	0.15
5/28	10.25	0.13	8/23	10.20	0.10
6/4	10.24	0.10	8/27	10.18	0.15
6/21	10.17	0.37	9/2	10.15	0.14
6/28	9.96	0.25	9/10	10.18	0.09

As the chart was plotted and analyzed, several changes were made. On 6/21, the technician was changed and an examination of the equipment and procedure was started. On 7/12, the equipment was changed and the original technician was placed back on the job. Calculate limits and plot a control chart for averages and ranges. Comment on the chart.[30]

8 One ball bearing was measured by one inspector 13 times with each of two vernier micrometers. The results are shown below:

Measurement Number	*Model A*	*Model B*
1	0.6557	0.6559
2	0.6556	0.6559
3	0.6556	0.6559
4	0.6555	0.6559
5	0.6556	0.6559
6	0.6557	0.6559
7	0.6556	0.6559
8	0.6558	0.6559
9	0.6557	0.6559
10	0.6557	0.6559
11	0.6556	0.6559
12	0.6557	0.6560
13	0.6557	0.6560

Suppose the true diameter is 0.65600. Calculate measures of accuracy and precision for each micrometer. What restrictions must be placed on the applicability of the numbers you determined?

9 Visit a local plant or school laboratory and prepare a statement summarizing the efforts made to control the error of measurement.

[30] Data taken from Kroll, *op. cit.*, pp. 1–15.

19 Vendor Relations

19-1 The General Problem of Vendor Quality

The vendor quality problem can be immense. Purchases from vendors are in any case a substantial part of the company budget. It is not unusual for the dollars of purchases to represent over 50 percent of net sales. These purchased goods contribute not only to the company's costs; they contribute also to the company's quality. For modern products this contribution can be decisive as to final product quality, for better or for worse.

The vendor quality problem has also been undergoing extensive change. In earlier centuries, vendors provided standard materials, natural or semiprocessed. The manufacturers then used these materials to make the finished products. Under such conditions, the quality problems were not severe, by today's criteria. Materials, while standard, were notoriously variable, and the buyer designed his products and processes accordingly. Both vendor and buyer guarded their trade secrets, so that there was little communication other than the specification, the contract, and the complaints.

Even in the twentieth century, there still remains a considerable residue of purchase of standard natural or semiprocessed materials. For such purchases the situation of two "independent worlds" for vendor and buyer is still very much the rule.

However, the twentieth century has also ushered in a totally new type of purchased goods—complex components and subsystems, highly engineered and exhibiting critical interfaces with other components and subsystems. Vendor relations for these new goods cannot be conducted under the plans in use in other centuries. Hence totally new plans of vendor relations have been evolved, for which the key feature is *interdependence* on a broad front, economic, managerial, and technological.[1]

Whether for simple or complex purchases, the ultimate objective in vendor quality relations is production of material which so adequately conforms to the buyers' requirements that there is no need for extensive acceptance or corrective procedures by the buyer. The activities needed to achieve this objective include:

1 Thinking through and publishing the company's policy of vendor quality relations
2 Using multiple vendors for major procurement items
3 Developing methods for identifying qualified vendors, and for eliminating those who are unable to meet quality requirements
4 Communicating essential and helpful information—designs, specifications, standards, standard practice, etc.
5 Communicating engineering changes promptly
6 Developing methods of detecting deviations promptly, especially through preproduction or trial runs
7 Providing for use of vendor quality data in lieu of incoming inspection
8 Developing methods of giving helpful assistance to vendors on their quality problems
9 Reviewing the performance of vendors through vendor rating or other plans and following up on the chronically poor vendors

The number of these activities to be conducted and the degree of formality to be used depend on a variety of factors: the nature of the goods purchased, the volume of purchases, the number of vendors, the extent of repeat purchase of the same goods, and the extent to which the purchases include collateral services provided by the vendor (research, design, subcontract management, etc.).

Table 19-1 shows the vendor relations activities usually appropriate for some situations commonly encountered.

19-2 Vendor Quality Policy

A limiting factor in vendor relations is that neither the vendor nor the buyer understands his responsibilities to the other. Some companies have prepared a statement of vendor quality policy.

An admirable example is that of Johnson & Johnson Company of

[1] For elaboration, see J. M. Juran, "Vendor Relations—An Overview," *Quality Progress*, vol. 1, no. 7, July, 1968.

Table 19-1 **Applicability of Defect Prevention Activities**

<table>
<tr><th rowspan="2">With</th><th colspan="2">And When Procurement Involves</th></tr>
<tr><th>Repeat Purchases of Same Items</th><th>New or One-shot Purchases</th></tr>
<tr><td rowspan="2">Few vendors and/or large volume and/or a complex item</td><td colspan="2">Vendor quality policy statement
Use of multiple vendors
Vendor quality survey
Exchange of standards and practices
Implementation of design changes
Vendor process control
Source inspection
Assistance to vendor on quality</td></tr>
<tr><td>Use of vendor quality data
Quality level agreements</td><td>Participation in planning
Preproduction evaluation
Pilot assembly lines</td></tr>
<tr><td rowspan="2">Many vendors and/or small volume and/or a standard item</td><td colspan="2">Vendor quality policy statement
Exchange of standards and practices
Implementation of design changes</td></tr>
<tr><td>Use of vendor quality data
Quality level agreements
Vendor rating</td><td>Vendor rating</td></tr>
</table>

Table 19-2 **Examples of Policy Statements**

Our Responsibilities to Our Suppliers:

1 To establish, through an understanding of Johnson & Johnson in its entirety, a definite conviction on the part of our suppliers that there exists a sound basis for a mutually profitable, long-range, friendly relationship.

2 To provide our suppliers with carefully written, completely detailed specifications that will establish, without question, what we require.

3 To develop with our suppliers the knowledge and the conviction that our most important requirement is the end quality of the product or service being purchased.

Our Suppliers' Responsibilities to Johnson & Johnson:

1 To supply materials or services to our specifications on a prompt delivery schedule.

2 To realize that it is not sufficient to accept returns willingly or negotiate disposition of materials not delivered to specification. A supplier should view such instances objectively and work constructively with us to correct the conditions that brought about the delivery of unsatisfactory material or services.

3 To look upon our association as a long-term working partnership and always be governed by the spirit of, "What is best for our partnership."

New Brunswick, New Jersey. Some *excerpts*[2] from the statement are shown in Table 19-2.

When broad quality policies are translated into specific actions, there is much variation among companies. The Vendor-Vendee Technical Committee of the American Society for Quality Control surveyed 209 companies. The response[3] to the 27 questions asked showed great variation. For example:

56.2 percent of the companies surveyed new suppliers prior to placing an order for most products.
42.0 percent periodically audited a supplier's program for most products.
42.7 percent required suppliers to have written inspection and test instruction sheets for most products.
57.1 percent accepted certified test results in lieu of testing for most products.

19-3 Means for Qualifying Sources of Supply

The evaluation of vendors as part of the vendor selection process can be performed in several ways.[4]

1 General reputation of the vendor. For large, well-known companies, there is such a thing as a general reputation (good or bad) and some use is made of it. For less well-known companies, the reputation is local and obscure. The effort to piece together what is the company's reputation is fully as great as the effort required to conduct a survey, and may be no more reliable. As matters stand, the general reputation of any but large well-known companies is a vague, imprecise basis for vendor evaluation before delivery.

2 Data from other buyers. On the face of it, the customers of this vendor should have good data on his actual prior performance. When one gets into it, this approach turns out to be not too helpful. The companies are generally unwilling to give out such data except to a data bank which keeps the data anonymous (and even then with reluctance). If what the companies buy (from this vendor) is the same product we plan to buy, they are likely our competitors, and hence unlikely to help us out. If what they buy from this vendor is something different from what we plan to buy, the data may not be pertinent.

[2] For the complete statement, see J. M. Juran (ed.), *Quality Control Handbook* (*QCH*), 2d ed., McGraw-Hill Book Company, New York, 1962, pp. 24-3 to 24-5.
[3] Report of the American Society for Quality Control, Vendor Control Subcommittee, May, 1966, available from American Society for Quality Control A&M Seminars, Imperial Reproductions, 105 Berlin Road, Cherry Hill, N.J. 08034.
[4] Juran, "Vendor Relations—An Overview," p. 13.

3 Data from a data bank. This device, which has been effective in the financial function, is just in process of development in the quality function. [This will be discussed in section 19-5.]
4 Vendor surveys. This will be discussed in the next section.

The actual decision in vendor selection is made by the company's purchasing department, which must consider all aspects of vendor relations. The more nearly the quality manager prepares vendor quality information in a way which is meaningful to the purchasing department, the more likely is this information to be given weight in vendor selection.

19-4 Vendor Quality Survey

A vendor quality survey is an evaluation of a vendor's ability to meet quality requirements. The results of the survey are used in the vendor selection decision or, if the vendor has already been chosen, the survey alerts the purchaser to areas where the vendor may require assistance in meeting requirements.

The survey can vary from a simple questionnaire sent to the vendor to a visit to the vendor's facilities by a team of specialists from the purchaser. The survey evaluates the vendor on:

1 Quality policies and practices. Policies are written or unwritten statements which define guidelines for the vendor's quality program. As guidelines they are really intentions which will be implemented to a variety of degrees. The survey has the problem of evaluating the policies and also determining the degree to which they are implemented.
2 Facilities. These not only include manufacturing facilities but inspection and test and any other facilities required to meet quality requirements for the product. The team may even request information on process capabilities to determine if the vendor can meet product tolerances. Samples of product may be checked with the vendor's gages and the buyer's gages to compare the gaging systems.
3 Procedures. This includes procedures for handling quality problems such as gage control, deviations from specifications, etc. These are often completed in a manual of quality procedures (see Chapter 29). Of course, the written procedure states what *should* be done. The survey faces the problem of determining if the procedure is really followed. (Sometimes it is not, for a good reason: the written procedure is out of date.) A survey of procedures may also include a sampling of inspection procedures to determine, for important characteristics, *if* a written procedure has been prepared and whether it is adequate.

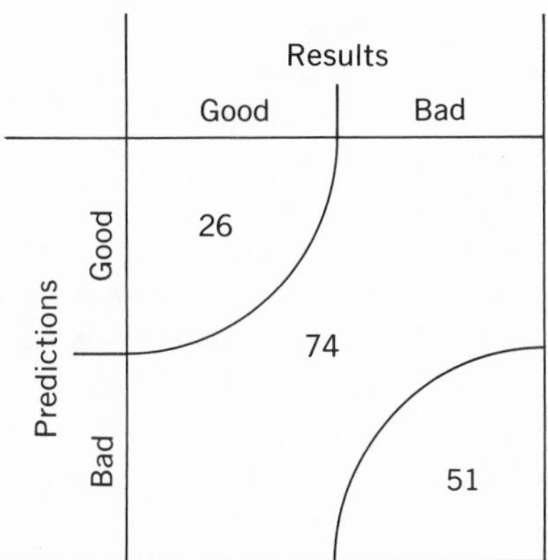

Figure 19-1 Ability of surveys to predict actual product quality.

4 Appraisal of the personnel. This difficult task evaluates the determination of the vendor's people to meet the specification and also their technical competence. The survey is weak in this area not only because of the subjective nature of the appraisal but also because of changing conditions due to turnover of key personnel.

Although the objectives of vendor surveys are sound, objectives are not always achieved. Brainard[5] presents a sobering example of failure. He compared the actual product quality of 151 currently active vendors to their ratings in the vendor survey made by the company. He concluded that the survey program was "completely ineffective":

1 There was no difference in product quality for vendors with acceptable ratings versus unacceptable ratings.
2 Company buyers did not show any preference for vendors with acceptable ratings.
3 The surveyors were not objective and could not reliably predict vendor quality performance (see Figure 19-1). Thus, 74 of 151 surveys incorrectly predicted product quality. As a survey costs several hundred dollars, a penny coin may have been an adequate substitute.
4 The significant elements of a quality system "were not required, were not measured, or were not measurable by the Company's system." Further, there was no relationship between the elements required of the vendor and his product quality.
5 The survey program did *not* result in reduced inspection.

[5] Edgar H. Brainard, "Just How Good Are Vendor Surveys?" *Quality Assurance*, pp. 22–25, August, 1966.

The company involved junked the survey program and substituted a simpler program. Prospective suppliers are asked to submit written information on test equipment, certifications, organization charts, and operating procedures. This permits objective evaluation and uncovers obvious deficiencies. In addition, the first article delivered comes with a form showing the actual results of vendor measurements. This indicates if the vendor is aware of all characteristics, is able to measure them, and if his results check with the company results.

The Brainard paper suggests that any company conducting vendor surveys should see how well the quality of subsequent deliveries correlates with the survey predictions. However, quite aside from the predictive value, these surveys provide objective evidence of facilities, procedures, and other tangible matters, including the attitude and responses of the key personnel. It is likely that the survey will remain as a regular use vendor relations tool for the foreseeable future. On balance, it seems that the surveys can be useful when applied to sophisticated products. Relatively standard products seldom require a vendor survey.

Until surveys can reliably predict performance, the best means of evaluating a vendor is his past performance. One means of doing this is a data bank.

19-5 Data Banks for Vendor Evaluation

In a data bank, vendor performance and survey data on quality from several buyers would be pooled and a summary made available to users of the data bank. Juran[6] pointed out that this has long been done in the financial field. The Dun and Bradstreet company collects credit data from organizations which deal with a debtor. The data are summarized and a report is sent to Dun and Bradstreet subscribers.

The parallel in the quality function is the pooling of quality data. This has already started in several[7] ways:

1 Pooling of data from several divisions within one large company. This may include the pooling of both quality performance data and survey data. One multidivisional company publishes a handbook showing the field failure rate and cost of remedying the failure for specific purchased components. Such information is valuable in negotiating new or renewal contracts with vendors.

2 Pooling of data by a government agency. Much effort has been expended in developing such programs, particularly for reliability data of new components (see Chapter 10).

[6] J. M. Juran, "Management's Corner," *Industrial Quality Control*, vol. 20, no. 11, pp. 65–66, May, 1964.
[7] *Ibid.*

3 Pooling of data by several companies within one industry. This has been tried for critical electronic components. A research organization provides a service in which its engineers thoroughly examine test reports from buyers who have purchased the name component. The data are pooled and a report is provided to subscribers of the service. The unique point is the recognition that the quality of the data varies greatly and must be carefully evaluated before digestion by data processing equipment.

The aerospace industry has made such extensive use of vendor surveys and approved vendor lists that some firms have pooled their information on vendors. An analysis revealed that many vendors had been surveyed and approved in the preceding year by as many as 15 of the participating aerospace firms. The firms have agreed to share their lists of approved vendors and provide backup data upon request. This has reduced the number of vendor survey teams each week. The program even provides for a member company to conduct a vendor survey in its geographical area for another member company, an example of industrial cooperation.[8]

19-6 Initial Quality Planning with Vendors

The planning must, by deeds and actions, recognize the interdependence of the vendor and buyer. This includes three areas: economic, managerial, and technological.

In the economic area, the concept of "value analysis" has urged purchasing agents to take a new view of their function. "Purchasing must be concerned with buying *value*—not just buying what is specified."[9] Applied to many twentieth-century products, designed for long life and continuous service, value analysis suggests that the purchasing department should be looking to the total cost of keeping these products in service over their designed life (rather than looking to the original purchase price) as a basis for making the decision to buy. As noted in section 4-9, this new concept requires that the costs of downtime, maintenance, spare parts, etc., associated with service failures be quantified. As a consequence, there arises a new form of economic interdependence. The vendor, no less than the buyer, must develop a continuing concern over the frequency and consequences of service failures, and must take action in a way which optimizes the buyers' costs.

The managerial area of interdependence is "cradle to grave," i.e.,

[8] "Aerospace Firms Pool Records and Reports on Suppliers," *Quality Assurance*, pp. 22–24, December, 1966.

[9] E. M. Trump, "New Profit Potentials from Purchasing," *Management Review*, pp. 4–9, August, 1965.

from initial planning through manufacture and usage of the product. The ingredients[10] of initial planning include:

1 Elaboration on the meaning of the specification, including the setting of standards for sensory qualities.
2 Defining key words and phrases to establish a common language.
3 Establishment of seriousness classification of defects [see Chapter 12].
4 Standardizing the methods of measurement and test.
5 Establishing compatible procedures and data systems.
6 Establishing sampling plans, acceptance levels and other criteria for action [see Chapters 12 and 13].
7 Avoiding duplication of gages, test facilities, inspection and test programs.
8 Establishing systems of lot identification and of preserving the order of manufacture.

These elements are often compiled into a written manual which defines procedures, audit plan, and organization plan.

Although there are general policies that would apply to all vendors, it would be foolish to have one set of detailed procedures for all vendors. For some companies, the spectrum of vendors might include a small vendor supplying standard nuts and bolts and a major vendor who must completely design and build a complex guidance system. The National Aeronautics and Space Administration has just this problem and has developed a number of documents defining quality program requirements. These documents include:

NPC 200-1A Quality Assurance Provisions for Government Agencies
NPC 200-2 Quality Program Provisions for Space System Contractors
NPC 200-3 Inspection System Provisions for Suppliers of Space Materials, Parts, Components, and Services
NPC 250-1 Reliability Program Provisions for Space System Contractors

The documents are listed here as a source of ideas for *all* companies. Chapter 13 also outlines some specifications for complete quality programs. Certainly some of the requirements go beyond the needs of conventional products. However, those who ignore (because "our product is different") the experience of the military and space agencies in handling vendor quality problems are missing a valuable source of advanced ideas.

More likely than not, the majority of vendors can be handled with

[10] Juran, "Vendor Relations—An Overview," pp. 10–11.

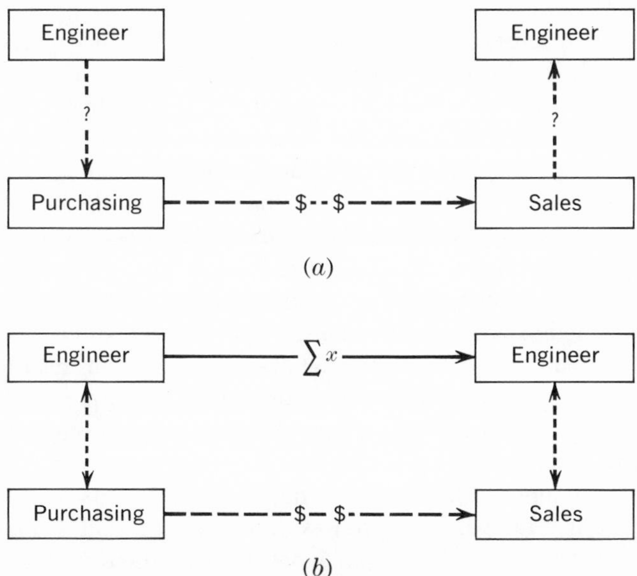

Figure 19-2 Relationships between engineering, purchasing, and sales departments. Source: P. U. Aasgaard, "Reliable Parts Procurement for Electronic Equipments," *Transactions of the National Convention of the American Society for Quality Control,* pp. 73–80.

a relatively simple program. However, there will also be the "vital few" vendors who may require an extensive vendor relations program.

Initial planning is often hampered by poor communications (Figure 19-2).[11] Figure 19-2*a* shows a situation where a purchasing department plays a commanding role: "The engineer must tell us what he wants, then we'll get it for him." An engineer transmits a set of requirements to his purchasing department.

Except for simple and standard items, some misunderstanding may arise because no specification will be perfectly clear and, further, the purchasing agent may not have the technical background to understand the requirements. However, the purchasing department is "responsible" for procurement and therefore transmits the technical requirements and handles the problems of price and other commercial factors by contact with the sales department of the vendor. That sales department then transmits the requirements to its engineer with more problems of misunderstanding arising.

In contrast is Figure 19-2*b* where it is recognized that technical

[11] P. U. Aasgaard, "Reliable Parts Procurement for Electronic Equipments," *Transactions of the National Convention of the American Society for Quality Control,* 1964, pp. 73–80.

matters must be handled directly by technical experts and commercial details by business experts. Jurisdiction for the complete transaction resides with the customer purchasing department (and vendor sales department), but technical discussions are encouraged as long as purchasing is kept informed. The purchasing and sales departments must be a party to *all* official relations between the two companies.

The same emphasis on collaboration is needed in the manufacturing and usage phases (see section 19-8).

The technological area of interdependence is particularly crucial for modern complex systems. Complex assemblies have always faced interchangeability problems. Many new systems also have serious problems of performance even though subsystems mate together dimensionally. Unknown interactions between subsystems suddenly appear. Basic causes are not obvious and manufacturers of different subsystems blame each other for the system problem. "To date our way of dealing with these mysteries has been to pull back the veil so that both buyer and seller can have a go at them in the development and design stages. Our experience has been that when we confront the mysteries with all the talent we can bring to bear, some of the mysteries disappear, and others can be cut down to size. Our experience has also been that if we do not conduct these reviews jointly, we will run into unwelcome surprises further down the road, with delays and costs out of all proportion."[12]

19-7 Defining Quality Requirements for the Vendor

Section 19-6 listed some key elements that must be agreed upon with the vendor, e.g., seriousness classification of defects and acceptable quality levels (see Chapters 12 and 17 respectively).

Seriousness classes and acceptable quality levels must be made part of the purchase contact. This can be done by incorporating them into the purchase order, inspection procedures, a separate quality specification, a quality manual, or the product specification itself.[13]

However, the setting of an AQL is not as important as other problems, e.g., interpretation of specification limits, methods of measurements, and standards for sensory characteristics.[14]

> ***Example*** The experience of a clothing manufacturer is pertinent here. One of its purchased items was ready for customer use; i.e., it was price tagged, bagged, and sent to stock for distribution. The

[12] Juran, "Vendor Relations—An Overview," p. 12.
[13] For further elaboration, see *QCH*, 1962, p. 24-14.
[14] *Ibid.*

vendor had an outgoing inspection and the buyer duplicated this with a 100 percent receiving inspection. About 6 percent of the items were not "first quality." These items were sold to employees as seconds and a claim made (and honored) against the vendor. The buyer initiated discussions with the vendor to eliminate the double cost of inspection. The discussions (typically) uncovered different views on what was acceptable but these were resolved. Finally, it was agreed that the buyer would pay the vendor to fold, tag, and bag the items. Except for a spot check, the receiving inspection was discontinued and the items are sent directly to stock.

Chapter 12 stressed the need for clarity in specifications. The need is even stronger for purchased materials because:

1 The distance between vendor and buyer makes it more difficult to correct unclear specifications. The people at the vendor's plant will simply assume that their interpretation is correct.
2 The use of sampling procedures in evaluating vendor product makes it critical that a correct decision be made on each unit of product. As explained in Chapter 17, the results of a small sample are used to "accept" or "reject" an entire lot. One or two inspection errors made in the sample may cause a wrong decision for the lot.

Example An example of the need to standardize inspection conditions occurred during the incoming inspection of yarn by a carpet manufacturer. The manufacturer complained to the yarn vendor about the yarn weight. This situation prevailed for many months and finally the vendor visited his customer to verify the test methods. The mechanics of the test methods checked. Next, an impartial testing lab was hired and verified the tests at the carpet plant. Finally, the mystery was solved. The supplier was spinning (and measuring) the yarn at bone dry conditions but the carpet manufacturer measured at standard conditions. During this period, $62,000 more was spent for yarn than if it had been purchased at standard weight.

19-8 Cooperation with Vendors during Execution of the Contract

The concept of interdependence means that the buyer and vendor are equally concerned with preventing defects or at least detecting them as early as possible. The monitoring of vendor performance aims at preventing defects and detecting defects as early as possible. The amount of effort may range from simple analysis of receiving inspection data to extensive formal surveillance systems. (Where a vendor must do considerable design and development, the surveillance may require periodic reviews of his reliability engineering efforts during design and development.)

One useful approach is a preproduction quality evaluation. This is an extension of the old concept of "first-piece inspection" which required that the first piece manufactured be inspected and accepted before production continues. However, the new approach stresses the active cooperation of buyer and vendor during the planning. The automobile industry illustrates the concept.[15]

> Every new model of automobile is quite an exercise in team work, since both buyer and vendor are yoked to a severe time table. The crux of it is joint venture on several fronts: engineers work directly with engineers; other specialists work directly with their opposite numbers. The manufacturing planning is a joint venture in providing an adequate process, machines, tools. The quality planning is a joint venture, down to avoiding duplication of gages. The first samples go through a prove-in procedure which includes assembly into prove-out vehicles. Through such joint planning and execution, not only is the preparation intrinsically better done; the personnel of the two companies are brought to work together face to face. The confidence born of such relationships can often outperform the most precisely drawn procedures.

During the production run, the buyer may station one of his inspectors at the vendor's plant. Particular circumstances may justify such "source inspection" but there are disadvantages. Source inspection is a means of detecting defects before the product is received at the buyer's plant.

Finally, cooperation between vendor and buyer must extend beyond manufacturing to the usage and service phase of the product life cycle. This means more than the buyer notifying the vendor of defective product.[16] The vendor may need detailed usage information in order to pinpoint the problem and correct it. In practice, the vendor is often not provided with this information or given access to it.

19-9 Writing the Quality Plan into the Contract

Increasingly, companies are putting their quality plans into the purchase contracts through "incorporation by reference." Reducing the quality

[15] Juran, "Vendor Relations—An Overview," p. 13.
[16] For elaboration, see *QCH*, 1962, pp. 24-24 to 24-27.

plan to a formal agreement is done typically through the following series of steps:

1 Both agree to the classifications of critical, major, and minor quality characteristics and to the acceptable quality levels for each. (Sometimes exceptionally important quality characteristics are specifically exempted from the agreement, and 100 percent conformance is required.)
2 The buyer affirms his intent to use a specific acceptance sampling plan and to accept lots which pass the plan.
3 The seller agrees to issue credits for lots rejected by the sampling plan and returned to him, or to reimburse the buyer for the cost of screening or repairing such rejected lots.
4 The buyer usually agrees to accept (and not return for credit) the defective items in accepted lots.
5 The seller agrees to supply specified inspection data with each shipment or to retain it for possible examination for a stated time period.
6 The buyer states his intent to waive incoming inspection when satisfied that seller meets requirements, reserving the right to return to full inspection if quality deteriorates. (Often, an audit-type sampling is used.)

The effect of a legal document is that, because it must be executed at high levels of each company, it gives legitimacy and permanency to procedures which are already in effect informally. The agreement, therefore, should not be regarded as the end object of the certification plan but rather as the final step in giving official status to the procedures for achieving complete vendor responsibility for quality, which is the essence of the plan.

It is customary to follow a definite pattern of preliminary steps before the signing of the papers:

1 *Discussion phase.* Quality control personnel of both parties meet together to review specifications and standards, to agree on changes to validate long-standing acceptable deviations, and to agree on tentative classification of characteristics, acceptable quality levels, measurement and gaging practices, visible-defect standards, and the like. In addition, the vendor's process control and/or final inspection methods may be reviewed and changes agreed on.
2 *Trial period.* A trial run of several shipments is made in which both parties operate to the new standards and exchange data. This ensures that the two parties are interpreting the specification in the same way and that the methods of measurement are producing identical results.
3 *Agreement signing.* High level officials of both companies affix their signatures to the legal document.

4 *Administration phase.* The plan goes into full formal operation, and actions are taken as prescribed by the rules. The records generally speak for themselves, augmented by occasional checkup visits to see that vendor's process controls remain in effect.

These agreements have some obvious advantages to both buyer and seller, particularly when the items involved are repeatedly purchased to the same specifications, or to specifications which differ only in detail from time to time. The buyer obtains lower inspection costs and greater assurance of quality and delivery. The seller achieves a stronger competitive position in recognition of his accomplishments in quality, plus greater assurance of customer acceptance by virtue of the meeting of minds on specifications and procedures. The willingness of the customer to absorb the costs of defectives in accepted lots relieves the supplier of certain nuisance costs and permits closer pricing. The spread of such agreements has also opened the way for quality managers to participate more fully in the writing of the purchase contracts. Informal vendor relations plans generally exhibit many features of the formal plans.

19-10 Uses of Vendor Quality Data

Data on the quality of a vendor's product can be used for several purposes:

1 Decisions for individual incoming lots
2 Detection of trends in quality from one vendor (perhaps based on monthly data)
3 Decisions on maintaining purchasing agreements with a vendor (perhaps based on annual quality summaries)

Purposes 2 and 3 can be incorporated into a formal vendor rating scheme as discussed in Chapter 20. Decisions on individual lots offer a special potential that deserves discussion.

Whenever a vendor is successful in meeting the customer's quality requirements, the way is open for using *vendor's* quality data to reduce the need for customer's incoming inspection. This is not a new idea. Metals, basic chemicals, and other raw materials have long been accepted for certain characteristics on the basis of analyses made by independent or even vendors' laboratories. The use of vendor generated data is particularly sensible in cases where the customer has insisted that a vendor use 100 percent inspection. Here, the customer is certainly paying for the inspection (in the selling price) and should make use of the data instead of paying twice by having an extensive receiving inspection himself.

The method is simple; it lies in *requiring the vendor's inspection data to be submitted with each shipment.* There is a period of necessary preliminaries to establish that such data are valid, honest, and reliable. During these preliminaries, both the vendor and vendee check the same shipments and exchange data long enough to clear up measurement and calibration differences, interpretations of the specifications, sampling biases, and so on. Convinced of the integrity of the vendor's data, the customer thereafter relies on the supplier's submitted data as the basis for lot acceptance, reducing his own product inspection to an occasional sampling to audit the inspection *decisions* of the vendor. Any serious discrepancy results in immediate return to full lot-by-lot acceptance by the customer's inspection.

One of the first instances of supplying of inspection data with each shipment was initiated by a vendor rather than being requested by a customer. The Hunter Spring Company of Lansdale, Pennsylvania, in the early 1940s started the practice of furnishing a frequency distribution with each lot of springs shipped.

19-11 Organization and Administration of Vendor Relations Activities

The purchasing function has the primary responsibility for procuring a product that meets *all* requirements (cost, delivery, quality). Although many purchasing departments have added technical capabilities in their staffing, they still must rely on the quality function for most of the technical aspects of evaluating and controlling vendor product quality. The interface between the functions will not always be completely harmonious because purchasing decisions must reflect far more that the interests of a quality function. One aid in obtaining a cooperative working relationship is a table of responsibilities. Table 19-3 shows a usual pattern of responsibilities.

In most companies, purchasing is the prime contact and the prime decision maker on vendor relations. Considerations of price, delivery, or multiple sources of supply may result in decisions to purchase from vendors about whose quality we are less than enthusiastic. Such decisions can rankle the quality specialists to a point that they accuse others of lacking quality-mindedness. Yet there is no escape from some degree of tradeoffs among specialties. The quality specialist must accept his role as a service agency in such cases, still reserving the right to go to higher authority when it seems that too much is being compromised.

Table 19-3 **TABLE OF RESPONSIBILITY FOR VENDOR QUALITY CONTROL**

	Functions or Departments Available for Participation			
	Product Design	*Purchasing*	*Laboratory or Inspection*	*Quality Control*
Qualification of vendors	—	X	X	XX
Negotiation with vendors on meaning of specifications	X	XX	—	X
Classification of defects for seriousness	X	X	X	XX
Establishment of acceptable quality levels	X	X	X	XX
Planning for inspection	—	—	X	XX
Inspection of incoming goods	—	—	XX	X
Waiver of nonconforming goods	XX	X	X	X
Quality surveillance of vendors	—	X	—	XX
Quality rating of vendors	—	X	—	XX

XX = prime responsibility
X = collateral responsibility

What the quality specialist should watch closely is the need for changes in the basic approach. The shift from purchasing standard commodities to purchasing complex subsystems is a revolution requiring a corresponding revolution in vendor quality relations. The company properly looks to the quality specialist to sense the need for such a revolution and to start the breakthrough sequence going to achieve it.

PROBLEMS

1 Visit the purchasing agent of some local institution to learn the overall approach to vendor selection and the role of vendor quality performance in this selection process. Report your findings.

2 Visit a sampling of local vendors (printer, merchant, repair shop, etc.) to learn the role of quality performance in their relationship with their clients. Report your findings.

3 Visit a local manufacturing company and create a table similar to Table 19-3.

4 Read the RPM case in the Appendix and answer the following problem.

You are Bjornson, the vendor QC manager, responsible for vendor quality control in RPM's car division. You have just attended a meeting with your boss, Schmidt, and his boss, Engblom, on the subject of vendor quality control.

At this meeting, all of you reviewed RPM's approach to vendor quality control. You are pretty familiar with this—you have come up through the ranks in this very department, and have been manager on this job for the last 8 years. But the review was a good thing, to put everything in perspective.

RPM's car division is a big buyer of hardware. It has 300 major vendors who supply 3,000 different parts. Another 600 vendors provide the lesser needs, serve as alternates, etc.

Years ago, RPM considered whether to follow a policy of working closely with vendors or to follow the hard line of relying on vicious competitive bidding and the associated threat of sudden death. RPM chose the former road. As a result, vendor relations have continuity and long standing. In any one year there are few new vendors, though the proportions bought from existing multiple sources may vary. RPM is convinced that it is able to keep its prices just as low by this policy of working closely with long-standing vendors as it would by use of the harder policy.

RPM's approach to vendor quality control has, of course, been affected by this broad policy. The stability of the vendor list has meant that the accumulated feedback of quality data is not lost through turnover in vendors. Despite this, quality control has maintained a strong incoming inspection. This has also meant buying costly gages, but the price has been paid rather than to put too much reliance on the vendor's quality control systems.

There has always been some visiting of vendors' plants by RPM people. But this has been primarily by manufacturing engineers and other technical people. When quality control people have gone along, it has been for educational reasons rather than for any surveillance or other form of quality control.

Within the last decade, some changes have come into RPM's approach to vendor quality control:

a The Damiano era brought in "scientific" sampling, including Shainin's Lot Plot plan. Some of this has endured.

b There has been a separation of inspection planning from inspections, so that QC engineers now prepare the more important inspection instruction sheets.

c A record system has been set up to record the results of incoming inspection.

d Some experiments are being tried in different schemes of vendor rating.

e The idea of using acceptable quality levels in purchase contracts has been sold to Purchasing. Progress has already been made in negotiating such levels with some vendors.

There are also some uncomfortable matters which were discussed at the meeting with Schmidt and Engblom:

a Quite a lot of product is being accepted outside of print tolerances. This has been going on for a long time. Some is due to the knowledge that the parts will work. Some is due to close scheduling and lack of inventory. You are all agreed that time is running out on this practice.

b About 25 percent of RPM's field failures are traceable to vendor failures. You are all agreed that this level must be cut sharply.

c There is clear evidence that the big car companies are heading in the direction of shifting from emphasis on incoming inspection to emphasis on vendor surveillance. You doubt that this is the road for RPM, because of the stable list of vendors. But Schmidt and especially Engblom are not at all sure they agree with you.

d The principle of a stable list of vendors has probably been overdone. There are about 40 or 50 vendors who should be dropped for quality reasons, but Purchasing keeps urging patience and collaboration. You all agree that the times are such that Purchasing could be bowled over if the facts were brought before the general manager. But you doubt the wisdom of this because you need the cooperation of Purchasing in a lot of ways, and you don't want to antagonize them.

Now you are back in your office and you have what is to you a big assignment: to come up with a comprehensive proposal of what should be done with respect to vendor quality control, as part of the broad program of improving reliability and generally modernizing the performance of the quality function.

Prepare your proposal.

20 Statistical Aids in Vendor Relations

20-1 Overall Role in Vendor Relations

The statistical aspects of vendor relations mainly concern the design of sampling plans for incoming inspection and the evaluation of vendor performance over a period of time. This chapter will emphasize (1) acceptance sampling plans that are particularly useful in the incoming inspection of vendor products and (2) the quantitative rating of vendors for product quality. The main discussion of acceptance sampling plans is in Chapters 17 and 24. Most of the plans in those chapters are readily adaptable to vendors' products. The plans in this chapter have been included here because they are particularly useful for incoming products.

20-2 Histogram Analysis

The frequency histogram is in widespread use as a tool for incoming inspection. A random sample is selected from the lot under consideration, and measurements are made for the selected quality characteristics. The data are charted as frequency histograms, and the analysis then consists of comparing the histograms to the specification limits and of drawing appropriate conclusions. As a general rule, at least 50 measurements should be taken to provide sufficient data to reveal the basic pattern of variation within the lot. Histograms based on too few measurements can lead to incorrect conclusions, because the shape of the histogram may be incomplete without the observer realizing it.

Figure 20-1 shows 15 typical histograms,[1] with comments on the significance of the distribution when compared to the product tolerance limits.

The patterns generated from variables data provide much more information than attributes data on the same units. For example, Figure 20-1*b*, *d*, *g*, and *i* warns of potential trouble even though all units in the sample are within specification limits. With attributes measurement, all the units would simply be classified as acceptable and the inspection report would have stated "50 inspected, 0 defective"—therefore no problem. One customer had a dramatic experience based on a lot which yielded a sample histogram similar to Figure 20-1*i*. Although the sample indicated that the lot met quality requirements, the customer realized that the vendor must have made much scrap and screened it out before delivery. A rough calculation indicated that full production must have been about 25 percent defective. The histogram enabled the customer to deduce this *without ever having been inside the vendor's plant.* (This is an example of how the "product tells on the process.") As the customer would eventually pay for this scrap (in the selling price), he wanted the situation corrected. The vendor was contacted and advice was offered in a constructive manner.

The comments associated with Figure 20-1 all assume that the distributions follow the normal curve. In the analysis it is necessary to check the validity of this assumption.

In using the disclosures of such histograms, their limitations must also be kept in mind. Since the samples are taken at random rather than in the order of manufacture, the process trends during manufacture are not disclosed. Hence the seeming central tendency of a histogram may be illusory—the process may have drifted substantially. In like manner, the histogram does not disclose whether the vendor's process was operating at its best, i.e., whether it was in a state of statistical control. A control chart analysis might be necessary to establish this.

In spite of these shortcomings, the histogram is a most effective tool for incoming inspection. The key to its usefulness is its simplicity. It speaks a language that everyone understands—comparison of product measurements against specification limits. To draw useful conclusions from this comparison requires little experience in interpreting frequency distributions, and no formal training in statistics. The experience soon expands, so that comparisons are made, e.g., histograms of products submitted by multiple vendors in response to the same specification, or of month-to-month shipments from the same vendor.

[1] G. R. Armstrong and P. C. Clarke, "Frequency Distribution vs. Acceptance Table," *Industrial Quality Control* (*IQC*), vol. 3, no. 2, pp. 22–27, September, 1946.

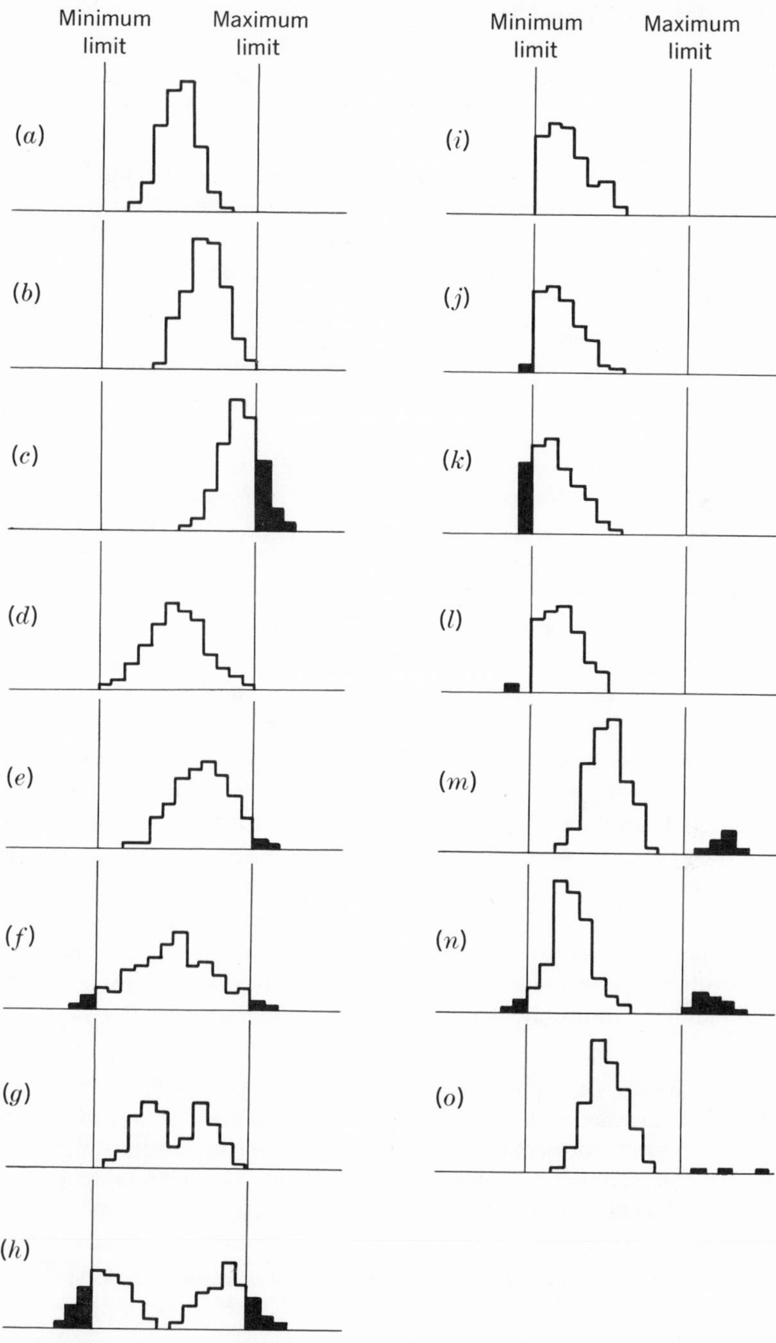

Figure 20-1 Distributions patterns related to tolerances.

20-3 The Lot Plot Plan[2]

This is a variables sampling plan, using the histogram analysis principle. A minimum sample of 50 units is taken at random and is measured with a variables gage precise enough to subdivide product variability into about 10 cells. The data are recorded on a special form (Figure 20-2), and calculations are made of $\overline{X}$ and σ. Then, under the assumption of a normal distribution, and by using a graphic "lot plot card" (Figure 20-2), it is readily possible to compute the percent defective and still other features of the lot. A decision to accept or reject the lot can then be made.

For example, to compute percent defective, the lot plot card is placed with the 50 percent point on X and the 0.1 percent point on a limit line (ULL) previously calculated as $\overline{X} + 3\sigma$. The intersection of the upper specification limit (USL) with the scale on the card predicts the lot percent defective above the USL (0.7 percent in this example). A similar procedure for the lower limit predicts that 2.4 percent of the lot will be below the LSL. [All this is, of course, an ingenious procedure for shortcutting the procedure for finding the area under a normal distribution curve (Chapter 6).]

The simplified mechanics of predicting the percent defective by making a lot plot of data should not overshadow the valuable information for quality improvement supplied by the histogram plot itself. Analysis of a plot (as illustrated in the last section) often helps in determining the corrective action needed. For this reason, a copy of the plot should be given to the operator responsible for controlling the characteristic. Comparisons of lot plots are also revealing, e.g., the plots of several vendors supplying the same part, the plots of one vendor supplying lots periodically, the plots from one vendor before and after material or other design changes, etc. Duncan[3] shows the operating characteristic curve for the Lot Plot plan based on a sample of 50. The following assumptions apply to the development of the curve:

1 Lots are normally distributed.
2 Nonnormal looking samples are assumed normal instead of using the special procedures required in the Lot Plot procedure.
3 The difference between tolerance limits is considerably greater than 6 standard deviations.

[2] Devised by Dorian Shainin. For a more detailed discussion, see J. M. Juran (ed.), *Quality Control Handbook*, 2d ed., McGraw-Hill Book Company, New York, 1962, chap. 17.
[3] A. J. Duncan, *Quality Control and Industrial Statistics*, 3d ed., Richard D. Irwin, Inc., Homewood, Ill., 1965, p. 438.

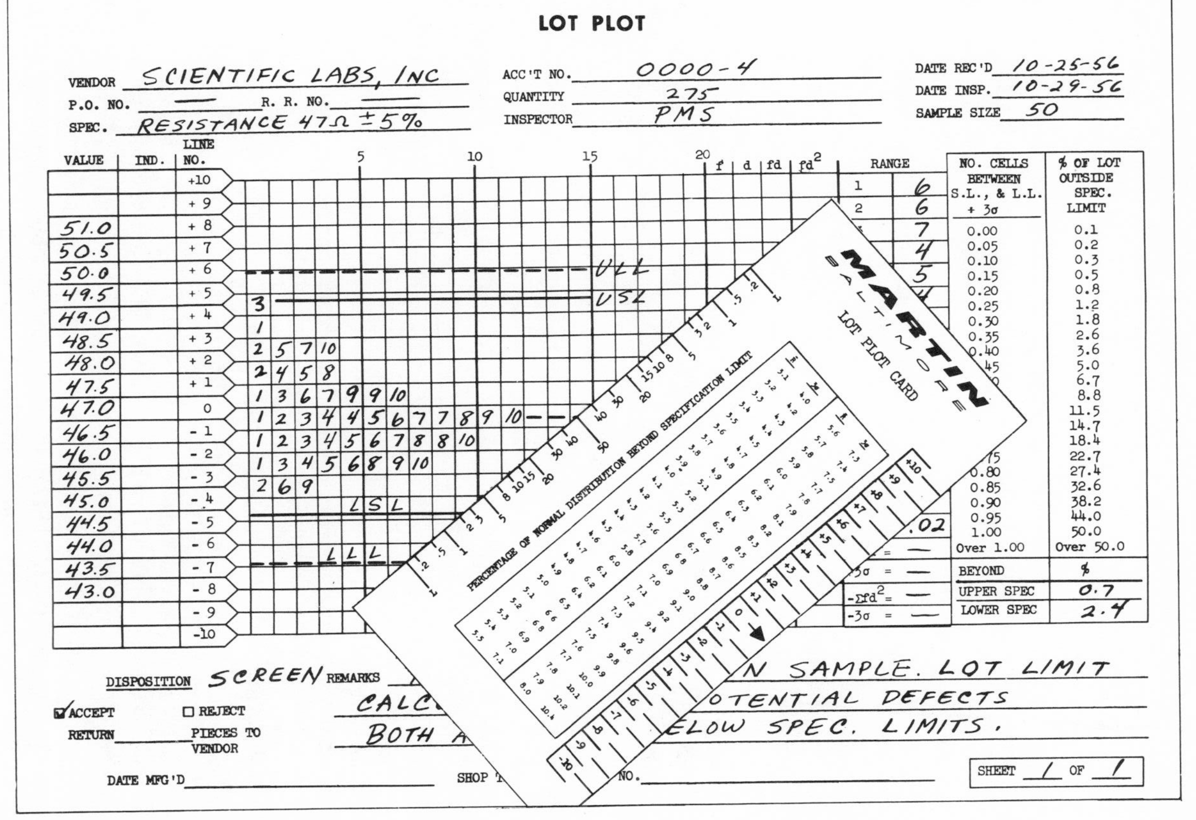

MARTIN BALTIMORE

PART NO.

NAME SWITCH

LOT PLOT

VENDOR SCIENTIFIC LABS, INC   ACC'T NO. 0000-4   DATE REC'D 10-25-56

P.O. NO. —— R. R. NO. ——   QUANTITY 275   DATE INSP. 10-29-56

SPEC. RESISTANCE 47Ω ± 5%   INSPECTOR PMS   SAMPLE SIZE 50

VALUE	IND.	LINE NO.	Plot
		+10	
		+ 9	
51.0		+ 8	
50.5		+ 7	
50.0		+ 6	- - - - - - ULL
49.5		+ 5	3 ——— USL
49.0		+ 4	1
48.5		+ 3	2 5 7 10
48.0		+ 2	2 4 5 8
47.5		+ 1	1 3 6 7 9 9 10
47.0		0	1 2 3 4 4 5 6 7 7 8 9 10
46.5		- 1	1 2 3 4 5 6 7 8 8 10
46.0		- 2	1 3 4 5 6 8 9 10
45.5		- 3	2 6 9
45.0		- 4	LSL
44.5		- 5	
44.0		- 6	
43.5		- 7	- - - LLL
43.0		- 8	
		- 9	
		-10	

Scale: 5 10 15 20; f d fd fd^2

RANGE: 1 — 6; 2 — 6; 7; 4; 5; .02

$+3\sigma$ = —; $-\Sigma fd^2$ = —; -3σ = —

NO. CELLS BETWEEN S.L., & L.L. + 3σ	% OF LOT OUTSIDE SPEC. LIMIT
0.00	0.1
0.05	0.2
0.10	0.3
0.15	0.5
0.20	0.8
0.25	1.2
0.30	1.8
0.35	2.6
0.40	3.6
.45	5.0
[illegible]	6.7
[illegible]	8.8
[illegible]	11.5
[illegible]	14.7
[illegible]	18.4
.75	22.7
0.80	27.4
0.85	32.6
0.90	38.2
0.95	44.0
1.00	50.0
Over 1.00	Over 50.0

BEYOND	%
UPPER SPEC	0.7
LOWER SPEC	2.4

DISPOSITION SCREEN REMARKS ... N SAMPLE. LOT LIMIT CALC... OTENTIAL DEFECTS BOTH A... ELOW SPEC. LIMITS.

☑ ACCEPT ☐ REJECT

RETURN ____ PIECES TO VENDOR

DATE MFG'D ____ SHOP ... NO. ____ SHEET 1 OF 1

MARTIN BALTIMORE LOT PLOT CARD

PERCENTAGE OF NORMAL DISTRIBUTION BEYOND SPECIFICATION LIMIT

+10 +9 +8 +7 +6 +5 +4 +3 +2 +1 0 -1 -2 -3 -4 -5 -6 -7 -8 -9 -10

Figure 20-2 Typical lot plot form and card.

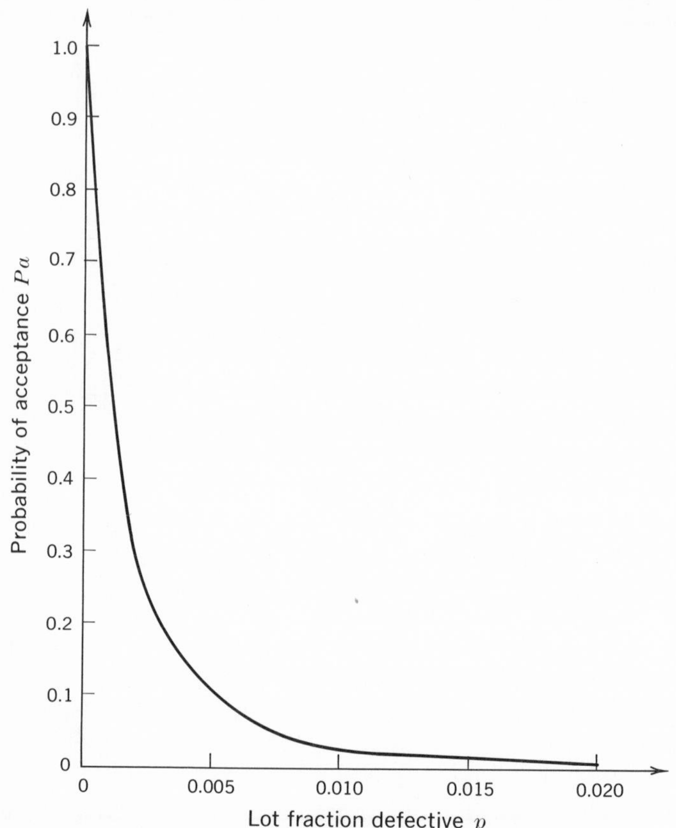

Figure 20-3 Operating characteristic curve for the lot plot method when lot distribution is normal end $n = 50$. Source: A. J. Duncan, *Quality Control and Industrial Statistics,* 3d ed., Richard D. Irwin, Inc., Homewood, Ill., 1965, p. 438.

The OC curve is shown in Figure 20-3. For some applications, the curve may be considered relatively steep. For example, there is only about a 10 percent chance of accepting a lot having a fraction defective of 0.005 (0.5 percent). However, it should be noted that the Lot Plot plan (and all variables sampling plans) can evaluate only one quality characteristic at a time.

There are special attributes plans for vendor product.

Discovery sampling is an attributes acceptance sampling plan that has proved useful in the evaluation of vendor lots. The plan recognizes that all quality levels are *not* equally likely to occur and provides plans that require smaller sample sizes than conventional attributes sampling plans.[4]

20-4 Evaluation of a Vendor's Design by a Reliability Prediction

Statistical aids in acceptance sampling concern the quality of the *finished* product submitted by the vendor. A more preventive approach, evaluation of the vendor's design *before* he starts production, is applicable for new products of relatively sophisticated design. For such products, there arises a need for coordination with the vendor in reliability prediction and planning. One technique for unifying this planning is to require that the vendor submit a complete numerical reliability prediction (see Chapter 10) of his design before going into production.

In such submissions it is usual to find that the vendor's and buyer's predictions will not be alike—the prediction techniques still include much judgment. However, the differences in prediction are worth discussion, since important misunderstandings can be discovered and cleared up.

The submission serves also to assure that the vendor understands the concept of quantifying reliability. One of the authors encountered a situation in which a vendor undertook to design and build a special air-conditioning system to a specified MTBF of 2,000 hours. The vendor, as part of his specifications to his subcontractors, specified a MTBF of 2,000 hours for each of their respective subsystems, thereby demonstrating his ignorance of the fundamentals of quantifying reliability. In the absence of a requirement for advance submission, much damage might have been done before his error was discovered.

20-5 Verification of Vendor Submitted Data

Significant economies in incoming inspection at a customer's plant can be achieved by using quality data from the vendor submitting the product. However, an investigation must first be made to ensure that the data submitted are valid, honest, and reliable. The vendor and customer should each check a number of shipments of product and should exchange their data to clear up measurement and calibration differences, interpretations of the specifications, sampling bias, and any other causes

[4] For further elaboration, see *QCH*, 1962, chap. 18.

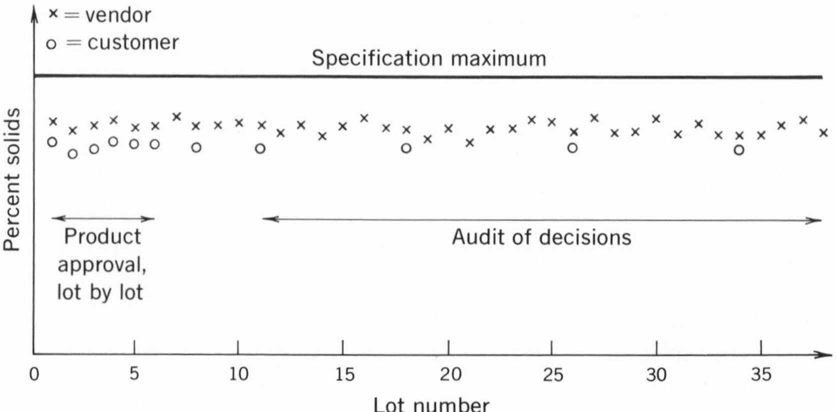

Figure 20-4 Concept of audit of decisions.

of discrepancies in the conclusions of the two inspection departments. When this is accomplished, decisions can be made on an audit basis.

The concept of audit of decisions is illustrated in Figure 20-4. For the first six lots, both vendor and customer measured the percent solids. The results show that:

1 All six lots were acceptable.
2 Both parties consistently agreed on the lot decision. The test results were slightly different but the vendor's test was more severe. However, the *lot decisions* were identical.
3 The vendor's process is capable of meeting the specification and the process variability is small enough to conclude that the process is sufficiently stabilized to be predictable using sampling.

Under these conditions, the customer can rely on the test results of the vendor to accept lots and can periodically audit the decision of the vendor by running duplicate tests. In Figure 20-4, audits were made on lots 8, 11, 18, 26, and 34. Thus, lot-by-lot inspection by the customer is eliminated by using the vendor results instead.

The same concept applies to the use of vendor data in the form of histograms, control charts, sampling plan results, etc. A special condition may arise if a customer uses the vendor's results of a sampling procedure. Suppose a procurement was made based on an acceptable quality level of 1.0 percent defective. A vendor produces lots in units of 1,000 each and uses a single-sampling type of plan. From MIL-STD-105D, a sample size of 80 units is required. The lot is to be accepted if 2 or fewer defectives are found in the sample. If the customer also took a sample of 80, the following results might appear:

	Vendor	*Customer*
Sample size	80	80
No. of defectives	2	5

The vendor would accept the lot but the customers would reject it. This could mean that there is a discrepancy between the inspection of the vendor and the customer. However, it is possible that the difference in sample results is due to sampling variation (see Chapters 11 and 17). If so, the vendor results are valid. Handbook H109[5] provides procedures and tables for evaluating the difference in sampling results.

20-6 Quality Rating of Vendors

Product quality submitted by vendors has always been evaluated and used as a factor in making purchasing decisions. Recently, the evaluation has been formalized by the use of vendor rating formulas which provide a quantitative measure of vendor quality. These ratings are primarily meant to provide an overall quality rating of a vendor for use in reviewing, comparing, and selecting vendors. Vendor rating is not a tool for making decisions on submitted lots.

To create a single numerical quality score is difficult because there are several inputs, each involving its own unit of measure:

1. The lot quality, expressed as lots rejected versus lots inspected
2. The parts qualities, expressed as percent defective
3. The characteristic qualities, expressed in numerous natural units, e.g., pounds per square inch, percent active ingredient, MTBF, etc.
4. The economic consequences of bad quality, expressed in dollars

Because these inputs differ in importance in different companies, the published rating schemes differ markedly in emphasis.

The early vendor rating formulas attempted to summarize into one number the quality of lots received over some time period. (The rating was usually on a 0-to-100 scale.) Some formulas were based on inspection dollar cost instead of on number of rejected lots. However, the

[5] *Statistical Procedures for Determining Validity of Suppliers' Attributes Inspection,* Handbook H109, Government Printing Office, May 6, 1960.

quality rating is only one of the inputs to decision making on choice of vendors. There are other inputs, principally prices and performance on delivery promises. The purchasing agents, who have responsibility for coordinating all inputs and who have the last word on purchasing decisions, have come up with vendor rating plans which take into account the subratings from all specialists, including the quality specialists.

The National Association of Purchasing Agents[6] has published three alternative plans:

1 *Categorical plan.* This is a nonquantitative system in which buyers hold a monthly meeting to discuss vendors and rate each as plus, minus, or neutral.

2 *Weighted-point plan.* Each vendor is scored on quality, price, and service. These factors are weighted 40, 35, and 25 percent respectively. A composite rating (on scale of 0 to 100) then is calculated for each vendor. Figure 20-5 illustrates the measurement criteria and the steps in this plan.

[6] *Evaluation of Supplier Performance,* National Association of Purchasing Agents, New York, 1964.

Figure 20-5 Weighted-point plan of vendor rating. Source: *Evaluation of Supplier Performance,* National Association of Purchasing Agents, New York, 1964.

	Supplier A	Supplier B	Supplier C
1 Lots received	60	60	20
2 Lots accepted	54	56	16
3 Percent accepted $\frac{1}{2}$	90.0	93.3	80.0
4 Quality rating 3×0.40	36.0	37.3	32.0
5 Net price*	0.93	1.12	1.23
6 $\frac{\text{Lowest price}^{\dagger}}{\text{Net price}} \times 100$	100	83	76
7 Price rating 6×0.35	35.0	29.1	26.6
8 Delivery promise kept	90%	95%	100%
9 Service rating 8×0.25	22.5	23.8	25.0
10 Total rating $4 + 7 + 9$	93.5	90.2	83.6

*Net price = unit price − discount + transportation

†Lowest price = minimum net price

	1 Net delivery price per unit A	2 Quality cost ratio $\frac{B}{H}$	3 Delivery cost ratio $\frac{C}{H}+D$	4 Service cost ratio G	5 Total penality 2 + 3 + 4	6 Adjusted unit cost (I × 5) + I	7 Number of units	8 Net value cost 6 × 7
Alpha & Beta Co.	$ 103.47	2.2%	1.2%	0.3%	3.7%	$ 107.30	1,000	$ 107,300
Delta Supply	98.11 ✓	4.7	1.9	1.0	7.6	105.57	1,000	105,570
Gamma Mfg.	98.38	3.9	2.3	1.0	7.2	105.46	1,000	105,460
Sigma Parts	99.59	1.3	0.9	0.4	2.6	102.18	1,000	102,180 ✓
Supplier "X"	—	—	—	—	—	—	—	—

A: Net delivery price per unit = unit price − discount + freight, etc.

B: Material quality cost = inspection cost + handling, manufacturing loss, spoilage, waste, etc.

C: Acquisition and continuity cost = cost of placing order + cost of receiving order

D: Promise kept penalty, percent = rank vendors in order of performance, then penalize each (except first) one quarter of one percent regressively

E: Competence and ability (numbers indicate rating scale) = research and development (0 - 15) + reputation (0 - 15) + technical ability of staff (0 - 10) + production capacity (0 - 10) + financial solvency and profit ability (0 - 10)

F: Attitude and special consideration = labor relation (0 - 5) + business approach (0 - 10) + field service (0 - 5) + warranty conditions (0 - 10) + communication of progress data (0 - 10)

G: Service cost ratio = [100 - (E + F)] x 0.01, except when E + F <60, then use 1 percent

H: Total value of purchase

Figure 20-6 Cost ratio plan of vendor rating. Source: *Evaluation of Supplier Performance,* National Association of Purchasing Agents, New York, 1964.

3 *Cost-ratio plan.* This plan compares vendors on the *total* dollar cost for a specific purchase. Total cost includes price quotation, quality costs, delivery costs, and service costs. The final "rating" is in dollars of net value cost. Figure 20-6 illustrates the steps in this plan. The net value cost is the product of the adjusted unit price and the number of units. The adjusted unit price incorporates three cost ratios.[7]

a The quality cost ratio reflects the relative cost of quality.

b The delivery cost ratio reflects the relative cost of placing and receiving an order. It also includes a "promises-kept" penalty based on a ranking of past performance of vendors.

c The service cost ratio reflects the technical, managerial, and field service competence of the vendor.

All three of these plans recognize quality in the rating of vendors but the rating is not restricted to product quality. An example of a formula based on quality alone is the Bendix vendor quality rating system.[8]

The formula for the Bendix system is

$$\text{Quality rating} = 70 + \left(\frac{\Sigma LR}{N} - 70\right)\sqrt{N} \tag{1}$$

where N = number of lots submitted during a given month

$$LR = \text{lot rating} = 70 - 10\,\frac{p - p'}{\sqrt{p'(100 - p')/n}}$$

where p = percent defective of the sample quantity inspected
p' = AQL percent
n = sample size

With Equation (1) each vendor starts out with a score of 70, which is considered to be average, and has points added or subtracted to this to yield his quality rating. The specific number of points added or subtracted in the formula is dependent on two factors: comparison of the percent defective with the AQL, and the number of lots submitted. The quality ratings thus obtained are interpreted as follows:

Above 90: quality is significantly better than the AQLs.
50 to 90: quality is acceptable.
Below 50: quality is significantly worse than the AQLs.

[7] For elaboration, see *ibid.*

[8] Robert G. Fitzgibbons, "The Bendix Radio Vendor Quality Rating System," *IQC*, vol. 11, no. 8, pp. 38–42, May, 1955. Also, Robert G. Fitzgibbons and Acheson J. Duncan, "Further Comments on the Bendix Vendor Quality Rating System," *IQC*, vol. 22, no. 1, pp. 21–22, July, 1965.

The literature contains many examples of other vendor rating formulas.[9] A basic decision must be made on whether the rating will be based solely on quality performance or on additional considerations such as cost and delivery.

True vendor ratings (for the purpose of making decisions on retaining or dropping vendors) are published infrequently, usually annually, semiannually, or quarterly. These ratings are not to be confused with monthly publications of "vendor performance" which serve mainly as product rating rather than vendor rating. However, some companies are not clear on this distinction, and try to make product rating data serve both purposes.

20-7 Using Vendor Ratings

Because we have not yet learned how to reduce all vendor quality data to a single index on which all can agree, it is necessary to use the ratings as a servant and not as a master for decision making. The single index hides important detail; the decision maker should understand what is hidden. The single index, being numerical, has a pseudoprecision, but the decision maker should not be deceived. He should understand the fringe around the numbers. The very purpose should be kept clearly in mind—product rating (for which the specification is usually the standard) should not be confused with vendor rating (for which other vendors may be the standard).

Vendor rating is an important defect prevention device if it is used in an atmosphere of interdependence between vendor and customer (see Chapter 19). This means that the customer must:

1 Make the investment of time, effort, and special skills to help the poor vendors improve.
2 Be willing to change the specifications when warranted. In some companies, 20 to 40 percent of rejected purchases can be used without any quality compromise. The customer must search for these situations and change the specifications.

Finally, in the cases of consistently poor vendors who cannot respond to help, the vendor rating highlights them as candidates to be dropped as vendors.

[9] See *QCH,* 1962, pp. 24-32 to 24-35. Also see J. L. Coburn, "Rating Supplier Quality Performance in a High Production Helicopter Program," *Transactions of the Twenty-first Annual Conference of the American Society for Quality Control,* 1967, pp. 127–137; Thomas J. Cartin, "Vendor Selection Using Expected Value," *Transactions of the Nineteenth Annual Conference of the American Society for Quality Control,* 1965, pp. 684–688.

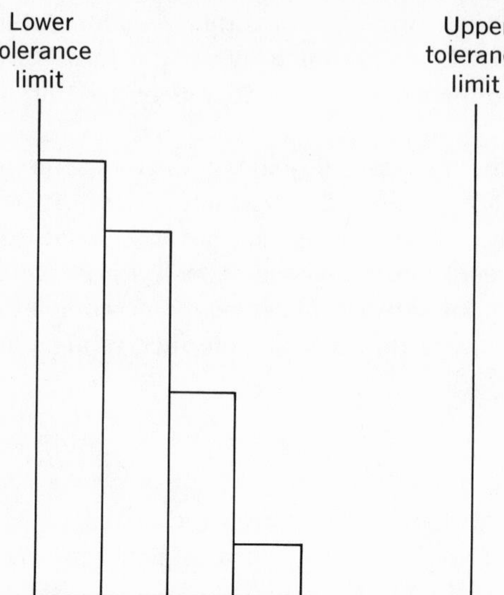

Figure 20-7 Sample histogram.

Figure 20-8 Sample histogram.

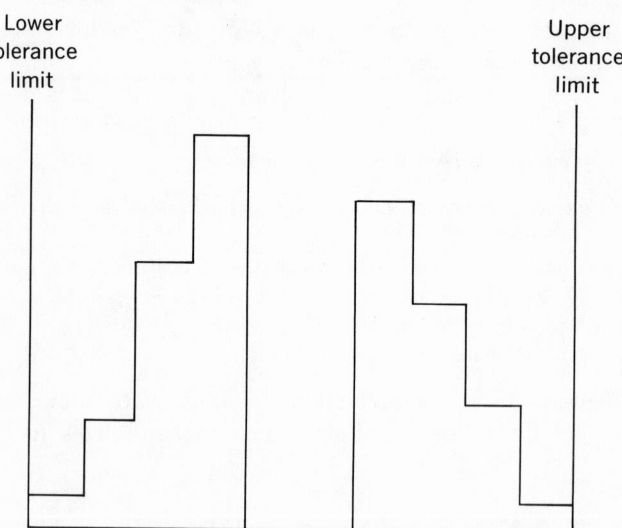

PROBLEMS

1 Apply the weighted-point plan of vendor rating (section 20-6) to compare three vendors for one of the following: (*a*) any product acceptable to the instructor, (*b*) an automatic washing machine, (*c*) a new automobile, and (*d*) a lawn mower.

2 Can you think of a situation *other than 100 percent screening inspection* that would result in the histogram shown in Figure 20-7?

3 Can you describe what caused the unusual histogram plot in Figure 20-8?

4 You have been asked to propose a specific vendor quality rating procedure for use in one of the following types of organizations: (*a*) a company acceptable to the instructor, (*b*) a large municipal government, (*c*) a manufacturer of plastic toys, (*d*) a manufacturer of earth moving equipment, and (*e*) a manufacturer of whiskey.
Research the literature for specific procedures and select (or create) a procedure for the organization.

5 Visit a local manufacturing organization and learn how it determines the quality of purchased items. Define the specific procedures used and what use is made of the information compiled.

21 Process Control

21-1 Controllability

The concept of self-control (section 1-12) requires that an operator know what he is supposed to do, know what he is actually doing, and have the means of regulating the process. These criteria for self-control make possible a separation of defects into various categories of "controllability," of which the most important are

1 *Operator controllable.* A defect is operator controllable if *all three* of the criteria for self-control have been met.
2 *Management controllable.* A defect is management controllable if any one or more of the criteria for self-control have *not* been met.

The theory behind these categories is that only the management can provide the means for meeting the criteria for self-control. Hence any failure to meet these criteria is a failure of management, and the resulting defects are therefore beyond the control of the operators. This theory is not 100 percent sound. Operators commonly have the duty to call management's attention to deficiencies in the system of control and sometimes they do not do so. (Sometimes they do, and it is management who fails to act.) However, the theory is much more right than wrong.

Whether the defects in a plant are mainly management controllable or operator controllable is a fact of the highest order of importance. To reduce the former requires a program in which the main contributions must come from the managers, supervisors, and technical specialists (see

Chapter 30). To reduce the latter requires a very different kind of program in which much of the contribution comes from the operators (see Chapter 22). The great difference between these two kinds of programs suggests that managers should quantify their knowledge of the state of controllability[1] before embarking on major programs.

Table 21-1 shows a controllability study in one department of a textile mill. Of the 37 kinds of defects, the top 6 (the vital few) account for 75 percent of all defects.

Table 21-1 **Controllability of Defects in a Textile Mill**

Defect	*Percent of Total Defects*	*Percent Operator Controllable*	*Percent Management Controllable*	*Percent Controllability Uncertain*
Leader end off-shade	17	0	17	
Start marks	16	8	8	
Broken ends	14	—	—	14
Creases	12	2	10	
Dirt	11	7	4	
Coarse filling	5	0	5	
All other (31 defects)	25	—	—	25
Totals	100	17	44	39

These top 6 have been classified by their source of controllability, i.e., operator, management, or uncertain. (The trivial many defects were not analyzed and hence were put into the uncertain class.)

For the defects analyzed, the management controllable greatly outnumber the operator controllable. Usually such knowledge is decisive in the grand strategy of formulating defect prevention programs.

One of the authors has made extensive studies in controllability and has found that the defects encountered are "by and large" about 80 percent management controllable.[2] This figure does not vary much from industry to industry, but varies greatly among operations. Other investigators, in Japan, Sweden, the Netherlands, and Czechoslovakia, have reached similar conclusions.[3]

While the available quantitative studies make clear that defects are mainly management controllable, many industrial managers do not know this, or are unable to accept the data. Their long-standing beliefs are that most defects are the result of operator carelessness, indifference,

[1] For a useful checklist of questions for determining controllability, see J. M. Juran (ed.), *Quality Control Handbook* (*QCH*), 2d ed., McGraw-Hill Book Company, New York, 1962, p. 11-6.
[2] Consulting experience of J. M. Juran.
[3] Private communications to J. M. Juran.

and even sabotage. Such managers are easily persuaded to embark on operator motivational schemes which, under the usual state of facts, aim at a small minority of the problem and hence are doomed to achievement of minor results at best.

21-2 Knowledge of "Supposed to Do"

This knowledge commonly consists of the following:

1 The product standard, which may be a written specification, a product sample, or other definition of the end result to be attained
2 The process standard, which may be a written process specification, a verbal instruction, or other definition of the "means to an end"
3 A definition of responsibility, i.e., what decisions to make and what actions to take

To a considerable degree, the definition of what is supposed to be done is made by the management hierarchy, after which there is a delegation to the nonsupervisors to do it. In some situations the management also delegates to the nonsupervisors the job of setting quality standards. This is quite common and appropriate in jobbing or maintenance work involving highly skilled craftsmen.

In still other cases the management *should* define the standards and responsibilities, but omits or avoids doing so. In such cases there is delegation by default. The nonsupervisors have no choice but to figure out as best they can what standards to follow and what responsibilities to assume. Often their conclusions are erroneous, since they do not have access to all the facts needed for wise conclusions.

Defining the Standards

The obstacles to operator knowledge of what he is supposed to do are legion. Singly or in combination these obstacles result in numerous quality failures.

The specification may be vague. For example, when fiber-glass tanks are to be transported in vehicles, the surface of the supporting cradles should be smooth. It was recognized that weld spatter would be deposited on the cradle surface, and so an operation was specified to scrape the surface "smooth." However, there was no definition of "how smooth," and many rejections resulted.

There may be conflicting specifications. The foreman's "black book" has had a long, durable career. Changes in specifications may fail to be communicated, especially when there is a constant parade of changes. In one instance, an inspector rejected product which lacked an angle cut needed for clearance in assembly. It was discovered that the inspector was using drawing revision D, the production floor had used revision B, and the design office had issued revision E just 3 days ago.

In another case, a new method of coding terminals and wire size of drawings was developed. The designers began using the new method on the drawings before the new system was explained to the production people, with much resulting confusion and guesswork.

There are also failures to provide operators with the means for interpretation. Lack of seriousness classification is widespread, so that the operator lacks complete knowledge of what is vital and what is trivial. To the same effect is lack of training in the "why" of the specifications.

The actions required to avoid such failures are well known. The information needed by the operator should in fact be compiled. It should be put into easily accessible form. The operator should be trained in the how and the why of the standards, especially in the key elements of seriousness. Conflicting "bootleg" standards should be abolished. A clear designation should be made of whom to consult for authoritative interpretation in case of question. There should be reviews of operator practice to see whether these means for providing knowledge of standards are in fact working out.

Defining Responsibility

The most fundamental need in any hierarchy is to define responsibility. When men lack this knowledge, they devote much of their energy to discovering what is to be their responsibility. The clearer it is, the more surefooted they become in carrying out their assigned duties.

Knowledge of "responsibility for quality" consists of knowledge of what decisions and actions one is expected to take. Unless the question "Who is responsible for quality?" is posed in terms of decisions and actions, there can be no clear answer.

The three most critical quality decisions on the factory floor are

Setup acceptance, i.e., the decision of whether a newly set up process is to commence operation

Running acceptance, i.e., the decision of whether a running process could continue to run or should be stopped

Product acceptance, i.e., the decision of what to do with the product

A useful tool for fixing responsibility for making these decisions is a responsibility matrix (Figure 21-1) showing, on the two axes, (1) what decisions are to be made and (2) who is available to make them.

The supervisors associated with the department under study convene for the purpose of arriving at a meeting of the minds on "Who

	Quality decisions on the factory floor		
	Process acceptance		
Who decides?	Setup acceptance	Running acceptance	Product acceptance
Setup man			
Operator			
Supervisor			
Patrol inspector			
Bench inspector			
Other			

Figure 21-1 Form to aid fixing responsibility for quality decisions.

should decide what?" As one step in reaching an agreement, they may first tally up on the matrix "Who *now* decides what?" From the study of the present pattern of decision making, they move to discussion of what *should* be the pattern.

Often this is done by "ballot." Each man executes a copy of the matrix in accordance with his views of who should decide what. All these views are then tallied on the blackboard to see what are the points of agreement and what are the areas of lack of agreement. The differences are then talked out to arrive at a meeting of the minds.

For example, a meeting of supervisors in a department making electronic parts tabulated "who now decides" as follows:

Table 21-2 **Results of Survey on Who Now Makes Decisions**

Who Is Available to Decide	*Who Now Decides*		
	*Setup Acceptance**	*Running Acceptance†*	*Product Acceptance‡*
Operator	3	2	
Setup man	8	2	1
Patrol inspector	4	5	6
Bench inspector	1		1
Production supervision	10	13	8
Inspection supervision	1		3
Production engineer			5
Quality control engineer		1	1

* Whether the machine is adequately set up to commence production.
† Whether the machine should continue producing.
‡ Whether the product should be accepted for further processing.

After discussion, they reached the following agreement:

	Setup Acceptance	*Running Acceptance*	*Product Acceptance*
Setup man	X		
Patrol inspector			X
Production supervision		X	

In defining these responsibilities, there is often need to impose limitations, i.e., criteria to be met by the man before he makes his decision. A mass producer of metal "tin cans" delegated the decision on setup acceptance to the setup mechanic, since setup of the complex machine was still mainly an art. However, experience showed that unless certain critical dimensions were properly regulated, serious quality troubles would result. Through engineering analysis, the tolerances needed were discovered. Then instruments and data sheets were provided to the mechanic, and he was required to meet a set of sampling criteria as a limitation on his decision to start the machine.

The foregoing relates to making responsibility clear. There is also the problem of making responsibility logical. As a corollary, there is a troublesome question: Should we look for universal patterns of responsibility which fit all companies, or must each pattern be tailor-made?

In the experience of the authors, there is no "right" way to organize. On the factory floor there are critical differences. In some processes it is "natural" for the same man to set up and operate. In other processes these duties are divided between two men. The separation between planning and execution for mass production shops differs from that used in jobbing shops or maintenance work. In a broader sense, each company is "different"; i.e., it consists of a combination of people whose background, training, traditions, cultural patterns, and goals are different from those of any other company. In consequence, it allocates responsibility for decisions and actions to meet the needs of this unique combination—a tailor-made pattern of responsibility.[4]

For example, in the screw machine departments of four companies, the patterns of quality decision making varied remarkably, despite the similarity in the machines and in the components produced by those machines. At one extreme, a company used inspectors at three stages of production: for setup approval, for running approval, and for product approval. At the other extreme, a company made no use of inspectors

[4] For an interesting "public" discussion, see J. M. Juran, "A Challenge to the Extinction of Homo Inspectiens," *Industrial Quality Control* (*IQC*), vol. 24, no. 6, pp. 298–299 and papers cited, December, 1967.

for decision making. The other two companies used intermediate forms.[5]

However, there has existed an aspect of the responsibility pattern which most companies have used during the twentieth century. Decisions on whether the process should run or stop have been made mainly by the production department. Decisions on whether the product is acceptable or not have been made mainly by the inspection department. Whether this pattern will endure is an open question, since the concept of "audit of decisions" has risen to challenge it. Under this concept, all inspection and all decisions, both on the process and on the product, are made by the production operator. However, an independent audit of these decisions is provided. Before this concept of audit of decisions can be made effective, it is first necessary to meet some essential criteria: reduction of management controllable defects; provision of adequate measurement and chart control; training of operators in advanced control methods; and, above all, achievement of a high degree of mutual confidence between the management and the men. It is a long, difficult process to meet these criteria, and so the concept of audit of decisions will be slow in adoption.[6]

21-3 Knowledge of "Is Doing"

For self-control, one must have the means of knowing whether he is conforming to standard. For some operations, the normal human senses are sufficient. The stenographer can see whether the typed letter follows the draft; the assembly operator can see whether the lock washer is in place. However, in most modern operations, the human senses are not adequate, and they must be supplemented by instrumentation.

> ***Example*** In highway construction, the men operating the earth compacting rollers are told to achieve a certain percentage of earth compaction. On some jobs they are given no means for measuring compaction. Only after the rolling is finished are they told whether the compaction is acceptable.

Where the operator himself is to use the instruments, it is necessary to train him: how to measure, what sampling criteria to use, how to record, how to chart, what kinds of corrective action to take. Motivating the operators to follow these instructions is so widespread a problem

[5] See *QCH*, 1962, pp. 25-21 to 25-27 and especially Table 4, p. 25-22.

[6] See, generally, J. M. Juran, "Inspectors—Headed for Extinction?," *IQC*, vol. 22, no. 4, pp. 198–199, October, 1965.

(see Chapter 22) that many companies go to lengths to minimize the need for operator action by providing instruments which require little or no operator effort to measure, record, and control.

When instruments are provided to operators, it is also necessary to ensure that these instruments are compatible with those used by inspectors and in other operations later in the progression of events.

On one construction project, the "form setters" were provided with carpenter levels and rulers to set the height of forms prior to the pouring of concrete. The inspectors were provided with a complex optical instrument. The differences in measurement led to many disputes.

Where the operator does not have access to the instruments, provision must be made for feedback of the essential data from someone who does the actual measuring. This brings additional people into the picture and creates some severe organizational and human relations problems. The longer the feedback loop, the more complex is the organization setup.[7] (See Figure 21-2.)

Feedback of Quality Data to Operators

The needs of production operators (as distinguished from supervisors or technical specialists) require that the data feedback:

Read at a glance.
Deal only with the *few important defects.*
Deal only with *operator controllable defects.*
Provide *prompt information,* both as to symptom and cause.
Provide enough information to *guide corrective action.* In a concrete mixing plant, the operators were told what were the final test results, but not in time to detect trends or to identify sudden changes due to material changes.
Provide a level of *interest.*

To meet these criteria for good feedback, use is made of modern communication technology: large departmental scoreboards; departmental loudspeakers or "bullhorns"; individualized signals (e.g., lights, flags) for specific processes, machines, or stations; computers for analyzing and summarizing quality data, and presenting the results on individualized instruments (these instruments can be wired into the private offices as well).[8]

Well-chosen feedback has both an informative and an incentive

[7] J. V. McKenna, "Managerial Control," *Mechanical Engineering,* pp. 662–664, August, 1955.

[8] John DiCicco, "Electronic Quality Surveillance and Instantaneous Quality Evaluation," *Transactions of the American Society for Quality Control,* 1964, pp. 171–179.

Staff organization and functions

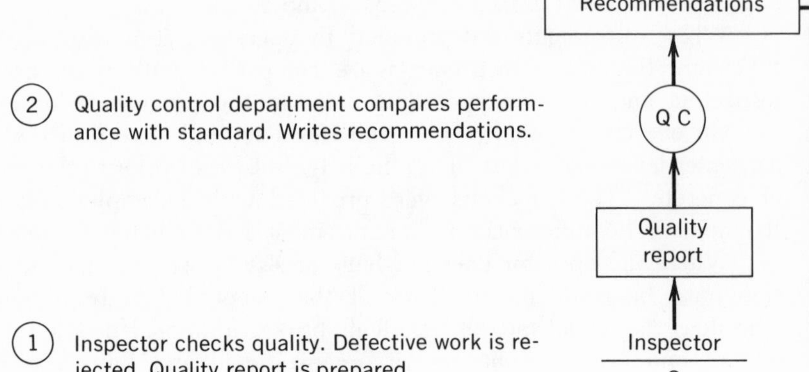
Recommendations
2 Quality control department compares performance with standard. Writes recommendations.
Q C
Quality report
1 Inspector checks quality. Defective work is rejected. Quality report is prepared.
Inspector
O

Staff organization and functions

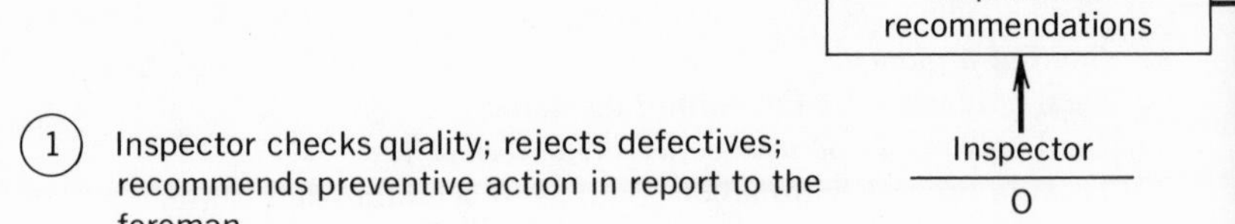
Report and recommendations
1 Inspector checks quality; rejects defectives; recommends preventive action in report to the foreman.
Inspector
O

Staff organization and functions

The functions of quality control are to provide the workman with control data, measuring instruments, and instructions.

Figure 21-2 Relation of staff functions to line functions in resolving control prob- (*c*) On-the-spot control.

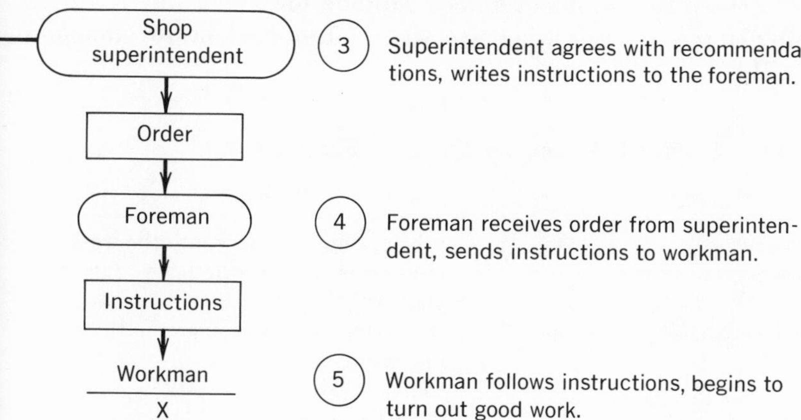
Line organization and functions
Shop superintendent
Order
Foreman
Instructions
Workman
X
3
Superintendent agrees with recommendations, writes instructions to the foreman.
4
Foreman receives order from superintendent, sends instructions to workman.
5
Workman follows instructions, begins to turn out good work.

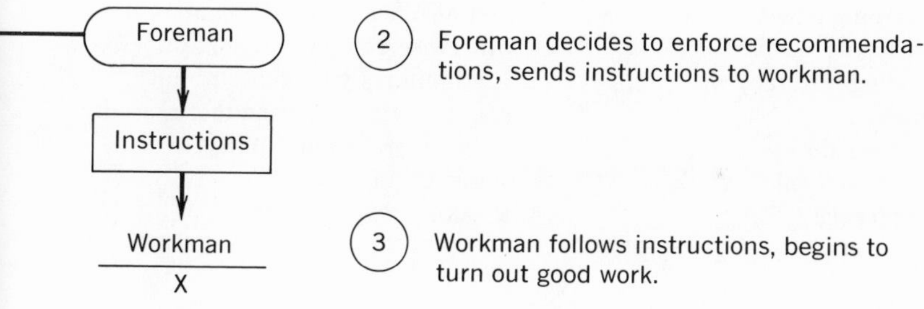
Line organization and functions
Foreman
Instructions
Workman
X
2
Foreman decides to enforce recommendations, sends instructions to workman.
3
Workman follows instructions, begins to turn out good work.

Line organization and functions

Workman
X

(1) Workman checks his own quality, decides when adjustments are necessary, makes them. Preventive action is immediate.

lems. (*a*) Control for complex problems. (*b*) Control for simpler problems.

effect on production and laboratory personnel. One investigation interviewed 106 workers to discover their attitude toward quality feedback in visible forms.[9] Their responses were categorized in 15 groupings (Table 21-3):

Table 21-3 ATTITUDE SURVEY ON QUALITY FEEDBACK

Category	*Example*
A. Variety	". . . it brings more variety into the work."
B. Judgment of bosses	". . . the bosses can see whether you're doing your best."
C. Suspicion against checking	". . . then you can have more faith in the inspector."
D. Necessity of control	". . . control is always necessary."
E. Figure stimulus	". . . those figures are nice to look at."
F. Insight	". . . you get a better idea of what you have to do."
G. Involvement	". . . you become more involved in the production process."
H. Eagerness for knowledge	". . . everybody wants to know how he has done."
I. Learning effect	". . . you also learn by it."
J. Neatness motive	". . . otherwise you go bungling on."
K. Feeling of duty	". . . it stimulates your sense of duty."
L. Speed	". . . also you start working more rapidly."
M. Responsibility	". . . to get more responsibility."
N. Improvement of work	". . . the work improves."
O. Assurance	". . . it takes away uncertainty."

The informative effect tells when corrective action is needed. The incentive effect influences behavior and attitude toward quality.

Charts as a Feedback to Operators

The criterion of "read at a glance" is normally met by use of indicating or recording instruments. However, in many situations, more sensitive detection can be provided through multiple measurements and through observation of trends. To meet these needs requires statistical processing of data plus a presentation of the statistics in their time-to-time relationship.

Charts can provide an excellent form of such sensitive feedback provided they are designed to be consistent with the assigned responsi-

[9] Thomas L. Stok, *The Worker and Quality Control*, University of Michigan, Ann Arbor, 1965, p. 73.

Table 21-4 **Operator Responsibility versus Chart Design**

Responsibility of the Operator Is To	*Chart Should Be Designed To Show*
1 Make individual units of product meet a product specification	The measurements of individual units of product compared to product specification limits
2 Hold process conditions to the requirements of a process specification	The measurements of the process conditions compared with the process specification limits
3 Hold averages and ranges to specified statistical control limits	The averages and ranges compared to the statistical control limits
4 Hold percent defective below some prescribed level	Actual percent defective compared to the limiting level

bility of the operator, as in the four situations listed in Table 21-4.

The foregoing is a logical relation of feedback tools to assigned operator responsibility. However, the advocates of statistical control charts have urged, ever since the 1940s, that because control charts have greater sensitivity, they should be used for feedback, even where the responsibility is to meet product limits for individual units of product. During the late 1940s and early 1950s these advocates had their way. Many companies adopted control charts on a massive scale—at every machine, on the factory walls, etc. The movement collapsed because, being technique oriented, it had not properly identified the responsibilities and needs of the operators before designing the tools responsive to those needs. Following the collapse, use of control charts largely disappeared from the factory floors, but remained, on a selective basis, in the laboratories and in the quality control department offices.

Meanwhile, the quality specialists have thought through a subtle distinction which previously was neglected. The purpose of the control chart is to detect change, whereas the responsibility of the operator is (usually) to meet specification limits. Hence, the need of the operator is not to detect change, but to detect a threat of failure to meet specifications.[10] Having grasped this distinction, the engineers came up with new control schemes and chart forms, "narrow-limit gaging," "modified limits," "precontrol," etc. (see Chapter 23). These schemes are designed to sound the alarm only when there is danger of failure to meet the specification.

Training of operators is a further prerequisite to use of charts for feedback. If the charts are to utilize statistical concepts (e.g., standard deviations) beyond the conventional knowledge of operators, consideration should be given to introducing the charts in stages. A first stage

[10] See, generally, J. M. Juran, "Cultural Patterns and Quality Control," *IQC*, vol. 14, no. 4, pp. 8–13, October, 1957.

can be the comparison of a sample of individual product measurements against specification limits. As experience is gained with use of such charts, there will also be grasped the idea that defects may be produced despite the fact that all the sample measurements fall within specification limits. Such understanding sets the stage for the introduction of a more sophisticated chart form, using medians or other statistics more sensitive than individual measurements.

Operator acceptance of chart feedback is influenced not only by solution of the problems of responsibility and training; the extent of work involved is also a criterion for acceptance. Much thought should be given to the design of the feedback in a way which makes minimal work for the operator: push-button charting, automatic charting, charting by someone else, etc. Among operators, the feeling is strong that their "regular" job is to produce, not to keep records or prepare charts.

A common pattern of responsibility for charting is seen in Table 21-5.

Table 21-5 **Responsibility for Charting**

Activity	*Operator*	*Inspector*	*Foreman*	*Quality Control Engineer*
Design of chart			X	XX
Measurements		XX		
Plotting of chart		XX		
Review of results	XX	X	X	
Action on results	XX	X	X	

XX = prime responsibility
X = collateral responsibility

21-4 Taking Regulatory Action

Regulatory action is the final step in closing the feedback loop. This requires, first, that production people be informed that action is necessary and, second, that the actions taken be effective.

Communication between inspection and production people takes many forms, depending on the general atmosphere of collaboration in the company and on the specific tribal relations between the two departments. Among the forms of feedback used are the following:

1 *Stop the machine.* In those cases where mutual confidence between the management and the operators is near the vanishing point, the inspectors are told to shut down the machines if the process is out

of control. This practice usually makes employee relations even more abrasive, since "property rights" of operators are invaded by the inspectors. In addition, the responsibility for adhering to specification shifts from the production to the inspection department.

2 *Ask the operator to stop the machine.* When the people at the very bottom of the company can be relied on to act on the feedback signals, it is evidence of a high level of cooperation between production and inspection.

3 *Ask the foreman to stop the machine.* In some companies the rule is firm that requests for operator action should come from the foreman and no one else. (The labor unions may be an additional force urging this.) Hence the inspectors are asked to deal with the foreman.

4 *Red tag the machine.* In still other companies the relationship between the two departments border on hostility to a degree that the feedback is impersonal. The inspectors tag the out-of-control processes and it is left to the production people to take it from there.

5 *Refuse to ship the product.* Where production takes no action on the feedback, inspection can exert great pressure by refusing to ship the product, even though the product is fit for use. Such refusal creates serious problems for production, since schedules are disrupted and floors become congested. If the product is fit for use, an appeal to the higher-ups will result in the inspectors being overruled, but the process of appeal also brings to the attention of higher-ups any production deficiencies which created the impasse.

It is seen that the language of feedback varies remarkably among companies. This variation results mainly from the cultural differences among the industrial societies in industry rather than from technological differences in the processes.

Under a state of self-control, regulatory action is taken by the operator because what he "is doing" no longer conforms to what he is "supposed to do." He heeds this alarm signal by changing the process, e.g., sharpening the tool, closing the valve, increasing the pressure.

Whether these actions really restore the process to a state which meets the specifications depends on a number of management controllable factors:

1 The process must be capable of holding the tolerances (see Chapter 15). In the absence of "process capability," the operator can act to minimize, but not to eliminate, defects.

2 The process must be responsive to the regulatory action in a predictable cause-and-effect relationship. Failing this, the operator becomes

frustrated and is driven to the alternatives of continuing to produce defectives or of stopping production.

Example In a process for winding grids for electron tubes, the means for centering the winding mechanism lacked provision for adjustment. The mechanic could cut and try, i.e., disassemble and reassemble the setup, but with no assurance that the setup would be an improvement. As a result it became common practice for the mechanics to avoid changes in setup.

3 The operator should be trained in how to use the regulatory mechamisms and procedures. This training should cover the entire spectrum of action: under what conditions to act, what kind and extent of changes to make, how to use the regulatory devices (sometimes the operators must be certified), why these things need to be done.

Example In one food plant, there were three qualified operators on a certain process. One of the three operated the process every week and became very proficient. The other two operators were considered relief men. The relief men were used only when the primary operator was on vacation or was ill, and thus the relief men never really operated the process often enough to become proficient. It was felt that any type of continuous training of the relief men was uneconomical, and agreements with the union prohibited the use of relief men except under the situations cited above. This problem is management controllable; i.e., additional training or a change in union agreements is necessary.

4 The act of adjustment should not be personally distasteful to the operator, e.g., require undue physical exertion.

Example In a plant making glass bottles, one adjustment mechanism was located next to a furnace area. During the summer months this area was so hot that operators tended to keep out of there as much as possible.

It is common to find that in the list of criteria for achieving a state of self-control, the attention to means for regulation is the weakest link in the feedback loop.

PROBLEMS

1 For any of the following processes, list the defects commonly encountered, estimate the frequency of each, and estimate the controllability of the principal defects. Tabulate your results and prepare your conclusions.

a A man playing golf
b A foursome playing bridge
c A regular on the football team
d A trucker driving in traffic
e A department store salesman waiting on a customer
f A stenographer transcribing notes
g A bricklayer on a construction job

2 Prepare a proposal for fixing responsibility for setup acceptance, running acceptance, and product acceptance for any process meeting the approval of the instructor.

3 In a certain operation involving the high volume manufacture of discrete pieces from continuous lengths of material, the following is the system of controlling quality on the manufacturing floor:

a The *operator* makes initial roll adjustments for each new length of material by examining sample pieces, and he subsequently continues to make such examinations and adjustments while the machine is running.

b The *patrol inspector* (reporting to the chief inspector) also periodically examines sample pieces directly from the machine and, if he thinks it necessary, asks the operator to make adjustments.

c Usually the operator complies with such requests, but on occasion he remonstrates. If he feels very strongly about it, he may appeal to his *foreman,* who may decide to let the machine run without making the requested adjustment.

d If this happens, the patrol inspector may appeal to the *chief inspector* (through his supervisor), who may then order the machine to be shut down. Such an order is complied with.

e All the pieces from the machine, in lots of 1,000, go to a *bench inspector,* who samples them according to a prescribed plan. He passes some lots to the next operation on the basis of a good sample and sorts those lots which fail the sampling plan if they contain particular types of defects (i.e., easily identifiable and removable). He routes other rejected lots (containing defects not so easily identifiable) to the next operation, marked to indicate the nature of defects to be removed.

f The next operation is final inspection, where a *final inspector* reexamines all lots 100 percent (accepted, sorted, and rejected lots) and is supposed to remove all types of defects, including some attributable to operations preceding the one we have been considering.

Using the following form, place the necessary check marks to indicate who has authority to make the decisions indicated.

	Operator	*Foreman*	*Patrol Inspector*	*Bench Inspector*	*Final Inspector*	*Chief Inspector*
Make setup						
Accept setup						
Operate process						
Accept process						
Correct process						
Accept lots						
Accept pieces						

Are the responsibilities clear-cut? Explain your answer.[11]

4 Study the system of performance feedback available to any of the following categories of "operator" and report your conclusions on the adequacy of this feedback for controlling quality of performance.

a The motorist in city traffic
b The student at school
c The supermarket cashier

[11] Course notes of Management of Quality Control Course, U.S. Air Force School of Logistics, 1959.

22 Motivation for Quality

22-1 Effect of Multiple Standards

The company has a variety of major goals, and it asks the people in the hierarchy, whether managers or nonsupervisors, to meet these goals. They are asked to meet not only standards for quality, but other standards as well—cost, expense, delivery, safety, etc. Often these standards are in precarious equilibrium; i.e., the company planning has been out of balance so that it is difficult to meet all the standards simultaneously. This difficulty is compounded when a "drive" is on (see below), in which case the standards can easily become mutually antagonistic.

Superimposed on the inherent equilibrium among the standards is a periodic shifting of management emphasis in response to changes in the economic climate. These changes trigger off a well-known sequence of "drives." An economic upsurge brings a demand for deliveries to an extent that customer service is impaired. To meet the delivery standards, overtime is authorized, resulting in violation of the cost standards. Deviations from specification are likewise authorized, resulting in violations of the quality standards. Then, as the delivery schedules are met, new "drives" are instituted to cut costs and to improve quality.

The upper managers are generally aware that they are responding to specific market forces. However, the rest of the organization, and especially the nonsupervisors, often lack this awareness, and hence are confused by the shifts in emphasis.

A further source of multiple standards is inherent in the complexity of modern industrial technology. To the operator, the quality standard actually consists of multiple quality characteristics, of unequal and usually unknown importance in a setting of numerous variables: the materials, tools, instruments, inspectors, expeditors, engineers, fellow workers, the union, and, of course, the boss. The multiple forces generated all converge on the operator, and he is faced with finding an equilibrium which is personally tolerable.

Finally, for each of the major company standards, the company has usually assigned staff specialists to observe the company's performance and to sound the alarms when there is deterioration. (In the quality function, the inspectors, testers, and quality specialists perform this signaling function.) Aspirations of these specialists become bound up with securing a high degree of conformance to their respective specialties. In consequence, the specialists become zealous and overzealous on matters of adherence to "their" standards. They develop positive means for urging good control: new control tools, better feedback of data, assistance in expediting. They also develop motivational tools (reward and penalty schemes) for stimulating good performance. In addition, some of them develop so fierce and narrow a loyalty to their specialty that they become personal about it; i.e., they accuse others of lack of quality-mindedness, lack of safety-mindedness, etc. These accusations can and do create serious antagonisms within the company.

22-2 Management Behavior toward Quality

We have seen, in section 21-1, that most quality failures are traceable to the doors of management. These management controllable failures are not merely the result of incompletely meeting the criteria for operator self-control; they extend to other matters which influence greatly the incidence of quality failures:

1 The technology, e.g., stable designs, adequate processes, precise instruments
2 The managerial process, i.e., sound policies, goals, planning, organization, manning, training, communications
3 The analysis provided to secure the facts needed for breakthrough and control
4 The decision making on day-to-day quality questions as they arise
5 The atmosphere created by the foregoing as evidence of management's dedication to quality versus the other standards

Management's effectiveness in doing all this has several major impacts on operator attitude and behavior.

1 Operators respect a managerial competence which clearly meets the criteria for self-control.
2 Operators watch management's decisions as the real index of how seriously managers regard the quality standards.
3 The extent of operator ideas for improving quality, and especially for more completely meeting the criteria for self-control, is strongly influenced by management's own behavior toward quality.

22-3 Managerial Quality-mindedness

Managers and supervisors are called on to make myriads of decisions affecting quality. Much of this decision making is by managers in the production and quality control departments: whether a process should run or stop; whether product should be shipped or not. However, managers in all other departments also make decisions affecting quality. Design engineering decides whether to change a design or not. Purchasing decides whether to buy from vendor A or vendor B. Marketing decides whether to sell a product for a borderline application. Personnel decides whether to schedule an operator training program. Finance decides whether to provide priority for processing quality control data on the computer. Manufacturing engineering decides whether to redesign a process.

While all departments have a voice in all major decisions, the departments set out above have the dominant voices in the specific decisions listed. In addition, top management has a voice of its own. Collectively, these decisions constitute the management's real emphasis on quality, no matter what the posters say.

The term "quality-mindedness" is often used to sum up a person's attitude toward quality. As a label for human attitude, the term is easily misleading. Many, many managers have been accused of lacking quality-mindedness because their pattern of decision making did not, in the eyes of the accuser, put quality in a high enough priority in relation to the other standards.

Actually, life would be easier for all managers if all standards could be met simultaneously. No one prefers poor quality, or for that matter, high costs, late deliveries, or unsafe plants. The manager who subordinates quality standards to other standards does so because of his interpretation of the forces which converge on him, and not because he likes bad quality. An accusation that he lacks quality-mindedness may be interpreted by him as an accusation that he likes poor quality, in which case the stage is set for recriminations and antagonisms.

The accusations of lack of quality-mindedness are not based merely on decisions dealing directly with product quality. The advocate of a technique may likewise accuse managers of lacking quality-mindedness because they have not accepted his program (for control charts, vendor rating, or whatever). In such cases the accuser is equating "quality" with some technique which he is advocating. They are not the same.

22-4 Operator Quality-mindedness

Operators also must make decisions on which standards to emphasize. Their collective actions likewise give rise to conclusions on their lacking or possessing quality-mindedness. The managerial hierarchy, in particular, observes operator behavior closely. In these observations, managers see that some or many operators are careless, indifferent, or ineffective in their quality performance. From observing this behavior, managers draw two widely different conclusions depending on whether the managers adhere to "Theory X" or to "Theory Y."[1]

Managers who adhere to Theory X believe that the cause of operator indifference is a decline in the moral fiber of affluent man; i.e., men are spoiled by the ease of life and lose their pride of workmanship, etc. Managers who follow Theory X conclude that the remedy for this decline in human pride in work lies in skillful use of motivation, i.e., the carrot and the stick.

Managers who adhere to Theory Y observe the identical phenomena of carelessness, indifference, or ineffectiveness. However, these managers believe that the observed indifference is the result of operators being assigned to those twentieth-century jobs which are repetitive, boresome, and often meaningless. In consequence, these managers believe that the remedy is to reorganize work so that operators' jobs are more challenging and meaningful.

These conflicting theories are not merely exercises in philosophic reasoning; the managers act on their theories. Table 22-1 compares the plans of operation of managers adhering to Theory X and Theory Y respectively.[2]

It is evident from Table 22-1 that Theory X managers rely mainly on systems, inspectors, incentives, and penalties to get results. Theory Y managers rely on the operators.

Some research has tended to show that operation under Theory Y is the more productive, but it is doubtful that conclusive proof has

[1] "Theory X" and "Theory Y" is the terminology created by McGregor in a classic discussion of management understanding of operator behavior. See Douglas McGregor, "The Human Side of Enterprise," *The Management Review*, November, 1957.

[2] J. M. Juran (ed.), *Quality Control Handbook* (*QCH*), 2d ed., McGraw-Hill Book Company, New York, 1962, p. 10-16.

Table 22-1 COMPARISON OF SHOP OPERATION UNDER THEORY X VERSUS THEORY Y

Plan of Operation under Theory X	*Plan of Operation under Theory Y*
Extensive use of piecework rates as an incentive to meet the standard	Less emphasis on piecework rates; greater use of supervisory leadership
Emphasis on wage-penalty clauses or disciplinary measures to punish poor quality performance	Emphasis on the "why" and "how" to improve poor quality performance
Reliance mainly on inspection personnel for tool control	Reliance mainly on production personnel for tool control
Reliance placed mainly on patrol inspectors to see that setups are correct	Reliance mainly on operators and setup men for correctness of setup
Reliance on patrol inspectors to stop machines which are found by inspectors to be producing defects	Reliance on operators to stop machines which are found by inspectors to be producing defects
Extensive use of formal inspection approval for piecework payment, movement of material, etc.	Limited use of formal inspection approval
Debates on the factory floor center on authority to shut down machines, and on motives	Debates on the factory floor center on the interpretation of specifications and measurements
Relationships between operators and inspectors tense, often hostile and acrimonious	Relationships between operators and inspectors businesslike, often good-natured
Upper-management criticism for high scrap losses directed at inspection as much as production	Upper-management criticism for high scrap losses directed at production
Operators exhibit no outward desire to do a quality job	Operators do exhibit the outward desire to do a quality job
Operators largely ignored as a source of ideas for improvement	Operators frequently consulted for ideas for improvement

been established. Meanwhile, research is continuing, because all managers regard the question as of great importance.

22-5 Subspecies of Operator Failure

Once the criteria for self-control have been met (by the management) it would seem that the rest is up to the operators. Like the motorist on the highway, they have been provided with knowledge of the posted speed limits, a speedometer, and controls for regulating the vehicle. Now it is up to them to be law-abiding citizens. However, an inquiry into the nature of operator errors soon discloses that the situation of the industrial operator is far more complex due to the multiple standards he must meet and due to the multiple forces with which he must contend.

The complexity has suggested to some investigators that any organized approach for reduction of operator errors must first classify the observed errors into logical subspecies. (To date the research on this has been sparse, and there remains much opportunity for creative study.) One of the authors[3] has proposed a classification as follows:

1 *Lack of skill.* The operator is *unintentionally* failing to comply. He is aware of the errors as he makes them, but he is *unable* to eliminate the errors—he is not skillful enough.
2 *Willful errors.* The operator is deliberately failing to comply. He could comply, but he has no intention of doing so, for reasons which are good enough for him.
3 *Inadvertence.* Not only is the error unintentional; the operator is even *unaware* that he has made the error.

Failures Due to Lack of Skill

In studying patterns of operator controllable defects, it is common to find that some operators make *consistently* fewer errors than others. The theories for the existence of these consistent differences vary widely; i.e., the superior operators are more gifted, attentive, skillful, quality-minded, etc. Not enough has been done to look for an explanation among the deeds rather than among the theories.

A (usually) rewarding approach is the following:

1 From data on operator performance, identify the consistently best results and the consistently worst results.
2 Study the work methods of the operators associated with these best and worst results. (At this stage, it is only the results, not the operators, which can be labeled "best" and "worst.") From this study, identify the differences in method.
3 From analysis and experiment, find which of these differences in work methods really account for the differences in results. These critical differences are the "knack" possessed by the "superior" operators.
4 Extend this knack to all operators through retraining or through redesigning the process to embody the knack on a foolproof basis.

Note that in such studies the analysis does not invent the solution. The analysis identifies the solution which has already been in use by some operators. (They often are not aware that they have a solution.) Once the nature of this solution becomes known to the managers, they

[3] J. M. Juran, "Operator Errors—Time for a New Look," *Quality Progress,* vol. 1, no. 2, pp. 9–11, February, 1968.

can improve on it and can take steps to provide the solution to all operators.

Willful Failures

These result from a variety of personal reasons, such as:

1 The operator is faced with what he regards as a hopeless problem to meet all the multiple, conflicting standards. Hence he violates one of them so as to meet the rest. (This parallels management's response to changes in the economic climate.)
2 The operator believes that the quality standard is unimportant.

Example In one textile mill, new women employees in the spinning room were failing to tie "weaver's knots" despite pleas or threats of supervisors. The personnel officer heard of this, and questioned one of the women, one whom he himself had hired. She was annoyed and ask him "What difference does it make?" Thereupon he escorted her to the weaving department and showed her how much trouble the weavers were having because of incorrect knots. On understanding this, the operator burst into tears, saying "Why weren't we told about this?" She was perfectly willing to tie weaver's knots when she understood the need for them but, lacking this information, had not been willing to do so merely because some foreman ordered her to do so.

(Managers likewise commit willful errors when they believe the quality standard is unimportant. In an electron tube factory there was a requirement that the assembled mounts be 100 percent visual inspected. The assembly department manager noted from the scrap reports that tubes made from uninspected mounts had no higher scrap levels than tubes made from inspected mounts. He therefore eliminated the specified 100 percent inspection, without advising the other departments he was doing so.[4])

3 The operator has a grudge against the company or against the boss. Violating the quality standard is one way of getting even.

There are, of course, still other reasons for willful errors. What they all have in common is that the operator knows he is in violation of the standard and that he intends to keep it up.

To date there has been virtually no research to discover the quantitative extent of willful errors and the quantitative extent to which the various personal reasons contribute to the total. There is need for such research.

The remedy for willful errors is "motivation." (See sections 22-6 to 22-10.)

[4] Consulting experience of J. M. Juran.

Inadvertent Failures

Human beings appear to be fallible.[5] With the best intentions, they make errors, and at the time are not aware they are making the errors. These errors are classified as "inadvertent."

Numerous investigators have studied the inadvertent errors of inspectors. They have found that inspectors find about 80 percent of all defects in the product submitted for inspection, and miss the remaining 20 percent. In contrast, there have been few published studies on the limit of human effectiveness in avoiding errors in production operations.

In the absence of consistent differences between operators (see Failures due to Lack of Skill, on page 458), managers have reduced the level of inadvertent errors through "foolproofing," i.e., designing the process so that errors are less likely or even impossible. Examples of foolproofing include:

Fixtures and indexes to ensure positive location of work pieces in the tools.

Interdependent operations to ensure that a sequence of operations is fully performed, and in the correct order. For example, in some mechanical processes, an earlier operation provides the locating hole needed to perform a later operation. Hence the latter cannot be performed unless the former was done first.

Redundancy to provide multiple alarm signals for multiple senses, e.g., simultaneous flashing of lights and ringing of bells superimposed on normal observation.

Remote control to make observation and action easy in difficult working conditions, e.g., observation by closed circuit television.

Magnification to make observation easier for the senses.

Tracers for identifying materials, e.g., painting code colors on tool steel.

Countdowns to check out the preparations for a major irreversible event, e.g., the elaborate countdown preceding the launching of a space vehicle.

These foolproofing devices are useful not only in dealing with inadvertent failures, but in the other types of failures as well.

22-6 Campaigns to Improve Motivation

Motivational campaigns have had a long, durable history in human affairs. The political rally, the military parade, the charity ball are

[5] This long-standing belief has been contested by some advocates of motivational schemes. (See section 22-8.)

examples of motivational devices for votes, patriotism, and contributions. Industrial companies have continued the tradition by conducting campaigns of all sorts, including those for better quality.

The premise behind a motivational campaign is that the employee has a useful contribution to make. He could reduce his own errors; he could point to deficiencies in the process; he could supply creative ideas for improvement. He fails to make these contributions because he regards them as unimportant, or because he is sulking, or because he is fed up with management's outward lack of interest in these things. The premise of the campaign is that a dramatic series of events will attract attention, open people's minds, and thereby open the way to a new level of action.

The "campaign" should be distinguished from other forms of company programs. A program to improve management controllable defects (e.g., design failures) requires policies, goals, plans, analysis, and action. However, the program requires this from relatively few people (some of the managerial hierarchy and technical specialists). Hence the program is conducted on a quietly analytical basis, without use of mass media for communication and stimulation of people. In contrast, a program to reduce operator controllable errors requires a contribution from a large number of people—thousands of them in even a middle-sized company. Here mass media play an important role.

Commonly, a campaign is designed to secure new action from nonsupervisors, usually shop operators, but sometimes from all nonsupervisors. In such cases the management hierarchy has the job of designing the campaign, giving and receiving communication, reviewing results, and stimulating action. However, some campaigns are designed more broadly to secure new action from supervisors and technical specialists as well as from nonsupervisors. In such cases the management is not merely directing the play; it is acting out some of the roles as well.

Enough campaigns have been conducted to make clear what are the essential ingredients and how they should be put together. It turns out to be a complex process.

The first step is *securing attention,* since many other programs, inside and outside the company, are competing for people's attention. The campaigns solve this problem by applying the known tools of salesmanship. Using these tools they:

- Set up mass meetings of employees, at which leaders in management, in the union, in the community, in the industry, etc., proclaim their support for the campaign
- Propagandize the campaign in all available media: bulletin boards, company newspapers, wall posters, giveaways (e.g., book matches, imprinted pens), letters to employees' homes

Set up exhibits showing scrap piles, customer letters (both complaining and laudatory), how the company's products are used

Collectively, these means are usually quite effective in securing employee attention.

Second, the campaign is faced with convincing the employee that the company quality and quality reputation are *important to his own well-being.* This is done in various ways. The attention-getting propaganda emphasizes that "quality makes sales, sales make jobs." The customer letters are to the same effect. Slogan and essay contests (with prizes) secure employee participation in proving the importance of quality to the employee. The winning slogans and contests, on publication, become effective propaganda in their own right.

Next, the campaign needs to convince the employee that there is *something he can do* to contribute to better quality, and to help him identify just what it is he can do. This identification must be personalized, i.e., it must convey a separate message to each operator in terms of his specific process and product. Getting attention can be done with general propaganda which is identical for all operators. However, identifying what action is expected of the operator must, to a large extent, be tailor-made. This tailor making is usually done by:

1 Identification of the "vital few" defects in each department, and the knack which can be used to avoid these defects. Through departmental posters, exhibits of good and bad work and foreman instruction, these vital few defects become the focal points of emphasis.
2 Departmental and individual scoreboards to measure progress and to serve as a basis for directing further stimulative action.

A well-designed campaign aims to secure a wide variety of action from operators. Some of this action is directed at operator controllable defects. Here the operator may be asked to:

1 Reduce willful errors by shifting his emphasis on meeting quality standards versus other standards.
2 Reduce errors due to lack of skill by accepting retraining to adopt the knack possessed by operators whose performance has been consistently superior
3 Reduce inadvertent errors by accepting changes in method designed to foolproof the operation

Other action is directed at management controllable errors. Here the operator is asked to help identify the failures of management to meet the criteria for self-control. This identification is through suggestions ("error cause removal" ideas; see section 22-8) and other forms of communication.

Finally, the campaign may invite the operator to participate in the analysis to discover causes of defects. The conventional suggestion system may stimulate some action along this line. However, the experience of the Japanese QC Circles (see section 22-7) suggests that a training program is necessary before operators will be very productive in their analysis.

22-7 Provision for Operator Participation

Applied to programs for quality motivation, the invariable breakthrough sequence consists of decision to embark on a campaign, goal setting, planning to meet the goals, choice of projects, analysis for causes, analysis for remedies, decision on action to be taken, execution of decisions, and revision of controls to hold the gains.

In practice, motivational programs vary remarkably in providing for operator participation in this sequence.

At one extreme is the "posters and suggestions" type of campaign, which provides virtually no management leadership. A propaganda campaign is set up, stressing the importance of quality and urging all hands to be more careful, to make fewer mistakes. The operators are left to figure out for themselves what they should do in response to the propaganda.

Often this propaganda is coupled with the prevailing suggestion system by offering awards for ideas on how to improve quality. As is usual in suggestion systems, provision is made for investigating the ideas through use of staff specialists, with review of their findings by the suggestion committee.

Mainly, these "posters and suggestions" campaigns accomplish little, since they are usually grounded on the false premise that most defects are the result of willful operator failures.

Example In a piston ring factory, the managers piled up a huge mound of scrap piston rings at the entrance gate, with a large sign proclaiming how much the pile had cost the company. At the same time a poster campaign urged operators to be more careful. Actually, over 80 percent of the defects were management controllable, and at that time neither the managers nor the operators had any idea of what to do to reduce the scrap. The main result of the campaign was that the posters became increasingly decorated with vulgar humor.

At the other extreme is the Japanese QC Circle[6] approach, in which operator participation can extend to every single element of the breakthrough sequence. This wide participation is made possible because of some unique favorable relationships existing in the Japanese culture between employer and employee. The possibilities are converted into realities through the essential step of training operators in how to set goals, conduct the planning and analysis, and follow through with action to achieve the planned goals and to hold them by revised controls.

Table 22-2 shows the opportunities for operator participation in error cause removal, for management controllable errors and for the various subspecies of operator controllable errors.

Table 22-2 **Operator Participation in Error Cause Removal***

	Possible Role of Operator in:			
		Operator Controllable Errors		
Activities in the Sequence	*Management Controllable Errors*	*Willful*	*Lack of Skill*	*Inadvertent*
Observe errors	X	XX	X	—
Theorize as to causes of errors	X	XX	X	—
Analyze to discover true causes	X	XX	X	—
Theorize as to remedies for causes	X	XX	X	—
Analyze to discover optimum remedy	—	XX	—	
Apply chosen remedy	—	XX	X	—

XX = operator's possible role is significant, even decisive.
X = operator's possible role can be useful.
— = operator's possible role is dubious.
* Juran, "Operator Errors—Time for a New Look," p. 11.

Table 22-2 is drawn to reflect usual American practice. In other Western countries, there would be differences in the operator role. In Japan, where operators have been widely trained in how to use the tools of analysis, the role of the operators is remarkably different.

The extent of this training in Japan has been astonishing. The QC Circle movement started in 1962. By mid-1969 over 200,000 of

[6] J. M. Juran, "The QC Circle Phenomenon," *Industrial Quality Control,* vol. 23, no. 7, pp. 329–336, January, 1967.

these circles had become active, meaning that over 2 million workers had been trained in such tools as the Pareto analysis, frequency histograms, Ishikawa cause-and-effect diagrams, and control charts. The projects worked out by these circles have reached to enormous numbers. The selected projects (the best, naturally) published in the Japanese journal *GEMBA TO QC* (*Quality Control for the Foreman*) are nothing short of admirable. Not only is the QC Circle movement drawing a fine-tooth comb through the intradepartmental quality problems of Japanese companies; it is also providing a generation of Japanese workers with the living laboratories and training needed to learn to solve industrial problems the rest of their working lives. What will happen when these workers become supervisors and managers can only be imagined, but the prospects are breathtaking.

A third form of campaign has been the "zero defects" family of programs. During the 1960s these programs attracted a good deal of attention, and it is well to understand why.

22-8 The "Zero Defects" Programs

In 1962 one of the divisions of the Martin-Marietta Corp. was producing a missile for a military customer. The division already possessed a fully established quality control program and had been delivering product within the "acceptable quality level" for the missile involved. However, the division undertook a program to improve quality even further. Part of this program[7] was motivational, and a slogan "zero defects" (ZD) was coined to publicize the ultimate goal. Because the program did achieve a superior result, the company and military officials publicized these results, together with the slogan "zero defects." The publicity was picked up by some national journals, and zero defects soon had a national favorable press. Attracted by this favorable publicity, and urged on by government and military officials, many top executives in companies in the defense and aerospace industries decided to embark on ZD programs. (These were conducted under a wide variety of names, including "zero defects.") These decisions were made partly for customer relations reasons and partly in the hope of improving quality.

The companies who pioneered in these programs (and who derived much favorable publicity thereby) also published technical papers describing what they had accomplished and what had been the content of the program leading to these results. Analysis of these publications tended to show that the programs consisted of:

[7] James F. Halpin, *Zero Defects*, McGraw-Hill Book Company, New York, 1966.

1 A motivational package aimed at reducing operator controllable defects. The contents of this package were such things as the big meeting rally, pledge cards, posters, attention getters, scoreboards.
2 A prevention package aimed at reducing management controllable defects. This package centered around "error cause removal" (ECR) suggestions to be made by employees for subsequent analysis and action by supervision. These suggestions were submitted to the supervisor on ECR forms which defined the probable error cause and proposed action and disposition. Procedures provided for prompt feedback to the worker.

As numerous companies got into these programs, much confusion and frustration set in. The pioneering companies generally had started with a known problem and designed a program to solve that problem. In relating their story they were less than complete in presenting a generalization of their program for the benefit of potential followers. There was little recognition of the need to distinguish between subspecies of operator controllable defects. It was generally assumed that under the stimulus of motivation, operators would somehow stop operator controllable defects. To a surprising degree it was urged that, through motivation, human beings could become errorless, i.e., could literally get down to zero defects.

In consequence, companies which adopted ZD programs under a top-management edict found themselves in many cases starting with an asserted all-purpose remedy when the problem was not even defined. The middle managers resented having such situations imposed on them by well-meaning Very Important People (top management, marketing management, and customers). Frustration and discord were widespread in such situations.

The ZD programs also introduced some innovations which are likely to become part of the permanent management tool kit. These include:

A well-defined, structured approach to launching a formal motivational program. (Structured approaches are not new, but none had ever attained such wide publicity.)
A concept of participation from *all* functions within the company, including the top management. (Previously, most programs had been directed solely at the factory personnel.)
Setting of quantitative goals for improvement.
A wide assortment of motivational techniques specially adapted to quality problems.
Formalized systems for error cause removal (really a specialized form of the long-standing suggestion systems).
Provision for reporting achievements against goals.

Every single one of these tools had been used here and there in older programs. However, the effect of the national (and international) publicity was the evolution of a standardized tool kit for dealing with a specific set of quality problems. The numerous failures were mainly the result of applying the tool kit without knowing whether the problems called for it. (The more zealous advocates of ZD were of no help here, since they urged ZD as a cure-all for any kind of quality trouble.)

22-9 Organization Structure for Motivation Programs

A simple poster campaign requires little formal planning or organization. Outside services are available to supply ready-made posters along with plans for using them. If the program is limited to poster publicity, this outside service plus an inside contact man (usually from industrial relations and/or quality control) can make all arrangements.

An elaborate campaign (such as a zero defects program) requires so much more planning that a formal organizational structure is created. This structure consists of (1) a committee to design the campaign and give it broad guidance, plus (2) a coordinator to carry out many of the details of planning and follow-up.

The committee is drawn from the various departments whose employees will be called on to take the ultimate action of error reduction. In addition, the committee includes membership from those staff departments whose services will be used extensively during the program. These include:

Personnel, which plays the dominant role in the communication media (bulletin boards, house organs, etc.)
Accounting, which keeps the score and issues the reports on results
Quality control, which supplies much of the investigative effort needed to study causes and to review error cause removal suggestions
Advertising, which provides some of the skills needed for selling ideas to people (in this case, selling to insiders)

The job of coordinator is often a full-time assignment, since such programs involve a great many details to be worked out and timetables to be kept concurrent. As the program reaches a maintenance state, less time is required for coordination, so that the job of coordinator may require less than full time. (In the maintenance phase, the committee meetings are less frequent, and the committee membership may be restructured to consist of lower-level managers than served on the original committee.)

22-10 Theories of Motivation

It was noted under Operator Quality-mindedness, section 22-4, that managers differ widely in their theories[8] as to the causes of operator lack of enthusiasm, and thereby as to the remedies to be applied. The motivational programs reflect the same differences. Theory X managers tend to emphasize increments of money (or threats of loss of money) as the means for motivating operators. Theory Y managers tend to emphasize nonfinancial incentives.

There is an interrelation between financial needs and nonfinancial needs. Financial needs are critical when people live below the subsistence level. Nonfinancial needs receive increasing attention as society becomes increasingly affluent.

In the late nineteenth century, the movement identified with Frederick W. Taylor[9] resulted in widespread adoption of wage-incentive systems based on the *quantity* of work performed versus a standard of a day's work arrived at by time and motion studies. (These incentive systems were generally effective in securing greater productivity, presumably because the workmen of that day had great difficulty supporting their families at nonincentive wages.) To guard against a deterioration of quality, these incentive systems commonly provided that an operator should be paid only for good work (or should repair or sort the defectives on his own time).

These "penalties" for poor quality resulted in many injustices, since operators were often penalized for defects beyond their control. The rise of industrial unionism in the mid-1930s brought these conditions to the surface through numerous employee grievances. The result was that the vast majority of companies abolished all penalty provisions for defects, even for those clearly operator controllable. These penalty provisions have, with rare exceptions, *not* been reinstated.[10]

[8] For a summary of various theories of motivation, see Saul W. Gellerman, *Motivation and Productivity*, American Management Association, Inc., New York, 1963.

[9] For a collection of Taylor's books and philosophy, see F. W. Taylor, *Scientific Management*, Harper & Row, Publishers, Incorporated, New York, 1947.

[10] A big deterrent is the fact that such reinstatement would have to be accepted by the labor unions, which have no enthusiasm for these schemes. In some states the prevailing legislation is a further deterrent. Section 103.455 of the Wisconsin Statutes makes it illegal for an employer to deduct wages for faulty workmanship unless the employee authorizes such a deduction in writing or unless the employer and an employee representative determine that the faulty work is due to worker's negligence, carelessness, or willful and intentional conduct, or unless the employee

Where quality still enters the wage formula it now does so as a positive incentive rather than as a penalty. For example, a trucking company engaged in transport of household goods and manufactured goods offers bonuses to employees for high quality performance in effectively packing breakable goods, adequately providing for full insurance coverage on shipments, and adequately handling service complaints.[11]

As American culture has become more affluent, managers have become aware that the wage-incentive schemes no longer have the "pull" they exhibited in earlier years. McGregor (footnote 1) has offered an explanation by suggesting that human needs[12] follow a definite order of priority list. It follows that the motivators should keep changing as the high priority needs are met (and hence are removed from the list). Using McGregor's list of priorities, the associated motivators would appear to be as follows (Table 22-3):

Table 22-3 **Hierarchy of Motivators**

Priority List	*Tools for Securing Motivation for Quality*
Physiological needs	Increments of money for better quality and for ideas
Safety needs	Appeals based on job security. "Quality makes sales; sales make jobs"
Social needs	Creation of team spirit and team activities
Ego needs	Appeal to pride of workmanship, to achieving a good score; trophies, publicity
Self-fulfillment needs	Appeal to creativity, working out of original ideas

This priority of human needs is greatly influenced by the prevailing national culture, which varies remarkably from one country to another and from one century to another. For example, Table 22-4 compares the motivational ingredients of the Japanese QC Circle system with those of most Western countries.

The willingness of Japanese workers to respond to a scheme of motivation so radically different from that used in the West is directly traceable to cultural differences.[13]

---

is found guilty in court. If such a deduction is made by an employer, the employer is liable for twice the amount of the deduction taken, in a civil action brought by the employee. Further, "any agreement entered into by employer and employee contrary to this Section shall be void and of no force and effect."

[11] Richard K. Smith, "Quality Incentives at North American Van Lines," *Transactions of the Twentieth Annual Conference of the American Society for Quality Control*, 1966, pp. 910–913.

[12] These needs and the concept of hierarchy were previously suggested by Maslow. See A. H. Maslow, *Motivation and Personality*, Harper & Row, Publishers, Incorporated, New York, 1954.

[13] Juran, "The QC Circle Phenomenon."

Table 22-4 **Comparison of Motivational Ingredients***

Elements of the Plan	*As Practiced in* Conventional Motivational Plans	*As Practiced in* QC Circles
Choice of projects	Left up to employee to identify his own project	Some projects identified by management; others identified by the QC Circle
Training in how to analyze a project	None provided	Formal training program provided. Out-of-hours; voluntary
Analysis of the project	By employee himself or with such aid as he can muster; otherwise, by formal suggestion which is analyzed by someone else	Analysis is by the QC Circle, out-of-hours; using training tools previously provided
Payment for time spent	None	Varies from no pay to full pay for hours spent
Payment for successful idea	Definite payment varying with value of idea	No payment; indirect effect on company profit and resulting bonus which uses one formula for all employees
Nonfinancial incentives	Opportunity for creativity and recognition; pride of workmanship	Opportunity for training; opportunity for creativity and recognition; membership in a group; response to company leadership

* J. M. Juran, "The QC Circle Phenomenon," *Industrial Quality Control*, vol. 23, no. 7, pp. 329–336, January, 1967.

PROBLEMS

1 For any process acceptable to the instructor, (*a*) identify the multiple standards faced by the operators, (*b*) list the principal operator controllable defects, and (*c*) classify defects as to whether due to lack of skill, willful, or inadvertent, and propose a program for reducing the error level.

2 For any motivational program acceptable to the instructor, analyze and report on (*a*) the methods for securing attention, (*b*) the methods for securing interest and identification with the program, and (*c*) the methods for securing action. Some suggested examples: a fund-raising drive, a traffic safety campaign, the Civil Defense program, a political campaign, and a keep-your-city-clean or similar campaign.

3 For any motivational program acceptable to the instructor, study the plan of organization used to achieve results, and report on the organization plan used.

4 Read the Metal Containers, Inc., case in the Appendix and answer the following problem.

You are Lafferty, the quality control manager. During your 2 months' survey, there were many conflicting statements regarding quality-mindedness. At company headquarters, the marketing people felt that the plant supervision was spotty—that there were strong and weak men when it came to quality. There was some feeling to this effect among the quality control supervisors as well.

At the plants, the plant managers and production supervisors had mixed views respecting their operators. The Chicago supervisors felt that their men couldn't be trusted and that the way to run that plant was through reliance on a good system, with penalties for poor quality. At the other extreme, the Detroit supervisors were pretty comfortable about their men, and had a lot of doubt about Chicago's approach.

Your own boss, Wallace, wanted you to be sure to get a good reading on quality-mindedness at the plants. Recently a consultant has been trying to sell Wallace a poster campaign to improve quality in the plants. Wallace has been holding him off to see what you come up with.

What do you recommend to the plant managers and to Mr. Wallace?

23 Process Control Techniques

23-1 The Concept of Dominant Systems

In all processes the input materials are subjected to a wide assortment of operational "systems" (machines, tools, operators, etc.) in order to achieve the desired end result. These operational systems are subject to variation, and it is regulation of these variables which is the basic problem of process control.

Although the process systems are numerous, it is common to find that one of them "dominates"; i.e., it is more important than all the rest put together. In designing the plan of process control it is most helpful to identify this dominant system. Four of the most usual dominant systems are

1 Setup dominant. The process has been engineered to so high a degree of reproducibility that, if properly set up in the first place, it will run the entire lot off successfully. Quality in such cases depends on validity of the original setup of the machine.
2 Machine dominant. The process keeps changing during the fabrication of the lot, requiring periodic check and readjustment.
3 Operator dominant. The operation has not been fully engineered, and the unengineered residue which depends on the skill and attention of the operator is the major cause of defectives.
4 Component dominant. The input materials or components are the main variable affecting the quality of the finished product.

Table 23-1 gives some examples of operations dominated in these four ways.

Table 23-1 **Process Dominance Categories**

Setup Dominant	*Machine Dominant*	*Operator Dominant*	*Component Dominant*
	Typical Operations		
Punching	Packaging	Arc welding	Watch assembly
Drilling	Staking	Hand soldering	Auto assembly
Cutting to length	Screw machining	Blanchard grinding	Other mechanical assembly
Broaching	Automatic cutting	Steel rolling	Plastics assembly
Die cutting	Volume filling	Turret-lathe running	Electronics assembly
Die drawing	Weight filling	Spray painting	Tube making
Molding	Paper making	Electronic "trimming"	Food formulation
			Vegetable packing
Coil winding	Wire enameling	Hand packing	
Labeling	Wool carding	Repairing	
Sheet-metal bending	Resistance welding	Adjusting	
		Inspecting	
Flame cutting		Card punching	
Heat sealing		Filing	
Printing		Order filling	
Mimeographing		Shoe lasting	
	Typical Control Systems during Manufacture		
First-piece inspection	Periodic inspection	Acceptance inspection	Vendor rating
			Incoming inspection
Lot plot	X chart	p chart	
PRE-Control	Median chart	c chart	Prior operation control
Narrow-limit gaging	$\overline{X}$ and R chart	Operator scoring	
	PRE-Control		Acceptance inspection
Attributes visual inspection	Narrow-limit gaging		
	p chart		
	Process variables check		
	Automatic recording		

The dominant system is not the sole source of defects for the process in question, but it is the main source. For example, a printing operation is setup dominated as to defects such as spelling or color of ink. It is operator dominated as to fingerprints on the sheets.

For each form of dominance there is an appropriate kit of control tools. Table 23-1 lists the most usual tools associated with the four forms of dominance presented.

Identifying the dominant system also has value in process improvement or breakthrough, since the biggest opportunity for breakthrough is in the dominant system.[1]

23-2 Setup Dominant Operations

When the setup is the dominant cause of defects, the process setup should be formally approved before production starts. An approval procedure should define who makes the decision, and the criteria to be used in making the decision.

The evaluation of a setup is usually made by the supervisor of the setup man, the production foreman, or an independent inspector. Whatever the choice in an individual plant, the responsibility must be clearly defined. There are several reasons for an independent inspector:

1 If many specifications and requirements are present, the potentially high number of errors of omission and interpretation can benefit from an independent check.
2 The independent inspector can be objective in insisting on meeting the specifications, since he personally does not have to correct the setup.
3 Where specialized judgment, interpretation of sensory standards (e.g., visual defects, color), and knowledge of customer idiosyncrasies are necessary, these can more easily be conveyed to a single inspector than to a number of setup men, particularly when many different products are involved.

Setup acceptance criteria have historically been based on rule-of-thumb methods, but a number of formal techniques are now available for choosing sample sizes and interpreting data for decision on setups. These are summarized in Table 23-2.

There are two possible mistakes that can be made in evaluating a setup—disapproving a correct setup or approving an incorrect setup. These can be defined in terms of quantitative sampling risks and sampling procedures developed to meet the risks.[2]

23-3 Machine Dominant Operations

In machine dominant operations, despite a satisfactory initial setup, there is a time-to-time or a random change which causes defects. Examples

[1] See J. M. Juran (ed.), *Quality Control Handbook* (*QCH*), 2d ed., McGraw-Hill Book Company, New York, 1962, pp. 26-12 to 26-21.
[2] C. J. Anson, "Procedures for Approving a Machine or Process Setting," *Quality Engineering*, vol. 25, no. 6, pp. 178–182, November–December, 1961.

Table 23-2 SETUP CRITERIA FOR MEASURABLE CHARACTERISTICS

Technique	*Application to Setup Acceptance*
Frequency distribution or lot plot (see sections 6-4, 20-2, and 20-3)	50-piece sample shows whether setup "aim" is correct and whether process capability (6 standard deviations) is within tolerance limits
PRE-Control (see section 23-4)	First 5 pieces judged against "PRE-Control" limits indicate whether aim and capability are adequate
Narrow-limit gaging (see section 23-4)	Gaging of first 5 pieces judged against special narrow limits provides basis for acceptance of aim and capability
Setup Criteria for Nonmeasurable Characteristics	
Attributes sampling (see Chapter 17)	Sample inspected visually is used to detect freedom from burrs, distortion, roughness, marks, and similar visual defects. Size of sample varies from 1 to 50 depending on stability of process and seriousness of defect

of the former are the wear of a grinding wheel, or the dilution of a solution. Examples of the latter are accidental tearing of sheet product, or occasional voids in castings. The more usual control tools include:

1 Average and range control charts, as in the polymerization example below. (For the statistical basis of the average and range chart see Chapter 15.)
2 Narrow-limit gaging and PRE-Control (see section 23-4).

Polymerization is the process of combining two or more molecules of the same substance to form a new compound. The process is widely used in the petroleum, synthetic rubber, and other industries.

From past data on the process (usually a minimum of 25 subgroups) control charts were set up for the three principal quality characteristics. The charts are shown in Figure 23-1. The left-hand side of the chart shows these data as "original process." Because these data showed wide variations, the variables were studied through statistical analysis. For example, the variables affecting softening point of the product were studied first by a multiple regression. Feed rate, polymerizer temperature, and sludge levels in the separator kettles were the factors correlated. The results of the first regression showed that all three of these factors had a significant effect on softening point and that together they accounted for 78 percent of the total product variation. Past perform-

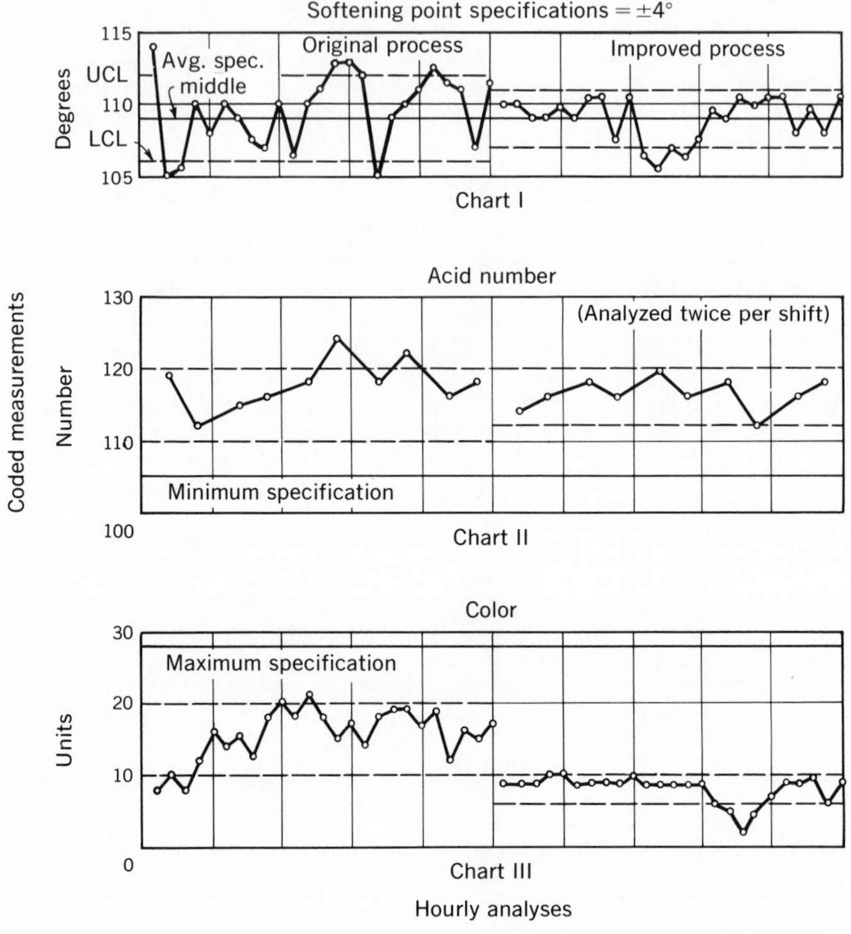

Figure 23-1 Master control chart polymerization process.

ance for these three variables was then plotted on action-control charts. These action-control charts suggested further steps to be taken with respect to each of these variables.[3]

In the example, the control chart indicated that assignable causes of variation were present in the process. The control chart is equally useful when it shows a process to be in complete statistical control. In such a case, the chart is saying that the process is operating with a *minimum possible variation* (chance variation). If a process is operating only with chance variation, but is still producing a high amount

[3] For further details, see *QCH*, 1962, pp. 23-34 to 23-37.

of defective product, then the chart at least helps to force the difficult decisions to be made. If a process is doing its best but is still producing defectives, then tolerances must be changed, a *fundamental* change must be made in the process, or a decision must be made to live with the defective product. Thus, the chart can help to end the frequent bickering between manufacturing and engineering people on whether the fault lies with "tight tolerances" or "careless workmanship."

It should be stressed that the statistical control chart distinguishes between chance variation and assignable variation (see Chapter 15). The ordinary control limits for sample averages *cannot* be compared to the specification limits, because control limits are based on sample *averages* and specification limits are usually based on *individual* items. The distribution of averages is always narrower than the distribution of individual items, and, therefore, control limits (for averages) cannot be compared to tolerance limits (for individual items). The transition from a control chart to an analysis of tolerance limits is explained in Chapter 15.

23-4 Narrow-limit Gaging: PRE-Control

The conventional control chart is designed to detect the presence of statistically significant changes. However, these changes are in many instances economically insignificant; i.e., they do not jeopardize conformance to specification. In such cases, the needed alarm signal is of another sort—a means for detecting those changes which might cause defectives. The model for such an alarm signal is shown in Figure 23-2.

The average of the process must be kept away from the specification limit by an amount of at least 3σ to avoid defects being made. This yields limiting values of the process averages as $\overline{X}_{\min}$ and $\overline{X}_{\max}$, respectively.

If the small samples always estimated the process average perfectly, then $\overline{X}_{\min}$ and $\overline{X}_{\max}$ would become the control limits. As small samples vary, the limits for $\overline{X}$ as observed in the small samples must be set still farther away from the product specification limits.[4] If the variation of averages of small samples is defined as $3\sigma/\sqrt{n}$, then the actual control limits will be inside the product specification limits by an amount equal to $3\sigma + 3\sigma/\sqrt{n}$.

An example of a practical application of the principle of modified limits is narrow-limit gaging (NL). In this technique, go-not-go gages

[4] This emphasizes protection to the consumer rather than to the producer. For further elaboration, see Acheson J. Duncan, *Quality Control and Industrial Statistics*, 3d ed., Richard D. Irwin, Inc., Homewood, Ill., 1695, pp. 431–433.

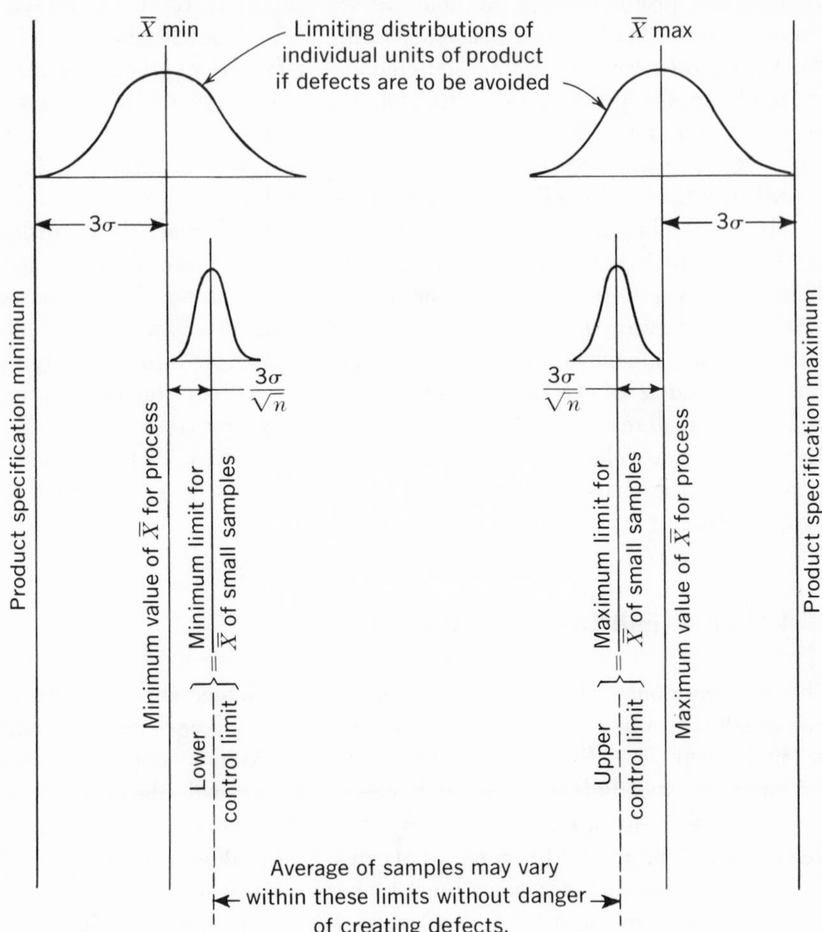

Figure 23-2 Modified control limits.

are prepared with limits stricter than the product specifications. Using these NL gages, a sample is taken from the process periodically. The number of units outside the narrow limits is counted. If this number exceeds the allowable number given by the NL gaging plan, the process is readjusted.

An example of the narrow-limit gaging approach is the PRE-Control technique. PRE-Control starts a process centered between specification limits and detects shifts that might result in making some of the parts outside a print limit. PRE-Control requires no plotting and no computations from sampled parts, and it needs only three pieces to give control information. The technique utilizes the normal distribution curve in the determination of significant changes in the aim or the spread of a production process that could result in the increased production of defective work.

The principle of PRE-Control is demonstrated by assuming the worst condition that can be accepted from a process capable of quality production, i.e., when the natural tolerance is the same as the print allows, and when the process is precisely centered and any shift would result in some defective work.

If we draw in two PRE-Control (PC) lines, each one-fourth of the way in from each print-tolerance limit (Figure 23-3), it can be shown that 86 percent of the parts will be inside the PC lines, with 7 percent in each of the outer sections. In other words, 7 percent, or 1 part in 14, will occur outside a PC line under normal circumstances.

The chance that two parts in a row will fall outside a PC line is 1/14 times 1/14 or 1/196. This means that only once in about every 200 pieces could we expect to get two parts in a row in a given outer band. When two in a row do occur, there is a much greater chance (195/196) that the process has shifted. It is advisable, therefore, to reset the process to the center. It is equally unlikely to get a piece beyond one given PC line and the next piece outside the other PC line. In this case, the indication is not that the process has shifted but that some factor has been introduced which has widened the pattern to an extent that defective pieces are inevitable. An immediate remedy

Figure 23-3a Assumptions underlying PRE-Control. *b* Location of PRE-Control lines.

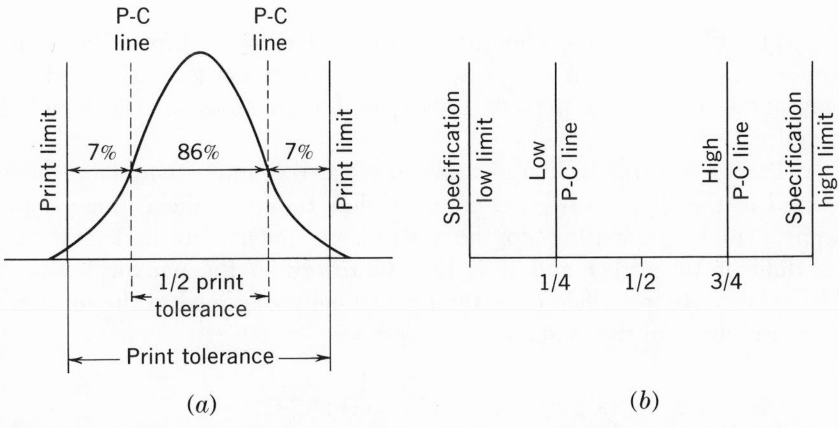

of the cause of the trouble must be made before the process can safely continue.

These principles lead to the following set of rules[5] that summarize the technique of PRE-Control:

1 Divide the specification tolerance band with PC lines at ¼ and ¾ of the tolerance, as in Figure 23-3.
2 Start job.
3 If piece is outside specification limits, reset.
4 If one piece is inside specification limit but outside a PC line, check next piece.
5 If second piece is also outside same PC line, reset.
6 If second piece is inside PC line, continue process and reset only when two pieces in a row are outside a given PC line.
7 If two successive pieces show one to be outside the high PC line and one below the low PC line, action must be taken immediately to reduce variation.
8 When five successive pieces fall between the PC lines, frequency gaging may start. While waiting for five, if one piece goes over a PC line, start count over again.
9 When frequency gaging, let process alone until a piece exceeds a PC line. Check the very next piece and proceed as in 6 above.
10 When machine is reset, five successive pieces inside the PC lines must again be made before returning to frequency gaging.
11 If the operator checks more than 25 times without having to reset his process, his gaging frequency may be reduced so that more pieces are made between checks. If, on the other hand, he must reset before 25 checks are made, increase the gaging frequency.[6] An average of 25 checks to a reset is indication that the gaging frequency is correct.

The PRE-Control technique indicates changes in process aim and process variation. It is simple to use, can use go-not-go gages, and can guarantee a specified percent defective if corrections are made when required.

Process control techniques that use narrow limits should be introduced to the shop with great care. Unless fully explained, these limits appear to be tightening up the tolerances. Such a misunderstanding is difficult to correct and it is best to record to the narrow limits in the quality control office until shop personnel are trained in the meaning and usefulness of the limits on the shop floor.

[5] See *QCH*, 1962, sec. 19, for further elaboration of PRE-Control.

[6] Practice has shown that it is proper to consider any single run between 21 and 29 not to differ from an *average* of 25.

23-5 Operator Dominant Operations

In the operator dominant operations, the primary cause of product deficiencies lies in the skill, decision making ability, or quality-mindedness of the individual operator. Defects usually occur at random times; the probability that a defect will occur at any particular time is equal to the "error rate" of the operator.

Error rates differ considerably from one operator to another, and these differences are measurable. These measures are a key element both in control and in improvement[7] of operator error rates. The control scheme is usually based on a running record of operator error which is then published in chart or time series form. The data are compared to a standard, thus providing a feedback of operator error data to the operators and the supervision. The standard for comparison may be

1 Historical performance
2 Performance of other operators
3 An improvement goal set by a customer or by the management

Motivation for meeting the standard can come from competition among operators, award or merit rating schemes based on quality performance, or threat of reassignment for inferior performances.

The choice of scoring or charting technique depends on the product and process, but one of the following is usually suitable:

1 Percent defective charts (p charts). Usually the inspection scheme is designed to serve both for product acceptance and for supplying operator control data. This method is used in attributes cases when the product is either good or bad.
2 Defects-per-unit charts. This method is used for more elaborate products in which multiple errors can occur on a single unit of product, e.g., a complex assembly, a large roll of woven carpeting. In such cases it is common to "headline" the frequencies of the vital few error categories.
3 Defects-per-opportunity charts. Each job is given a rating, prior to production, based on its complexity, or on the "number of opportunities for a defect." In electronics assembly work, for example, the rating is the sum of all components, solder connections, hardware insertions, etc., for the operations to be performed by a given opera-

[7] Analysis of reasons of differences in operator error rate leads to discovery of the causes of low error rates (aptitudes, training, experience, knack, motivation). These discoveries can then become the basis for breakthroughs in operator error rate. For example, see Chapter 22.

tor. The operator's error rate is then the actual number of defects found in a sample divided by the total number of opportunities for a defect (i.e., units inspected times opportunities per unit for that job). This denominator puts all operators on the same basis, so that fair comparisons can be made between them.

4 Defects-per-characteristic charts. For mechanical parts or assemblies, "opportunities for a defect" may more conveniently be defined as the total number of quality characteristics for a given operation. This is used as the denominator. Such a scoring method is readily adapted to cumulative "pyramiding" for higher-management reporting.

Example Figure 23-4 shows a percent defective chart for an operator dominant forging process. The operator has standard samples of visual defects at his hammer and is responsible for informing the lead man when he cannot meet the standard of visual defects—dies are then replaced. Data for the chart (which is posted near the forging hammer) are determined by sampling each operator's work after it leaves the cold-press operation. Each lot is identified by a lot number and operator number.

Figure 23-4 *p* chart for forging defects.

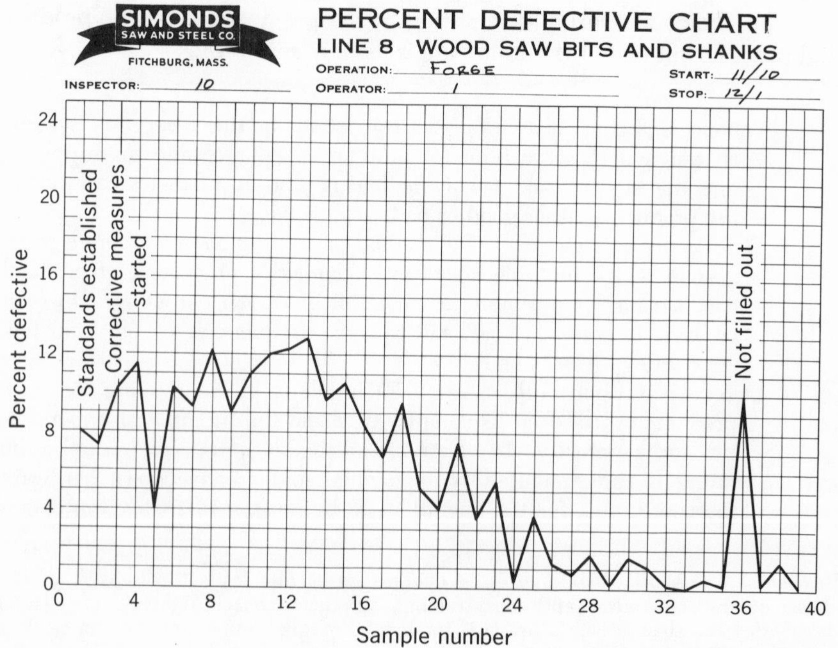

Example Another example involved two conveyorized assembly lines for assembling finished parts into sewing machine cabinets. The quality of *incoming* parts to the assembly area was plotted on a number of defect charts. These c charts showed the number of defects per unit (three parts) and the nature of those defects. This information on incoming quality was compared with the number of defects per unit of three *assembled* cabinets found on final inspection. The *difference* in defects per unit found on incoming parts and on assembled cabinets was a measure of the quality of work being done by the assembly department.

Data were accumulated on each model cabinet and were presented by top management to the managers of the finishing, assembly, and inspection departments for them to set quality goals and a timetable for accomplishing these goals. The goals and program were then explained to the production workers.

Quality scoreboards were posted showing incoming quality of parts and assembled quality of parts on each model as well as the average cumulative total of defects per cabinet, both incoming and after assembly for each style of cabinet. Colored disks were used to indicate that quality was within the goal (green), about equal to the goal (yellow), or worse than the goal (red).

Competition developed between departments and between groups within each department working on different models. The quality control group analyzed these charts daily and investigated and corrected the causes of recurring defects. For example, a revision of conveyor-maintenance procedures was effected as a result of these investigations.

The results over a period of 5 months are shown in Figure 23-5. The $\bar{c}$ on incoming cabinets (parts) had been reduced from 8.5 to about 3.5 (slightly above the goal 3.0) and the $\bar{c}$ on assembled cabinets reduced from 11.5 to approximately 5.5. A considerable saving of repair costs on assembled cabinets resulted and final product quality was improved because less repair work was needed.

It should be noted that the last two control charts did not have statistical limits. The concept of statistical control limits *does* apply to such charts (see Chapter 15) but the results were achieved without the statistical refinement. When a p or c chart is installed on a process a rather dramatic and relatively quick improvement in quality is sometimes noted. A review of such charts may show that the improvement took place *before* sufficient data were available to calculate the control limits. When this is so, it should be realized that the improvement in quality could *not* have been due to the use of the *statistical* limits because they were not available before the quality improved. This means that some other factor must have been responsible for the improvement in quality. Particularly for operator dominant operations,

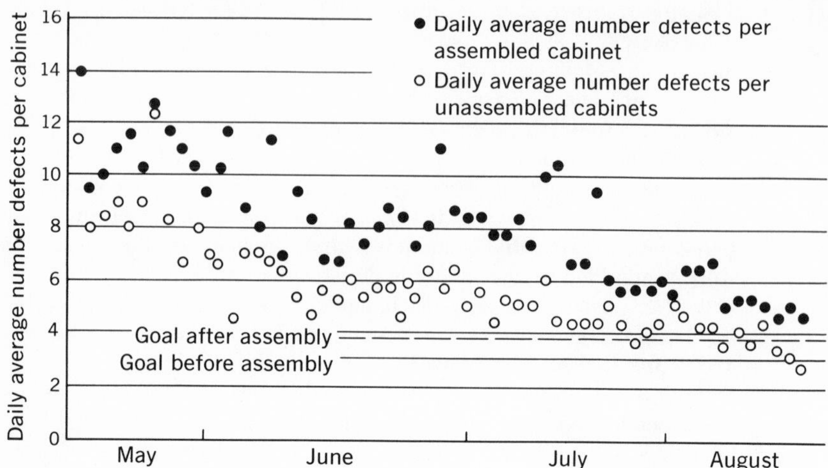

Figure 23-5 Statistical quality control program brings reduction in defects on cabinets on a conveyorized cabinet assembly line.

the improvement in quality may mean that the chart provided the operator with tangible evidence that management is truly concerned with product quality and may thereby have instilled the quality-mindedness necessary in the operator to achieve an improvement.

23-6 Component Dominant Operations

In component dominant operations, the prime cause of poor quality is in the input materials to the operations. Defects may occur in epidemics (a lot of wrong material or out-of-specification material) or they may occur at random (a lot containing partially defective material). Assembly and subassembly, many chemical and food processing operations, and packaging operations are typical examples.

The solution lies in establishing control systems for the earlier operations producing the material. For in-plant operations, the control system needed will depend on whether the particular operation producing the defect is setup, machine, or operator dominant, all of which have been treated in this section. For vendor material, a vendor control system employing incoming inspection, corrective-action follow-up, vendor rating, and similar features may be needed (see Chapters 19 and 20). For internally produced material or components, a temporary system of lot-by-lot acceptance of components before the component dominant operation may be beneficial. The cost of the extra inspection must be evaluated against the advantages:

1 Saving due to reduction of defects from the component dominant operation
2 Demonstration to the component-producing operation personnel that their output is not acceptable
3 Clarification of the nature and frequency of specific component defects as a basis for corrective action.
4 Improved morale among the operators of the component dominant operation who may have struggled so long with problems beyond their control that their own quality problems do not seem worth tackling.

23-7 Automation of Process Control

The use of electronic data processing equipment on quality data has started to yield some degree of automation in process control. As data processing equipment has become more sophisticated, advances have also been made in the basic measuring equipment itself. Thus, all varieties of special gages are now available. The combination of automatic measuring devices and data processing equipment (and, where necessary, devices to transmit information from the measuring equipment to a form that the data processing equipment will accept) appears to make likely major breakthroughs for quality data systems.[8] The American Society for Quality Control has established a special technical committee on "quality information equipment."

Bishop[9] summarizes the potential as follows:

> Three significant areas of development offer potential assistance in stepping up the pace of quality activity. These are: (1) in-line high-speed or continuous-reading instrumentation, (2) in-line and time-shared analogue and digital computers, and (3) automatic feedback control. The results of advances in these areas are already manifest in many companies in numerous industries.
>
> For instance, in-line instrumentation is in wide use for determining such properties as product thickness, density, moisture content, and electrical resistance and capacitance. A number of process control centers include computers for real time calculation[10] of means, vari-

[8] A discussion of the concept of quality information equipment can be found in "Quality Information Equipment—The Tools that Make Total Quality Control Work" by Bernard Sussman, *Industrial Quality Control* (*IQC*), vol. 21, no. 1, p. 10, July, 1964. A more extensive reference on this subject can be found in chap. 9 of *Total Quality Control* by A. V. Feigenbaum, McGraw-Hill Book Company, New York, 1961.

[9] Albert B. Bishop, "Automation of the Quality Control Function," *IQC*, vol. 21, no. 10, p. 509, April, 1965.

[10] Instantaneous calculation as soon as data are available.

ances, correlation functions, and other statistics. Automatic sorting and process control applications are too numerous to mention, but are particularly prevalent in the electronics, tobacco, steel, paper, chemical, bearing, and plastic industries.

A typical application of computers and data processing equipment is the analysis of quality data in automotive assembly.[11] This system is similar to others in that it not only digests a large amount of detailed information, but also transmits this information to operating personnel within a short time period; e.g., "an electronic typewriter in each superintendent's office reports the status of quality approximately twenty microseconds after a defect is detected."

The system[12] in an appliance manufacturing plant consists of a digital computer and associated equipment, a dispatch office, data links to nearly 1,500 operator stations and nearly 3,000 production machines, and a paging system. As a result of the up-to-the-minute flow of information, it is possible to generate "alarm printouts" such as the following:

Machine 602 is reported down—no reason given.
Machine 806 has been down 20 minutes due to tool troubles.
Machine 1400 has produced 5,623 parts and should finish run in 10 minutes.

In addition, a wide variety of reports can be prepared such as on the status of individual machines, on reasons for downtime on machines, and on breakdown of rejected pieces by machine.

The future will probably bring increased emphasis on automation of quality data systems, particularly for process control use. An important initial step is to make sure that the process is capable (see Chapter 15) of meeting tolerances. As the hardware of sophisticated data systems becomes available, thorough feasibility studies will be needed to make sure that the equipment does provide the service that the quality system requires and to make sure that the economics of the equipment are justifiable. Care must be taken to adapt the available hardware to meet the needs of the quality system, instead of adapting the quality system to meet the capabilities of the available hardware. The more sophisticated the hardware, the more likely it is that there will be a period of debugging before the system is finally able to achieve the

[11] John J. DiCicco, "Dynamic Quality Control," *IQC*, vol. 22, no. 5, p. 235, November, 1965.

[12] W. E. Arnold, "Quality Control and Digital Control Computers," *Transactions of the Nineteenth Annual Conference of the American Society for Quality Control*, 1965, p. 329.

original objectives. Errors in choosing and installing automated process control devices can be expensive, so that alternatives must be carefully evaluated[13] in the manufacturing planning phase (see Chapter 14).

23-8 Quality Systems Audit

To provide a framework for the audit[14] of activities affecting *manufacturing quality,* it will be worthwhile to refer back to the three fundamentals of self-control and the corollary fundamental of attitude toward quality (see section 1-12). For each fundamental, some examples of activities that might require audit are given:

The operator must know what he is supposed to do:

1 Completeness and adequacy of specifications
2 Classification of characteristics including the identification of critical parts
3 Issuance and control of drawings and revisions
4 Written manufacturing procedures with provisions for the issuance of revisions
5 Written inspection and test procedures with provisions for the issuance of revisions

The operator must know what he is doing:

1 Operator sampling of product during manufacturing
2 Independent inspection of product
3 Maintenance of manufacturing and inspection logs to record all changes and unusual situations occurring during manufacturing
4 Use of quality charts when required
5 Feedback of inspection results to operator

The operator has the means for regulating what he is doing:

1 Certification and recertification of operators and inspectors
2 Certifications of machine and test equipment (including the quantitative determination of process capabilities)
3 Review of initial tooling setup and periodic review during manufacturing
4 Calibration and control of test equipment

[13] James A. Curry, "Automatic Production Recording—Yield—and Quality Control," *IQC*, vol. 17, no. 8, p. 12, February, 1961. Also, N. B Angelo, "Process Computers—Which One for the Job?," *IQC*, vol. 22, no. 4, p. 167, October, 1965.

[14] Marguglio presents an excellent discussion of the principles of executing quality surveillance. See B. W. Marguglio, "Quality Systems Audit," *IQC*, vol. 20, no. 1, p. 12, July, 1963.

5 Control of environmental conditions (temperature, dirt, etc.) when important
6 Certification and control of the storage, packing, and shipping operations during manufacturing
7 Use of seals when required to prevent accidental or purposeful tampering with completed units of product

The operator must have the state of mind which makes him so desirous of meeting the specification that he will make use of the knowledge of performance and the means for regulating the process:

1 Orientation program to inform the operator on the use of the product, on the importance of quality, and on the specific actions he must take to adequately control quality
2 Feedback to operator on the results of the analysis of product failures
3 Operator stamp system to identify the product made by each operator
4 Record of action taken or not taken on feedback of product failures
5 Employee suggestion file to see the extent of suggestions dealing with quality
6 File of personnel actions dealing with quality reprimands or with employee grievances respecting quality

These comments are oriented toward the surveillance of the *manufacturing* aspects of quality. See Chapter 28 for a broader discussion of surveillance. Also see section 14-11 for a discussion of process audit or surveillance.

PROBLEMS

1 Visit a local plant and report on the techniques used to control the variation of key dimensions of a specific product.

2 Visit a local plant and study several important manufacturing operations. Classify each operation as setup, machine, operator, or component dominant. (Try to find operations illustrating each of these dominance categories.)

3 Select *one* of the activities listed in section 23-8 and define the specific steps you would take to audit that activity.

4 The percent of water absorption is an important characteristic of common building brick. A certain company occasionally measured this characteristic of its product but records were never kept. It was decided to analyze the process with a control chart. Twenty-five samples of four bricks each yielded these results:

Sample Number	$\bar{X}$	*R*	*Sample Number*	$\bar{X}$	*R*
1	15.1	9.1	14	9.8	17.5
2	12.3	9.9	15	8.8	10.5
3	7.4	9.7	16	8.1	4.4
4	8.7	6.7	17	6.3	4.1
5	8.8	7.1	18	10.5	5.7
6	11.7	9.1	19	9.7	6.4
7	10.2	12.1	20	11.7	4.6
8	11.5	10.8	21	13.2	7.2
9	11.2	13.5	22	12.5	8.3
10	10.2	6.9	23	7.5	6.4
11	9.6	5.0	24	8.8	6.9
12	7.6	8.2	25	8.0	6.4
13	7.6	5.4			

a Plot the data on an average and range control chart with control limits.

b During the manufacturing process, water is added to a clay paste to produce a workable mass. The amount of water added depended on the feel of the clay paste to the "pug mill" operator. Several process changes were made after the original data were taken. A key change was the installation of a flowmeter to better control the quantity of water added. Twenty-five new samples of four bricks each were taken with the revised process:

Sample Number	$\bar{X}$	*R*	*Sample Number*	$\bar{X}$	*R*
1	6.7	4.0	14	9.6	6.9
2	7.7	8.4	15	11.1	2.9
3	8.0	4.0	16	13.2	13.2
4	10.9	8.3	17	8.7	8.7
5	8.7	2.2	18	4.7	4.7
6	8.2	4.1	19	6.1	6.1
7	9.9	3.7	20	4.8	4.8
8	11.1	8.0	21	2.3	2.3
9	10.5	6.1	22	4.2	4.2
10	7.3	2.7	23	4.5	4.5
11	8.8	2.2	24	2.0	2.0
12	10.8	6.3	25	2.9	2.9
13	10.4	7.7			

Plot these data on an average and range control chart with limits calculated from the new data. Comment on the charts for both sets of data.

5 The following data were collected for a control chart plot of the net weight of cosmetic cream X. Each sample consisted of four packages and the total data covered sampling of 8 days of production.

Sample Number	$\overline{X}$	*R*	*Sample Number*	$\overline{X}$	*R*
1	3.34	0.029	17	3.23	0.002
2	3.28	0.000	18	3.28	0.000
3	3.20	0.030	19	3.26	0.000
4	3.30	0.058	20	3.28	0.029
5	3.21	0.000	21	3.56	0.030
6	3.29	0.120	22	3.62	0.000
7	3.25	0.062	23	3.38	0.000
8	3.24	0.000	24	3.23	0.000
9	3.22	0.000	25	3.41	0.000
10	3.24	0.030	26	3.27	0.000
11	3.34	0.048	27	3.33	0.000
12	3.23	0.030	28	3.24	0.020
13	3.47	0.000	29	3.23	0.035
14	3.25	0.030	30	3.35	0.004
15	3.35	0.000	31	3.29	0.000
16	3.27	0.032			

Plot the data on an average and range control chart with control limits. Comment on the large number of out-of-control points.

6 The following data represent the number of defects found on each sewing machine cabinet inspected:

Sample Number	*Number of Defects*	*Sample Number*	*Number of Defects*
1	8	14	6
2	10	15	4
3	7	16	7
4	7	17	5
5	8	18	8
6	6	19	6
7	9	20	4
8	8	21	5
9	4	22	7
10	7	23	4
11	9	24	5
12	6	25	5
13	5		

Plot a control chart with control limits. Comment on the chart.

7 A sample of 100 electrical connectors was inspected each shift. Three characteristics were inspected on each connector but each connector was classified simply as defective or acceptable. The results follow:

Sample Number	*Percent Defective*	*Sample Number*	*Percent Defective*
1	4	14	4
2	3	15	4
3	5	16	5
4	6	17	3
5	7	18	0
6	5	19	3
7	4	20	2
8	2	21	1
9	5	22	3
10	6	23	4
11	4	24	2
12	3	25	2
13	3		

a Plot a control chart with control limits. Comment on the chart.
b If the inspection results had been recorded in sufficient detail, what other type of chart could also have been plotted?

8 An average and range chart based on a sample size of five has been run with the following results:

	Averages	*Ranges*
Upper control limit	78.0	8.0
Average value	75.8	3.8
Lower control limit	73.6	0

How large an increase in the overall process average would have to occur in order to have a 30 percent chance that a sample average will exceed the upper control limit?

24 Finished Goods Acceptance

24-1 Acceptance Inspection: General

Every fabricator must decide, on completion of his work, whether the resulting product is fit for use. In earlier days, the village tanner, the temple architect, and the royal cook all made this decision. Today the same decision must be made by the neighborhood cobbler, the large factory, the large service company.

The act of deciding fitness for use is commonly called "inspection." It consists of examining the product and judging whether to "accept" it as meeting the conditions of use. (Hence the term "acceptance inspection.")

In the small shop of today (as traditionally) the craftsman or master performs the acceptance inspection based on his personal knowledge of the conditions of use. The inspection occupies only a minority of his time; mostly he is engaged in production. However, in the large shops of today (as traditionally) the workman often is confined to a narrow operation and does not understand precisely the relationship of what he does to the conditions of use.

Of course, he is usually given some approximation: a product specification, a sample, a process specification. However, these do not fully reflect the various conditions of use. In addition, he is often subjected to multiple standards to an extent which makes it difficult for him to meet all standards simultaneously. In consequence, there has arisen a practice, in the larger companies, of making inspection a full-time job category, and of delegating the product acceptance decision to the inspector.

(This practice is actually of ancient origin. The ancient Egyptians used full-time inspectors on construction jobs. In Figure 2-1, one of the workmen is dressing a stone block while an inspector measures the flatness with a piece of string.)

Whether performed as a full-time job category or as incidental to production, "inspection" consists mostly of interpreting the specification, examining the product, judging conformance, deciding on acceptance, disposing of the product, and recording the necessary data.

With elaboration of technology, the term "inspection" has come to acquire the narrower meaning of *visual* examination. Other forms of examination include mechanical gaging, electrical testing, chemical analysis, etc. As a result, the job category of examining the product includes inspectors, gagers, chemists, metallurgists, testers, etc. (Some "inspectors" have nothing to do with quality of product, e.g., police inspectors, fire inspectors.)

24-2 The Modern Acceptance Function

The modern large company conducts this function under a system of delegation shown in Table 24-1.

Table 24-1 THE MODERN ACCEPTANCE FUNCTION

Activity	*Delegated to*
Interpretation of specification	Inspection or inspection supervisors
Examination of product	Inspectors or instruments
Judgment of conformance	Inspectors
Acceptance of product which conforms to specification	Inspectors*
Review and disposition of nonconforming product	Material review board
Recording the data	Inspectors or instruments

* A small minority of companies have created the conditions which make it possible to delegate the work of product acceptance to production operators, subject to an "audit of decisions." See J. M. Juran, "Inspectors—Headed for Extinction?," *Industrial Quality Control*, vol. 22, no. 4, pp. 198–199, October, 1965.

24-3 Interpretation of the Specification

Specifications are prepared by comparatively few men, each well aware of the conditions of use. In contrast, specifications must be interpreted by many operators and inspectors, most of them ignorant of the conditions of use. This gap is bridged in one of several ways:

1 Training the inspectors to understand the conditions of use.
2 Training the inspection supervisors to understand the conditions of use and reserving to them the duties of interpreting specifications.
3 Preparing detailed standards and manuals to aid all concerned in interpreting the specifications. Usually this is done by quality control engineers acting as inspection planners.

Example In an optical goods factory the term "beauty defect" described several conditions which differed widely as to fitness for use. A scratch in the focal plane made the product unfit for use. A scratch on the large end of a pair of binoculars was not serious functionally, but was visible to the user, and hence not acceptable. Other scratches were neither functional nor visible to the user, and hence not important. Only through planning analysis were these distinctions made clear and woven into the procedures. (Also see the discussion on sensory qualities in Chapter 13.)

4 Standardizing the conditions of inspection to simulate the conditions of use. This principle is widely used in inspecting such products as electrical appliances, floor coverings, decorations, etc.
5 Classifying quality characteristics for seriousness (see below).
6 Inviting the designer to give an interpretation, often in a meeting of a material review board.

In many companies there is a general situation of unrealistic tolerances loosely enforced (see Chapter 12). In such cases there is a chronic problem of securing interpretations and waivers from the designers unless a basic solution is worked out.

24-4 Seriousness Classification of Defects

Defects are unequal in their effect on fitness for use. Those who understand fitness for use have always considered this inequality during decision making. However, in modern times when many inspectors and operators lack adequate knowledge of fitness for use, alternative methods must be provided to convey to them what is the seriousness of defects. One such method is a formal plan of seriousness classification of defects. The resulting classification is then used widely in the company: in specification writing (see Chapter 12); in vendor relations (see Chapter 19); in manufacturing planning, production, and inspection; in check inspection and executive reporting (see Chapter 28). Because of this multiple use, preparation of the plan of classification is made by an interdepartmental committee which:

1 Decides how many classes or strata of seriousness to create (usually three or four)
2 Defines each class
3 Classifies each defect into its proper class of seriousness

Definition for the Classes

Several sets of definitions have been developed, of which one is set out below.[1] Table 24-2 on page 498 is a summary of the Bell System plan (which has influenced the design of many other plans).

Increasingly, the seriousness of a defect is being determined from the user's viewpoint—the effect on his continuity of service, on his sensibilities.

Classifying the Defects

This tedious job is not only an essential part of the planning; it yields a welcome by-product in clearing up vague meanings, and in discovering misconceptions and confusion among departments. This confusion comes to light in large part because the classification plan has several purposes:

1 Setting or revising tolerances in product specifications[2]
2 Design of sampling plans (see Chapter 17)
3 Rating quality of outgoing product (see Chapter 28)

The discipline of meeting this variety of needs is wholesome for clarifying the understanding of all concerned.

24-5 Acceptance of Systems

Complex systems generate some special problems in acceptance inspection. Such products must be checked for visual and dimensional characteristics and for proper operation. The operational test may simply be a test to see if each system functions properly when all subsystems

[1] For other examples, see J. M. Juran (ed.), *Quality Control Handbook* (*QCH*), 2d ed., McGraw-Hill Book Company, New York, 1962, pp. 8-11 and 12-18.
[2] Paul E. Allen, "Evaluating Inspection Costs," *Transactions of the National Convention of the American Society for Quality Control,* 1959, pp. 585–596.

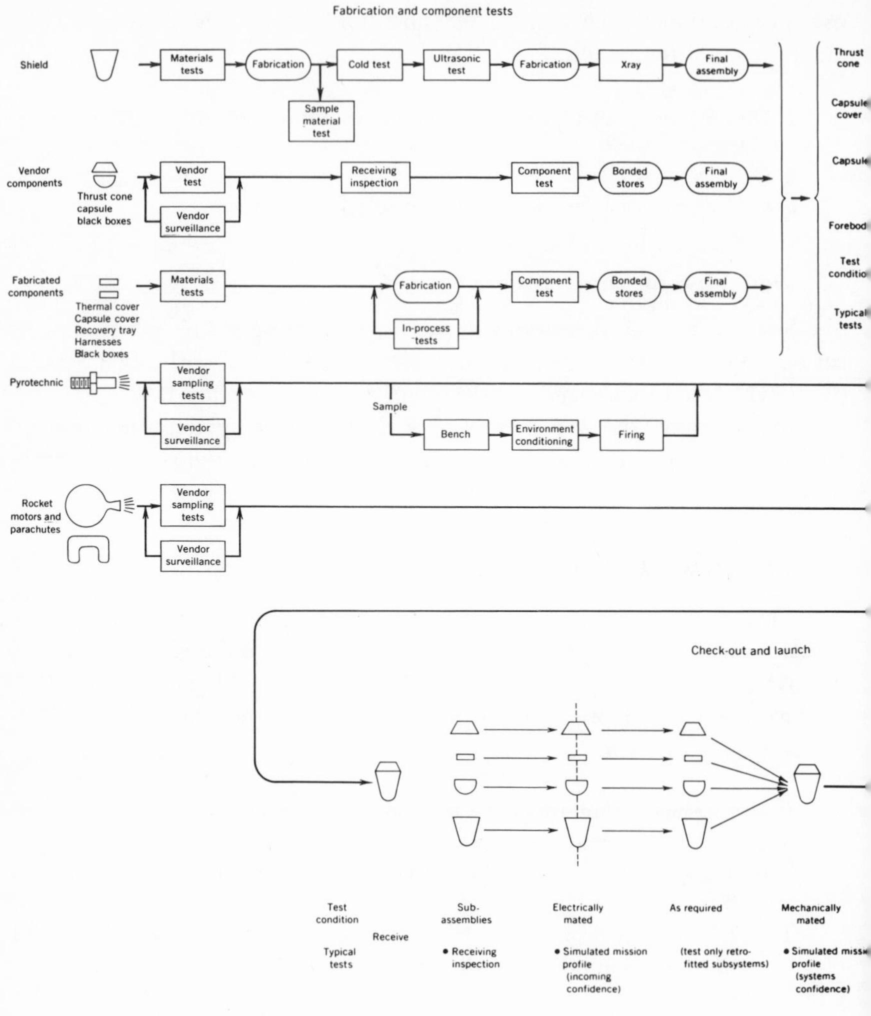

Figure 24-1 Acceptance test flow for a reentry vehicle. Source: M. Wilson, Electric Co., Reentry and Environmental Systems Division, Mar. 15, 1961.

are assembled. However, the test may be more extensive and involve operating each system (or a sample) under environmental conditions such as vibration, shock, humidity, etc. When run as part of an acceptance test, the levels of environmental tests must be carefully selected to prevent degradation of any part of the system.

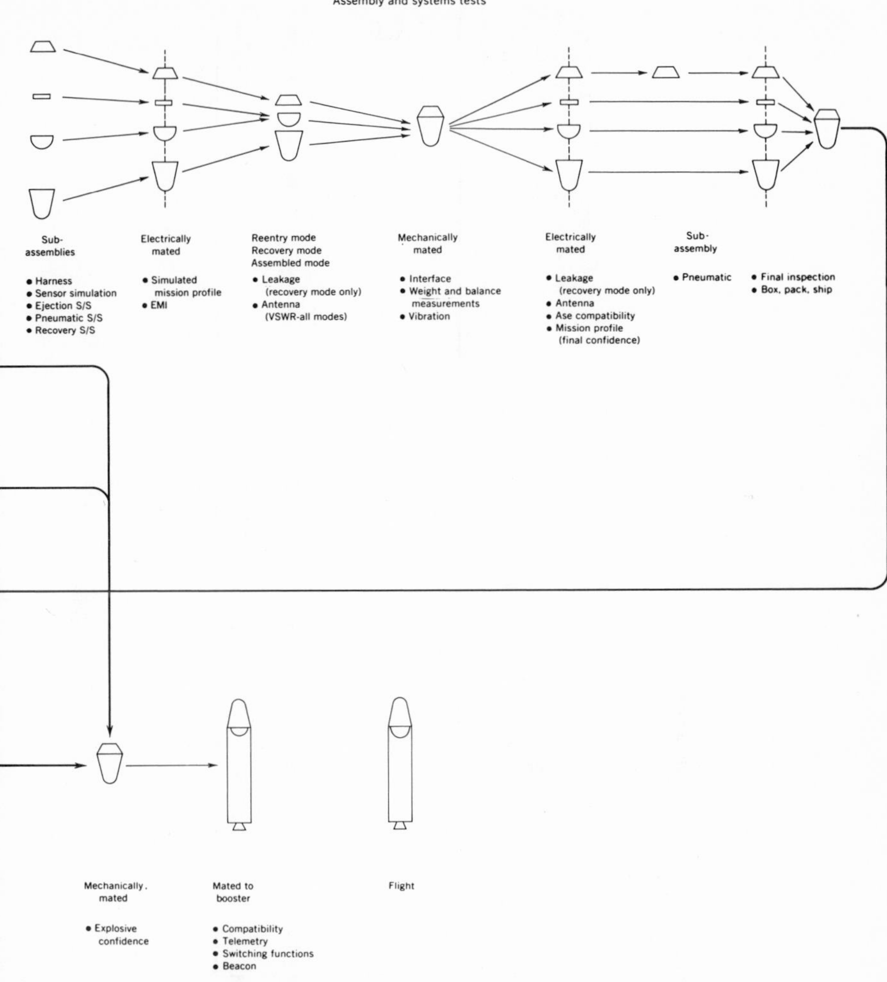

identify the many echelons of testing required to maintain the depart-
at the materials level and follow through to system level and on to ultimate flight test.

Malfunctions in Missile and Space Equipment, Document No. R61SD1, General

The development of a systems acceptance test requires a great deal of knowledge about the design of the system and its operation under actual usage conditions. The specifics of the test partly depend on what inspections and tests were conducted at lower levels of hardware such as components and subsystems. Figure 24-1 illustrates the steps

Table 24-2 **Basic Definitions—Classification of Defects**

Defect Class	*Demerit Weight*	*Cause Personal Injury*	*Cause Operating Failure*	*Cause Intermittent Operating Trouble Difficult to Locate in Field*	*Cause Substandard Performance*	*Involve Increased Maintenance or Decreased Life*	*Cause Increase in Installation Effort by Customer*	*Appearance, Finish, or Workmanship Defects*
A	100	Liable to	Will surely*	Will surely				
B	50	—	Will surely† Will probably	—	Will surely	Will surely	Major increase	
C	10	—	May possibly	—	Likely to	Likely to	Minor increase	Major
D	1	—	Will not	—	Will not	Will not	—	Minor

* Not readily corrected in the field.
† Readily corrected in the field.

in a complete acceptance test program for a reentry vehicle.[3] Note that the systems tests include conventional inspection, operational, and environmental tests. Also note that testing starts at the material level and extends to systems tests during the checkout and launch phase of the product.

24-6 Judgment of Conformance

The enormous, growing size of our economies requires myriads of judgments of conformance to be made by people who lack knowledge of fitness for use (and by others who have that knowledge). To aid in making these judgments, a whole kit of tools has been developed and improved:

Improved specifications and clearer definition of sensory qualities (see Chapter 12)
Seriousness classification of defects
Lot acceptance criteria, to define lots, sample sizes, acceptance numbers, etc. (see Chapter 17)
Material review boards (see section 24-8)
Written procedures for disposition of goods, identification, documentation, etc.

Use of these tools is guided by still another tool—the plan of delegation of who may decide what. Disposition of *conforming* material is delegated to the bottom of the organization; disposition of *nonconforming* material is reserved to the higher-ups. A further division of responsibility is based on the seriousness classification as in the following typical (but not universal) example:

Type of Defect Exceeding AQL	*Authority of Inspection*
Critical or major	No authority to ship nonconforming product. May request waiver from design department, or from material review board
Minor A	May ship after consultation with design department
Minor B	May ship on own judgment

In the background of all this delegation is the vital concept of "law and order"; i.e., the men must carry out their delegated responsi-

[3] From General Electric Co., Missile and Space Vehicles Division.

bilities but must not go beyond them. A common violation is for the inspectors to accept borderline product without reporting it. It is analogous to the police officer giving a borderline violator a "break." (Sometimes the violation is the result of the supervisor's brushing off previous cases of reported borderline product.) The result is an erosion of the discipline of conformance until some catastrophic failure brings an investigation which discovers the practice.

24-7 Acceptance by Production Departments

Most finished goods acceptance is performed by inspection departments. However, much acceptance of goods in process (usually movement of goods between fabrication departments) is done by production departments. In addition, there are companies in which the final inspection and test departments report to a production manager rather than to an inspection manager (but subject to an inspection audit).

Modern practice is to require that the inspection plan, whether

Figure 24-2 Plan for disposition of machined parts.

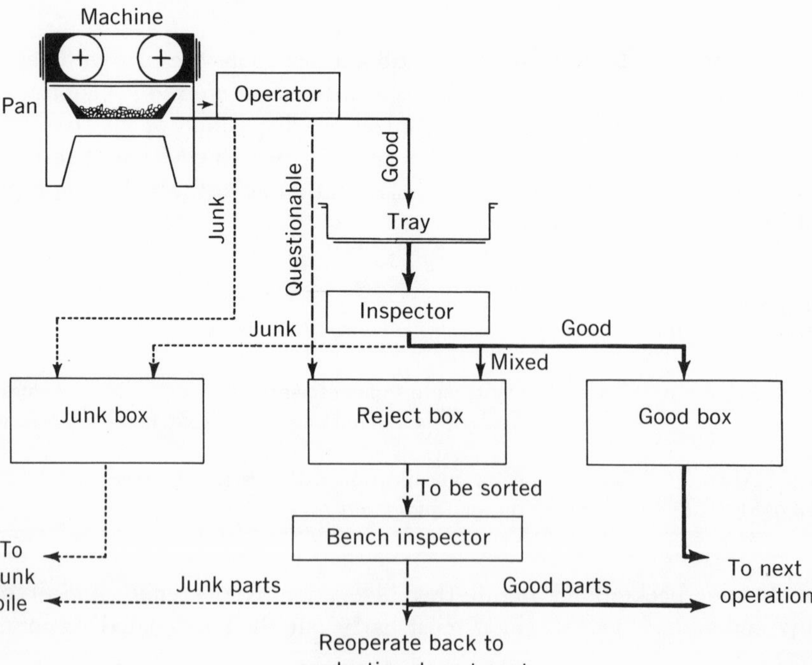

to be carried out by the inspection or the production department, is to be designed by the quality control planners. If the planning is well done, the question of jurisdiction over the inspectors shrinks in importance.

It is quite common to find acceptance decisions being made by *both* operators and inspectors. A department subject to patrol inspection is an example of this. In such cases it is well to coordinate the work so that each has access to the measurements of the other. In addition, the rules of jurisdiction over product disposition must be worked out with care. (Important "property rights" arise in the factory as well as outside, affecting floor space, tools, machinery, product, and everything else.) An example of a coordinated plan is shown in Figure 24-2. The operator classifies a "pan" of product as junk, questionable, or good. An inspector makes the same classification for the product called "good" by the operator. A bench inspector then classifies product in the reject box as junk, rework, or good.

The question of who physically controls the product during the acceptance procedure must be settled. Elaborate systems (some necessary, some unnecessary) have been developed to define the jurisdiction of production and inspection departments.[4]

24-8 Disposition of Nonconforming Product; Material Review Board

Disposition of nonconforming product is usually outside of the delegated responsibility of the inspector. A review procedure is provided. At one extreme, the inspection supervisor is delegated the responsibility of review and disposition. At the other extreme, a formal committee known as a material review board (MRB) is set up to conduct the review and make the disposition. A complete review procedure usually includes:

1 A *deviation report,* prepared by the inspection department. This is a factual statement of the condition of the product, identifying the lot(s), the type and extent of the defect, the probable cause, etc. Copies of the report are sent to interested departments, including production control.

2 A *request for waiver,* prepared by an investigator, often a member of the staff quality control department. This request includes suggested means for avoiding recurrence as well as proposed means for disposition of the product.

[4] See *QCH,* 1962, pp. 8-21 to 8-22.

3 *Means for segregation* of the questionable material. This varies from the red tag all the way to a separate hold desk or hold area for getting all dubious material out of the regular stream of production. In any event, serial numbers should be used on the papers to ensure that there is no unauthorized usage of the product.
4 *Means for decision making* on waiver requests. These may be the regular chain of command, or informal meetings of departments interested in the particular material, or a formal committee such as a material review board (see below).
5 *Recorded decisions.* These may be in the form of changes of specification, or numbered written waivers. (Sometimes the waiver request is signed by the material review board members.)

The concept of a material review board recognizes that decisions on the possible salvage of defectives requires the participation of several parties, such as representatives of the customer, of the engineering department, and of quality control. The sequence of events is generally:

1 Nonconforming material is discovered either by production or inspection.
2 Inspection (or production) writes a "hold" form to identify the material.
3 The quality control representative reviews all cases and disposes of many.
4 Customer representatives review remaining cases and dispose of many.
5 The remaining cases are then shown to the full material review board for one of the several decisions:
   a Accept "as is." This usually requires unanimous approval of the board.
   b Rework.
   c Scrap.
6 Physical disposition is made on the basis of the decision in step 5.
7 Hold forms are, in theory, analyzed for repeat cases so that action can be taken to prevent future defectives.

A common criticism of the MRB procedure is the emphasis on disposition of current nonconforming product. Often, the board devotes much time to arguing about the disposition of specific units of product and devotes insufficient time to correcting the causes of nonconformance. This may require setting up some specific organizational mechanism to follow up on all board decisions with recommendations for corrective action.

The final disposition of nonconforming material will be to rework, scrap, or downgrade to a lower grade. This is an economic decision that depends on costs, the time required, and related factors.[5]

[5] See *QCH*, 1962, p. 8-23.

24-9 Accounting for the Cost of Rejections

"Rejection" may mean:

1 Nonacceptance
2 Return of all the goods
3 Return of the defectives only
4 Nonacceptance unless certain conditions are met

Suppose a vendor ships material which is 10 percent defective. There is no argument that the 10 percent defectives are to be returned to the vendor for credit, but who should pay the cost of sorting the good from the bad? The decision should minimize the sum of all costs.

To save shipping charges and time, it may be better to sort the good from the bad at the customer's location than to ship the entire lot of product back to the vendor's plant. The actual sorting can be done by either the vendor's personnel or the customer's personnel, and the *cost* of sorting paid by the vendor. This follows the concept of minimizing the sum of all costs to the parties. The lower the sum, the easier it will be to agree on how to divide the charges between the parties.

24-10 Feedback of Inspection Data

A key element of control is the need for a person to know how he is doing. This need occurs at all levels and therefore a carefully planned system is needed for feedback of inspection data at all levels.

Example An example of such a system[6] is seen for a research and development[7] job shop manufacturing electronic modules. Inspection data was classified into 12 major defect categories:

1 Assembly
2 Broken or damaged
3 Electrical
4 Fabrication
5 Missing and incorrect
6 Potting
7 Procedural
8 Processes (general)
9 Soldering

[6] H. Robert Daw, "Systematic Procedure of Trouble Spotting," *Industrial Quality Control* (*IQC*), vol. 21, no. 9, pp. 443–449, March, 1965.
[7] See *QCH,* 1962, pp. 29-12 to 29-19, for a similar system developed for mass production operations.

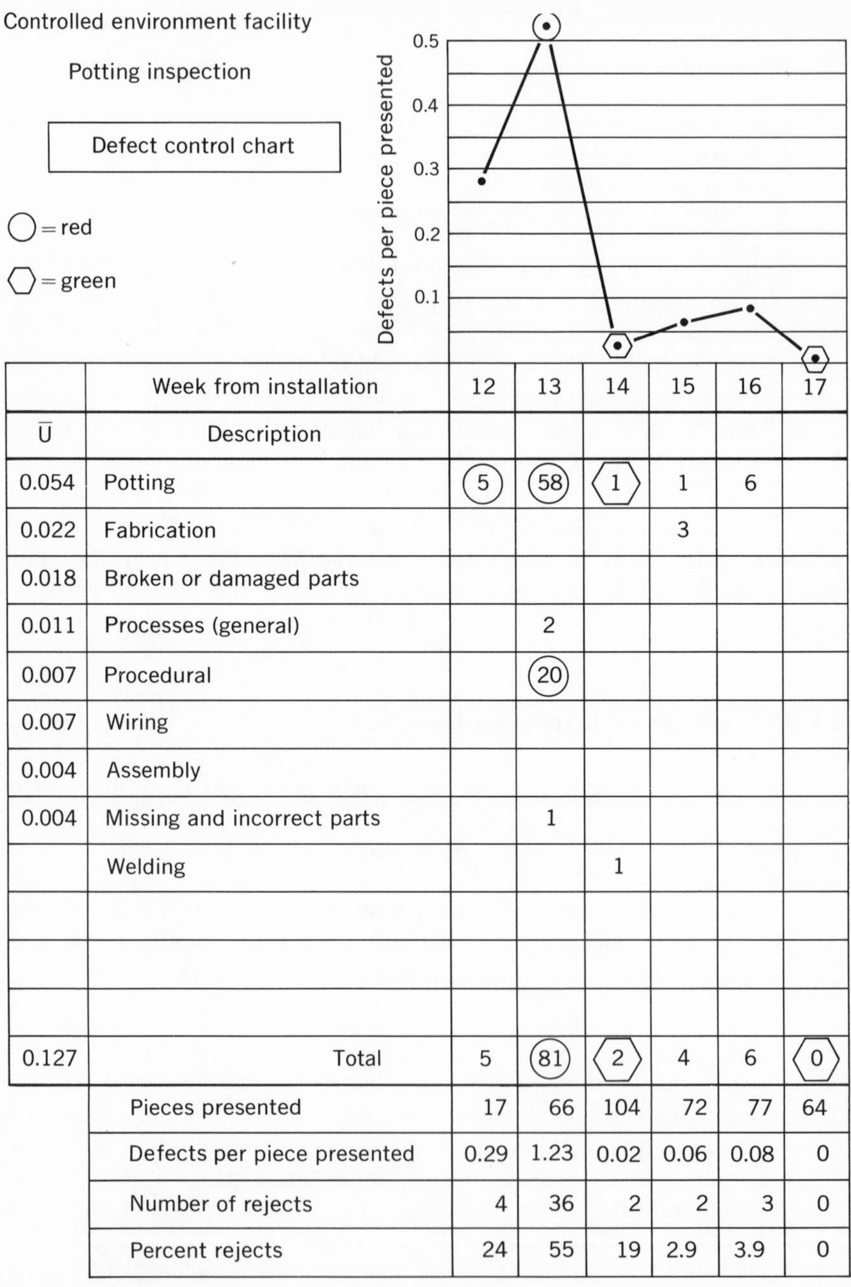

	Week from installation	12	13	14	15	16	17
$\bar{U}$	Description						
0.054	Potting	(5)	(58)	⟨1⟩	1	6	
0.022	Fabrication				3		
0.018	Broken or damaged parts						
0.011	Processes (general)		2				
0.007	Procedural		(20)				
0.007	Wiring						
0.004	Assembly						
0.004	Missing and incorrect parts		1				
	Welding			1			
0.127	Total	5	(81)	⟨2⟩	4	6	⟨0⟩
	Pieces presented	17	66	104	72	77	64
	Defects per piece presented	0.29	1.23	0.02	0.06	0.08	0
	Number of rejects	4	36	2	2	3	0
	Percent rejects	24	55	19	2.9	3.9	0

Figure 24-3 Potting inspection ($U = c'/100$). The circled numbers show out of control conditions based on the table for number of defects, red on the high side and green on the low side. Source: H. Robert Daw, "Systematic Procedure of Trouble Spotting," *Industrial Quality Control,* vol. 21, no. 9, pp. 443–449, March, 1965.

10 Surface finish
11 Welds
12 Wiring

For each category, inspection results are posted weekly. Figure 24-3 shows the chart for potting defects for the twelfth through the seventeenth week after installation of the data system. For the thirteenth week, 81 defects were found in 66 pieces presented, or there were 1.23 defects per piece. This was compared to a previous process average $\bar{u}$ of 0.127 and the difference was found to be statistically significant. The value of 1.23 (which exceeds the scale limits on the chart) is circled to indicate red, for significantly poor quality. The values for the fourteenth and seventeenth weeks are highlighted to indicate green, for significantly good quality. The chart also defines the "vital few" defects in the thirteenth week as due to potting and procedural operations.

Figure 24-4 shows how this basic information is pyramided into a series of charts for levels of management, i.e., manufacturing foreman, manager of shop operations, and manager of manufacturing. Finally, Figure 24-5 shows supplementary charts issued to shop supervision to help identify defect problem areas by operator number. Again the vital few problem areas are highlighted: operators 4, 7, and 11 account for 104 of 133 defects, and potting, procedural, soldering, and welding defects account for 120 of the 133 defects.

Defect prevention requires data and data cost money. A system for adequate feedback of inspection data to all action levels is an investment whose payoff is reduced quality costs.

24-11 Acceptance Sampling Plans for Life and Other Time Characteristics

Published sampling plans are available for use in evaluating performance against a requirement in terms of life, mean time between failures (MTBF), or its reciprocal, the failure rate. In general, the same plan is applicable to any of these time indices. Tables are available providing attributes plans and variables plans based on both the normal and exponential and the Weibull distributions. Martin[8] summarizes some of the key plans.

Sampling plans for evaluating reliability indices have focused atten-

[8] Cyrus A. Martin, "Rationale and Use of Military Sampling Handbooks," *Proceedings of the Annual Symposium on Reliability*, 1968, Institute of Electrical and Electronics Engineers, Inc., pp. 385–390.

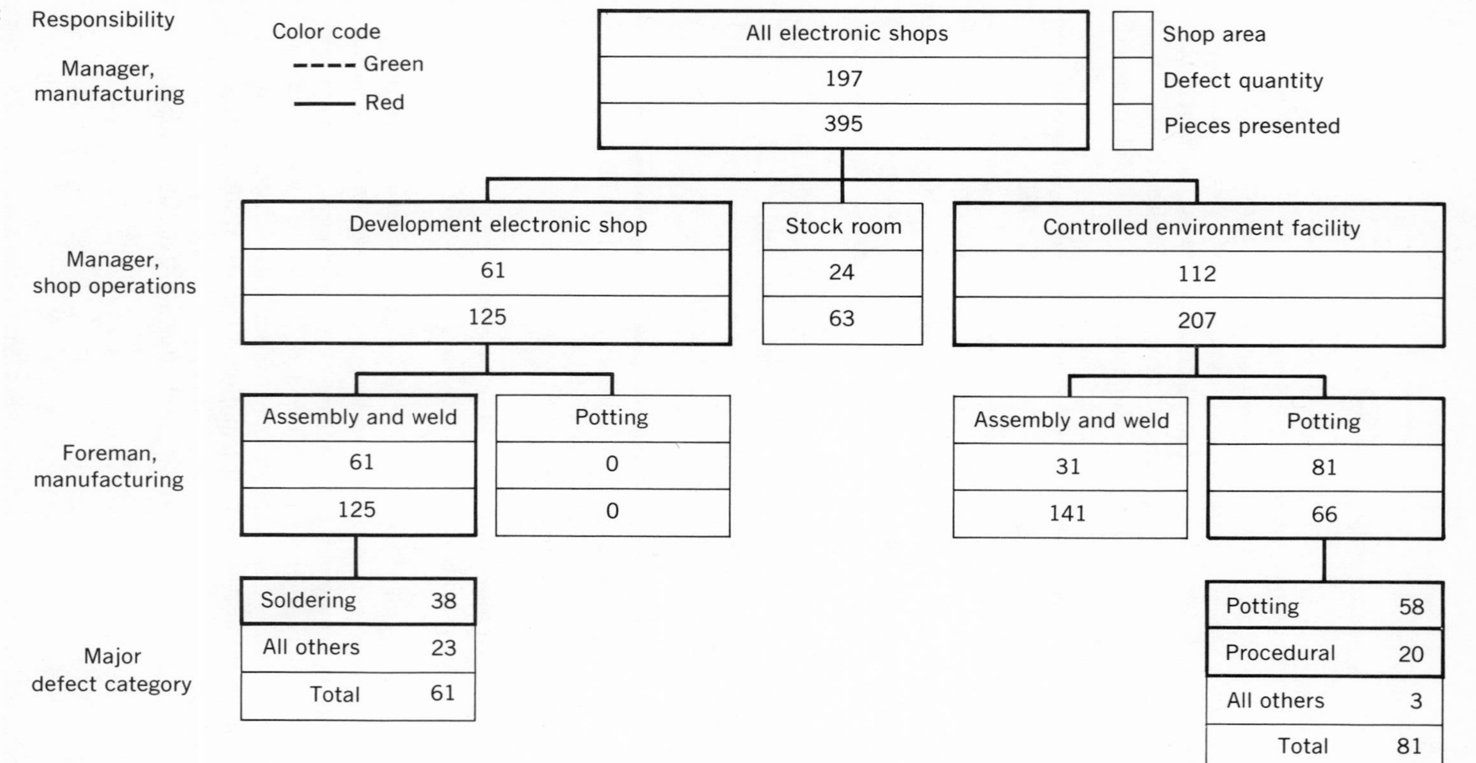

Figure 24-4 Pyramid chart—week +13. Source: H. Robert Daw, "Systematic Procedure of Trouble Spotting," *Industrial Quality Control,* vol. 21, no. 9, pp. 443–449, March, 1965.

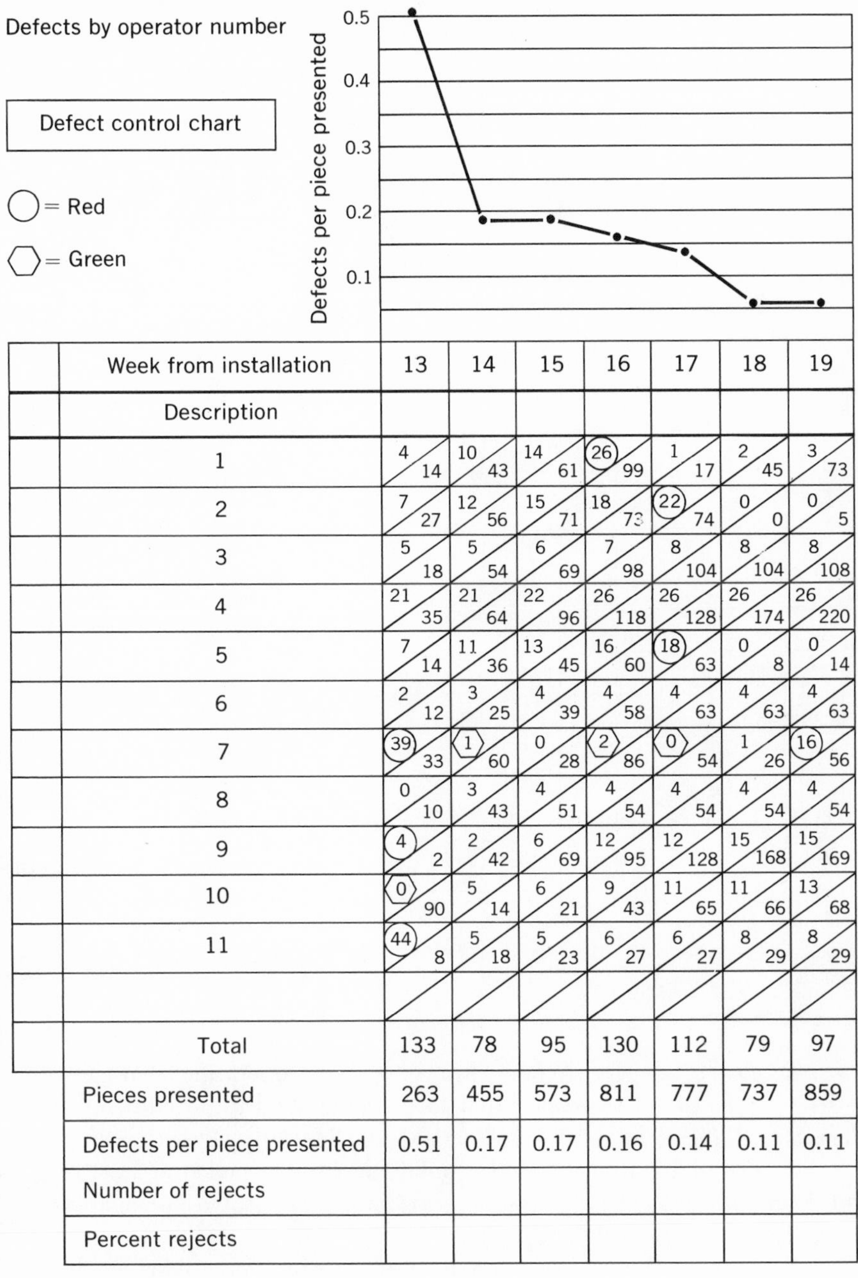

	Week from installation	13	14	15	16	17	18	19
	Description							
	1	4 / 14	10 / 43	14 / 61	26 / 99	1 / 17	2 / 45	3 / 73
	2	7 / 27	12 / 56	15 / 71	18 / 73	22 / 74	0 / 0	0 / 5
	3	5 / 18	5 / 54	6 / 69	7 / 98	8 / 104	8 / 104	8 / 108
	4	21 / 35	21 / 64	22 / 96	26 / 118	26 / 128	26 / 174	26 / 220
	5	7 / 14	11 / 36	13 / 45	16 / 60	18 / 63	0 / 8	0 / 14
	6	2 / 12	3 / 25	4 / 39	4 / 58	4 / 63	4 / 63	4 / 63
	7	39 / 33	1 / 60	0 / 28	2 / 86	0 / 54	1 / 26	16 / 56
	8	0 / 10	3 / 43	4 / 51	4 / 54	4 / 54	4 / 54	4 / 54
	9	4 / 2	2 / 42	6 / 69	12 / 95	12 / 128	15 / 168	15 / 169
	10	0 / 90	5 / 14	6 / 21	9 / 43	11 / 65	11 / 66	13 / 68
	11	44 / 8	5 / 18	5 / 23	6 / 27	6 / 27	8 / 29	8 / 29
	Total	133	78	95	130	112	79	97
	Pieces presented	263	455	573	811	777	737	859
	Defects per piece presented	0.51	0.17	0.17	0.16	0.14	0.11	0.11
	Number of rejects							
	Percent rejects							

(*a*)

Figure 24-5a Defects by operator number. *b* Defects by manufacturing operator. Source: H. Robert Daw, "Systematic Procedure of Trouble Spotting," *Industrial Quality Control*, vol. 21, no. 9, pp. 443–449, March, 1965.

Week + 13

Operator number	DEFECT CATEGORY												Total defects	Total pieces inspected
	Assembly A	Broken or damaged B	Electrical C	Fabrication D	Missing and incorrect E	* Potting F	* Procedural G	Processes (general) H	* Soldering K	Surface finish L	* Welding M	Wiring N		
1	1					1	1				1		4	14
2	2				1						4		7	27
3	1										4		5	18
4*					1	17	3						[21]	35
5	1												7	14
6				1	1						6		2	12
7*						30	9						[39]	33
8													0	10
9											4		4	2
10													0	90
11*	1			1					38		2	2	[44]	8
Totals	6	0	0	2	3	[48]	[13]	0	[38]	0	[21]	2	[133]	263

(*b*)

Figure 24-5 (Continued).

tion on two concepts: confidence level and the operating characteristic curve. Chapter 11 explained that to demonstrate a required mean time between failures at a specified confidence level requires that the test results lead to a lower one-sided confidence limit which is at least equal to the required MTBF. Tables are available that provide plans to demonstrate reliability in this manner.

Table 24-3 shows a portion of a table providing plans at the 90 percent confidence level.[9]

[9] M. Sobel and J. A. Tischendorf, "Acceptance Sampling with New Life Test Objectives," *Proceedings of the Fifth National Symposium on Reliability and Quality Control*, New York, 1959, Institute of Radio Engineers, pp. 108–118.

Table 24-3 MINIMUM SIZE OF SAMPLE TO BE TESTED FOR A TIME t TO ASSURE A MEAN LIFE OF AT LEAST μ_0 WITH CONFIDENCE P^* WHEN c IS THE ACCEPTANCE NUMBER

	t/μ_0			
c	1.0	0.1	0.01	0.001
0	3	24	231	23,026
1	5	40	390	38,898
.				
.				
.				
5	13	97	927	92,747
.				
.				
.				
10	22	161	1,541	154,068
.				
.				
.				
14	29	212	2,013	201,280

The table shows the number of items to be tested each for a time t and the allowable number of failures c to demonstrate with 90 percent confidence that the reliability requirement μ_0 has been met or exceeded. For example, a sample of 40 units each tested for $t = 100$ with 1 or fewer failures will provide demonstration of a 1,000-hour MTBF at the 90 percent confidence level. The table assumes a constant failure rate. This allows tradeoffs of sample size with test time.

The advantage of a demonstration plan based on a specified confidence level is that proof is provided in terms of a probability value that the requirement has been met. However, there are disadvantages:

1 A large sample size is often required. In the case above, 4,000 hours of test time would be required *assuming that only 1 or fewer failures occurred* (which is unlikely).

2 The alternative to a large sample size is for the true reliability to be much better than the requirement. For example, suppose a contractor wanted to demonstrate a 1,000-hour MTBF by testing 24 units each for 100 hours and allowing 0 failures (thus demonstrating the units at the 90 percent confidence level). Calculations show that the true reliability would have to be about 46 times the requirement if the contractor wants to be 95 percent sure of actually passing the test. This is not what a producer desires, nor is it what his customer wants just to provide contractual demonstration of a requirement.

Chapter 17 explained the concept of the OC curve. It is a plot of the probability that a given quality level will be accepted by a sampling plan versus the value of that quality level.

For most plans based on the OC curve approach the sample size required is usually smaller than that based on the confidence level approach. However, for plans based on the OC curve approach, the acceptance decision does *not* always mean that the requirement has really been achieved. In fact, it is possible to obtain an accept decision based on a sample reliability result which is lower than the requirement.

Confidence level and the probability of acceptance given by an OC curve are different concepts and the numerical values are *not* even approximately equal. Occasionally, "confidence" or "confidence level" is used to describe a probability of acceptance point of an OC curve. This is usually *incorrect. When the sample size and number of failures are known,* a relationship between confidence level and probability of acceptance can be defined.[10]

The remaining plans in this chapter are based on the OC curve approach because of its apparent usefulness for most acceptance sampling situations.

24-12 Application of Standard Attributes and Variables Plans to Time Characteristics

When time is a performance characteristic of a product, and is specified in numerical terms, an attributes plan such as MIL-STD-105D can be used. In this case, it is not necessary to assume a specific probability distribution.

> ***Example*** The minimum life requirement on a battery is 20 hours. The acceptable quality level has been defined as 2.5 percent defective. What single-sampling plan should be used to accept or reject lots each containing 500 batteries?

Solution: From Table 17-4, the sample size code letter is determined as H. The plan is read from Table 17-5 as

Sample size: 50 Acceptance number: 3

The plan directs that a sample of 50 batteries be tested for 20 hours. If 3 or fewer fail before 20 hours, the lot is accepted. Otherwise the lot is rejected. The sampling risks incurred in this plan are described by the operating characteristic curve included in MIL-STD-105D.

If the Weibull distribution is applicable and an estimate of the

[10] D. J. Cowden, *Statistical Methods in Quality Control*, Prentice-Hall, Inc., Englewood Cliffs, N.J., 1957, pp. 523–525.

shape parameter is available, procedures[11] have been created for using MIL-STD-105D plans for evaluating life in terms of other than simply a minimum life requirement. Three alternative criteria are possible: mean life, hazard rate (similar to failure rate), and reliable life (life beyond which a specified portion of the population will survive). These procedures also provide for terminating ("truncating") the testing at a preassigned period of time. The tables in TR7 can meet a wide range of cases because plans are provided for Weibull shape parameters from ⅓ to 4.0.

If it can be assumed that the time characteristic follows a normal distribution, a plan may be selected from a set of variables plans such as MIL-STD-414 (see Chapter 17).

> ***Example*** The life of the battery mentioned in the previous example follows a normal distribution. The standard deviation is unknown. What form 1 type of variables sampling from MIL-STD-414 should be used?

Solution. From Table 17-6, the sample size code letter is determined as I. The plan is read from Table 17-7 as

Sample size: 25 Acceptability constant: 1.53

The plan directs that 25 batteries be placed on a life test until all fail. The average life is computed and the quantity $(\bar{X} - L)/s$ compared with the acceptability constant k. Suppose the average life of the sample from a lot is 22 and the standard deviation is 4. Then $(\bar{X} - L)/s$ is $(22 - 20)/4$, or 0.5. As this is less than 1.53, the lot is rejected. The sampling risks incurred are described by the operating characteristic curve included in MIL-STD-414.

24-13 Example of a Reliability Sampling Plan

Tables[12] have been prepared that provide sampling plans for use with reliability requirements that have a time base. These plans assume an exponential distribution. The concepts of sampling risks are as applicable to these plans as to all sampling plans. The distinctive feature of these plans is the detailed procedures that have been developed to recognize the special problems connected with time measurement. Three types of plans[13] are available:

[11] *Factors and Procedures for Applying MIL-STD-105D Sampling Plans to Life and Reliability Testing,* TR7, Government Printing Office, 1965.

[12] *Sampling Procedures and Tables for Life and Reliability Testing,* H108, Government Printing Office, 1960.

[13] See *QCH,* 1962, pp. 13-111 to 13-118, for elaboration.

1 Life tests terminated upon occurrence of a preassigned number of failures. Here n units are tested until r failures occur and the failure times are compared to criteria given in the tables.
2 Life tests terminated at a preassigned time. This is discussed below.
3 Sequential life testing plans. Here, n units are placed on test and results are recorded until sufficient data are available to reach a decision. Periodically, the accumulated results are compared to criteria given in the tables.

Another example of a reliability sampling plan are the tables[14] which evaluate reliability in terms of mean life and assume a Weibull distribution. These tables provide sampling criteria to meet specified sampling risks related to mean life (μ) when the Weibull shape parameter is known and when the test time on each unit is set at t. The tables provide plans for shape parameters ranging from ⅓ to 5.0. An example will illustrate these plans.

Example A sampling plan is desired which will have a 10 percent probability of accepting a lot with a mean life of 1,000 hours and a 95 percent probability of accepting a lot with a mean life of 2,000 hours. Each unit is to be tested 500 hours. Past experience indicates that the Weibull shape parameter is 2.5.

Solution: Table 24-4 is an excerpt from one of the tables in TR3. Entering the table with $(t/\mu) \times 100 = (500/1{,}000) \times 100$ or 50,

[14] *Sampling Procedures and Tables for Life and Reliability Testing Based on the Weibull Distribution*, TR3, Government Printing Office, 1961.

Table 24-4 Excerpts of Sampling Plans for $\beta = 2\frac{1}{2}$

$(t/\mu) \times 100$ Ratio for Which $P(A) = 0.10$ (or less)

c	100	50	10	1.5
0	4	20	1,002	115,000
	(20)	(10)	(2.2)	(0.33)
1	6	34	1,692	194,000
	(37)	(18)	(3.8)	(0.58)
2	9	46	2,314	266,000
	(45)	(23)	(4.7)	(0.71)
3	11	58	2,905	334,000
	(51)	(25)	(5.2)	(0.80)
.	.	.	.	.
.	.	.	.	.
15	37	189	9,258	1,060,000
	(73)	(36)	(7.3)	(1.1)

the third column is examined for a parenthetical value of $(t/\mu) \times 100 = (500/2{,}000) \times 100 = 25$. The criteria can then be read as a sample size of 58 and an allowable number of failures of 3. This means that 58 units must each be tested for 500 hours. The lot is accepted if 3 or fewer failures occur; otherwise, the lot is rejected.

These tables provide many alternative plans (all meeting one set of sampling risks) for making tradeoffs between test time per unit and number of units tested.

24-14 Sampling for Bulk Product

The previous discussions of acceptance sampling have been restricted to the sampling of discrete units of product such as an electronic component or a mechanical assembly. Acceptance sampling procedures for *bulk* product present special problems.

Bulk product may be gaseous, liquid, or solid in form, e.g., cement or corn syrup. There are fundamental differences between acceptance sampling for bulk product and for discrete products. First, in bulk product, there is usually a continuous mass of material that can be subdivided into smaller quantities. For example, a bolt of cloth can be divided into several pieces. Discrete product usually cannot be subdivided; i.e., to subdivide an electronic component would yield several items which are different from the original component. Second, bulk product frequently involves a physical mixing during production. This mixing or blending results in a physical averaging of the quality of the product. Frequently several samples of the material are taken, blended together into a single composite sample, and test is made on a portion of or on all of the composite sample. The act of blending creates problems in obtaining a measure of the variation within a composite sample.

Third, the quality characteristic being investigated is usually the mean value. The variability of the product is, of course, important, but the mean value is usually the key criterion. Some bulk sampling problems are due to a failure to *clearly* identify the quality characteristic. For example, a numerical specification must be clearly identified as applying to *average* values or *individual* values. If the numerical value is not clearly defined, then one runs the risk of having it defined to suit the needs of the person currently using the specification.

As most bulk sampling problems are unique to a particular product and plant, it is difficult to prepare standard sampling tables for bulk sampling. The approach here will be to indicate some of the basics involved in designing a specific plan.

Basic to the development of a plan for bulk product is the identification of the fundamental sources of variation of the product. Consider the sampling of tank cars of a chemical.[15] Each tank car will be considered a batch (Figure 24-6). From each tank car three sub-batches are chosen, and two samples are taken within each sub-batch. Finally, on each of the two samples per sub-batch, two analyses are run.

The fundamental sources of variation are (1) variation between batches (or tank cars), (2) variation within batches (variation between the three sub-batches), (3) variation between samples, and (4) variation due to analytical error.

Before a sampling plan can be derived, a quantitative measure of each of the above four sources of variation is necessary. This requires that a carefully planned test program be conducted and the results analyzed to divide the total variation into the four sources above. The analysis of variance technique (see Chapter 11) is useful in making this subdivision of the total variation.

The subdividing of the total variation into component sources is important from both the statistical and engineering viewpoints. For example, if the variation between sub-batches is small, then it is likely that this variation is not significant to producer or consumer and, therefore, a sampling plan need not detect large differences between sub-batches. However. if the variation between sub-batches is large, then such differences may be significant to either the producer or the consumer. If such differences are to be detected, this will influence the number of samples to be taken and the manner in which the samples are selected from the tank cars. In addition, if differences between sub-batches must be detected, then the common approach of blending samples together to form a composite sample may hide a difference between sub-batches. These are merely *examples* of situations that can arise in the design of a bulk sampling plan. These examples illustrate why bulk sampling plans must be created to fit each case instead of relying on sampling tables to provide the plans.

The literature[16] provides further information on designing bulk sampling plans. However, the major steps that should be considered are

1 Define a unit of product and the fundamental sources of variation which are possible in the product.
2 Define a method of obtaining actual data in order to measure the sources of variation.

[15] A. R. Crawford, *Statistical Quality Control and Industrial Statistics*, Esso Research and Engineering Co., 1956, p. 33.

[16] For example, see R. S. Bingham, Jr., "Bulk Sampling—A Common Sense Viewpoint," *Transactions of the National Convention of the American Society for Quality Control*, Washington, Mar. 8, 1961. For an example of a specific plan, see vol. 11, no. 5, pp. 3–5, February, 1955.

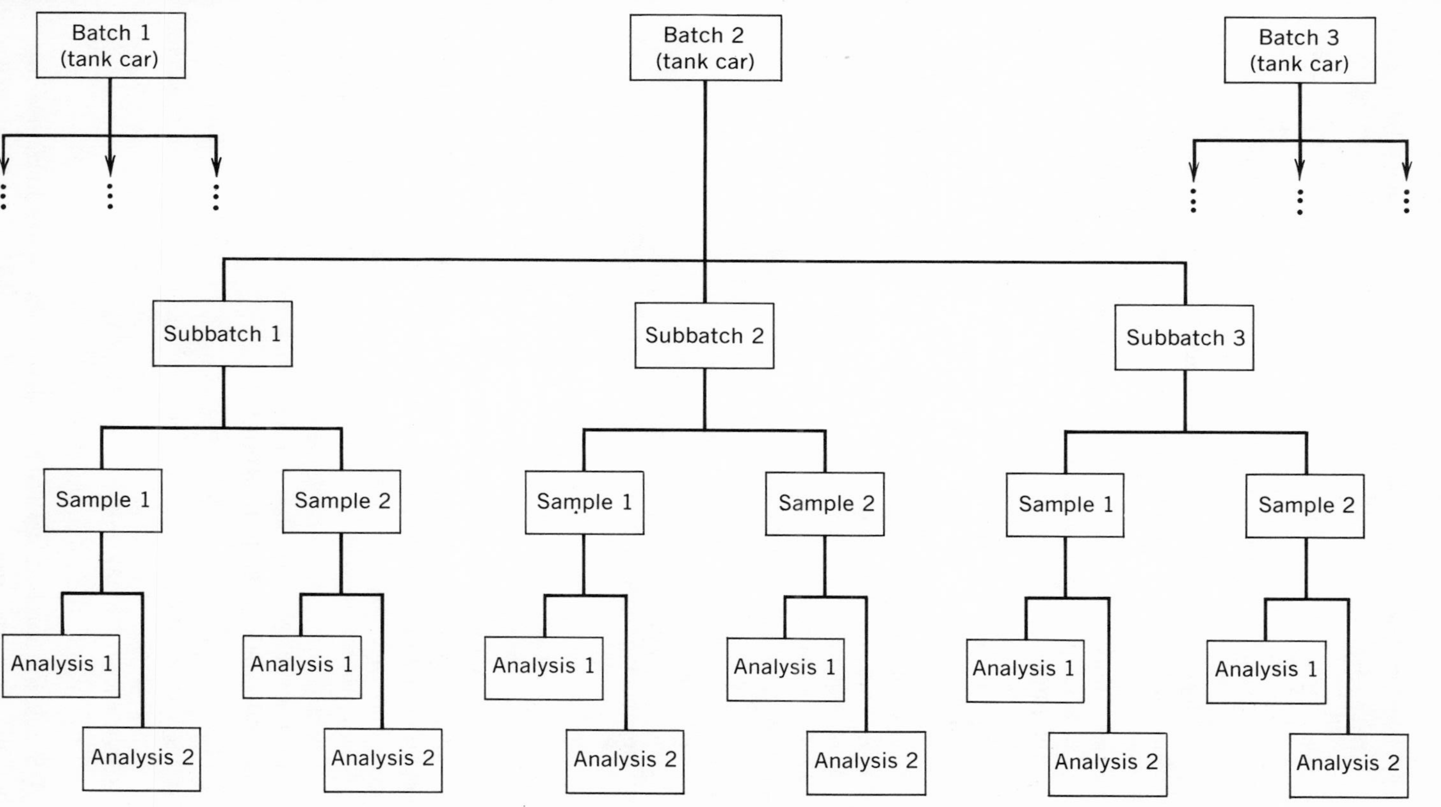

Figure 24-6 Sampling of tank cars of a chemical.

3 Obtain the data and perform an analysis to provide quantitative measures of the amount of variation due to each of the fundamental sources.

4 Define the sampling risks which are to be permitted in the sampling plan. As in conventional sampling plans, there are two risks: the risk of rejecting good product and the risk of accepting bad product. This will force a careful definition of what is meant by both bad product and good product. For bulk sampling, this is *probably* in terms of average values for an entire population.

5 Investigate the cost factors concerned with obtaining samples and running tests. The final determination of the amount of data needed in a sampling plan will involve a choice of selecting many samples of product and running one or just a few tests on each sample or selecting just a few samples of a product and running many tests on each of the few samples. The final decision here will depend on (*a*) the cost of obtaining one sample, (*b*) the cost of running one test, (*c*) the variation between samples, and (*d*) the variation due to measurement error.

6 Investigate the problems involved with the physical selection of the sample. This involves consideration of such factors as the selection of the sample on a periodic or random basis, the locations (for example, within a tank car) in which samples should be taken, whether a composite or blended sample should be selected, etc.

7 Analyze the information from all the steps above and develop a specific plan in terms of (*a*) number of samples, (*b*) number of tests per sample, (*c*) procedure for selecting samples, and (*d*) method of evaluating the data to accept or reject the product.

PROBLEMS

1 Select a problem in sensory qualities acceptable to the instructor (see section 13-7 for examples).

a Define precisely what is meant by the quality characteristic in question and propose some possible methods of developing an instrument to measure the characteristic in question.

b Propose a general plan for experimentation and test to design and prove in the instrument.

2 For any product or service acceptable to the instructor:

a Prepare a list of defects encountered during acceptance inspection and during use.

b Design a plan for seriousness classification of defects and apply the plan to the list in question.

3 Study the method used for dealing with nonconformance in any of the following processes: arrests for traffic violations, grievances under

the collective bargaining agreement, protests over tax assessments, and any other process acceptable to the instructor.

Report the plan in use, including the equivalents of inspector, designer, supervision, material review board, etc.

4 An electronic component has a specified mean life of 1,500 hours. It is desired to demonstrate with 90 percent confidence that this specification has been met. Assuming that the life of the component is exponentially distributed, propose four alternative test programs.

5 In problem 4, what would be the smallest number of components which could be tested (based on the table provided in the chapter)? What would be the acceptance criteria for this size test program? What is the probability that the product would meet these criteria if the actual mean life in the lot is 1,500 hours?

6 A fuel pump has a minimum life requirement of 600 hours. Pump life is normally distributed. Pumps are shipped in lots of 1,000 units. The acceptable quality level has been defined as 1 percent. Propose a variables sampling plan.

7 The specified mean life of a 24-channel communications set is 1,000 hours. A sampling plan is desired to meet the following conditions:

Test each set in the sample for a maximum of 200 hours. Failures are not to be replaced.

Have a 90 percent probability of accepting lots whose product has a mean life of 1,000 hours.

Have a 90 percent probability of rejecting product having a mean life of one-third of that specified.

a Assuming an exponential distribution, define the sampling plan.[17]

b Summarize the effect of test time on sample size for the probability risks stated above by plotting a graph of test time per set (horizontal) versus sample size (vertical). Comment.

8 A mechanical assembly has a specified mean life of 5,000 hours. Past data on similar assemblies show the Weibull slope parameter to be 2½. An acceptance sampling plan is desired. Lots which meet the requirement should have a 95 percent chance of being accepted. Lots having a mean of only 1,000 hours should have only a 10 percent chance of being accepted.

a Define a sampling plan with the smallest possible sample size.

b Plot a graph of mean life having a 95 percent probability of acceptance (horizontal) versus sample size (vertical) required. Comment.

[17] Refer to *QCH,* 1962.

25 Control of Product Quality during Customer Usage

25-1 Quality Problem Areas during Customer Usage

The act of final product acceptance may be regarded as terminating the manufacture phase of the product. Following this act, and before customer usage of the product, there take place a number of preusage phases: packing, shipping, receiving, storage. Finally, there are the usage phases: installation, checkout, operation, maintenance. Table 25-1 shows these phases,[1] categories affecting them, and the authors' opinion on significant problem areas.

The significant problem areas shown in the matrix are only meant to indicate relative emphasis. In specific cases, review of the complete matrix will help to pinpoint significant areas. This chapter will discuss elements of the matrix that are applicable to a wide range of products.

It should be stressed that industry has always recognized the customer usage phase and has performed the tasks listed in the matrix. Thus, the concepts are not new. However, quality problems during usage have recently received more public attention and there is now a movement to refine the methodology in this area. Following the pattern in reliability engineering, the refinements will probably include qualitative and quantitative techniques.[2]

[1] Based on the results of a task force of the Electronic Industries Association.
[2] See the following pages from the *Annals of the Sixth Annual Reliability and Maintainability Conference*, 1967, Society of Automotive Engineers: S. J. Kasper, "Factors Degrading Reliability in the Use Phase," pp. 284–291; W. L. Hadley, "Reliability from Factory to Target," pp. 356–363.

Table 25-1 QUALITY PROBLEM AREAS DURING CUSTOMER USAGE

	Activity Phases						
Variables Affecting Phases	*Ship*	*Receive*	*Store*	*Install*	*Align and Check out*	*Operate*	*Maintain*
Personnel		*		*	*	*	*
Organization						*	*
Support equipment and tools						*	*
Major equipment	*	*	*	*	*	*	*
Information (procedures, specifications, etc.)				*	*	*	*
Materials and spares			*				*
Facilities/environment			*	*		*	
Packaging design	*						

* Potentially significant problem areas.

25-2 Shipping, Receiving, and Storage

The work of achieving a product fit for use can be nullified through some seemingly simple follow-on operations:

Identification

The very act of inspection verifies the identity of the product. Unless the product is then and there clearly identified by code stamps, tags, etc., the subsequent stock handlers, packers, and shippers can easily confuse the identities, with resultant trouble for all concerned. In recent years there has also been a proliferation of the need for "traceability," i.e., the ability to identify all product made with components coming from a particular heat number, recipe, batch number, etc. This need arises in case of recalls of automobiles, pharmaceuticals, or other products involving human safety and health. For such products the problem of identification requires much new planning, both as to the physical goods and as to the documentation.

Protection

Packages must now be designed to protect products against shock, moisture, oxidation, temperature, and still other environmental changes. Design, construction, and test of these packages create problems which can compete in difficulty with many product quality control problems.

Shipment

This introduces a long list of new perils to the product: handling in transport, environments during travel and storage, damage during unpacking, and still others. Planning defenses against the known perils is not enough, since there are other risks due to ignorance, carelessness, blunder. Good feedback from prior experience with similar products can be helpful in minimizing damage to the product. The shipping phase includes the preparation for shipment, loading, transportation, and unloading. Baker[3] reports 54 handling operations merely in *preparing the shipment* of a transmitting tube (out of a total of 888 handling operations starting with raw materials). Each handling is a potential source of damage.

Environmental conditions during shipment are another problem, e.g., the effect of temperature on frozen food, photographic film, and drugs. Temperature, humidity, vibration, and shock encountered in shipment may be *more* severe than the levels used by the designer based on the usage of the product. In such cases, the design levels must be based on those met in shipment, or some form of protection from the extreme shipping environments must be provided. Failure to provide for this can cause a subtle and more serious problem. Shipping environments may partially damage a product without any external indication. When the item is placed in service it may prematurely fail due to the overstressing it was subjected to during shipment.

Heine[4] traces the shipping damage problem on household ranges and reveals a typical pattern. The damage was over 8 percent but was regarded as inevitable due to rough transportation. Many opinions were available but facts were lacking. Finally, distributors were asked to submit specific damage information with every railroad carload of ranges. Testing procedures were then set up to measure vibration, shock, and impact. Within the *first few hours* of testing, certain problem areas were revealed. It is instructive to note that the corrective measures were not simply better packaging. The damaging was reduced to less than 2 percent by making nine product design changes, six processing changes, and six packaging changes.

The storage of finished goods involves the initial placement into storage, maintenance during storage, and issuance from storage. The federal government was sufficiently concerned with the problem to develop a "Storage Quality Control" program. Skolnik[5] discusses this and shows the basic approach in Figure 25-1. It is presumed that the ac-

[3] Bronson B. Baker, "Product Reliability through Integrated Packaging and Handling," *Institute of Electrical and Electronics Engineers, Transactions on Product Engineering and Production,* PEP-7, no. 3, pp. 1–11, July, 1963.

[4] R. E. Heine, "How We Reduced Our Shipping Damage," *Finish,* pp. 71–79, May, 1951.

[5] Samuel S. J. Skolnik, "A New Department of Defense Approach to Quality Control of Stored Material," *Transactions of the American Society for Quality Control,* 1963, pp. 219–229.

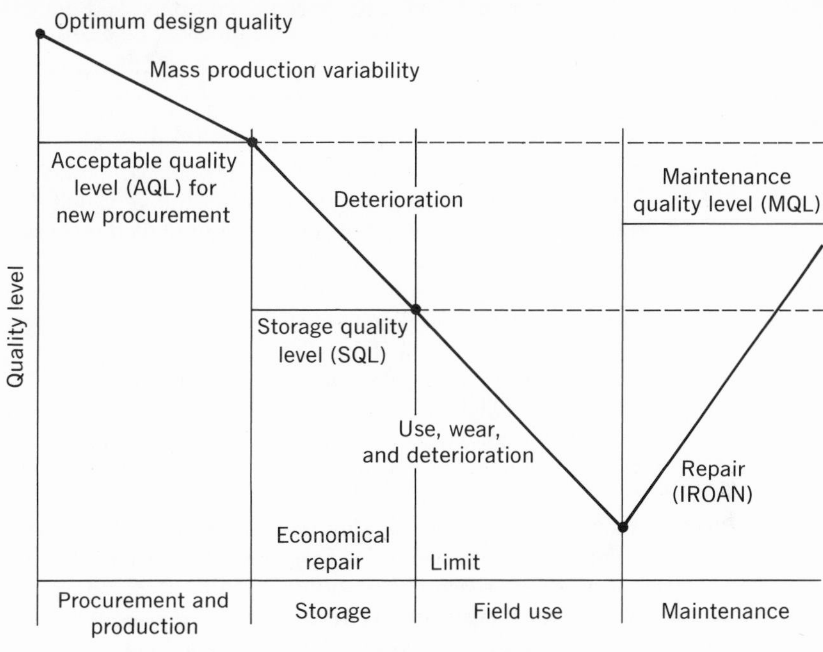

Figure 25-1 Total quality control quality progression. Source: Samuel S. J. Skolnik, "A New Department of Defense Approach to Quality Control of Stored Material," *Transactions of the American Society for Quality Control,* 1963, pp. 219–229.

ceptable quality level will deteriorate during storage to a lower level (storage quality level) which must exceed field requirements. A further deterioration takes place during field use. After extended use, the economics of repair or discard must be considered. The program usually tries to raise the quality to a level (MQL) somewhere between the AQL and SQL. When storage can have a significant effect on a product, the specifics of a program may include:

1 Control of the storage environment to minimize degradation
2 Minimizing degradation due to improper handling during storage
3 Determination of "shelf life" of products (storage life at which quality may have deteriorated below acceptable limits)
4 Feedback of technical data on storage effects to improve future designs

25-3 Installation, Alignment, and Operation

These phases of the product life cycle contain many opportunities for damage or deterioration of the product. For some products, the preparations for actual use of a product can become complex and include such factors as

1 Facilities for housing the product. This not only includes sufficient space but may also include special equipment to continuously control temperature, humidity, vibration, or other phenomena. When the effect of environmental factors on product operation is not fully understood, there is a real danger of treating the installation problem as minor and emphasizing minimum cost at the expense of necessary facilities.

2 Facilities for housing auxiliary equipment necessary for the operation of the primary product. Such "support" equipment may include complex hardware for aligning, checking out, and maintaining the product. (The complete installation problem for one of the major missile systems was so enormous that the prime contractor set up a separate "activation division" with its own vice-president.)

3 Instructions for installing and checking out the product. Written instructions for complex products are forever subject to omissions and mistakes because of sheer complexity. The effect on quality can be highly dramatic. One of the failures in America's early space shots was traced to the omission in checkout procedures of a certain adjustment on the product. The need for the adjustment was known and had been made on previous shots by a technician on the launch crew, but he had failed to record the step in procedures. On a certain launch, this technician was not part of the launch crew and the adjustment was not made. The launch was a complete failure. Such problems are not restricted to complex products. The following examples of poor installation instructions have been recorded by one of the authors:

   Example A rather expensive blackboard was purchased, installed, and used with ordinary chalk. However, the writing could not be completely erased even with great pressure on the eraser. Inquiry revealed that the board must first be rubbed with a thin coating of chalk and then erased before any writing is done. Without the coating (which is to be applied by the customer) the board will be ruined because the writing will not erase. There were no instructions about the need for coating.

   Example A metal bookcase was shipped unassembled with detailed assembly instructions. The assembler was confused by the presence of a critical screw hole that was not mentioned in the instructions.

   Example A television converter contained detailed instructions for assembly with the television set. The converter contained one external wire which was not mentioned at all in the instructions.

Some similar examples can be cited for operating instructions which are supposed to define how to use the product:

Example The main gas line on a home furnace is not noted on any of the diagrams in the instructions and is not labeled on the furnace. Also there are no "off" or "on" markings on the control.

Example A can of latex paint did not contain any instructions for thinning. A woman decided the paint needed thinning and used paint thinner. She painted a wall and the result was terrible. She learned later that the paint was to be thinned with water. (The cost of quality to the paint store was the cost of repainting the wall.)

These examples of extremely simple products are mentioned to emphasize that quality degradation can be caused by poor instructions even for simple products. In these cases, preventive action is simple. With more complex products, the prevention tasks will require more effort.

Another area is the problem of product failures or deterioration due to human mistakes made during the operation of the product. It is often tempting to use customer misuse as a cause of poor performance. However, mistakes made during operation sometimes are the fault of the designer for not having investigated the physical limitations of humans and designing around them. An automobile manufacturer once received complaints about the driver's seat on a truck. Analysis revealed that "the operator for the bus had to be at least 6 ft. 7 in. tall and not more than 110 pounds if he were to fit under the steering wheel, get his feet on the pedals, and still be able to get his head high enough to see the road."[6]

Other operator errors may be due to inadequate training. As products have become more complicated to operate, the training problem has become severe in some industries. For example, the emphasis on sophisticated weapons has created enormous training problems because of the constant obsolescence of weapons plus the turnover of military personnel.

25-4 Operational Conditions and Maintenance

Experience on two early missile programs indicates the effect of "operational conditions" on quality. The data present the number of signifi-

[6] A. D. Wilder, "Let's Produce Equipment a B.A. Graduate Can Use," *Annals of Reliability and Maintainability*, vol. 4, pp. 799–802, Spartan Books. Inc., New York, July, 1965.

cant reliability problems encountered. A "problem" here is not simply a product failure but a recurring condition that affected reliability. For missile X, 16 problems were due to operational conditions, while missile Y had 24. The reasons are shown in Table 25-2.

Table 25-2 **Problems due to Operational Conditions**

Cause	*X*	*Y*
Inadequate operating or maintenance procedures	6	13
Inadequate maintenance equipment	4	3
Product deterioration due to environment	3	1
Human error during maintenance	1	4
Inaccessibility to repair	1	
Inadequate packaging	1	2
Defective spare parts	—	1
	16	24
Percent of all reliability problems due to operational conditions	23%	10

Aside from preventive maintenance, maintenance to correct a failure involves:

1 Verification that a failure has occurred
2 Diagnosis to determine the nature of the failure
3 Corrective action
4 Verification to assure that the product now operates properly

As products become complex, maintenance programs require significant investments in personnel and equipment. Details such as adequate maintenance instructions and spare parts of sufficient quantity and quality can have a snowballing effect for complex equipment if not thoroughly planned and controlled.

25-5 Configuration Control

The complexity of some products can result in certain field problems that require the use of formal systems for controlling the status of the hardware itself. Although the problem is not unique to military products, certain conditions have resulted in an emphasis on complex military products.

Ideally, a complex product is thoroughly evaluated and changes are made during design and development so that the design released to production is fixed. Of course, this is often not the case and changes are made after production has started. Some recent military programs have gone even further by using the "concurrency principle" to reduce the time required to place new weapons in the field. This principle recognizes the possible need for further quality development after the initial units have been released to the field. Thus, design changes will be forthcoming and will require "effectivity" decisions (defining in which units of hardware the changes will be incorporated). If there are many changes and if most of them are to be made to units already in the field, the customer is confronted with a maze of questions:

1 Can and/or should all units in the field receive the change from an engineering point of view?
2 What are the maintenance problems involved in making the change in the field (personnel, facilities, time, etc.)?
3 How should changes in operating and maintenance instruction be made?
4 What should be done with spare parts that become obsolete by the change?
5 Will the change require any retraining of operator or other personnel?

The collection of activities needed to administer these changes for complex military products are called "configuration control." The objective is to incorporate changes into equipment in a manner that provides the customer with equipment that will function properly and also be maintainable in the field. A configuration control program usually has three elements:

1 A system for defining and identifying every element of the product. This not only includes all physical elements (to the lowest level) but also supporting documentation such as drawings and maintenance and operating instructions. Here then is a huge but necessary paperwork operation.
2 A system for reviewing and making decisions on proposed changes. Particularly as a program goes into production, the performance or reliability improvements of engineering changes must be carefully weighed against the cost of making any change. As the decisions affect many functions, a "configuration control board" is often set up to review all proposed changes. The board may have representatives from engineering, quality control, field engineering, purchasing, manufacturing, and the customer. Usually, the chairman has the authority to make a decision on the change (which decision may be appealed through normal channels). The quality control representative may often be in the position of justifying the importance of a change to overcome the cost and schedule disadvantages that are often present.
3 A system for keeping track of the exact status ("configuration") of

every unit of hardware. A formal accounting-type system is required because changes may continually be made. Every change to be made must be compatible with existing hardware from an engineering point of view. Thus the "configuration" of each unit of hardware must be recorded and reviewed in deciding which units will receive the change—again, a large but necessary paper work system.

25-6 Feedback of Usage Data from Customers

Industry has always recognized the need for customer feedback on product quality. In some cases, there is a formal system involved. This is particularly so for products having some type of guarantee. Here, the manufacturer may have a complex data processing system on product claims. This then serves as a key source for defining quality problems. Such information is vital to any company and the analysis is discussed in Chapter 27.

The point to be made here is that those quality problems reported by customers may *not* be a true picture of quality. Consider the case of a resistor which had a high failure rate in practice but which was no problem to the manufacturer because there were few complaints. Investigation revealed that the resistor was quite easy to replace and the high failure rate was accepted as inevitable by the customer. (The resistor could have been detected as a problem if spare parts sales had been analyzed by the manufacturer.) Another case involved welding defects under a motor mounting. These were sanded down in the field and never reported to the manufacturer. Imagine the cost of field repair to the customer. Such examples can be ignored by reasoning that if the customer does not complain, then the manufacturer should not waste effort on them. (Also, there are other unsolved quality problems that the customer *is* reporting.) However, this reasoning (although appealing in the short run) is not consistent with the concept of quality as fitness for use.

The manufacturer vitally needs two items of customer information:

1 *All* the quality problems encountered by the customer. This includes problems that are excluded under formal written guarantees. It also includes problems that occur after the guarantee period has expired. Thus, this goes beyond the legal commitment of the manufacturer (which has probably been influenced by competition). However, the solution of quality problems that are not covered under guarantees is a potential source of increased sales income, particularly if competitors have similar problems.

2 Relative importance of quality problems as viewed by the customer. To a manufacturer, the number one problem may be component X because it is incurring the highest dollar amount of warranty claims. However, the relative importance of problems from the viewpoint of the customer depends on many possible factors—effect on downtime, availability of spares, supply of maintenance personnel, etc. Many of the factors involve customer operations and thus the information is not ordinarily available to the product manufacturer.

These two items of information are *not* easy to obtain but their criticality in assuring proper emphasis in a quality program dictates that an effort be made to obtain the information. The difficulties involved are sometimes used as an excuse for not collecting the information but, at a minimum, the information can be obtained on a sampling basis.

PROBLEMS

1 Visit any transportation service involving loading, storage, crating, unloading, transport, etc. (acceptable to the instructor). Some possible examples: a trucking terminal, a steamship pier, an air freight terminal, a commercial warehouse, the shipping area of a factory, the parcel room of the post office, and a rail freight yard.

Study the processes in use and their likely effect on the qualities of the products undergoing these processes.

Report your findings and conclusions.

2 Visit some manager who is exposed to problems of installation and operation of consumers' hardware, e.g., the buyer in the home appliance department of a department store or the service manager of a factory making home appliances.

Secure information on major troubles encountered by users in installation and operation, and also on what steps are being taken to minimize these troubles.

Report your findings and conclusions.

3 For some product (approved by the instructor) study the forms of feedback used by the manufacturer to inform himself on fitness for use, e.g., customer complaints, salesmen's reports, servicemen's reports, sale of spare parts, etc.

Report your findings and conclusions.

4 For some product (approved by the instructor) study the operating and/or maintenance instructions. Report your findings and conclusions.

5 Read the RPM case in the Appendix and answer the following problem.

You are Day, the reliability manager. You have reviewed the present system of securing field failure information, and it is about as follows:

RPM has about 2,000 dealers scattered over the country. Collectively they sell all of RPM's 180,000-car output (not at all evenly, since the top 20 dealers account for about 5 percent of the total sales, and the top 100 dealers account for 15 percent of the total sales).

RPM's arrangement with its dealers requires them to keep a record of troubles encountered on customers' cars during the guarantee period. This record is supposed to include the serial numbers, the odometer reading, date, description of the troubles, the labor and material used to fix the troubles, etc. These records are used in several ways, including (*a*) as part of the system of checking the dealers' claims against RPM and (*b*) as an input to the quality data system.

The records have been generally unsatisfactory to Quality Control. The descriptions of the defects are seldom complete enough to be a guide to action or to be summarized to discover concentrations of troubles.

There is also a problem of shipping the samples. The dealers are supposed to ship to RPM samples of the hardware involved in any trouble which they have encountered for the first time. Their performance on sending in samples is very poor. Here are some conclusions from a report made by a team of RPM men (from marketing, quality control, and engineering) who had been sent into the field to see what could be done:

> The real interest of the dealer is first to get credit from RPM, and second to get the customer off his neck. The dealer's interest in paper work and samples is primarily to be able to establish proof of the work he did, so that he can get his credit.
>
> Very few dealers understand the importance of data feedback for quality purposes. Most of them really don't want to understand. They regard it as a nuisance. They also don't know where to send the samples, but they would prefer to hold the samples anyhow, to help get credit. The language used to describe defects is remarkably different around the country.
>
> We are convinced it is out of the question to get more than a small

minority of the dealers to follow our instructions on data feedback and samples.

Notwithstanding this bleak history and the gloomy report, you are faced with finding a way to get data which will provide RPM with the following essential feedback:

a A *very* prompt notification of troubles encountered on new models, so that the fixes can be made before too many cars are shipped

b Identification of chronic failures so that the processes, designs, or components can be changed

c A measure of field failures adequate for executive reporting on the program of improving reliability

What do you propose to accomplish these purposes?

26 Customer Relations

26-1 The Customer Viewpoint

Customer needs are for "services": nutrition, shelter, transportation, recreation. To meet these needs, customers may buy a "product" or hire the service. For example, to obtain transportation, the pedestrian buys shoes, the vacationist hires a trailer, and the rail traveler buys a "ride."

In early human history, human needs were met by natural, unfabricated materials plus human energy. With the advance of technology, other forms of energy were harnessed to fabricate natural materials and to create unnatural products. These developments, glacially slow in early centuries, are now coming thick and fast, so that there is a wide range of choice of means for meeting human needs. In a market-based economy, this choice is made by the consumer, not the producer.[1] (The company that makes millions of cars, or whatever, must still sell them one by one.) Because the consumer dominates the marketplace, the manufacturing and service companies should take steps to become informed on such facts as:

What are the present qualities needed by users?
What new qualities might users be induced to buy?

[1] This is not universally true. In times of shortages of goods, the consumer takes what he can get, and even bids up the price in the process. In centrally planned economies, the planners, not the users, make the choice, though the planners try to consult the users before making the decision. However, in the normal operation of a competitive economy, the "anarchy of the marketplace," i.e., the collective effect of the decisions of many users, is decisive as to what is bought and sold.

What are the competitive facts as seen by the user?
How does the user make his decisions?

Despite the obvious value of such information, most companies do not become adequately informed, for several reasons:

1 They lack awareness of the opportunities to increase income through more precise information on users' habits and thinking. These opportunities include such matters as what are the users' costs (see section 4-9); what work does the user now do which might be done better at the factory through some systems redesign; what are the users' future plans; what are the trends in users' tastes.

2 They fail to organize their search for knowledge in a way which will identify the opportunities. The "natural" field intelligence service is the field sales force. These salesmen are highly sensitive to signals about quality troubles which might lose sales; often they overreact to such signals. They are also sensitive to the possibilities of increasing sales through improving quality of design and quality of conformance. However, they are seldom able to quantify these signals. In addition, they are seldom sensitive to the opportunities inherent in the other forms of increasing income, e.g., systems redesign, new contract forms, changes in reliability, etc.

   Generally, identification of the opportunities requires a team approach. A team of managers, from all key departments, discusses and agrees on what information is needed from the marketplace to guide decision making in product design, sales contracts, service contracts, quality control programs, etc. Quite often it will be found that to secure this knowledge is beyond the capacity of the field sales force. Hence use is made of a special service known as market research. Such a service, staffed with skilled specialists, does most of the work of securing the data without preempting the time of the regular departments.

3 The companies fail to act on the available knowledge. Managers harbor many axiomatic beliefs and departmental goals which may drive them to take action not in the best company interests. In the power tools case (see section 4-9) the improved reliability of the company's products warranted a price increase. The marketing manager resisted increasing the prices because of his axiomatic belief that "standard" products must be priced to meet competition. In the breakfast cereal industry, the sales appeal is made to children as well as to adults. One result is that more complaints from users deal with deficiencies in the toys than with deficiencies in the food. (Charlie Brown will not tolerate a defective truck.) Yet the production and quality managers feel the toys are a nuisance which must be tolerated; the real need is to do a good job on the food. The general managers are also susceptible to axiomatic beliefs. A manufacturer of hairnets concentrated on solving marketing, produc-

tion, and quality problems of his "product." He lost his business because someone else developed a chemical spray which could hold women's hair in place without a hairnet. He had really been in the business of selling a service for holding women's hair in place.

26-2 Market Research

The systematic approach for discovering the facts and opportunities in the marketplace is known as "market research." Knowledge about fitness for use is only one of the kinds of market knowledge needed; others include market sales potential, competitor practice, secular and seasonal trends in sales, relation of sales to variables in the economy (population, region, etc.), and many more.

In small companies, market research is carried out personally by the managers, with occasional aid from a consulting service. In large companies the amount of market research needed grows to a size requiring a special market research department.

A wide variety of tools are used by market researchers in their quest for information on quality and related matters. These tools include:

1 The printed questionnaire with illustrations and checklists designed for securing many answers without occupying much of the user's time.
2 Physical display of products to secure a ready choice of user preference. At one extreme is the booth in a busy department store at which free coffee is offered to secure data on the preference of A versus B. More sophisticated is the traveling show of carpet samples to secure user preference data. A manufacturer of toys invites teachers to bring their class of young school children to play in a room full of toys while men behind one-way mirrors time how long the various toys hold attention. (Later they analyze the toy breakage as a means of increasing the useful life.) More elaborate is the annual industry show at which new models of industrial products, processes, and materials are displayed to prospective industrial buyers.
3 Consumer panels to whom products are consigned for consumption or use. A large food company uses a panel of several hundred consumers to whom samples of food (new formulations, competitors' products, etc.) are sent for preference evaluation. One automobile company, considering a revolutionary design of a turbine driven car, made up 200 of them and loaned them, 3 months at a time, to hundreds of motorists to secure the usage data needed as an input to further decision making.
4 Consumer (or dealer) advisory panels to meet periodically with company officials to discuss quality and other user problems.

Market research should also be distinguished from product research. The former is done to discover what quality characteristics are needed or can be sold to users. The latter is a problem in science and technology—how to create goods and services which can provide these needed quality characteristics.

26-3 User Information in Product Redesign

Chapter 1 described one cycle of the quality spiral as starting with information on the product qualities desired by the user and ending with feedback from users to suggest product improvements. A new cycle of the quality spiral then starts with research leading to product redesign.

The redesign may be to reduce field failures which are the subject of complaints. This impetus for redesign is strong and action usually comes fast.

Redesign to change the quality of design will evolve more slowly unless the manufacturer aggressively seeks out possible redesigns that have market potential. Risks are lowest when redesigns are based on user experience and desires. There are many types of user desires for product redesign, but several categories that can be defined are

1 Additional functional performance (e.g., horsepower in automobiles and size of homes)
2 Additional performance in other product "abilities" (e.g., longer life of transistor radio batteries, faster drying time of paint, and ease of maintenance in certain electronic assemblies)
3 Transfer of work from user site to factory (e.g., self-lubricating bearings, ready-mixed concrete, and frozen dinners)
4 Ease of installation and use (e.g., window air-conditioners, adhesive wallpaper, and frost-free refrigerators)

Each of these examples has resulted in significant sales from redesigns initiated by user feedback on a product. The countless number of other examples is a tribute to the creativeness of designers and the responsiveness of aggressive manufacturers to meet user needs.

26-4 Development of a Positive Quality Reputation

It is possible to fail in business even though quality of product is good. However, it is not possible to stay in business if quality is poor, unless one has a monopoly. The products which have endured for centuries, like Wedgwood pottery, Damascus steel, or Cordovan leather, all are firmly grounded on a strong quality base.

This positive reputation reaches its highest importance in selling consumers' goods. The consumer, who can neither understand the intricacies of a technical design nor test a product for conformance to that design, learns to rely on companies whose goods have given him satisfaction. Market research has demonstrated over and over again that such a buying habit, once established, is not easily shaken.

A company making an entire "line" of products, such as household utilities, must make the whole line to a consistently good quality. An owner of a refrigerator will be influenced by its performance in deciding whether or not to buy a washing machine from the same company, and so on.

One of the authors sends a notice of "superior or inferior quality" to manufacturers. The notice is a two-part form with space for providing sufficient detail to be helpful to the manufacturer. One part of the form is used to report an instance of superior quality; the second part of the form provides details on inferior quality. The answers to letters of inferior quality are always polite and offer replacement of the item. Sometimes the answers are humorous in failing to view the problem as a customer would:

A major food company sent the following reply to a complaint (made on the above form) about moldy pancake syrup:

"You will be glad to know that the moldy syrup is not harmful. Often it can be reconditioned. To do this simply pour the syrup into a saucepan and when the mold rises to the top skim it all off. Then bring the syrup to a boil and store in a sterile, scalded jar."

A high class toy manufacturer was sent a complaint (made on the above form) about the string on a toy top breaking after being used twice. The manufacturer replied by asking for the part number of the string.

A dramatic result of a customer complaint is the example (see section 4-9) of the customer who received $985 for the time spent in taking a defective product back to the dealer for repair.

Not only for consumers' goods, but also for raw materials and semifinished goods used by producers of finished goods, the problem of conformance still remains largely in the hands of the producer. The consumer must rely on the producer or expend much effort himself to check up somehow. A producer who can develop a reputation for conformance and for making good any failure to conform has achieved much in the way of adding value to his product.

A good quality reputation is a treasure. It is a matchless tool for competition by helping to obtain a better price as well as greater volume of sales. The quality reputation also has high advertising value. Finally, the quality reputation is a morale builder because people derive much satisfaction from being on a winning industrial team.

Drawing conclusions from both consumers and retailing executives, the *New York Times*[2] stated:

"Retailers have grown complacent about handling customer complaints. They haven't paid the same attention to this problem as to the original sale. And they haven't given the same caliber of executive supervision to the need to improve customer services as they have to merchandising buying and store operations."

One illustration given is the rate of returns due to faulty furniture as one piece out of seven (13.7 percent) as compared to 9 percent on gifts (on which people notoriously change their mind). One executive summarized his concern about reputation: "Of course, we are liberal on our return policies, and sometimes take a beating because of it, but that is the way over the years that we have built the faith and confidence of our customers."

26-5 Advertising of Quality

For centuries, merchants have made skillful use of adjectives and pictures to attract buyers. For example, cigarettes and beverages are touted as less irritating, smoother, milder, better. The tools of the advertisers are usually words, not facts, but some manufacturers are now trying to determine facts on quality that are appropriate for consumer advertising. Table 26-1 lists some examples.

Table 26-1 QUALITY FACTS IN ADVERTISEMENTS

Product	*Quality Facts Included in Advertising*
Cigarettes	Nicotine and tar contents
Springs	Histogram of spring characteristic
Tractor	Repair times for typical components
Capacitor	Acceptable quality level values
Car rental	Offer of $50 free rental for any complaint
Luggage	Lock is guaranteed forever

In part, the advertising of quality is done by salesmen. These men must be trained in the quality characteristics of the product they are

[2] Nov. 14, 1965.

to sell, and of competing products also. Such training should include visits to the design, production, and inspection functions to permit the clearest visualization of how the qualities get into the product.

The cliché that a good quality product will advertise itself by word of mouth has been accented in recent years by the product ratings published by private testing organizations (see Chapter 4). A top rating given to two washing machines by such a testing company (X) caused an executive of the manufacturer to state: "X put us in the washing machine business."[3] Ironically, the manufacturer did not spend $1 on advertising this rating because it is prohibited by X.

Consumers can also go to extremes to advertise their displeasure. A dramatic example concerned the lack of service provided by a national auto insurance company in settling a claim involving damage to a home caused by an automobile. The homeowner painted a derogatory picture (of the company) on the *entire side* of his home.

26-6 Guarantee of Quality

Obviously life is simpler for both producer and consumer when the product is fit for use. However, there are enough cases of failure to urge the user to demand prior assurance that the producer will protect him against failures through restoring service and paying the damage. This prior assurance is the quality guarantee.

The quality guarantee is not new (see section 2-8 for an example from 429 B.C.).

Ideally, the guarantee should provide for putting the user into the position he would have been in had the product not been defective. This is difficult to carry out in practice because the responsibility for failure and the anticipated consequences of failure are both blurred in many instances. The guarantee is therefore limited as to:

1 *Responsibility,* so that the producer will not suffer loss due to damage by the user or by a third party. Many producers administer this provision by giving users the benefit of the doubt in borderline cases.
2 *Expense,* so that the producer will not become the victim of escalation of costs. For example, a piece of tool steel costing $100 may contain a crack which is not discovered until $1,000 in die work has been done on the tool steel. Under the usual form of guarantee, the producer's liability would be limited to the replacement value of the tool steel.

[3] Philip Siehman, "U.S. Business' Most Skeptical Customer," *Fortune,* pp. 157–229, September, 1960.

3 *Time,* so that failures after a reasonable period of use will not be held against the factory.

A typical guarantee is the following by a well-known maker of gas ranges:

> We warrant each new———range manufactured by us, to be free from defects in material or workmanship under normal use and service, our obligation under this warranty being limited to exchanging at our factory any———or parts, within one year from date of delivery to the original purchaser of this range, and which our examination discloses to our satisfaction to be thus defective; this warranty being expressly in lieu of all other warranties, express or implied, and of all other obligations or liabilities on our part, and we neither assume nor authorize any other person to assume for us other liability in connection with the sale of——— ranges.
>
> This warranty will not apply to any range which has been subject to accident, abuse, or misuse.[4]

From the producer's viewpoint, a quality guarantee can mean a large risk. Some household appliances are guaranteed against factory defects for 5 years, and thereby 5 years' production can simultaneously be subject to the guarantee. A single type of defect, not detected in the factory, might contaminate many months' production before developing trouble in the field.

Quality-type guarantees have been increasingly used as a marketing tool. Some examples:

1 Longer warranty periods for products such as appliances.

2 Quantification of quality where never done before. Contracts[5] on some products now include numerical definitions of reliability, maintainability, or availability. The increase in rental of product has brought with it demands by customers for guarantees of performance, particularly when equipment downtime may be an economic production loss to the user of the equipment. Guarantees can even be stated in economic terms. For example, a diesel generator set was sold to generate power for a manufacturing company as an alternative to purchasing the power from a utility company. The contract guaranteed that the cost of generated power would be less than the cost of purchased power.

[4] For a discussion of the information that should be included in a guarantee, see J. M. Juran (ed.), *Quality Control Handbook (QCH)*, 2d ed., McGraw-Hill Book Company, New York, 1962, pp. 1-28 to 1-29.

[5] For a discussion of the hazards of including (or not including) reliability statements in contracts, see Ross H. Johnson, "Contracting for Reliability," *Proceedings of the Eighth National Symposium on Reliability and Quality Control,* 1962, pp. 181–184.

The concept of quality guarantee has grown to a point where it has become an important element of competition. In consequence, companies should be alert to take the initiative in offering guarantees. This requires that they first study the failure data to predict the likely effect of alternative forms of guarantee and to take the action needed to reduce their exposure. In contrast, if a competitor takes the initiative with a new guarantee, the companies which try to meet this competitively may endure severe losses while they are trying to catch up on analysis and corrective action.

26-7 Quality Labeling and Certification

Producers provide labels and certificates with their products for three general purposes:

1 *Product information.* The informative label (or leaflet) describes the product and its ingredients, how it should be used and maintained, and what cautions to observe. By implication, the manufacturer warrants that the product conforms to label.

Example Garments made of a certain fabric contain a "wear dated" tag guaranteeing a year of wear. (The guarantee is provided by the chemical company that manufactures the synthetic fiber—not by the garment manufacturer.)

Example Many organizations include a code in labels to identify containers, raw materials, or final product. This can be helpful in sampling inspection, inventory control (e.g., to ensure "first in, first out"), and investigation of complaints. Coding may take such forms as color, numbers, or small notches on the edge of a label.[6]

2 *Sales promotion.* Producers properly urge users to buy the product for the qualities claimed on the label. However, in their zeal, the producers have from time immemorial exaggerated their claims, and this practice goes on to this day. (The legal term is "puffing.") There is now a trend, through government intervention, to insist on "truth in packaging." For example, olives are graded by size, there being eight grades. The smallest is "standard"; i.e., there are no "small" olives. Above standard are medium, large, extra large, mammoth, giant, jumbo, colossal, and supercolossal. The extent of such exaggeration is very wide, particularly in the advertising, and there is enough deceit to enable reformers to make out a case for government intervention.

[6] Robert E. Southworth, "Using Codes for Packaging Control," *Management Review*, pp. 66–69, August, 1962.

Other forms of labels, while factual, nevertheless have an incidental use in promoting sales.

Example The Harris Tweed Association marks its cloth with an insignia consisting of an orb surmounted by a Maltese cross. For a tweed to be marketed as Harris it must be 100 percent pure virgin wool produced in Scotland, spun, dyed, and finished in the Outer Hebrides, and handwoven by the islanders at their own homes.[7]

3 *Product assurance.* Information designed to give the user added assurance that the product meets the claims is commonly by "certificate" (i.e., make sure). The certificate is usually provided by an independent laboratory.

Example Underwriters Laboratories (originally created by insurance companies to aid in fire prevention) develops and publishes standards of materials, products, and systems related to fire protection, burglary protection, hazardous chemicals, etc. The agency also makes independent tests of manufacturers' goods to determine compliance with these standards. Goods which comply are "listed." Because the insurance company rates are lower when listed goods are used, the listing is important to the marketability of the goods.

Example The British Hallmark system (instituted in 1300) provides an independent test of fineness in gold or silver products. The mark identifies the manufacturer, the fineness (e.g., 14 carat gold), the testing office, and the year of test.

Example In some countries, certain classes of products may not be exported out of the country without a certificate from an independent laboratory. The purpose is to protect the national quality reputation.

Example Some forms of drugs (those injected into the human body) may not be released by producers until there has been an independent test and approval by a laboratory of the Food and Drug Administration.

Such forms of product assurance by certificate are likely to grow so long as human beings continue to build health and convenience around the quality of products and services.

26-8 Product Liability

The twentieth century has witnessed a remarkable growth in the concept of product liability—the responsibility of the manufacturer for injuries

[7] Veronica Thomas, "The Wizard of Plockropool," *Atlantic Monthly*, p. 124, June, 1966.

which his products inflict on the public. This responsibility is based not on an agreement between the parties involved, but on court decisions growing out of the laws of negligence. The emerging theory seems to be that a manufacturer who puts unreasonably dangerous products on the market should be held responsible if someone gets hurt.

In earlier years the injured person had redress only if he himself had bought the dangerous product. In addition, he had to prove that the manufacturer had been negligent. Now the courts tend to find an implied warranty which renders proof of negligence unnecessary. In addition, they tend to regard this implied warranty as running with the goods, so that the rule of "privity" (being a party to the purchase transaction) no longer applies. The consequence is an emerging situation of liability without proof of fault.

The growth in product liability claims is traceable in part to the evolution of law as stated above. In addition, the extent to which human beings are putting their lives and health at the mercy of manufactured products (automobiles, drugs, etc.) has grown enormously. Finally, there has been a growing consciousness of the potentialities inherent in product liability claims. There is money in it, for victims and for their lawyers, since some jury verdicts run to hundreds of thousands of dollars.

This growth in product liability claims is causing companies to examine broadly their exposure to these claims and what to do about it.[8] The remedies run across the entire company:

Adopting broad policies of product safety
Designing and making products to be safe and foolproof
Avoiding exaggerated claims in advertising
Training salesmen, servicemen, and users in proper use of the product
Securing field feedback on usage and on failures
Adequate insurance against catastrophic claims
Prompt, thorough investigation of claims
Remedial action to prevent recurrence of known causes of failure

Product liability is a problem that indeed requires those in the quality function to go far beyond the technical aspects of the problem.

26-9 Consumerism

The users of "hardware," e.g., machine tools, tractors, TV sets, etc., are faced with two very different problems depending on whether the

[8] See, for example, *Products Liability and Reliability,* Machinery and Allied Products Institute, Washington, 1967. Also, James D. Ghiardi, "Corporate Liability for Defective Products," *Quality Progress,* vol. 2, no. 5, pp. 8–11, May, 1969.

apparatus is still in guarantee or not. In the former case, the users have some clear rights by prior arrangement which are commonly respected. In the latter case the user is on his own. If the apparatus fails, he must find someone who can restore service, and then arrange to get the repair work done.

The "large user" is in a favorable position to secure good maintenance. Backed up by competent buyers and engineers, the large user is usually able to take advantage of his economic power to secure good quality work, promptly and at competitive prices.

The small user is in no such situation. When his automobile or TV set fails, he must find a repair shop and get service restored. Lacking personal knowledge of the labor and parts needed, the user has little choice but to take what he gets and to pay the bill. This use of complicated hardware by "ignorant" users has grown enormously, but the service industry has not kept pace. It has not provided enough shops which can give skilled, prompt, and efficient service. (In such seller's markets even the integrity of the shops deteriorates notoriously.)

The resulting situation is one in which millions of users of apparatus find themselves unable to secure adequate redress when they are out of service. The guarantee period is small in relation to useful life (10 percent or so). Hence during most of the useful life, the users have no assistance from the manufacturers or the merchants; they are on their own and must deal with the repair shops. Even during the guarantee period, there is an uncomfortable extent of irresponsibility.

In democratic societies, whenever huge numbers of people are faced with common problems for which the existing solutions are inadequate, there arises new leadership offering new solutions. This new leadership comes from reformers, intellectuals, politicians, etc., both the sincere and the insincere. Either way the result can be to mobilize the grievances of the many into political action.

The phenomenon called "consumerism" is precisely of this nature. Huge numbers of consumers have encountered problems for which they feel existing solutions are inadequate. The resulting pressure has created new leadership, and political forces are forming which will try to achieve by legislation what the marketplace has failed to do.

PROBLEMS

1 Prepare a plan of market research to discover the quality needs of users, and what qualities might be sold to users, for any of the following categories of product or service: children's footwear, pedestrian protection from rain, peeling fruit, writing letters, soft-boiled eggs, illumination for reading, and anything else acceptable to the instructor.

2 Study the methods used for advertising quality for any of the following categories of product or service: beverages; cigarettes; automobiles; household appliances; airline travel; moving picture entertainment; electronic components, e.g., resistors; and mechanical components, e.g., bearings. Report on the extent to which the advertising (*a*) identifies specific qualities, (*b*) quantifies the extent to which the product possesses these qualities, (*c*) engages in exaggeration or deceit, and (*d*) appeals to various human traits, e.g., vanity, greed.

3 Study the guarantee published for some product or service (acceptable to the instructor). Compare the terms of the guarantee with the "Guides against Deceptive Advertising of Guarantees" (*QCH,* 1962, pp. 1-28 and 1-29). Report your findings and conclusions.

27 Statistical Aids in Customer Relations

27-1 Significance of Field Complaints

Historically, customer "satisfaction" has usually been measured negatively by some type of complaint index. These will be discussed in this chapter, but first the question: Are complaints an adequate measure of customer satisfaction? Often, they are not.

Most product defects are *not* made known to the manufacturer. One study[1] found that only one out of every three dissatisfied buyers of a washer or television set complained to the manufacturer (although they all complained to their friends). Further, it appears that every appliance that fails soon after it is purchased results in publicity that turns *ten* potential customers against the brand.

Whether a complaint is made on a defect depends on several variables:

1 The economic climate. Complaints tend to rise in a buyer's market and fall in a seller's market.
2 The temperament of the individual customer. Individuals react differently to defects—the same product generates complaints from some customers but not from others.
3 The seriousness of the defect as seen from the customer's viewpoint. The scope of products here is broad. A serious defect may be a faulty braking system in a vehicle or a missing toy in a cereal box.
4 The value of the product. Complaint rates are influenced by the unit value of the product (see section 27-2).

[1] Herbert E. Klein, "Golden Key to Production Profits," *Dun's Review and Modern Industry*, p. 50, April, 1963.

All this means that the complaint rate may be a poor index of product quality because the rate can be affected by variables that are unrelated to quality. A reasonable guide is this: A high complaint rate means customer dissatisfaction but a low rate is *not* proof of customer satisfaction.

Comparison of the causes of customer complaints with the causes of factory rejections can reveal facts of critical importance for future manufacturing and inspection planning. In one instance, 81 percent of the field returns of an optical product were for mechanical breakage. Factory data revealed no such problem because there was *no* systematic factory test for this defect. Factory data did have a headliner; that is, 65 percent of factory rejects were for minor appearance faults, but there were no field returns for these faults.

In another example, a summary of defects on rubber garments revealed the following percentages:

	Field Complaints	*Field Returns*	*Factory Rejections*
Torn, location A	58	77	
Torn, location B	33	19	
Scuffs			28
Blisters			13

The conclusion is revealing—factory evaluation does *not* reflect user desires. In this example, a drop in sales was blamed on poor "quality." The president ordered an improvement and the factory promptly tightened standards on the key defects causing factory rejection (visual defects such as scuffs and blisters). Visual defects were reduced, costs were increased, and the user continued to receive garments which tore during use. The facts were available but lack of coordination led to the wrong action.

In these two examples, the path of necessary action is rather obvious but this is so only because of the data collected. The key point illustrated is the need to periodically collect data to compare the causes of factory rejections and user complaints.

27-2 Effect of Unit Price and Time on Complaint Rate

Data suggest that unit price of a product may be the largest single factor in determining complaint rates.

For example, a company making shotgun shells found that the number of misfires in the outgoing product is about fifty times the number of misfire complaints. For this product (priced at about 10 cents) the ratio of complaints to serious defects is 2 percent. For data on razor blades (selling at about 3 cents), the ratio of complaints to serious defects was under 1 percent.

Ratios on some other products are shown below:

Product	*Unit Price*	*Ratio of Complaints to Serious Defects*
An article of clothing	About $5	About 10%
A small electrical appliance	About $30	About 25%
A large electrical appliance	About $200	About 50%
A motor vehicle	About $2,000	About 70%
A large engineered facility	About $50,000	100%

For low unit prices of product, the complaint rate can greatly understate field difficulties; the complaint rate may need to be inflated from twenty to over one hundred times to arrive at the actual defect rate. For low priced products the measure of outgoing quality must come from check inspection. As low unit prices usually mean large-volume production, the cost of check inspection will be relatively low, since the sample needed is only a small fraction of the total product. For high priced products, the number of units is few, and the cost of check inspection becomes prohibitive. However, the ratio of complaints to defects tends to approach unity, so that the complaint rate tends to reflect product quality more closely, thereby minimizing the need for check inspection.

The effect of time on complaint rates is influenced by the nature of the failure. A useful classification is

1 "Infant mortality" failures, i.e., misapplication, design weaknesses, factory blunders, shipping damage. These tend to show up early in service—usually within the stated or implied guarantee period.
2 "Accidents" arising from misuse in service, i.e., sporadic overloads, poor maintenance. Some of these take place within the guarantee period or the time span during which complaints are commonly entered.
3 Wear out, i.e., failure in old age after giving acceptable service life. These seldom generate complaints.

It is important to record the cardinal dates associated with the product—the date of manufacture, the date of sale, and the date of going into service. These dates can disclose useful knowledge about the product and can aid in analysis and prevention (see section 25-2).

Dates of manufacture (normally month or week) are marked on the product through coded serial numbers, through date stamps of invisible, indelible ink, and through other ways. Date of sale is far more difficult to secure on a detailed basis. Where serial numbers are used, arrangements can be made with selected dealers to note the date of sale on a large enough sample to determine the usual shelf time. Other methods are periodic inventories of dates of manufacture of goods on dealers' shelves and the use of serially numbered guarantee cards to be filled out by the customer or dealer. Date of going into service is normally secured only for large units of product by using a guarantee card or similar device to record the date.

27-3 Complaint Indices

Some typical complaint indices are

1 Service calls per 1,000 units out on warranty. When calculated for a time period such as a month this is referred to as a "failure rate."
2 Number of field complaints related to total production during the same period or simply the absolute number of field complaints per time period.
3 Dollar value of claims paid related to value (or volume) of production or simply the absolute value of claims per time period.

These are examples of indices used to separate the vital few from the trivial many problems. Such indices provide little recognition to the degree of dissatisfaction *as viewed by the customer.* Thus, although the dollars paid by a manufacturer to repair or replace a product are an important cost item to *him,* they do not reflect the customer cost of the repair or replacement, e.g., the value of lost production while the product was under repair (see section 4-9 for examples). Sometimes the effect on the customer is subtle. Suppose a fleet of earth moving equipment is being used to remove earth from one location and transport it to another. The fleet may include a "motor grader" to maintain the "haul road" in good condition for tractors and other vehicles that will use it. Now suppose the motor grader fails and is out of commission. If the failure is covered by warranty, it will be fixed and part or all of the repair cost will be covered by the manufacturer. However, suppose no other graders are available and the condition of the haul road deteriorates due to lack of constant conditioning by a grader. This can mean that the other vehicles in the fleet using the road will have to work at reduced speed. Such a reduction in efficiency is critical to an earth moving contractor because it increases his cost per yard of earth moved. Thus, the repair cost (even if paid by the manufacturer) may be much less important than the loss in production.

Complaint indices need to recognize customer effect in addition to effect on the manufacturer. Such indices must be created uniquely by each company, but factors that should be considered in evaluating different failures include:

1 Total repair time (including waiting for parts and/or technicians)
2 Immediate effect of failure (whether or not the failure results in immediate shutdown)
3 Total production downtime
4 Maintenance labor cost
5 Other effects on customer operations

Some of this information is accurately known only by the customer. Some information may vary greatly depending on customer procedures. There are many reasons why it will be difficult to obtain this type of information. However, the key point is that the customer *is* experiencing these effects of failures. The difficulties of measurement might better be viewed as an opportunity to achieve a competitive advantage by solving the measurement problem and acting on the results.

27-4 Obtaining Customer Information on Quality

Relatively little has been done to develop sensing systems to obtain the customer view on quality. Ideally, the sensing system should report the following types of information:

1 Number of failures (or other instances of trouble) experienced by the customer during and after the warranty period. The expiration of a warranty may end legal obligations but the customer continues to evaluate product performance during its *entire* life.
2 Sufficient detail on failures to permit accurate analysis for failure causes.
3 Effect of the failures on customer operations in terms of downtime, value of lost production, etc.

The difficulty of obtaining this type of information raises the possibility of using a sampling approach rather than trying to cover the entire customer population. Concentrating effort to obtain a higher quality of information in a small sample may yield more useful results than over a larger group.

The approach[2] at Buick Motor Division will serve as an example. Buick receives failure information from four sources:

1 Warranty reports
2 Summary problem reports from each of the 27 zone managers throughout the country
3 Individual product reports from dealer service people
4 Special investigations

A system was *additionally* imposed to obtain actual failed parts within a few days of the failure. The system covers a sample of only 37 dealers who sell a total of about 6 percent of the sales volume. The system provides for these dealers to send the failed parts (from all the cars they sell) to a central Buick location. With the part is included detailed usage information. These parts are then analyzed by a reliability engineer and an in-depth analysis is made to accurately determine failure cause.

Schilling and Davies[3] describe a "quality assessment" system used for grinding wheels. The program has four objectives:

1 Determine causes for quality deterioration at the point of consumption
2 Compare the quality of similar products made in different plants within the same company
3 Indicate potential changes necessary to meet market demands
4 Compare product quality with competition

The system is based on the continuous purchase of company products on the open market. The product is tested to represent field use and appropriate analyses made. The scope of the program is indicated by the 2,400 tests per month conducted and the 600 control charts maintained.

27-5 Processing and Analysis of Customer Complaints

Quality complaints require two levels of investigation, one oriented to the complaining customer and the other oriented to the product.

[2] J. R. Gretzinger, "Buick's Reliability Program," *Industrial Quality Control,* vol. 21, no. 9, pp. 449–453, March, 1965.
[3] Edward G. Schilling and Harold J. Davies, "Quality Assessment—A Missing Link," *Transactions of the American Society for Quality Control,* 1964, pp. 279–286.

The first level of investigation includes:

1 Restoration of service. Where service is interrupted, the first need is to reestablish service. This may require temporary repairs, replacement parts, or replacement of an entire unit.
2 Financial adjustment. Prompt financial adjustment, if justified, is a second essential step toward reestablishing customer relations.
3 Restoration of customer goodwill. There is still a problem of soothing ruffled feelings. A product failure is an annoyance at best, but the irritation remains despite the restoration of the status quo in service and in finance.

The second level of investigation includes:

4 Discovery of whether the condition is isolated or general. If isolated, little more need be done. If general, there is need for:
5 Analysis for causes. This is the investigation, both in the field and in the laboratories (the autopsy), to discover the causes of the failures.
6 Remedy. This is the "fix," which may require changes in design or in usage. It may also require recall of product in the field, sometimes on a heroic scale.
7 Summarized reporting for executive information and action.

The first of these levels of investigation is commercial in nature; the second, technological. It is usually difficult for one man to conduct both. As the company grows, and especially if the product involves advanced technology, increasing use is made of two different individuals to conduct these two levels of investigation. The former is from sales or sales service; the latter is from the technical department. Very large companies tend to create a special complaint bureau to conduct the second level of investigation.

It is also common practice to help sales and sales service personnel by preparing a manual for handling complaints and claims. The manual normally contains:

The company's statement of policy in handling complaints and claims.
Samples of complaint forms, properly filled out. (These forms differ greatly among companies because of differences in product, usage, etc.)
Detailed instructions for securing essential data, including explanations of *why* these data are needed.
Description of usual defects, usual causes of each, and related information for use in dealing with customers.
Statement of responsibility of salesman (or investigator) with respect

to settling claims, recommending settlement, making repairs, classifying causes, etc.

Glossary of terms.

Code letters and numbers used in systematic classification.

Chronic field complaints may require experiments in which changes made in the factory are tried out in the field.

Heine investigated the cause of porcelain breakage on Norge kitchen ranges. To do this he developed a diagram[4] on which the distributor could record the location, in the railroad car, of the damaged range. In addition, he provided an exploded diagram of the range itself on which the distributor could mark the exact point of porcelain damage.

The returned diagrams were summarized monthly into master charts which told at a glance where the concentration of crate damage was, how the steel bands were holding up, and where on the range itself the damage was concentrated.

Analyses resulted in changes in crate design, in range design, in car loading, and in car routing. Transportation damage was reduced from 9 to 3 percent over a 2-year period, at a negligible increase in cost.

27-6 Cumulative Returns Analysis

Because failure rates change with length of product usage, analysis of *cumulative* returns of charges is most revealing. Such analysis is helped greatly if the product is "dated" to show when it was made, sold, or installed (see section 27-2).

The various dates are useful not only in disposing of complaints and claims for product of a slow-perishable character (candy, photographic film, etc.), but they also can help in predicting the failure rate of various product designs.

Example An example of the use of product dating concerned a certain component of women's clothing. About 2 percent of the product shipped was being returned as having torn in service (despite a statement on the container that the product was *not* returnable for tearing in service). The price of the product made it appear (see section 27-2) that actually about 20 percent of the product was tearing in service—a serious situation.

An analysis was made of the returns from the 20,000 units of product shipped during December, 1951. For *each* month from January, 1952, through December, 1953, a count was made of the number

[4] See J. M. Juran (ed.), *Quality Control Handbook,* 2d ed., McGraw-Hill Book Company, New York, 1962, pp. 12-12 and 12-13.

of units (shipped in December, 1951) that were returned in the current month. The monthly count was made cumulative and finally expressed as a percent of the 20,000 units shipped.

Meanwhile, the research department had evolved a new product which went into full-scale production in February, 1953, during which month 20,000 units of product were shipped. For the first 12 months, the cumulative percent of returns of these 20,000 units was recorded in the same form as the old design.

Even more striking is the comparison when "months following manufacture" are used so that both charts start from the same point (see Figure 27-1).

By such analysis the company was informed *within the first few*

Figure 27-1 Comparison of cumulative return based on months since month of manufacture.

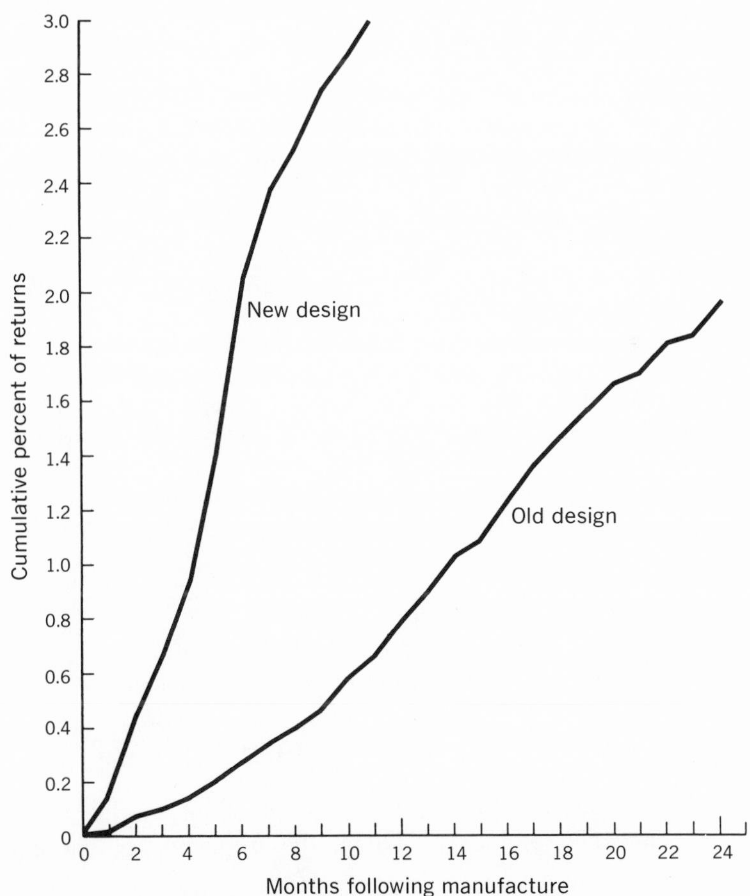

months of the life of the new design that the new design had, as to the problem of tearing, made the situation worse rather than better.

The cumulative percent principle has universal application, even for large equipments.

Example The following example concerns equipments which are used in steel mills, paper mills, and cement mills, and which sell in the price range of $50,000 to $250,000 per unit. The equipments failed frequently in service, and repairs were paid for by the equipment manufacturer under the guarantee he made to the mills that bought the equipment.
The equipments were redesigned to eliminate the failures, but there was controversy over whether the redesigns solved the problems. However, since cost data were not separated by types of design, the monthly repair charges did not disclose whether the new designs were superior to the old designs.
The problem was solved through use of a cumulative repair charges chart.
For each equipment, information was recorded on the date of installation, the original installed cost, and the monthly repair charges since installation.
The analysis proceeded as follows:

1 For each equipment there were computed the *cumulative* repair charges since installation, month by month.
2 For all equipments of the same design, it then became possible to add the cumulative repair charges since installation, month by month, despite the fact that the installation dates differed.
3 The sums of the cumulative repair charges were then computed as a percent of the original installed cost.
4 The values of the cumulative repair charges as percent of the original cost were then plotted. Figure 27-2 shows such a study.

The apparatus prevailing in 1948 resulted in repair charges amounting to 4 percent of the original cost.
A redesign late in 1948 turned out to be unfortunate but another in the latter part of 1950 was effective as seen in the 1951 product, which in the first 12 months of service accumulated charges of less than 1 percent of the original cost (see Figure 27-2).

A variation of the effect is the "hours of service" concept used in maintenance programs. There, the frequency of check of equip-

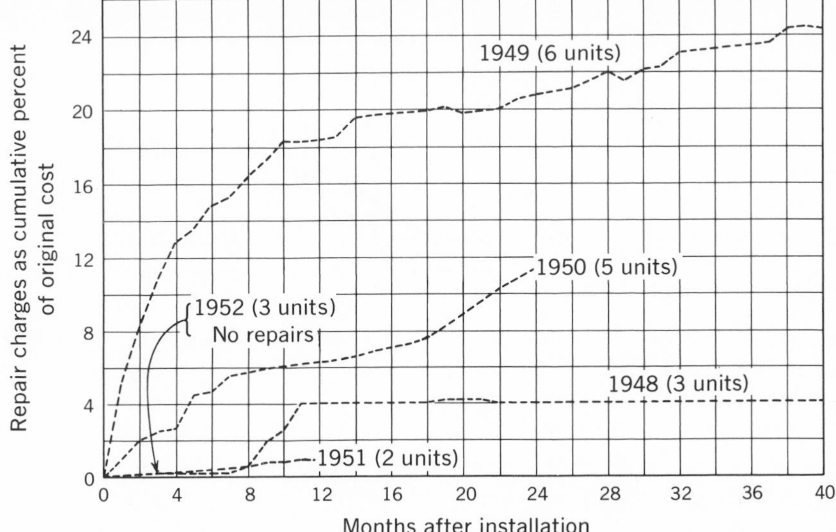

Figure 27-2 Study of design effectiveness of large equipments through analysis of cumulative repair charges.

ment in service is based on extent of actual use; e.g., "hours flown" is the basis for scheduling overhaul of aircraft components.

27-7 Use of Probability Paper for Predicting Complaint Level

The foregoing examples were based on *arithmetic* scales for time and for complaint rate to make the analysis. With these arithmetic scales, the cumulative rates showed up as curved lines. In many cases it is possible to choose nonarithmetic scales which will convert the cumulative complaint curves into straight lines. These straight lines make it much easier to predict, early in the marketing life of the product, what to expect in the future. Use of "probability paper" (see section 6-9) is a common method for converting the cumulative complaints or failures into straight lines.

The approach[5] is a melding of statistical analysis and engineering judgment and can be summarized in the following steps:

1 Select a typical company product for which complete warranty data are available. For example, the complete data for an automobile

[5] B. H. Simpson, "Reliability Prediction from Warranty Data," *Proceedings of the Automotive Engineering Congress, Society of Automotive Engineers*, 1966.

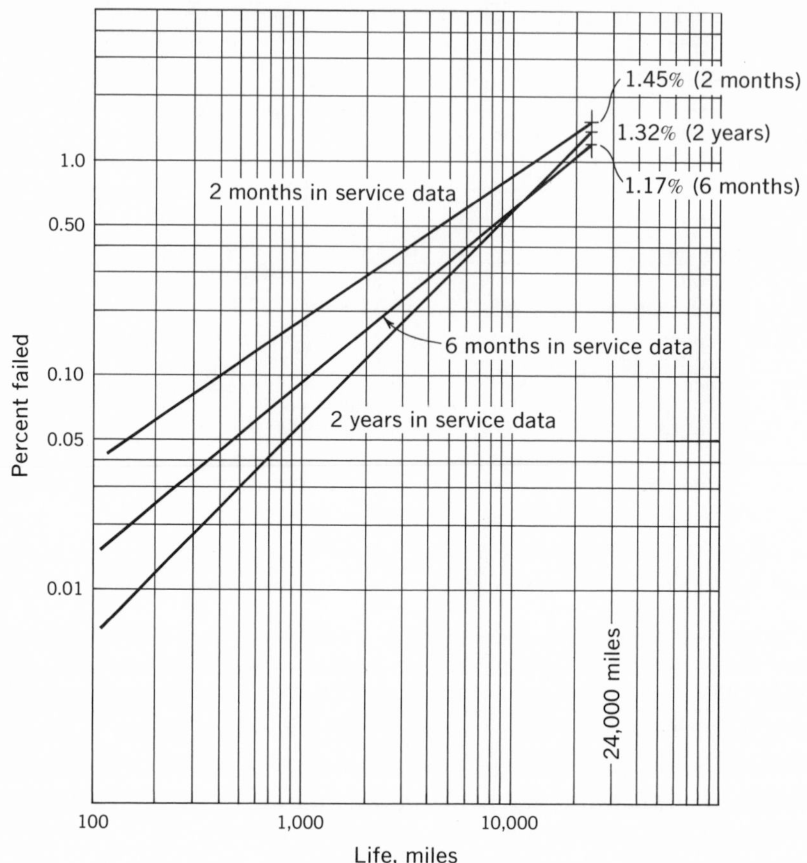

Figure 27-3 Projection with various time in service. Source: B. H. Simpson, "Reliability Prediction from Warranty Data," *Proceedings of the Automotive Engineering Congress, Society of Automotive Engineers,* 1966.

component having a 24-month or 24,000-mile warranty would be the record of all failures that occurred during the complete warranty period.

2 Select an early portion (for example, 2 months) of the data and plot it on probability paper. If the plot approximates a straight line, draw the line and extend it forward to the full warranty period of 24,000 miles (see Figure 27-3). The top line is based on warranty data for a group of cars that all had been in service for 2 months. The graphs plot the cumulative percent of the group that had failed at a specified mileage. In calculating this percent, it is necessary to have mileage distribution data for cars which have been in service for the period involved, that is, 2 months.

Knowing the total number of cars in service for 2 months and the percent having achieved a specified mileage, the number of cars eligible for repairs can then be determined for each mileage. The repair claims are then matched with these data and the cumulative percent failed is calculated.

3 Plot the complete set of data for the full warranty period and compare the actual percent failure at the end of warranty to the predicted percent obtained by extrapolation in step 2.

4 Repeat steps 2 and 3 for several periods of early data. Figure 27-3 shows predictions based on 2 months and 6 months of data and the actual data for the full 2-year period.

5 Draw conclusions on the validity of the approach. Such conclusions would include:

a Whether the prediction based on *any* amount of early data is sufficiently accurate for evaluating potential problems.

b How much early data is required. (About 6 months of data appears to be necessary based on results reported by Simpson.[6])

c Whether extrapolation seems reasonably valid.

6 If the conclusions from step 5 are positive, then apply the procedure to new products.

Extrapolation beyond the limits of actual data is always questionable. However, if the comparison of actual and predicted checks well for some products and if judgment suggests that the extrapolation is valid, then the approach would seem to justify its consideration as a useful working tool. It should be stressed that decisions on the "vital few" warranty problems deserving major attention *must* and *will* be made by someone. Although the approach suggested here is not rigorous, it surely seems to be a major step beyond intuitive decisions.

27-8 Design of Competitive Tests

The producer who has a policy of meeting or exceeding competitor quality cannot judge his conformance to that policy unless he secures data on competitor products. These data seldom become available in the normal course of conducting business. Instead, positive steps must be taken to secure these data. In addition, if a comparison is to be made of "our product versus competitors'," the comparative data must be collected in a way which avoids bias.

For example, the Home Company[7] annually compares its product with that of competitors for each quality characteristic. Table 27-1 illustrates the results in terms of relative ranking of companies. The results

[6] *Ibid.*

[7] Actual instance, name fictitious.

Table 27-1 **Comparison of Product X Home Company Versus Competitors**

Characteristics of the Product	Ranking of the Companies				
		Competitors			
	Home Co.	A Co.	B Co.	C Co.	D Co.
P	3	2	1	4	5
Q	2	4	3	1	5
R	1	3	2	5	4
S	1	5	3	4	2
T	5	2	3	1	4
U	4	1	2	5	3

not only tell which characteristics may need improvement (T and U) but also highlight that the company is the leader on R and S. Characteristics R and S may deserve advertising emphasis or, alternatively, they may be a source of cost reduction if they are of relatively low importance to the user.

In collecting competitive data, it is important to secure data as seen by the user. For example, a company making abrasive cloth (for polishing metal) secured samples of competitors' products, tested them in the laboratory for various qualities, and found that its own product competed well. However, what was important to the user was something quite different: "Cost per 100 pieces polished." To the user, abrasive cloth was a "supply" which served only to perform a production operation (polishing hoods). The measure of usefulness was "cost per 100 hoods," and the user compared this cost for the competing sources of abrasive cloth. Under this comparison, the company's product was the least valuable to the user, no matter what the laboratory tests showed.

PROBLEMS

1 Propose one or more complaint indices for the manufacturer of one of the following products:

a Any product acceptable to the instructor
b Household refrigerators
c Passenger automobiles
d Corn processing by-products sold primarily to brewers of beer and pharmaceutical companies (for making medicine capsules)
e The totality of products sold by a large department store
f Jet engines for passenger planes

2 Select a product which you or a friend owns and which has been tested and reported on in one of the consumer magazines.

a Comment on the adequacy of the tests made to evaluate competing brands.

b Compare the opinions of the user to the evaluation published in the magazine.

3 Visit a local manufacturer and report on the procedures used to process complaints and to summarize complaint information for executive action.

4 Warranty data on an ignition switch show that 0.15 percent have failed by 3,000 miles, 0.25 percent by 6,000 miles, and 0.40 percent by 12,000 miles. Predict the percent failure at 24,000 miles and at 50,000 miles. State the assumptions necessary.

5 Past data on a certain type of windshield wiper motor indicate a Weibull slope of about 0.50. A goal of no more that 0.5 percent failures by 24,000 miles has been set. What percent failure observed at 6,000 miles would indicate that the goal would probably not be met? State the assumptions necessary.

28 Quality Assurance

28-1 Nature of the Assurance Function

"Quality assurance" as used in this book means an evaluation of quality activities, conducted by independent evaluators, for the purpose of giving supervisors and managers an added confidence that all is going well. This independent review goes by many names: control, audit, survey. The process of conducting the review and of reporting the results to give this added confidence is called (in this book) quality assurance. The department which carries out this process is sometimes called Quality Assurance (department). There are also other uses of the term "quality assurance" which make it difficult to understand what is meant unless one examines the deeds being done to understand their significance.[1]

It is useful to look sideways at the finance function to understand the elements of quality assurance. Every company has means for financial control: general accounting, cost accounting, etc. In addition, an independent audit is conducted to provide assurance that (1) the design of the control plan is such that, if followed, it will correctly reflect the company's financial condition, and (2) the plan is being followed.

Generally, "quality assurance" includes the same role as applied to the quality function. In addition, the term commonly includes analy-

[1] See, for a discussion of this and related terminology, J. M. Juran, "Activities and Labels; Functions and Names," *Industrial Quality Control* (*IQC*), pp. 248–250, November, 1967.

sis of evidences of failure of control and participation in remedial action. Specifically, the activities usually include:

Analysis of field complaints (see Chapter 27)
Measure of field quality performance
Quality rating of outgoing product
Quality survey or audit
Executive reports on quality

This chapter discusses the latter four items.

28-2 Measures of Field Performance

These include:

1 The measure of field complaints (Chapter 27), which represent the most dramatic evidence of negative performance.
2 The measure of returns for credit. These are a form of silent complaints and should enter the picture of field quality performance. The department which is responsible for preparing the summary of complaints should also be responsible for the summary of returned goods.
3 The information on orders for spare parts. Many companies look on spare parts orders as a good source of revenue, forgetting that these orders are a big expense and annoyance to the customer. Customers often become shockproof with respect to spare parts, and accept such maintenance as a fate. However, the design and manufacture of maintenance-free products are just emerging as a problem to be solved, and products which need spare parts will be unable to compete against those which do not.
4 The reactions of the customers themselves. Some of these reactions come through the sales force (often a biased communicator). Others are sought out through questionnaires, surveys, advisory councils of customers, and still other structured approaches.
5 The record of customers lost. Companies watch to see which customers stop buying. A review of the dossiers of such ex-customers will disclose whether the loss was due to quality factors.

Aside from the foregoing, there are commonly available the financial data on credits for returns, payment of claims, charges to guarantee accounts, and related costs.

28-3 Quality Rating of Outgoing Product

A most practical measure of field quality performance is through measurement of the product before it is shipped to the user. This measure of outgoing quality is referred to as "rating," and the process of collecting the data is known as "check inspection."

Creation of a plan of quality rating is normally undertaken by an interdepartmental committee which considers and makes decisions with respect to a whole series of matters:

1 The purpose of the rating. The ultimate purpose is to obtain knowledge on how well the product performs for the user. The best source for this knowledge is actual service performance. The difficulties of obtaining service data usually result in making the rating at an earlier stage (or stages). Some possibilities are

Stage of Measurement	*Difficulties of Using This Stage*
Performance in service	Creating a reporting system that can and will be used by many users
Quality as received by users	Does not reflect use of product
Quality as received by dealers	Does not reflect use or effect of user storage
Quality at time of shipment from plant	Does not reflect use or effect of shipping or user storage
Quality after acceptance by inspectors	Does not reflect use or effects of packing, or shipping, or user storage

Each of these stages can be viewed as an approximation of the final use stage. In practice, the rating is often performed at the time of shipment and thus is a valuable independent check on inspection and packing and shipping operations. The selection of the stage of measurement depends on the purpose of the rating and on the difficulty of obtaining the information.

2 Collection of data. As the purpose of a rating does not include the formal acceptance of product, the amount of data collected is often arbitrarily a balance between cost of large samples and the unreliability of small samples. An example[2] of a specific formula is

$$\text{Monthly sample size} = a\sqrt{2n}$$

where n is the monthly production and a is a constant that depends on the complexity of individual units of the product and on the degree of variation in design characteristics of a class of product to be rated. The formula is empirical and was designed to meet the specific needs of one company. Where a company has selected a sample size simply on the basis of convenience and cost, it can evaluate the statistical adequacy of the sample size by calculating

[2] H. F. Dodge and M. N. Torrey, "A Check Inspection and Demerit Rating Plan," *IQC*, vol. 13, no. 1, pp. 5–12, July, 1956.

confidence limits for typical sample results, repeating this for alternative sample sizes, and then making a judgment on whether the proposed sample size achieves the desired balance between cost and statistical validity.

In selecting a sample, the units should be representative of the month's production, both as to the types of product made and as to days of the month. The units in a sample must be chosen at random from a total production quantity (e.g., a lot) produced under reasonably similar conditions. This is important for small samples to obtain a representative picture of all product.

3 Definition of defects. This requires collaboration among sales, engineering, and quality control personnel, in order to reflect factors such as engineering specifications, sensory standards, packing and shipping specifications, and field use experience.

As it is desirable to have an overall quality rating for the product, the different degrees of seriousness of defects must be defined (see section 24-4) and numerical weights ("demerits") assigned to each class of defect. The numbers are arbitrary but *trends* in product quality can still be detected.

4 Unit of measure for the rating. The overall rating is the number of demerits for the defects found divided by the number of units of product in the sample. The result is in terms of demerits per unit. If desired, the index can be converted to another scale; e.g., a rating of 100 could mean perfection while barely acceptable product (say 4 demerits per unit) could score a 60 rating. This could be accomplished by the following conversion chart:

Demerits per Unit	*Quality Rating*
0	100
1	90
2	80
3	70
4	60
5	50

Conversion scales can be completely arbitrary as long as the results make sense to the users of the ratings.

5 Standard for comparison. The usual standards for judging current quality are (1) the company's own past history and (2) competitor quality.

The company's past quality rating is established by calculating the rating for a representative "base period." This period should not include such factors as spottiness, introduction of new products, and changes in design. A base period should not be changed until it is no longer representative of the designs being manufactured, or of the processes. The calculation for a base period is shown below:

		Found during Base Period	
Defect Class	*Demerit Value*	*Number of Defects*	*Total Demerits*
A	100	50	5,000
B	75	370	27,750
C	25	1,400	35,000
D	5	450	2,250
			70,000

Total units inspected during base period = 75,400

$$\text{Demerits per unit} = \frac{70{,}000}{75{,}400} = 0.928.$$

The standard value for comparison is 0.928.

Using history as a standard runs the risk that the company will perpetuate an uneconomic performance. The economic soundness of the standard can be judged by rating competitors' products by the same rating plan.

A further refinement in the presentation of ratings would be a control chart[3] with statistical control limits for signaling significant departures from standard.

6 Correlation of outgoing quality data with actual user data. All the previous items assumed that the manufacturer's definition of what is a defect and the relative importance of different defects was an accurate reflection of how the user would rate the product when he inspected it upon delivery. However, there may be serious lack of agreement on these points between the manufacturer and the user. Several different customers may not agree among themselves. Where it is possible to do so, it is a worthwhile investment to check the correlation between the manufacturer's results on outgoing quality and the user's corresponding results ("incoming quality") for the identical units of product.

28-4 Overall Survey of a Total Quality System

The terms "quality," "survey," and "audit" refer to the appraisal of the quality system of an entire plant, or of one product, or of one activity. A survey may be conducted on a scheduled periodic basis or only when needed by the existence of symptoms of a quality problem. Chapter 19 discusses the use in evaluating a vendor's entire quality system. This section will discuss the evaluation of a company's own system. The

[3] See J. M. Juran (ed.), *Quality Control Handbook (QCH)*, 2d ed., McGraw-Hill Book Company, New York, 1962, p. 12-22.

next section discusses the evaluation of one or more activities in a total system.

The quality survey is analogous to an accounting audit. The audit checks the books, the bookkeeper, and the accounting system. The quality survey evaluates the product, the inspector, and the system for achieving product quality. In both instances, the audit is functionally independent of the activity being audited.

A complete quality survey involves bringing together data from a number of investigations:

1 Examination of customer needs and the adequacy of design specifications in reflecting these needs.
2 Examination of detailed manufacturing specifications for fidelity to design specifications, completeness, and clarity.
3 Examination of vendor product specifications and monitoring procedures.
4 Examination of customer quality complaints and adequacy of corrective action.
5 Examination to judge the adequacy of equipment, data collection systems, and decision making processes used by line personnel.
6 A first hand inquiry into the performance and understanding of all personnel having quality responsibilities.
7 Examination to judge the adequacy of the quality cost and other information provided to operating and top management for evaluating quality performance.
8 Examination of the scope and organization of the program to assure that all personnel understand the actions they must take to achieve the quality needs of the customer.[4]

All this takes time. This is the price one pays for the realism of facts over the drama of opinions. The task can be lessened by:

1 Using skilled fact finders to collect the detailed data in accordance with a general plan[5] and then have a survey committee review the findings, study broader matters, and issue the report.
2 Separating the "vital few and trivial many" product quality characteristics. The auditor (an individual or a team) can examine in detail the planning for the vital few; he can sample the planning for the trivial many. This sampling principle can be extended (see below).
3 Using a sampling approach for the entire survey. If a broad picture of the quality effort for many products is desired, then some type

[4] J. M. Juran, *Management of Inspection and Quality Control,* Harper & Row, Publishers, Incorporated, New York, 1945, p. 193.
[5] An excellent reference on this is *Evaluation of a Contractor's Quality Program,* Quality and Reliability Assurance Handbook H50, Government Printing Office.

> of sampling approach is imperative. The selection of the sampling units must be tailor-made for each situation but provisions should be made to include basic sources such as different products, plants, stage of product development, quality system activities, and customers. Just as acceptance sampling inspection sometimes produces a better overall result than 100 percent inspection, audit sampling may yield better overall results than a 100 percent audit.[6]

The sampling type of audit assumes that the plant has a day-to-day quality control program that includes many tasks (possibly including detailed audits) to achieve proper quality. This is analogous to a final acceptance sampling inspection which assumes that many tasks have been executed (possibly including 100 percent inspection) on a daily basis to achieve proper quality. The sampling gives an overview of the entire picture.

In preparing survey reports, top management needs a summary. An extensive quality survey of a giant multidivision consumers' goods company was summarized as follows:

> In essence, we find that:
> ——— is well organized to take care of day-to-day quality problems and is doing a good job on these.
> ——— is not well organized to identify and solve the long-range quality problems. As a result, a formidable list of important long-range quality problems is stacked up waiting for solution.
> Solution of these long-range quality problems would yield important cost reductions and provide improved understanding and control of quality.
> In essence, we *recommend* that.
> ——— should attack this list of long-range problems by:
>
> 1 Strengthening the quality function in each division.
> 2 Providing, at the corporate staff level, a manager of quality services, to help the divisions become self-sufficient in solving the long-range quality problems.

When really important matters are at stake there is no point in mincing words. For example:

> Considering the size of the ——— operations, and considering the universal conviction among the executives that the quality of product is of vital importance, I can say flatly that the facts available

[6] For a discussion of this point and an imaginative application of sampling in an audit, see W. R. Purcell, "Sampling Techniques in Quality System Audits," *Quality Progress*, vol. 1, no. 10, pp. 13–16, October, 1968.

to them are hopelessly inadequate to permit executive control of the quality function with anything like the control available on profit, cost, delivery schedule, and other essentials of industrial life.

A great risk in conducting a survey is that the omissions or errors of top management are not identified. Even when they are identified, it takes courage to tell management of them. The Japanese are trying to involve their managements. In Japan, a companywide or "president's audit" is conducted in many companies, usually on a semiannual basis. The audit team consists of the company president as chairman, plus several assistants, with outside consultants as observers. The team reviews:

1 Proposed quality objectives for the 6 months ahead
2 Attainment status of objectives for the previous 6-month period (results, training, standardization, etc.)
3 Status of solution of important problems
4 Status of administration of the quality control system (the conventional review of the quality plans and of the execution of the plans)

Since this is a high level audit, the inputs come up from lower-level audits conducted at plants, sales offices, and corporate office departments. At each level appropriate subjects are chosen for review. The audit team members discuss their findings and report their commendations as well as their recommendations.

For companies using such a network of audits, which bring in the personal review of the managers, including the top men in the company, there is less reliance on formal quality control systems and greater reliance on personal supervision.[7]

28-5 Quality Audits of Specific Activities

Some companies[8] maintain an independent audit group to define problem areas in sufficient detail to generate corrective action. Several principles are essential:

1 The activities subject to audit should include any that affect quality regardless of the internal organizational location. Examples might

[7] For elaboration, see S. Mizuno, "Quality Systems in Japan," *Reports on Statistical Application Research,* Japanese Union of Scientists and Engineers, vol. 15, no. 1, 1968.

[8] For example, see Benjamin W. Marguglio, "Quality Systems Audit," *IQC*, vol. 20, no. 1, pp. 12–15, July, 1963.

include definition of quality requirements on drawings and supplier documents, operator or inspector performance on a specific operation, adequacy of maintenance instructions, etc. Further, the audit scope should reflect ultimate user needs rather than only formal contractual requirements.

2 The auditor must have sufficient technical skill to uncover detailed facts to carefully define the problem and give some indication of the causes. It is not critical that the audit have a proposal for corrective action but it is critical that the audit furnish enough detailed facts so that the affected function can determine what action is necessary. Only then will the audit be viewed as a constructive service to the function which was audited.

3 The human relations problems must be recognized as potentially explosive. People simply do not like to be audited. First, the purpose of the audit should be *thoroughly* explained (shortcutting here to save time is inviting trouble). Second, the audit results should be shown to the function *before* the report is issued (no surprises in dramatic reports). Finally, the manager of the activity audited should clearly understand that he makes the decision on corrective action. It is the auditor's job to overcome the human relations problems. His promises are suspect until they are verified by his deeds and actions.

As an example, the following summarizes the results of an audit conducted on manufacturing and test equipments with respect to their calibration and certification:

1 Approximately 2 percent of the———equipments did not bear evidence of being calibrated for the current period.
2 Thirty-five additional———equipments have not been calibrated because they are not included as part of the control system for calibration as should be the case.
3 Criteria have not been established for determining which equipments are to be calibrated and which are not.
4 The referenced procedures are incorrect and inadequate.

These results were supported with five additional pages of detail from which the functions concerned could decide on corrective action. Such detail is the crucial *service* provided by the audit.

28-6 Control Subjects for Executive Reports on Quality

Although the facts needed by executives vary widely, there are some essential quality control subjects which are common to most companies:

1 Measures of customer satisfaction or dissatisfaction with the product, i.e., complaint rate, returns rate, gain or loss of customers due to quality, customer costs during and beyond warranty period, etc.
2 Direct measures of outgoing quality of product, i.e., results of quality rating, data on specific qualities, company's quality versus competitor quality, etc.
3 Losses due to defects
4 Cost of acceptance
5 Cost of prevention
6 Measures of how the overall system of quality is being carried out, i.e., quality survey or audit
7 Status of the solution of major quality problems

There are certain principles which fit any problem in executive reporting. These include:

1 There should be agreement on just what needs to be controlled (the control subjects), e.g., complaints.
2 There should be agreement on what is a sensible unit of measure for each control subject, e.g., number of complaints per million packages.
3 There should be means of securing facts on actual performance.
4 There should be agreement on a standard of performance.
5 The facts should be summarized for comparison with the standard.
6 The reports should read at a glance and should be presented as a coordinated instrument panel.

To design the executive control panel, a skilled investigator, e.g., the quality manager, personally interviews the key people to secure their ideas on what are the needed control subjects. When he summarizes the results of these interviews, it will be found that some control subjects are important to most of the executives. Other control subjects have only limited appeal. From this summary the investigator drafts a plan for a proposed report which meets the consensus of needs. This draft is then reviewed at a meeting of executives, modified, and approved. An example of such a draft is seen in Figure 28-1.

The unit of measure must be in terms useful to the level receiving the data. Reports to lower levels must emphasize *things* (defects, pounds, etc.). The higher levels need summary measures with *money* as the common denominator. Examples of the former are quality rating in terms of demerits per unit and warranty claims in terms of service calls per 1,000 units out on warranty. Examples of the latter are incoming inspection costs in dollars per dollar of goods purchased and warranty claims in terms of dollars.

Control subject	Unit of measure	Type of standard	Source of information	Frequency	Distribution
Customer complaints	Service calls per 1,000 units on warranty	Historical	Formal report	Monthly	Sales, service, factory, top management
	Cost of service calls due to quality failures	Historical	Formal report	Monthly	Sales, service, factory, top management
⋮	⋮	⋮	⋮	⋮	⋮
Cost of spoilage in factory	Dollars of rework and scrap	Engineered	Formal report	Monthly	Factory, top management
⋮	⋮	⋮	⋮	⋮	⋮
List of principal assembly defects	Number of defects	Plan	Formal report	Weekly	Factory

Figure 28-1 Plan of executive reports.

28-7 Securing the Facts

Fact and opinion are often not the same.

The proprietor of a small company can get production facts by simply counting the product in the shipping area. The manager of a large plant must ask the production superintendent, who asks the general foreman, who asks the foreman, who asks some group leader.

These two ways of finding out the amount of production should yield identical answers. In practice they do not. The difference is due to (1) *misunderstanding* of what is asked as the question goes down the line, and of what is replied as the answer goes up the line, and (2) a *coloring* of the answer. The fact is that production is only 200 units, which is less than yesterday's 250 units. However, another 60 units are complete except for some minor repairs and therefore the foreman reports the production as 260 units.

Information can be distorted by passing through many levels. This can be avoided by securing the facts directly from the original source rather than through the numerous levels. The original source is the mechanical instrument or the stream of papers which records the events of industrial life.

The case of the mechanical instrument is simple. If the executive requires knowledge of what is the time of day, or what is the steam pressure, then this information can be "piped" into his office as well as to the foreman's wall. The managerial instrument is the same in principle but is much newer in practice. Information on volume of production and on cost of defects can be "piped" into the executive's office as well as to the foreman's bench.

But what is the originating source for the pipeline? For example, production volume can be measured at several points. Volume moving past a station of a conveyor line can be registered by a mechanical counter. Volume of good production after inspection comes from inspection records. Volume of production packed for shipment comes from the packing slips. Volume of production shipped comes from the invoices or from bills of lading. One of these sources of information must be chosen as official.

28-8 Setting a Standard of Performance

There are four usual standards of comparison: (1) historical, (2) engineered, (3) market, and (4) plan. The historical standard uses past performance as a basis. This appeals to practical men as realistic but it may perpetuate poor performance. The engineered standard is hopefully based on what performance can be achieved, but this is often difficult to establish. The market standard is extremely useful from a competitive viewpoint, but the information may not come easily. The plan type of standard is a goal that seems realistic in light of past performance and activities planned for the future,

The standard will not be easy to set (but the discussions to set them will be fruitful). Unless the executive has some standard for comparison, he will be grappling with phantoms. Reporting results without a standard is like recording a golf score without a "par."

28-9 Summarizing Managerial Facts

Some engineers prepare reports in great detail so that the executive "can get anything he wants" out of the report. Most executives will get *nothing* out of such reports because they do not have the time

Figure 28-2 A summary quality report for a major product.

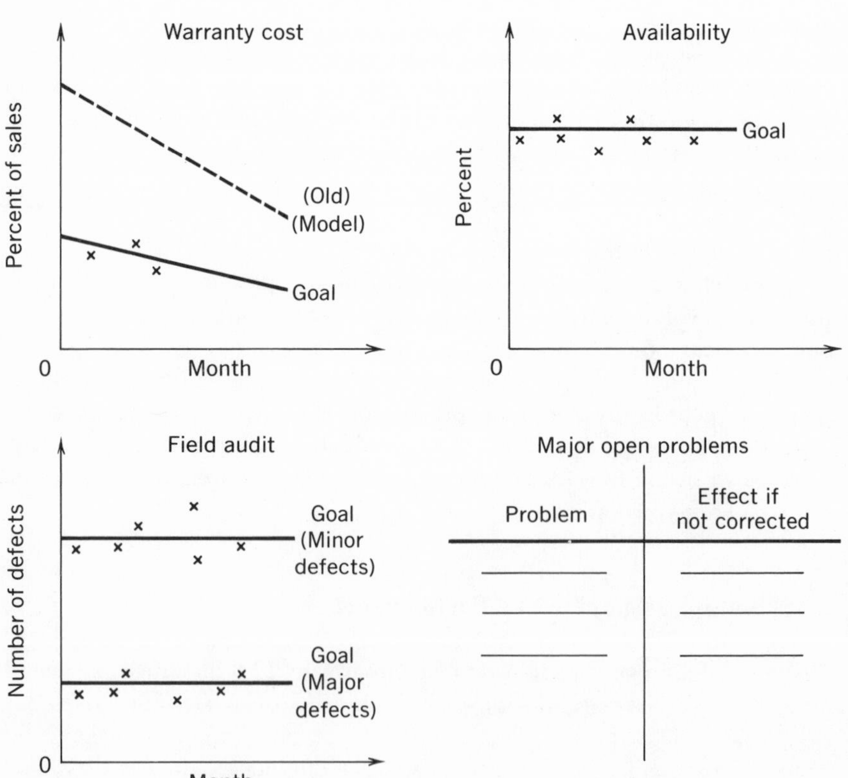

to sift out the vital from the trivial. Executive reports must not only be summarized; they must also include a standard for comparison, read at a glance, and present a coordinated story.

Section 5-6 presented examples of reporting the cost of defects. Figure 28-2 shows a report that summarizes several types of quality information for one major product. A coordinated reporting system requires:

1 A time schedule for various reports so that they can be reviewed regularly during executive meetings.
2 Grouping of related information to permit rapid review and comparison.[9]
3 Uniformity in reporting to minimize the time required to understand new indices or types of charts.
4 Highlighting of trends (with future projections if warranted) and conditions requiring executive action. Proposals for action should, of course, be available.

The quality problems of a company are many, but the serious problems are few. These vital few must be identified and the information provided to help executives decide on a course of action.

PROBLEMS

1 Draft a quality survey for use in a total quality system applicable to any product or service approved by the instructor. The draft should make clear the approach to be used in (*a*) review of customer needs and of adequacy of design specifications in reflecting these needs, (*b*) adequacy of manufacturing specifications, (*c*) review of customer complaints and of associated corrective action, (*d*) review of adequacy of methods of manufacture and process control, and (*e*) review of adequacy of product acceptance procedures.

2 For any company or institution acceptable to the instructor, design a system of executive reporting on quality, including (*a*) choice of control subjects, (*b*) units of measure, (*c*) form of standard to be used, (*d*) plan for securing facts on actual performance, (*e*) plan for comparing performance against standard, and (*f*) format for a coordinated instrument panel.

[9] For an example, see *QCH*, 1962, p. 12-31.

29 Quality Data Systems and Quality Manuals

29-1 Quality Data Systems: Scope

In the past, quality data systems have been concerned mainly with in-plant inspection data. This is, of course, a vital part of the quality data system. However, as products become complex and data from design tests and field usage become important, the scope and role of the data system expand (see Figure 29-1). One of the first considerations, then, in the planning of a quality data system is to define its scope. Some of the possible inputs for a quality data system are

1 Product design tests
   Examples are development test data, data on parts and components under consideration from various suppliers, and data on the environment that the product may encounter.
2 Data on procured parts and materials
   Examples are receiving inspection data, data on tests conducted by a supplier at his plant, and data on tests conducted by an independent laboratory on a procured item.
3 Process inspection data
   These data cover the entire in-plant manufacturing inspection system from the beginning of manufacturing up to final inspection.
4 Final inspection data
   These data are the routine data at a final inspection.
5 Field data
   Examples are data from a company proving ground, and warranty and complaint information obtained from the customer. It may also include "configuration control" data, defining exactly what hardware is included in each serial number of equipment sent to a customer.

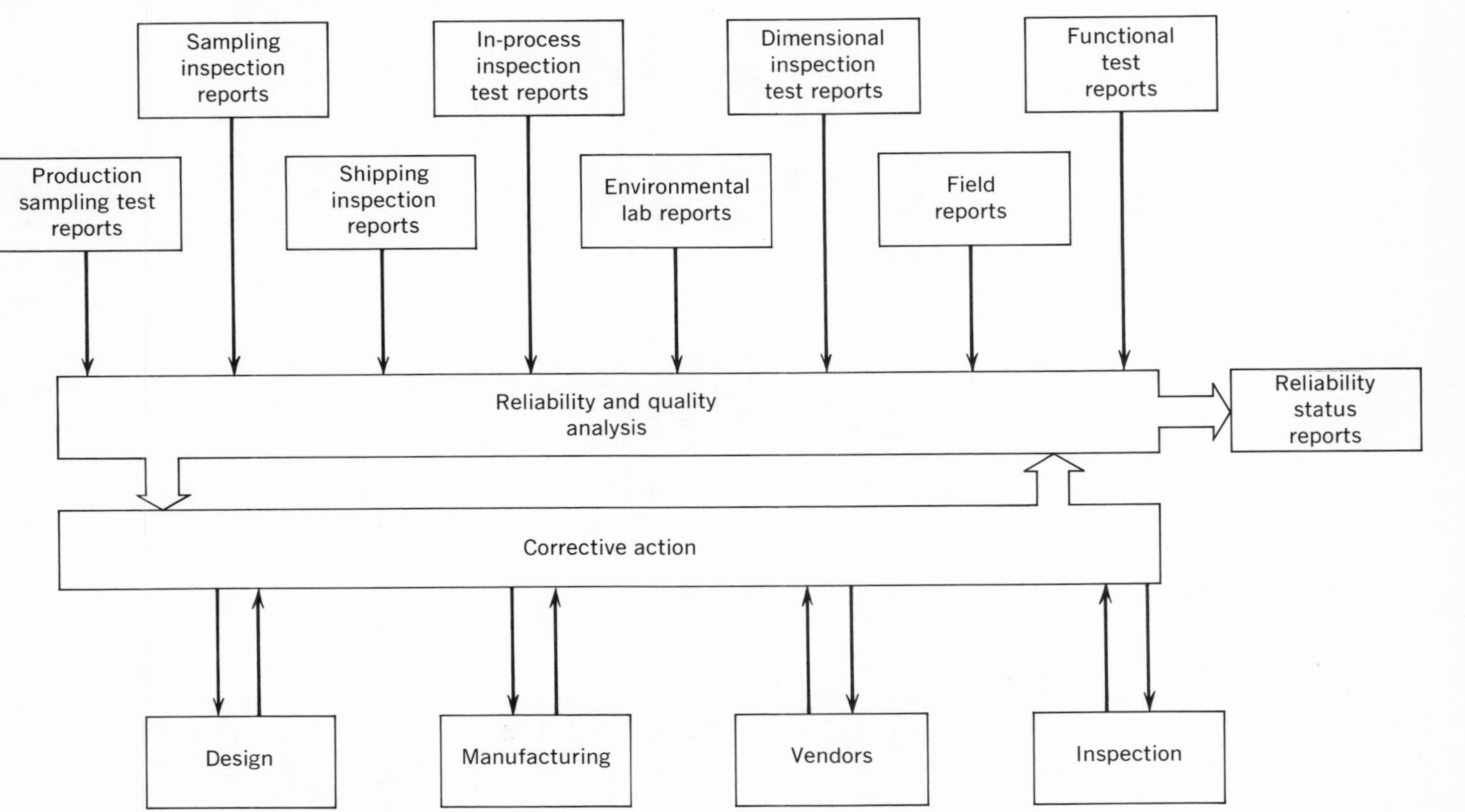

Figure 29-1 Quality and reliability data usage.

Thus, the scope of a quality data system may vary from a simple system covering in-process inspection data to a broad system covering all data applicable to the overall effectiveness of a product.[1]

29-2 Quality Data Systems: Design Principles

A data system must be tailored to meet the needs of each individual plant. The literature provides detailed examples of forms and procedures for various applications. However, the following principles are generally applicable:

1 Plan the system to receive information in almost any form imaginable. Although most of the information will be received on special forms, the system should make it possible to receive and process information by means of a telephone call, letters, or other media.
2 Provide flexibility for meeting new data needs. A cardinal example of this is the failure reporting form that must be revised periodically, because someone suddenly discovers a critical need for an additional item of information to be recorded (see section 29-6).
3 Provide for eliminating collection of data no longer useful, and for eliminating reports no longer needed. This requires a periodic audit of the use (or lack of use) of the data and reports.
4 Issue reports that are readable, timely, and have sufficient useful detail on problem areas to facilitate investigations and corrective action.
5 Prepare summary reports covering long periods of time to highlight potential problem areas and show progress on known problems.
6 Test the proposed system with a small pilot installation in one area.

Moore[2] discusses some of these principles in further detail.

29-3 Importance of Accurate Quality Records

Quality data are an input for decisions on product quality and also serve other purposes. The spectrum includes:

1 Measurement of reliability
2 Detection of potential quality problems
3 Analysis of chronic quality problems

[1] In the latter connection, see *Data Collection and Management Reports,* U.S. Air Force Systems Command, Weapon System Effectiveness Industry Advisory Committee, January, 1965.
[2] W. N. Moore, "Collection of Inspection Data through the Use of Computers," *Industrial Quality Control,* vol. 20, no. 12, p. 10, June, 1964.

4 Selection of components for future designs
5 Selection of vendors
6 Determination of piecework payments
7 Definition of quality guarantees
8 Assignment of machines to jobs
9 Release of designs to production
10 Determination of quality costs
11 Measurement of inspector accuracy

The impact on such a wide variety of company operations means that the data must be reasonably accurate.

Even when the initial planning is thorough, useful data may not result. For example, some companies find that a shockingly high percentage of forms for describing product failures are of little or no help in determining the cause of failure. Equally dramatic is the situation where defectives disappear and no rejection form is prepared, thereby making it impossible to obtain a true picture of the total quality loss. Getting good data is a never-ending task. Certainly, forms and data sheets must be carefully designed to promote ease of use and neatness. People must be *thoroughly* briefed on why the data are important and on how they should prepare the data to make them useful. People must be given the time and the means to prepare good data. Such actions are often good *intentions,* but the follow through may not exist. (A good way to find out is to periodically audit data generation methods and the resulting written results. Such an audit can pinpoint data system weaknesses.) Some companies have found it necessary to assign engineers to assembly, test, and inspection areas to make sure that the failure reports contain the detailed information required for later diagnosis of failure causes.

The problem of obtaining accurate data is even more acute in the field. Obtaining good information on *all* field failures is generally impossible. Only a portion of the failures are documented in any form. Descriptions of failures that are reported are often lacking in technical depth of information. A more realistic approach is to *sample* the failures and obtain completely detailed data on the sample. This will require joint planning with customers, training of personnel, and probably some form of compensation (see section 27-4).

29-4 Initial Planning of Paper Work Systems

The initial planning should start with the creation of a flowchart for paper work based on a flowchart for the design, fabrication, and use

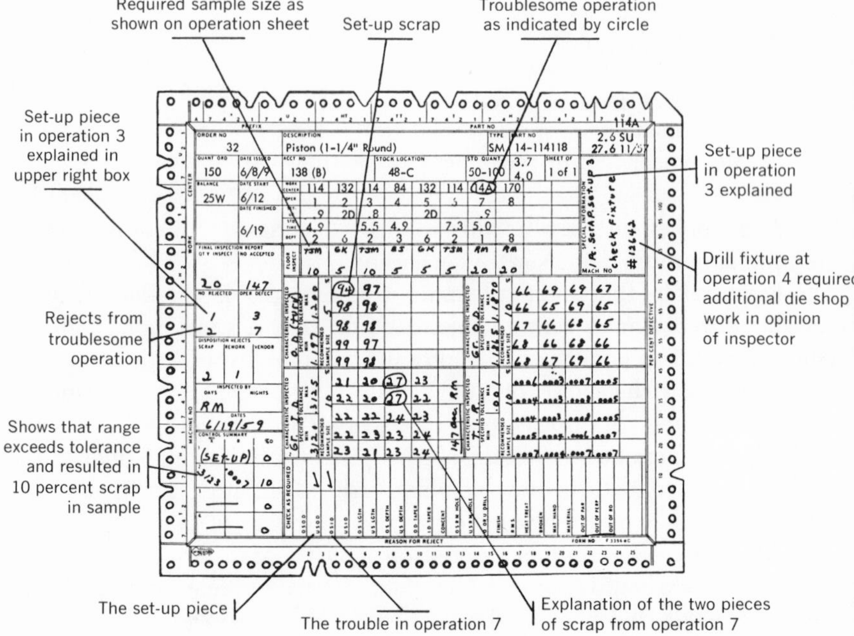

Figure 29-2 Example of edge-punched card. Source: Harley Wehrwein, "A Six Step Quality Control Program," *Industrial Quality Control,* p. 21, December, 1960.

of the product itself. Section 14-6 presents an example[3] of a flowchart for paper work. Flowcharts not only help in designing the paper work system, but they are also invaluable later on in explaining the reasoning and the importance of the various elements in the system to all functions concerned.

The flowchart provides an overall plan for the hard work of developing the detailed procedures and forms for executing the paper work system. What procedures are required depends on the product and the scope of that quality system (see section 29-1). These procedures spawn many forms for the recording of data.

The planner should keep in mind the *analysis* of data that will follow. (Care must be taken that the planning is not restricted to forms and analysis for *existing* data because real breakthroughs often require creating *new* data never before available.) Sometimes recording forms can be designed to make the analysis easier; e.g., if data can be recorded

[3] For another example, see J. M. Juran (ed.), *Quality Control Handbook (QCH)*, 2d ed., McGraw-Hill Book Company, New York, 1962, p. 21-3.

directly in frequency distribution form, the analysis is partly finished when the data are recorded.

As the data or analysis gets more complex, some form of mechanization may be needed. One modest means is the use of edge-punched cards (see Figure 29-2). These cards can be used both for the recording and for the analysis of data. The data recorded are coded by notching various holes. Sorting of many cards for various categories of data is done by inserting a long needle through the appropriate hole. The needle picks up the unnotched cards which present data for one category. The example in Figure 29-2 stores all information for one job. The method of analysis permits detection of problems for categories such as specific parts, assemblies, operations, machines, and inspection equipment. The card also can store information on machine capability, specifications, and vendor quality.

Another means of analysis is, of course, electronic data processing equipment. The far-reaching potential of such equipment for process control is discussed in section 23-7. The next section illustrates the application to the analysis of assembly data.

29-5 Example of the Use of Data Processing Equipment

The use of data processing equipment will be illustrated by an analysis of quality data from assembly operations. Figure 29-3 summarizes the steps:

1 Assign code numbers for all possible defects in the product inspection procedures. A "double coding" system minimizes the number of items in the defect code list and makes it easier to find code numbers which apply to particular defects found. In such a system, the first several digits identify the part involved (e.g., screws) and another set of digits identifies the trouble (e.g., loose).

2 Designate classes of defects such as major, minor, and incidental and assign demerit values, for example, 15, 5, and 1 respectively.

3 Record the assembly inspection results on a finished products analysis card. This card contains the product serial number and accompanies the unit as it proceeds to each inspection station. (Alternatively, the inspector could fill out a card for each defect noted.)

4 After the last inspection, keypunch (K.P.) the analysis cards and punch a card for each defect on the analysis card. (The first card contains a special punch in one column to provide a count of the total number of units inspected during the period.)

5 Accumulate the cards until the end of the period. The machine then sorts the detail cards, match-merges them with defect master cards,

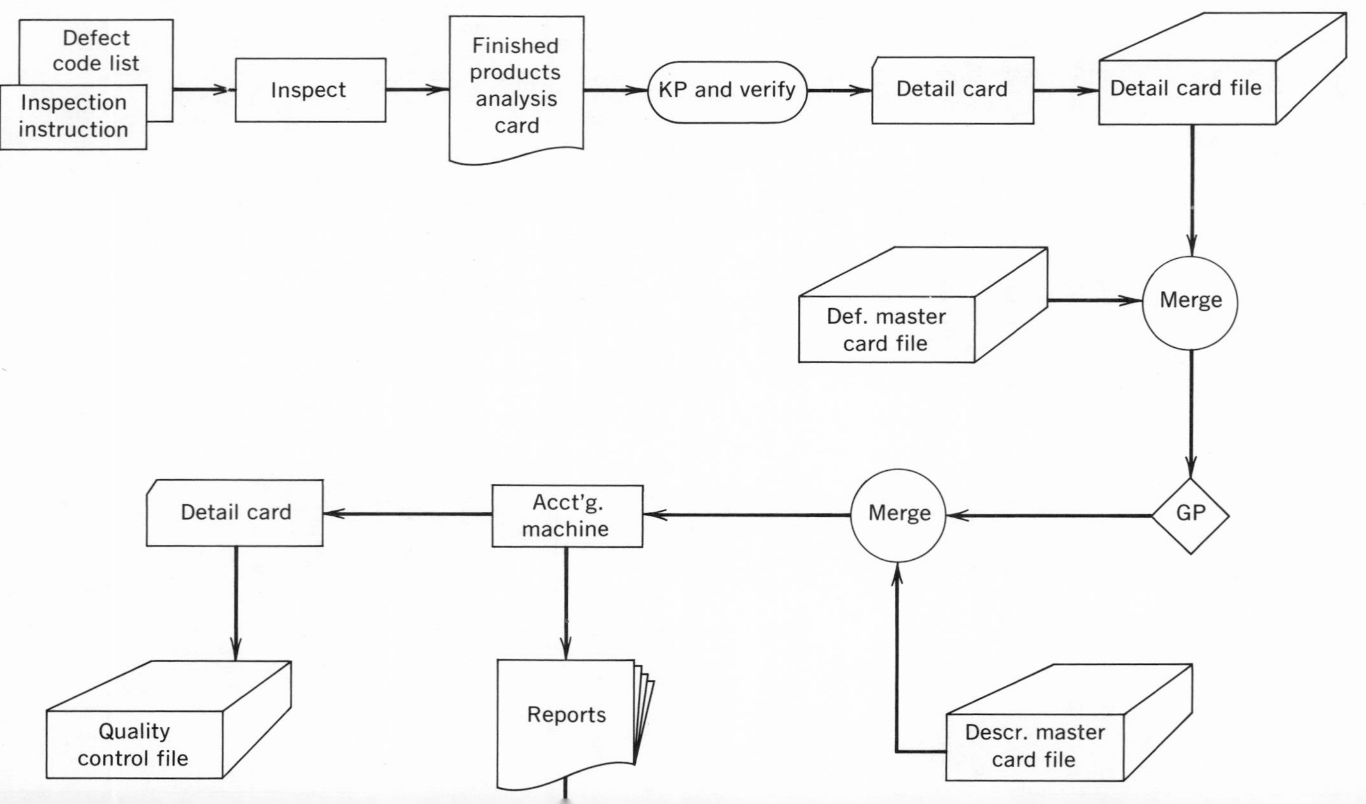
Defect code list
Inspection instruction
Inspect
Finished products analysis card
KP and verify
Detail card
Detail card file
Def. master card file
Merge
GP
Merge
Descr. master card file
Acct'g. machine
Detail card
Quality control file
Reports

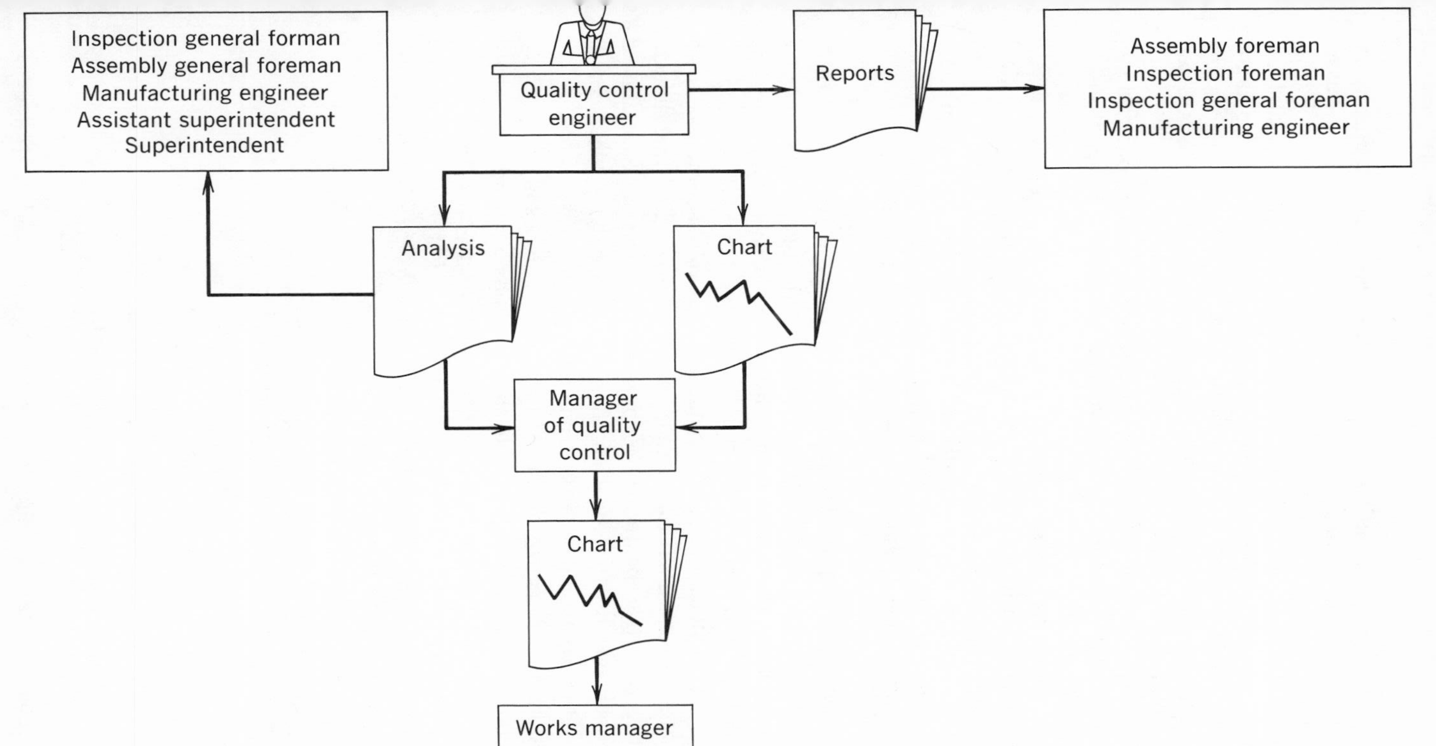

Figure 29-3 Flowchart for quality evaluation in assembly.

interisperse gang punches (G.P.) the class of defect and defect name, and groups them behind the master unit-description cards. An accounting machine then prepares reports by defect number, by department responsible, and by inspector number (Figure 29-4):

a The report by defect number contains product type, operation number, defect number, defect name, class of defect, trouble code, the total number of defects by class, the total units inspected, and the total defects found.

b The report by department responsible contains similar information, except that it indicates the department responsible, rather than the operation number.

c The report by inspector number contains the product type, operation number, inspector number, number of units inspected, and total number of defects by class.

6 Copies of the reports are distributed by the quality control engineer to the inspection foreman, the inspection general foreman, the manufacturing engineer, and the assembly department foremen.

Data processing equipment also makes it possible to prepare from the same punched cards additional reports by operation number, by trouble code, and by class of defect. Reports of this nature pinpoint the specific defects and areas which need corrective action. A comparison from period to period indicates whether remedial action has been taken and points up the need for more intensive effort if the same defects appear on the reports repeatedly. Comparisons of sets of data should include a check on the statistical significance, particularly when small amounts of data are involved. (See Chapter 11.)

The report by defect number (Figure 29-4) is used to prepare a demerit chart (Figure 29-5) which shows trends in product quality from period to period. Each point on the chart represents the average number of demerits per unit calculated by totaling the demerits and dividing by the total units inspected. For example, in December 466 units were inspected with these results:

Class	*Demerit Value*		*Number of Defects*		*Demerits*
I	15	×	2,594	=	38,910
II	5	×	3,817	=	19,085
III	1	×	1,718	=	1,718
			Total		59,713

$$\text{Average number of demerits per unit} = \frac{59{,}713}{466} = 128$$

This report goes to the manager of quality control and to the works manager.

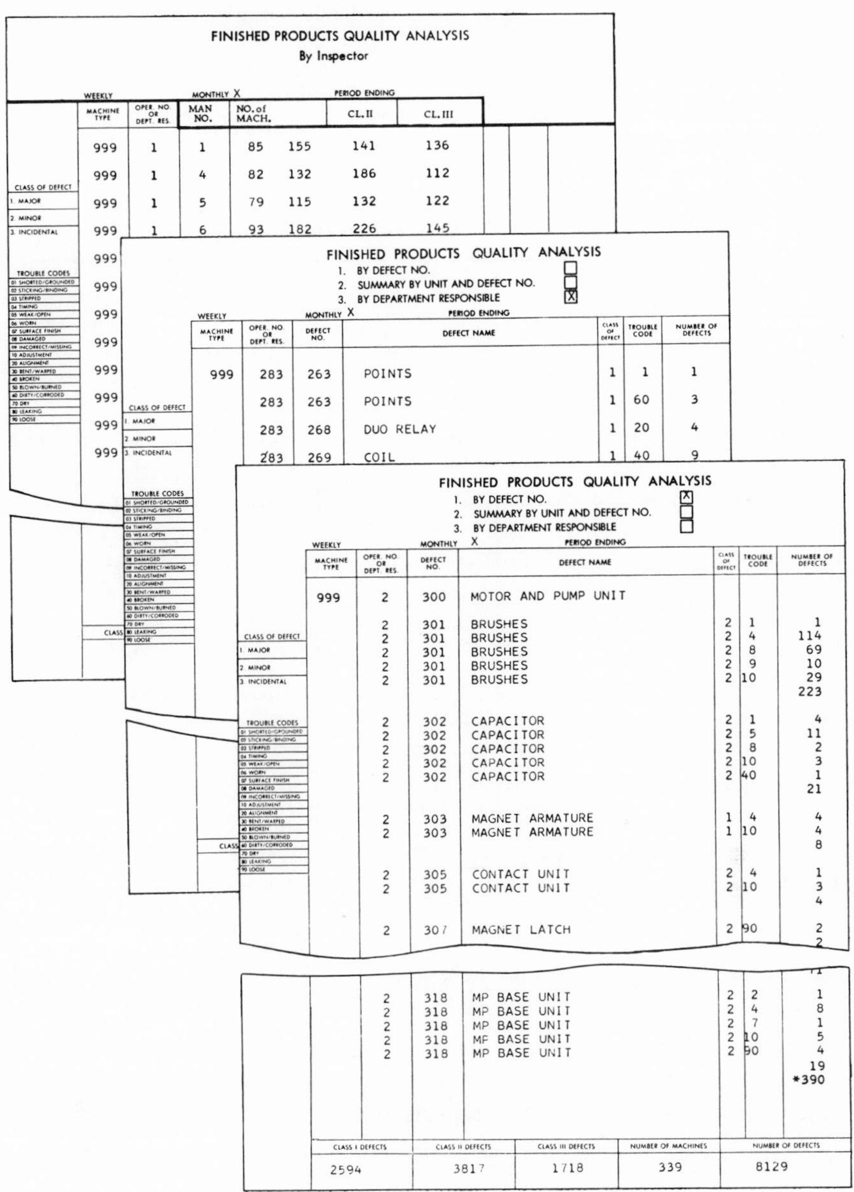

FINISHED PRODUCTS QUALITY ANALYSIS
By Inspector

WEEKLY MONTHLY X PERIOD ENDING

MACHINE TYPE	OPER. NO. OR DEPT. RES.	MAN NO.	NO. of MACH.		CL. II	CL. III
999	1	1	85	155	141	136
999	1	4	82	132	186	112
999	1	5	79	115	132	122
999	1	6	93	182	226	145
999						
999						
999						
999						
999						
999						
999						
999						

CLASS OF DEFECT
1. MAJOR
2. MINOR
3. INCIDENTAL

TROUBLE CODES
01 SHORTED/GROUNDED
02 STICKING/BINDING
03 STRIPPED
04 TIMING
05 WEAK/OPEN
06 WORN
07 SURFACE FINISH
08 DAMAGED
09 INCORRECT/MISSING
10 ADJUSTMENT
20 ALIGNMENT
30 BENT/WARPED
40 BROKEN
50 BLOWN/BURNED
60 DIRTY/CORRODED
70 DRY
80 LEAKING
90 LOOSE

FINISHED PRODUCTS QUALITY ANALYSIS
1. BY DEFECT NO. ☐
2. SUMMARY BY UNIT AND DEFECT NO. ☐
3. BY DEPARTMENT RESPONSIBLE ☒

WEEKLY MONTHLY X PERIOD ENDING

MACHINE TYPE	OPER. NO. OR DEPT. RES.	DEFECT NO.	DEFECT NAME	CLASS OF DEFECT	TROUBLE CODE	NUMBER OF DEFECTS
999	283	263	POINTS	1	1	1
	283	263	POINTS	1	60	3
	283	268	DUO RELAY	1	20	4
	283	269	COIL	1	40	9

FINISHED PRODUCTS QUALITY ANALYSIS
1. BY DEFECT NO. ☒
2. SUMMARY BY UNIT AND DEFECT NO. ☐
3. BY DEPARTMENT RESPONSIBLE ☐

WEEKLY MONTHLY X PERIOD ENDING

MACHINE TYPE	OPER. NO. OR DEPT. RES.	DEFECT NO.	DEFECT NAME	CLASS OF DEFECT	TROUBLE CODE	NUMBER OF DEFECTS
999	2	300	MOTOR AND PUMP UNIT			
	2	301	BRUSHES	2	1	1
	2	301	BRUSHES	2	4	114
	2	301	BRUSHES	2	8	69
	2	301	BRUSHES	2	9	10
	2	301	BRUSHES	2	10	29
						223
	2	302	CAPACITOR	2	1	4
	2	302	CAPACITOR	2	5	11
	2	302	CAPACITOR	2	8	2
	2	302	CAPACITOR	2	10	3
	2	302	CAPACITOR	2	40	1
						21
	2	303	MAGNET ARMATURE	1	4	4
	2	303	MAGNET ARMATURE	1	10	4
						8
	2	305	CONTACT UNIT	2	4	1
	2	305	CONTACT UNIT	2	10	3
						4
	2	307	MAGNET LATCH	2	90	2
						2
	2	318	MP BASE UNIT	2	2	1
	2	318	MP BASE UNIT	2	4	8
	2	318	MP BASE UNIT	2	7	1
	2	318	MF BASE UNIT	2	10	5
	2	318	MP BASE UNIT	2	90	4
						19
						*390

CLASS I DEFECTS	CLASS II DEFECTS	CLASS III DEFECTS	NUMBER OF MACHINES	NUMBER OF DEFECTS
2594	3817	1718	339	8129

Figure 29-4 Reports on finished products quality.

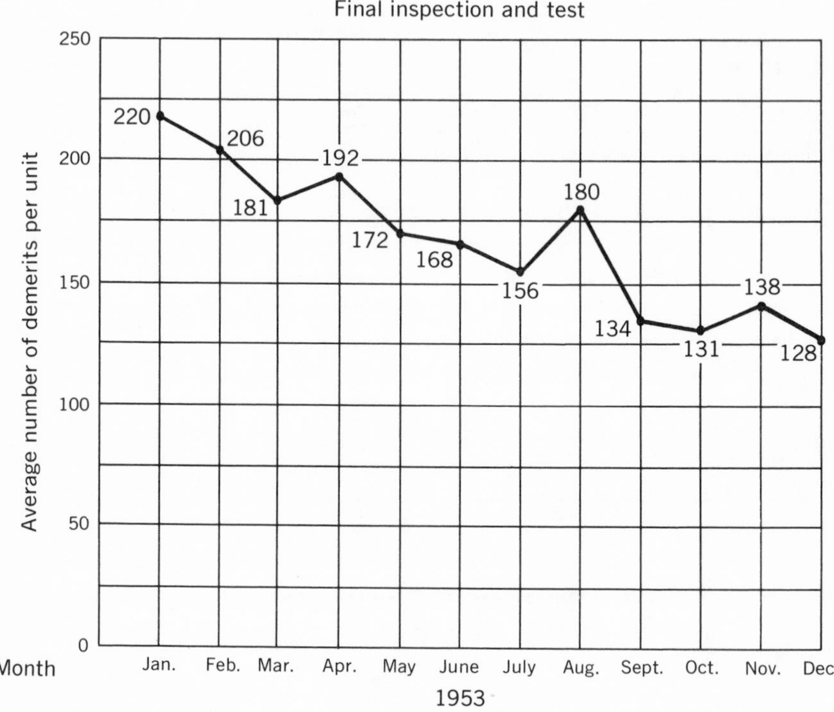

Figure 29-5 Chart of demerits per unit.

Another report which can be prepared is a list of the items appearing with the greatest frequency and of the items requiring corrective action most urgently. All the reports pinpoint specific defects, identify areas requiring action, and show the effects of remedial action.

29-6 Reliability Data Systems

A reliability data system usually connotes a system for collecting data on failures and operating time on complex products. There are three basic sources of this type of data—in-plant testing, field testing, and operation by the ultimate user.

The data system is initiated by a failure report form. Many varieties[4] of such forms are in existence but they should be tailor-made for each company. This form includes information on:

1 Complete identification of the failed item
2 Operating time
3 Description of the trouble
4 Immediate action taken and recommended further action
5 Time to restore the system to operational status
6 Findings of an analysis of the failure
7 Conditions found during repair
8 Parts replaced
9 Repair time

A useful feature of the form is the inclusion of numbered blank spaces in anticipation of additional data needs in the future. As the need arises, the spaces can be easily labeled. This avoids a complete redesign of the form every time an additional item of information must be collected.

The information on failure report forms is the raw material for analysis for two distinct purposes: short-range corrective action and longer-range evaluation of components and systems. In the short range, failures must be analyzed for criticality and repetitiveness, decisions made on failures requiring action, failure causes diagnosed, responsibility defined, and follow-up made to verify that the action worked. For some products, this is formalized into a "corrective action system" with an organization structure spanning several functional areas, a paper work system, and heavy use of data processing equipment for quantitative recording and analysis of failures (see section 9-12). In the longer range, failure reports and operating time information are the basis for the measurement of reliability. Overall reliability can be measured and compared to requirements. In addition, reliability levels achieved and failure modes discovered can be documented at the component level for use in the design of new systems.

29-7 Purposes of a Quality Control Manual

A quality control manual is a document containing company policies and procedures affecting product quality. It serves as:

1 A reference for the policies and procedures as well as the reasons behind them. The manual provides proof that policies and procedures have been thought out and documents the reasoning to help those who must execute the plans.

[4] For an example, see *QCH*, 1962, p. 20-34.

2 A textbook for training. The widest training use is for the inspection and quality control personnel, but the training extends also to production supervisors, engineering personnel, and others.
3 A precedent for future decisions. The manual codifies past practice and agreements; e.g., quality standards can appear in the manual.
4 An aid to continuity of operations despite employee turnover. Without a manual, personnel changes can result in a change of practice, sometimes drastic. The manual helps to stabilize practice and to conduct operations based on "laws, not men."
5 A reference base against which current practice can be audited.

Some companies have required their vendors to prepare quality control manuals as an aid to vendor relations. The results have been mixed. In many cases, the vendor has benefited greatly from thinking through and documenting his procedures. In other cases, unnecessary formalization and documentation of practice has resulted in more expense than benefit from the manual. Blanket requirements that are arbitrarily made on all products and vendors are inviting unnecessary expense in the form of documentation that is used only to impress the organization that required it.

29-8 Evolution of the Manual

The manual has much in common with a code of laws. For example, the motor vehicle code came into existence because there was a need for it. The horse and buggy days required few traffic rules, but the greatly increased number of vehicles, their greater speed, their heavy loads, their interstate range, etc., generated the traffic codes of today. In like manner, there is little need for a quality control manual in circumstances of few quality characteristics and low precision, but as these multiply, the manual begins to evolve. The idea of a manual is born when someone concludes that an orderly means must be created out of the assortment of memoranda, notices, and other bits of written and unwritten practice. The size of the company is not the main factor in creating the need for a manual. The extent and complexity of the quality practices determine the need. The use of manuals in other areas (e.g., drafting standards in engineering or an employee relations handbook in personnel) will influence the ease with which the quality control manual concept is accepted.

29-9 Development of the Manual

The manual may have to start in a modest fashion using existing policies, procedures, instructions, and rules as building blocks. The steps necessary to incorporate an idea into the manual are

1 Reduction to writing of the idea
2 Review by the originator and procedures writer and preparation of a formal policy or procedure for the manual
3 Approval
4 Issuance
5 Maintenance

The review (step 2) provides technical ideas from the procedures writer and fosters a sense of participation for the originator. The procedure is usually written by a planning section in the quality control department (in conjunction with a company systems and procedures department if one exists). The procedure usually contains:

1 Title and identifying number
2 Effective date (and expiration date, if any)
3 Section replaced in the current manual
4 Table of contents
5 Purpose or objective
6 Scope
7 Summary (in flowchart or narrative form)
8 Detailed procedure including forms
9 Glossary of terms
10 Authorizing signature

All departments affected by a procedure should formally approve it before it is issued. Securing approval from the various departments takes time, but it may generate important ideas. Equally important is the effect of participation—a better chance that the procedure will really be followed.

Responsibility for issuance and maintenance of the manual should be assigned to one and only one department, e.g., the quality control department or the systems and procedures department. This avoids confusion due to conflicting procedures, failure of maintenance, etc. The responsibility includes obtaining necessary approvals, duplicating and distributing approved drafts, and issuing emergency procedures.

Maintenance of the procedures is important, because nothing can render the manual ineffective so quickly as the accumulation of obsolete procedures. Maintenance includes a periodic review of the manual books to see if the latest edition of all procedures is included. The review also includes an audit of actual practice versus the manual (see section 29-11).

The manual should be printed and bound for durability, neatness, and easy maintenance. Copies should be numbered, and a list kept showing to whom the books were issued so that revised procedures can be easily distributed.

29-10 Content of the Manual

What policies and procedures are necessary must be decided by each company. The following list is *not* meant as complete but shows areas[5] that are often included:

1 Information about the manual itself
2 Sections dealing with administration of the quality function
3 Sections dealing with quality and inspection planning
4 Sections dealing with design
5 Sections dealing with vendor relations
6 Sections dealing with process control
7 Sections dealing with finished goods control
8 Sections dealing with field performance
9 Sections dealing with general test methods
10 Sections dealing with measurement
11 Sections dealing with specific products
12 Sections dealing with specific processes
13 Sections dealing with personnel
14 Sections dealing with government contract procedures
15 Sections dealing with control and costs of the quality function

In deciding on the content of the manual, thought should be given to using a policy in place of a procedure. Sometimes it is *not* desirable to prepare a detailed procedure on accomplishing a task, but it is important to define in writing a policy which (1) states that a certain task must be conducted, and (2) provides general *guidelines* on how the task should be conducted, without defining it to the detail of a procedure. This is the case when the task involved has one or more detailed elements that are continually changing due to normal company operations.

29-11 Audit of Actual Practice Versus Manual

Supervisors are responsible for implementing procedures but there are forces acting that often result in the need for independent audit of actual practice versus the manual. The supervisor does not have the

[5] This is a revision of the list in *QCH*, 1962, pp. 5-5 to 5-7. The reference includes a list of detailed items.

time for organized follow-up on adherence to procedures and naturally relies on his men to follow procedures. Sometimes, conflicting pressures force the supervisor to knowingly deviate from approved procedures.

For these and other reasons, a periodic independent audit of practice versus the manual is often useful. The results of an audit are varied. Sometimes, the formal procedure is shown to be inadequate or even wrong. Sometimes, the procedure is fine but personnel are unable to execute it because of lack of knowledge, lack of adequate equipment, or other reasons. Sometimes, the procedure has been revised but the new edition has not yet been distributed. In some companies, these audits are a part of the broader quality audit or survey (see section 28-4).

PROBLEMS

1 For any product or service acceptable to the instructor, propose a list of subjects for procedures in a quality control manual.

2 For any of the following, obtain the necessary information and draw up a flowchart for the data collection and analysis: (*a*) an operation acceptable to the instructor, (*b*) final inspection results at a plant, (*c*) goods returned to a plant, (*d*) traffic fines, (*e*) complaints at a department store, (*f*) automobile accident insurance claims, (*g*) performance deficiences by an athletic team, and (*h*) loss of utility service to homes.

Make recommendations or changes or additions.

3 Prepare a list of the *specific* items of information to be included on a form for reporting quality deficiencies for one of the following: (*a*) a product acceptable to the instructor, (*b*) a residential home, (*c*) airline service, (*d*) service in a department store, and (*e*) a passenger automobile.

30 Diagnosis Techniques for Quality Improvement

30-1 The Breakthrough Sequence in Quality Improvement

The term "quality improvement" may have several different meanings:

1 Restoring out-of-control processes to a state of control (restoring the status quo). Much of this was discussed in Chapters 3 and 23.
2 Achieving new records of performance through use of the breakthrough sequence (as discussed in Chapter 3).
3 Planning new processes and products in a way which will launch them on levels of performance superior to those attained by present processes and products. (Such quality planning is discussed in Chapters 9 and 14.)

This chapter is mainly concerned with (2) above. Chapter 3 stated the steps in the breakthrough sequence as (1) breakthrough in attitudes, (2) discovery of the vital few projects, (3) organizing for breakthrough in knowledge, (4) creation of the steering arm, (5) creation of the diagnostic arm, (6) diagnosis, (7) breakthrough in cultural pattern, (8) breakthrough in performance, and (9) transition to the new level. This chapter will concentrate on the diagnosis phase, and many of the techniques discussed apply to all three cases stated above.

30-2 Quality Symptoms, Problems, Projects, Causes, and Remedies

The invariable sequence of events for breakthrough (Chapter 3) includes the process of diagnosis. This is a subsequence which begins with symptoms and ends with remedies.

A quality *symptom* is any observed evidence of quality failure: a defect causing unfitness for use, a deviation from specification, an error in procedure, etc. These symptoms signal the presence of varying degrees of "problems" faced by the organization.

A quality *problem* is a potential task resulting from the existence of quality symptoms. The problem may be so minor that nothing will be done about it—it will be endured. It may be left to voluntary, informal solution by someone in the hierarchy. (This is usually the case in restoring the status quo.) Or it may be chosen to become the basis of an organized, formal problem solving activity—a project.

A quality *project* is a problem selected for solution through a formally organized problem solving procedure.

A *cause* is the reason for the existence of symptoms and is discovered by diagnosis, i.e., examination of the theories of causation and discovery of the real cause(s). Diagnosis is the process of discovering the true origin of symptoms.

A *remedy* is a means discovered for eliminating or neutralizing the cause of symptoms. Discovery of remedies involves a second level of diagnosis.

Companies are faced with myriads of quality symptoms, and thereby with numerous quality problems to be solved by people at all levels. Selection of formal projects for solution through the breakthrough process is done via the Pareto principle of the vital few and the trivial many (Chapter 4). The criterion for creating a project is "return on investment," i.e., the belief that the effort expended in diagnosis and remedy will be more than recovered by the improvement in results.

(A further concept in diagnosis techniques is that of the quality *syndrome,* which may be defined as a group of symptoms so uniquely interrelated as to point convincingly to a common cause. This concept is used in medical research, but as yet there has been limited use of it in diagnosis for quality improvement.)

30-3 General Approach to Management Controllable Problems

In diagnosing for causes, a main fork in the road is whether the defect is operator controllable or management controllable (see Chapter 21). Reviewing, a problem is operator controllable if means have been provided so that the operator knows what he is supposed to do, knows what he is doing, and can regulate the process. If all these conditions are present, the problem is operator controllable and the approach to solving the problem is discussed in Chapter 22. If any of the conditions are missing, the problem is management controllable and the approach is discussed in this chapter.

Management controllable problems can be studied by a structured pattern of diagnosis shown in Table 30-1. A dissectable characteristic is one that can be measured between manufacturing steps. For example, a shaft diameter measured with a micrometer during a series of turning and grinding operations illustrates a dissectable characteristic which is measured by variables. The number of chips on a lens during a series of operations illustrates a dissectable characteristic measured by attributes. A nondissectable characteristic is one that can be measured only at the completion of all the operations which contribute to the quality characteristic, e.g., the electrical output of a vacuum tube.

Table 30-1 PATTERN OF DIAGNOSIS FOR MANAGEMENT CONTROLLABLE DEFECTS

Dissectable characteristics—variables (section 30-4)
Process capability analysis (section 30-5)
Dissection of streams (section 30-6)
Graphic analysis (section 30-10)
Dissectable characteristics—attributes
Analysis of past data (section 30-8)
Dissection of streams (section 30-7)
Defect concentration studies (section 30-9)
Nondissectable characteristics (section 30-11)
Correlation studies
Diagnosis through experiments

30-4 Diagnosis for Dissectable Characteristics

Chapter 15 introduced the concept of process capability as a tool in the *planning* phase of manufacturing. The aim there was to *prevent* quality problems by assigning products to processes which are capable of meeting the tolerances. In this chapter, process capability analysis is discussed as a means of correcting an existing problem. In all cases, the "process" is the combination of machine, material, method, and operator. All these contribute to the final variation in the product.

Process capability analysis can involve a variety of techniques[1] suitable for different problems. The simplest of all is to take a sample of consecutive pieces from a process, measure the characteristic under study, and plot the individual measurements in chronological order on a chart containing the tolerance limits. The example on the watch part

[1] See J. M. Juran (ed.), *Quality Control Handbook (QCH)*, 2d ed., McGraw-Hill Book Company, New York, 1962, sec. 11.

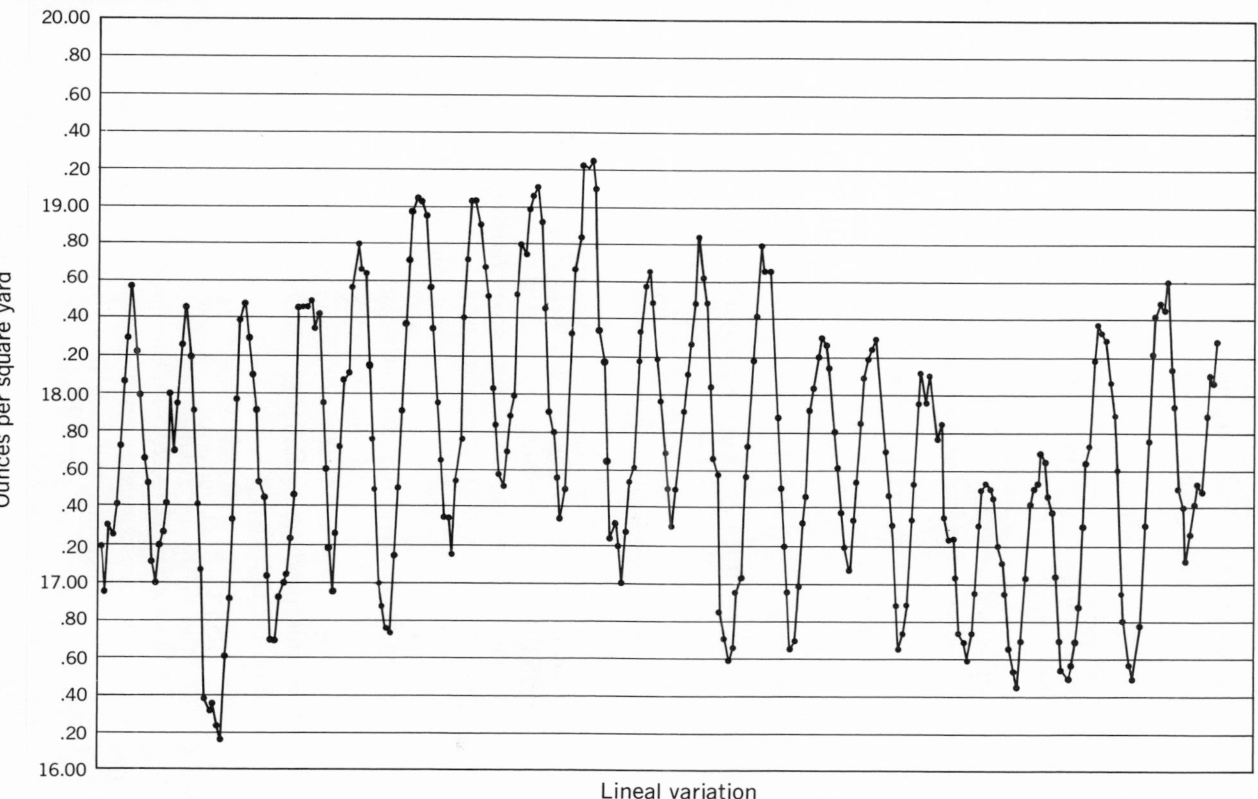

Figure 30-1 Process capability study—calendering rubber sheets.

in section 15-2 illustrates this and should be reviewed. Figure 30-1 shows another example[2] of how this simple charting technique vividly portrays cycles, trends, and piece-to-piece variation in a process. This simple chart can show:

1 The ability of a process to meet tolerances
2 The rate of change in a characteristic due to tool wear or some other "natural" cause
3 Where to set the process in order to get the longest run of good product without resetting
4 How often to inspect and adjust the process

This chart is useful in practice because everyone can understand it. Even if additional techniques are needed, the simple chart will at least help everyone to understand the problem.

30-5 Elements of a Complete Process Capability Analysis

A more complete analysis consists of five elements:

1 The specification tolerance
2 Determination of whether the process average is "centered" midway between the tolerance limits
3 Measurement of the inherent (piece-to-piece) variability of the process
4 Measurement of the actual variability over a period of time
5 Causes of the difference between inherent and actual variability

With the above information, alternative courses of action can be considered (see Table 30-2) to eliminate the defective work. All these alternatives have economic implications. Widening a tolerance may reduce the value of a product. Changing a process may require a complete overhaul of equipment.

[2] *QCH,* 1962, pp. 11-20 to 11-23, summarizes the study. The original reference is Anthony Oladko, "Developing a Calendergraph," *Rubber Age,* September, 1949. This provides a good discussion of process capability analysis in terms of advantages that would apply to any industry.

Table 30-2 COURSES OF ACTION

Process Capable of Meeting Tolerances	*Process Not Capable of Meeting Tolerances*
1 Recenter process aim	1 Widen tolerances
	2 Make a basic process change
	3 Sort defectives

An effective way of making the analysis is by means of a control chart and a frequency distribution (see Chapter 15). A number of samples are taken over a period of time. Each sample consists of consecutively made pieces (to use as a basis for measuring piece-to-piece variability.) The analysis proceeds:

1 Calculate the average $\bar{X}$ and range R of each sample.
2 Calculate the grand average $\bar{\bar{X}}$. This measures the centering (aim) of the process.
3 Calculate control limits and plot average and range control charts. This measures the stability of the process, i.e., the extent to which it changes with time.
4 Calculate process capability as ± 3 standard deviations based on the *within*-sample variation. As shown in Chapter 15, this is $3\bar{R}/d_2$. This measures the piece-to-piece variability of the process.
5 Record all the individual values into a frequency distribution and histogram.
6 Calculate the actual variability or spread as ± 3 standard deviations of the total data in the frequency distribution.

Figure 30-2 shows some typical results. Note that the control limits on sample averages are narrower than the tolerance limits on individual values because averages vary less than individuals. If the process is found to be stable (as indicated by the control chart being within control in the top half of Figure 30-2), the spread and capability will be approximately equal. (Spread is the variability among pieces produced over a period of time, i.e., 6σ on a histogram.) If the process is unstable (out-of-control situations in the bottom half of Figure 30-2), the spread will be greater than the capability. If the *capability* (piece-to-piece variability) is within tolerance limits, then finding and eliminating the causes of the out-of-control points will remove the time-to-time variability, reduce the spread, and solve the problem. However, as the last figure shows, if the capability is outside the tolerance limits, then only a basic process change or widening the tolerances will completely solve the problem. Thus, the combination of control charts of average and range, plot of individuals, and histograms have complemented each other in the diagnosis of causes and remedies.[3]

30-6 Dissection into Streams: Variables

A "lot" of product as received for inspection is usually quite a mixture (see Figure 30-3). It is composed of several main "streams" originating

[3] See *QCH,* 1962, pp. 11-26 and 11-27, for further elaboration.

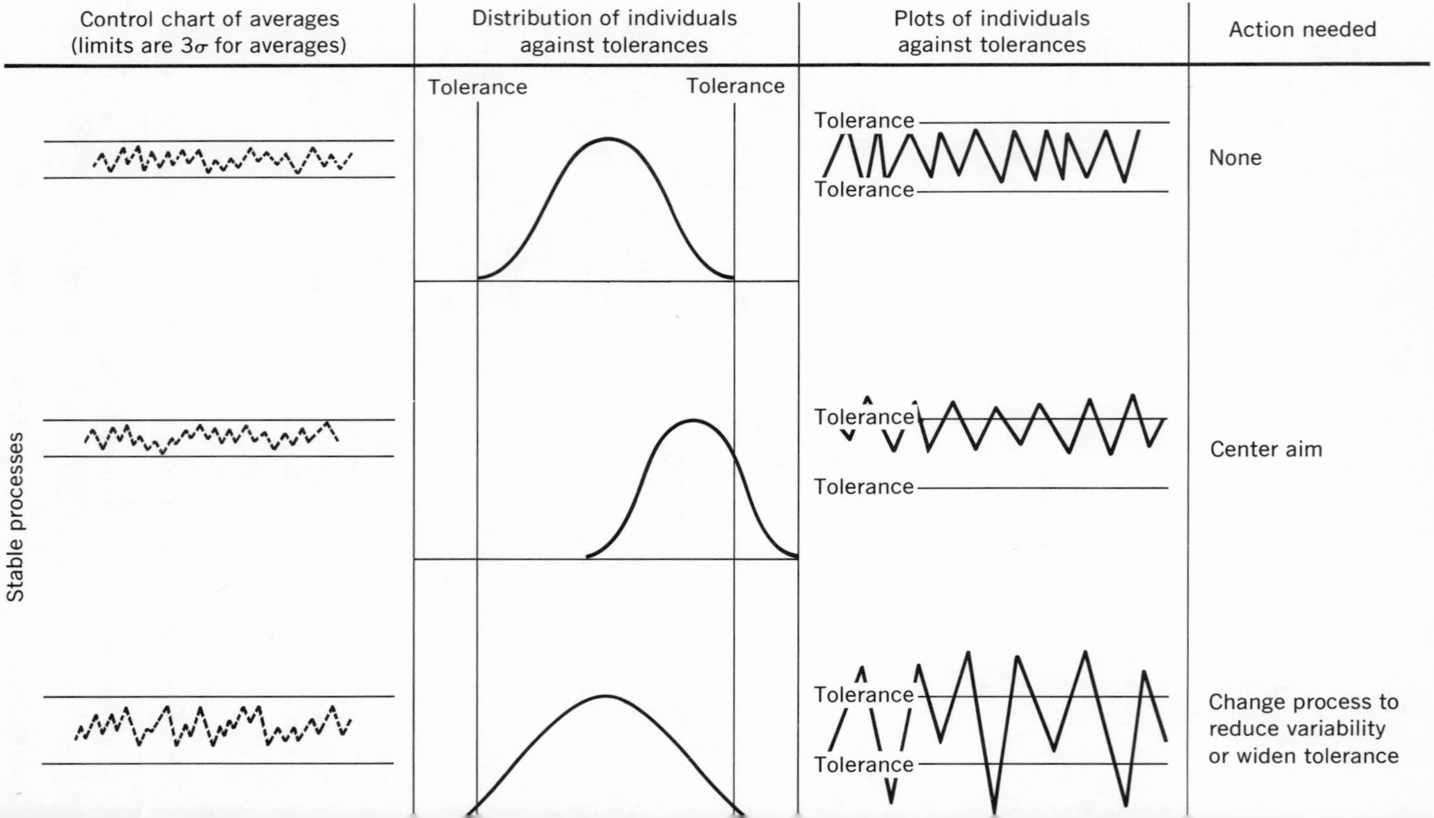

	Control chart of averages (limits are 3σ for averages)	Distribution of individuals against tolerances	Plots of individuals against tolerances	Action needed
Stable processes		Tolerance / Tolerance	Tolerance / Tolerance	None
			Tolerance / Tolerance	Center aim
			Tolerance / Tolerance	Change process to reduce variability or widen tolerance

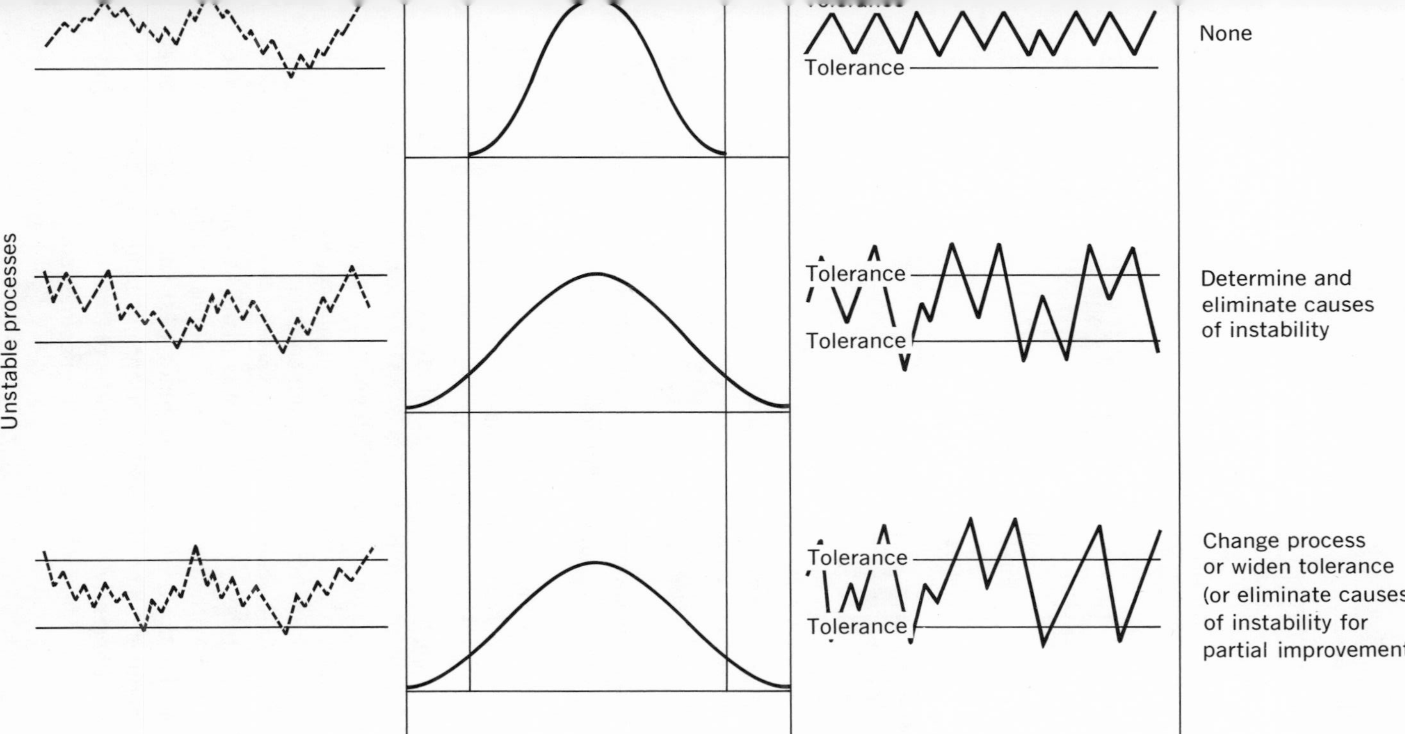

Figure 30-2 Results of process capability analysis.

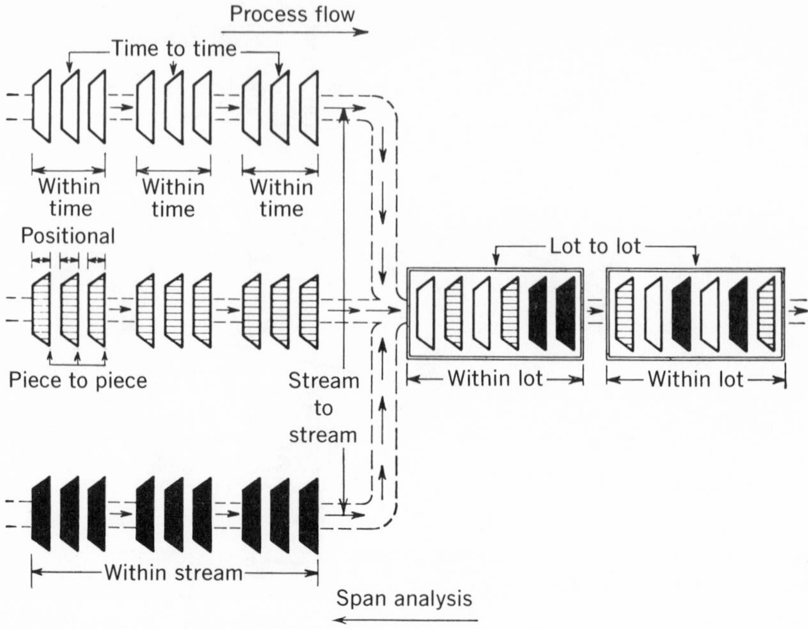

Figure 30-3 Span capability analysis for discrete piece process.

from different machines, operators, material batches, etc. The qualities of each stream may be strongly influenced by differences in these multiple origins.

Each stream, in turn, may exhibit time-to-time differences. A solution progressively becomes more dilute (or concentrated); the tool wears; the operator tires. Within any short time interval there are piece-to-piece or specimen-to-specimen differences because of local variations in machines, materials, and men. There are even within-piece differences: eccentricity of a shaft; polish on a finished surface.

In one example,[4] there was disagreement among various supervisors as to the cause of high rejections for dimensional control (length) of certain electrical components: unrealistic specifications, poor workmanship, incapable machines, etc. A study was made to compare the product histogram data with the product specifications. Figure 30-4 shows the results for the three identical machines ("streams") used to meet production schedules. The bimodal distributions made it clear that the machines were inherently capable. Study of tool change procedures helped to explain the bimodal distributions encountered.

[4] See *QCH*, 1962, p. 25-5, for another example.

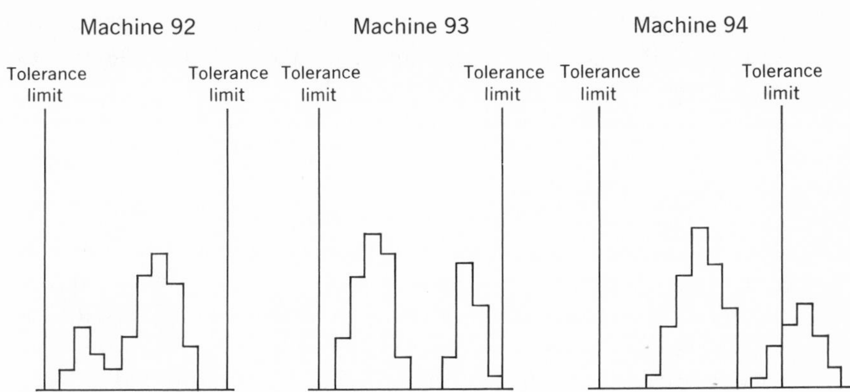

Figure 30-4 Performance of three identical machines making the same part.

30-7 Dissection into Streams: Attributes

The dissection of streams applies to problems of attributes as well as variables. For example,[5] razor "caps" exhibited a high rejection rate due to unpolished areas missed by the polishing wheel. The unpolished areas seemed to occur in random patterns. The caps were conveyed to the polishing wheel in an endless chain of 138 holding fixtures (that is, 138 streams). The dissection of streams was made by an experiment to test defect concentration by fixture number. Successive pieces from the same fixture were identified: i.e., pieces no. 1, 139, 277, 415, etc., all were processed in fixture no. 1. The previously "random" patterns

[5] See *QCH*, 1962, pp. 11-41 to 11-42, for other examples.

Figure 30-5 Mysterious polishing patterns were easily explained.

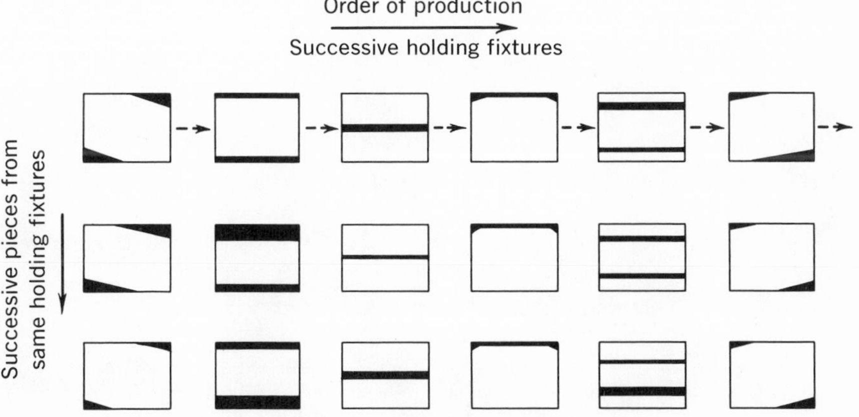

emerged as a stable pattern of variation—each different pattern was produced by a different holding fixture (see Figure 30-5). Regrinding and realignment of the fixtures solved the problem.

30-8 Analysis of Past Data

In the early stages of diagnosis, data from the recent past can help in locating overall defect concentrations and in pointing up the further study needed. Summarizations of data to determine the "vital few" have already been discussed (see section 5-5).

An example of the use of existing data concerns pitted castings in a foundry. It was theorized that a smaller "choke" cross section of the pattern prevented loose lumps of sand from being swept into the casting by the onrushing metal.

Past data could help to pretest this theory without conducting any experiments. When choke dimension was plotted against percent scrap (Figure 30-6), a correlation was evident. Subsequent experimentation revealed an optimum choke dimension of 0.050 inch.

The point is that a theory was tested and a correlation established, *all without experiment*. Experiments were needed to find the optimum, but the total amount of experimentation was small because of the use of past data.

Sometimes listing types of product, *in order of percent of defective*,

Figure 30-6 Pit scrap versus choke thickness.

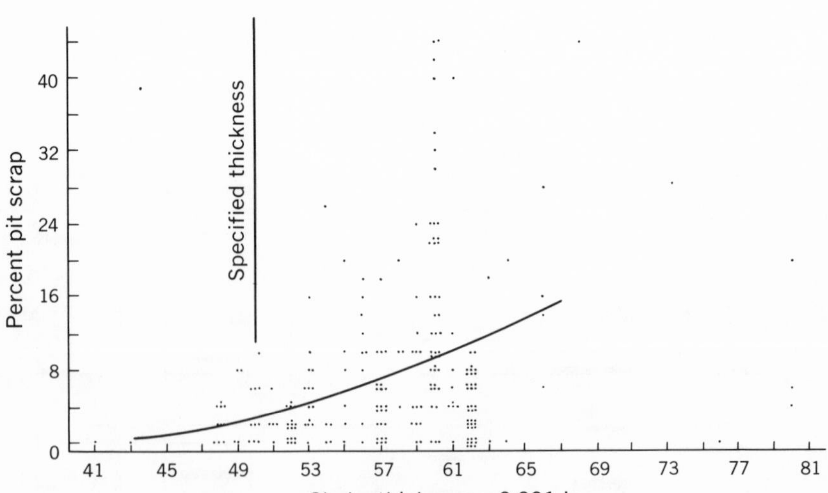

will disclose obvious correlations. For example, an automobile parts plant averaged 22.1 percent defective for an important characteristic. Table 30-3 lists the types of product involved in order of percent defective. Someone noted that types of product containing swaged tubes seemed to be high on the list, and so the list was marked to show which types were swaged. The result was dramatic—the worst seven types were all swaged, the best seven were all unswaged. Investigation

Table 30-3 **Rankings of Percent Defective by Product Type**

Type	*Percent Defective*	*Swaged (Marked X)*	*Type*	*Percent Defective*	*Swaged (Marked X)*
A	52.3	X	M	19.2	X
B	36.7	X	N	18.0	X
C	30.8	X	O	17.3	
D	29.9	X	P	16.9	X
E	25.3	X	Q	15.8	
F	23.3	X	R	15.3	
G	23.1	X	S	14.9	
H	22.5		T	14.7	
I	21.8	X	U	14.2	
J	21.7	X	V	13.5	
K	20.7	X	W	12.3	
L	20.3				

showed that the swaging operation introduced serious variations partly because the specifications did not limit a certain vital coaxial relationship between the unswaged and the swaged parts of the tubes.

30-9 Defect Concentration Studies

Dissection by streams (sections 30-6 and 30-7) can reveal where defects concentrate. Defect concentrations may also occur on the same physical area or location on a succession of product pieces. When this is the case, the remedy is usually obvious; e.g., the tool designer recognizes the fault in the die when all the cracks are in the same location on the punch press part. However, a defect concentration may elude detection because it is intermittent, or unexpected, or simply clouded by the presence of product from other streams by the time the inspection is made. Special data may be needed.

Collection of the basic data is simple. The inspector is provided with a sketch of the product and marks the location of the defect on the sketch. The data are summarized in one diagram or on a three-dimensional model of the product.

For example,[6] sport guns required extensive refinishing operations due to handling damage on the polished gun stocks. The defects on 100 damaged stocks were summarized by painting dots on one unfinished stock used as a three-dimensional sheet. Each dot represented one defect and was placed on the "data sheet" in the same position as the defect on the damaged stock. On completion, the "data sheet" exhibited over 200 dots (some stocks were damaged in more than one place). However, the dots were highly concentrated in just two areas—the resting points of the guns in their mobile storage racks. Investigation showed that in many racks the protective leather pads had fallen off without being replaced. The remedy was simple once the defect concentration was revealed.

Table 30-4 **Rubber Boots. Recapitulation of Leak Test Inspection**

Style of Boot	1	2	3	4	5	6	*Total*	*Percent of Product*	*Percent of Leaks*
Boots tested	120	4,060	2,242	728	1,485	1,009	9,653	100.0	
Number of leaks at location:									
A	17	402	241	79	306	309	1,354	14.0	40.6
B	4	24	158	23	48	94	551	5.7	16.6
C	2	65	67	6	107	158	405	4.2	12.2
D	1	102	86	0	0	0	189	2.0	5.8
E	0	20	42	0	0	2	64	0.7	2.0
F	0	74	39	81	114	90	398	4.1	12.0
G	0	73	34	71	101	82	361	3.7	10.8
H	·0	0	12	0	0	0	12	0	
Total defectives	24	960	679	260	676	735	3,334	34.4	100.0
Percent defective	20.0	23.6	30.3	35.7	45.5	72.8	34.4		

A manufacturer of rubber boots had the problem of 34 percent leaky boots. Table 30-4 shows the data collected. Almost half the defects were concentrated in one of eight locations on the boots and this concentration cut across all six styles of boots.

30-10 Graphic Analysis

There is much opportunity for use of graphic analysis to aid and to dramatize study of variability. For example, Figure 30-7 (a "Multi-Vari" chart) compares three processes which are incapable for three different reasons:

[6] See *QCH*, 1962, pp. 11-43 to 11-46, for additional examples.

Left-hand case I might be the result of out-of-roundness of diameter; the middle case II might be the result of an operator controllable piece-to-piece variation (tightness of clamping); the right-hand case III, the result of supply material batch differences.

This example shows just one of the many[7] ingenious graphical approaches that have been developed.

30-11 Diagnosis for Nondissectable Characteristics

Diagnosis for nondissectable characteristics is complicated by the fact that the characteristic can only be measured after the completion of the operations that determine the characteristic. Sometimes use can be made of the diagnostic techniques for dissectable characteristics, but a more fruitful approach is to convert the nondissectable characteristic to a dissectable one by:

1 Measuring related properties during the progression of the product, e.g., viscosity as a measure of polymerization, hardness as a measure of tensile strength.

2 Creating a new measuring instrument, e.g., the air-flow instrument for measuring uniformity of textile sliver.

3 Using existing instruments in new places. In a paper making process, the material batches were cooked in a sealed digester with a fixed cooking cycle. A breakthrough was made when a valve opening was cut into the digester to permit drawing off samples for test. By

[7] See *QCH*, 1962, pp. 11-32 to 11-39, for other examples.

Figure 30-7 The Multi-Vari chart.

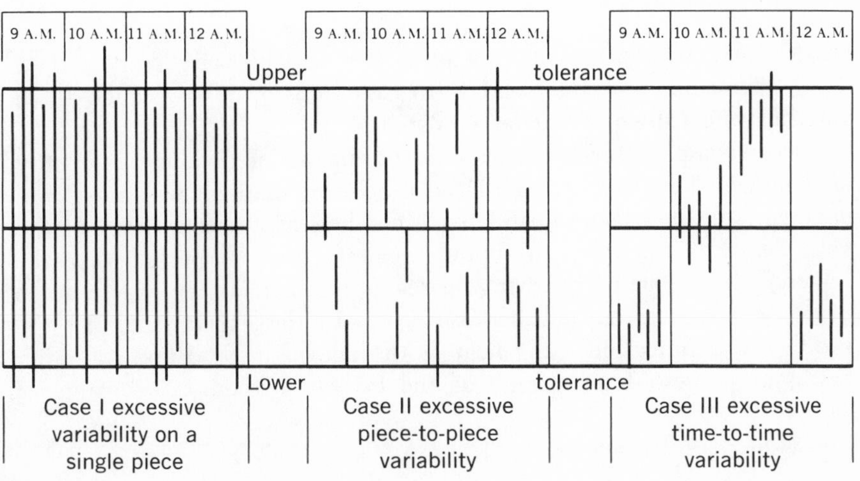

correlating test data with properties of the subsequent product, a variable time cooking cycle was established.

In critical welding of pipe there were numerous defects due to "voids" which then underwent a costly grinding out and rewelding process. The various theories of causation could not be tested since the X-ray test for voids was not made until numerous beads and layers had been welded to form the joint. By conducting the X-ray test after each bead and each layer, the precise cause of the voids was established.

4 Running a parallel pilot plant. One way is to take samples from various stages of production and process them in a small pilot plant setup in a laboratory. The results provide a "control" against which regular production can be compared.

If it is not possible to convert the characteristic from nondissectable to dissectable, two other approaches are available. First, past data can be analyzed to search for correlations between quality results and the levels at which process variables were run. However, this requires detailed production and process records. If such records are available, they may contain a gold mine of information on causes and remedies of quality problems. The tapping of the mine may require powerful statistical techniques such as multiple regression. The analysis may get complicated but there may be a big payoff without conducting any new experiments. The second approach requires that experiments be run to uncover the causes and remedies of the problem. Section 11-12 should be reviewed to furnish the background on the classical and modern methods of experimentation.

30-12 Experiments for Evaluating Dominant Variables

In this type of experiment, one or a few variables are suspected of being dominant, and the need is to determine just how much these variables affect the quality characteristic.

In its simplest form, the experiment consists of dividing a quantity of homogeneous material in half and processing one half with the dominant variable at one level and the other half at a second level of the variable. For example,[8] an optical manufacturer suspected that the "takeoff" operation of removing lenses from their holders was a main cause of scratches on polished surfaces. In the experiment, half the product was processed using heat as the takeoff method and the other half was processed using chill as the takeoff method. The extent of

[8] See *QCH*, 1962, pp. 11-49 to 11-51.

scratches after takeoff showed no significant difference for the two processes, thus eliminating both takeoff operations as suspects for the dominant cause of scratches.

Another example[9] of this approach is the component search pattern experiment. This is useful when there is prior knowledge of a difference between two versions of a product and the problem is to determine which component is causing the difference.

Suppose two identical pumps are giving different outputs. The pump designated as "Hi" gives an output of 6 gal/min and the "Lo" pump gives an ouput of 4 gal/min. Each pump has 40 components of which the critical ones have been determined as A, B, C, E, and H. The problem is to determine which of these 5 components is causing the difference in output.

The approach and results in the component search pattern technique are summarized in Figure 30-8. In run 1, component A from the "Lo" pump (A_L) was put in the "Hi" pump and run with the remaining components (R_H) from the "Hi" pump. The result of (6) indicates that A is *not* causing the difference in output. In run 2, component B from the "Lo" pump (B_L) is run in the "Hi" pump. The result of (3) indicates that B has an effect on output. Similarly, test run 3 indicates that C has an effect. Test runs 4 and 5 indicate that component D and the remaining components do not have an effect.

The results are summarized in Figure 30-9. It is concluded that

[9] E. S. M. Group, "Increasing the Efficiency of Development Testing," *Annals of Reliability and Maintainability,* vol. 4, 1965, Spartan Books, Inc., New York, pp. 796–798.

Figure 30-8 Summary of component search pattern. Source: E. S. M. Group, "Increasing the Efficiency of Development Testing," *Annals of the Fourth Annual Reliability and Maintainability Conference,* 1965, Spartan Books, Inc., New York, pp. 796–798.

Test run	Result, gpm	Analysis
A_L R_H	6	~~A~~
A_H B_L R_H	3	B+
A_H B_H C_L R_H	2	C+
A_H B_H C_H D_L R_H	6	~~D~~
A_H B_H C_H D_H R_L	6	~~R~~

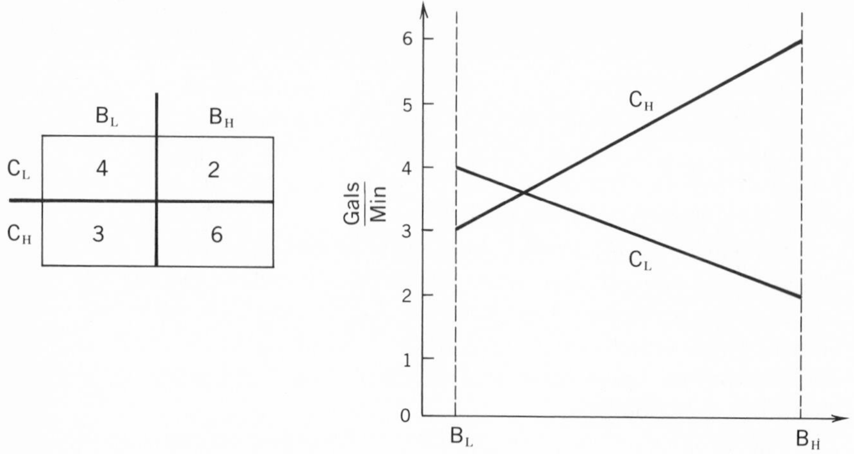

Figure 30-9 Results of experiment. Source: E. S. M. Group, "Increasing the Efficiency of Development Testing," *Annals of the Fourth Annual Reliability and Maintainability Conference,* 1965, Spartan Books, Inc., New York, pp. 796–798.

an interaction (see Chapter 11) between B and C exists and that each of these components has an effect.

The approach is applicable whenever it is possible to subdivide an assembly into parts, a process into steps, or a total environment into separate environments.

Experiments which evaluate the effect of previously defined dominant variables appeal strongly to practical people. The results will confirm or deny existing theories and will provide solid numerical results by concentrating test effort on just a few variables.

30-13 Exploratory Experiments

The exploratory experiment searches for the dominant variables instead of previously defining them. A product is followed through various processes and an extensive collection of measurements is made on materials, process variables, environments, and product characteristics. It is hoped that analysis of all the data will lead to the dominant variables by uncovering correlations between materials, process, or environments and the quality characteristic.

Examples of exploratory experiments extend to quality problems on a wide variety of products.[10]

[10] For examples involving adhesive tape, paper, steel sheet, and carpet yarn, see *QCH*, 1962, secs. 11, 30, 34, and 35 respectively.

	Characteristic	How measured	M, S, R or V
		The input lenses	
1	Curvature	W gage	V
2	Visual condition	Visual	M
3	Lens type and power		S
4	Grinding cycle	Timer	M
5	Blocking conditions		M
6	Surface texture	Profilometer	M
		The tools	
1	Curve of presser	W gage	S
2	Curve of polisher body	W gage	S
3	Curve of polisher lap	Contour projector	M
4	Thickness of felt before and after	Micrometer	M
5	Diameter of polisher body	Scale	S
6	Ooze on polisher	Visual	M
7	Source of felt		S
		The process	
1	Condition of upper pin	Visual	M
2	Condition of bushing	Visual	M
3	Polishing time	Timer	S
4	Machine setting	Protractor	S
5	Machine number, spindle number		S
6	Operator		S
7	Date, time		M
8	Process performance	Visual	M
		The finished product	
1	Presence and location of unpolish	Microscope, visual, arc lamp	
2	Visual condition	Visual	
3	Targets	Arc lamp	
4	Curve	W gage	

Figure 30-10 Detailed plan of experiment.

A well-organized exploratory experiment has a high probability of identifying the dominant causes of variability. However, there is a risk of overloading the experimental plan with too much detail. A check on overextension of the experiment is to require that the analyst prepare a written plan to be reviewed by the steering arm. This written plan must define:

1 The characteristics of materials, process, environment, and product to be observed.

2 The control of these characteristics during the experiment. A characteristic may be
   - *a* Allowed to vary as it will, and measured as is (M)
   - *b* Held at a standard value (S)
   - *c* Deliberately randomized (R)
   - *d* Deliberately varied, in several classes or treatments (V)

3 The means of measurement to be used (if different from standard practice).

Figure 30-10 provides an example of such a plan. If the plan shows that the experiment may be overloaded, a "dry run" in the form of a small-scale experiment is in order. A review of the dry run experiment can then help decide the final plan.

30-14 Graphic Analysis and Experimentation

An example of experimentation combined with simple graphic analysis is seen in a study of quality of car painting by Hayashida.[11] One of the key characteristics was durability. A measure of durability is the ability to retain the initial glossiness. The factors contributing to this are depicted in a cause-and-effect diagram[12] shown in Figure 30-11.

In the Ishikawa diagram, principal operations or subdivisions of the activity are represented by the main diagonal lines. Attached to

[11] Hiroomi Hayashida, "An Approach to the Quality of Car Painting," *Transactions of the Twentieth Annual Conference of the American Society for Quality Control,* 1966, pp. 580–588.

[12] Originally developed by Prof. Kaoru Ishikawa of Tokyo University.

Figure 30-11 Characteristic diagram of paint durability (gloss retention percentage). Source: Hiroomi Hayashida, "An Approach to the Quality of Car Painting," *Transactions of the Twentieth Annual Conference of the American Society for Quality Control,* 1966, pp. 580–588.

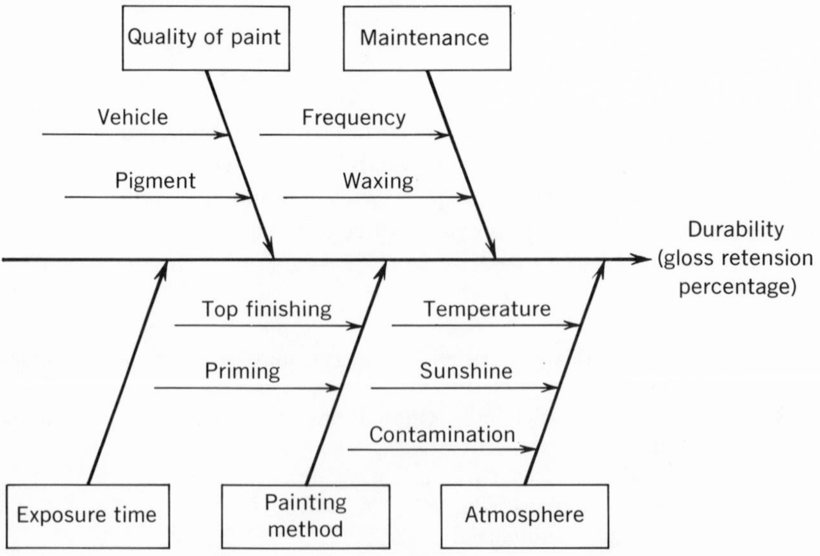

these lines are other lines representing subactivities or variables. This process of subdivision is carried on until all variables are reflected.

The diagram permits viewing the variables in perspective and in relation to each other, with resulting improvement in communication and planning. Analysis of the diagram resulted in a decision to run an outdoor weathering test in the form of a factorial experiment (see Chapter 11) as outlined in Table 30-5. The results were graphed and

Table 30-5 **Factors and Levels (Three-way Layout)**

Factors	*Levels*
Paint material (A)	Nine kinds (A1–A9)
Exposed location (B)	Toyota, Sapporo, Amagasaki, and Sumoto (B1–B4)
Waxing (C)	Not applied (C1) and applied (C2)

some examples are included here. It was concluded:

1 Durability widely differs from one paint to another (Figure 30-12).
2 Waxing has an effect, but not on every paint material (Figure 30-13).

Figure 30-12 Effect of "A" (paint material). Source: Hiroomi Hayashida, "An Approach to the Quality of Car Painting," *Transactions of the Twentieth Annual Conference of the American Society for Quality Control,* 1966, pp. 580–588.

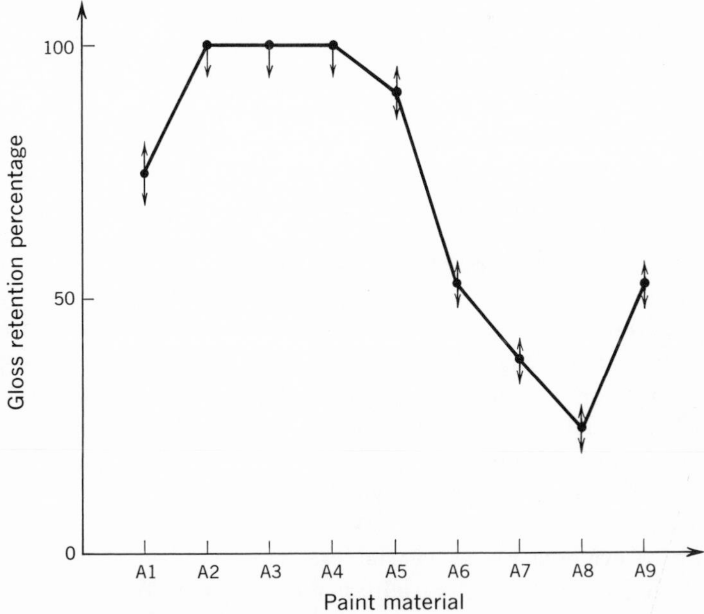

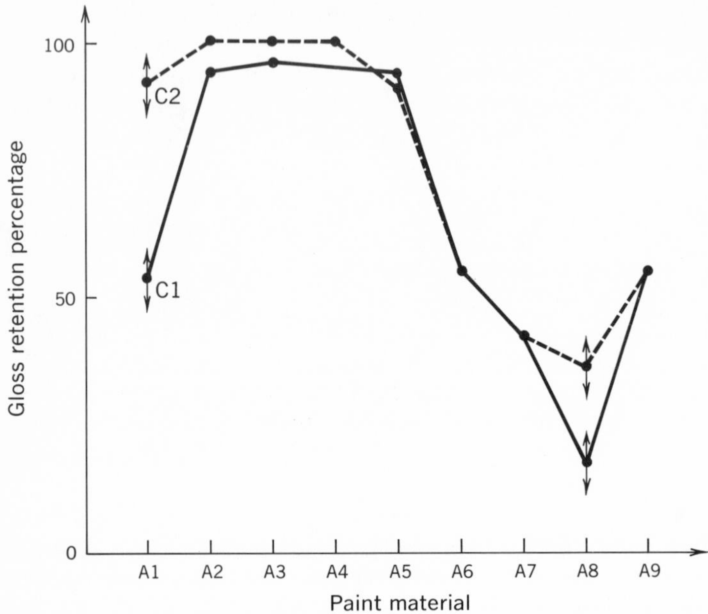

Figure 30-13 Effect of "A" × "C" (paint material and waxing). Source: Hiroomi Hayashida, "An Approach to the Quality of Car Painting," *Transactions of the Twentieth Annual Conference of the American Society for Quality Control,* 1966, pp. 580–588.

3 Testing in different locations did not have much effect. (It was decided to standardize on the Toyota district.)

Subsequent changes were made in paints A and B and the results are summarized in Figure 30-14.

30-15 Evolutionary Operations (EVOP)

Experimentation is often regarded as an activity that can only be performed under laboratory conditions. To achieve maximum performance from some manufacturing processes, it is necessary to determine the effect of key process variables on process yield or product properties under shop conditions. Laboratory experimentation to evaluate these variables does not always yield conclusions that are completely applicable to shop conditions. When justified, a "pilot plant" may be set up to evaluate process variables. However, the final determination of the effect of process variables must often be done during the regular

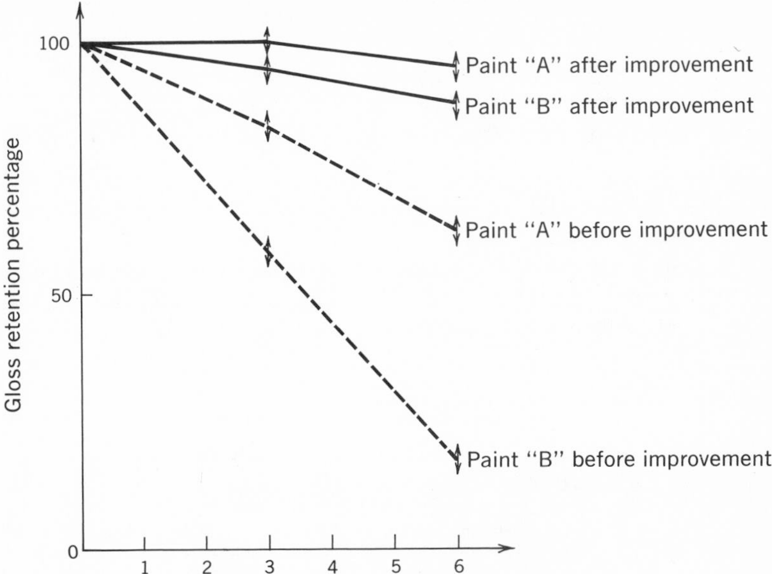

Figure 30-14 Results of the outdoor weathering test before and after improvement of the paint material. Source: Hiroomi Hayashida, "An Approach to the Quality of Car Painting," *Transactions of the Twentieth Annual Conference of the American Society for Quality Control,* 1966, pp. 580–588.

production run by informally observing the results and making changes if these are deemed necessary. Thus, informal experimentation *does* take place on the manufacturing floor. To recognize this and provide a methodical approach for process improvement, Dr. G. E. P. Box developed a technique known as Evolutionary Operations (EVOP):[13]

> The thesis of EVOP is that every processed lot can contribute information about the effect of one, two, or three processing variables on process yield or product properties of interest. By *slightly* displacing each of the variables under study from its normal standard operating value (first higher, then lower according to a fixed pattern) the risk of making large quantities of non-conforming product is minimized. In high volume production, each repetition of the pattern serves to decrease the uncertainty of the results. When "significant" variables are identified, the *normal* value is adjusted to make the most of each. The procedure can then be repeated

[13] Richard S. Bingham, Jr., "EVOP for Systematic Process Improvement," *Industrial Quality Control,* vol. 20, no. 3, p. 17, September, 1963.

for further improvement with these variables, or others may be considered.

The approach can be illustrated[14] for evaluating the effect of certain alloys on hot tensile properties of Zircalloy-2 metal. The steps are

1 Select the "vital few" variables. Suppose that tin and chromium were key variables.
2 Determine the levels within which each variable can be set without producing defective material. For example, in Table 30-6, additions of the two alloys were chosen.

Table 30-6 OPERATING CONDITIONS FOR ALLOYING INGREDIENTS

Test Number	*Percent Ingredient Added* *Sn*	*Cr*
1	1.45	0.10
2	1.40	0.07
3	1.50	0.13
4	1.50	0.07
5	1.40	0.13

3 Define the test criteria for evaluating results. For example, one criterion might be the ultimate tensile strength at 600°F.
4 Estimate the number of process cycles needed to detect a specified improvement in the process. For example, test of hypothesis concepts (Chapter 11) showed that eight cycles would be needed to detect a 905-psi change in strength.

An example of a test cycle and the results is shown in Figure 30-15. The numbers presented in Table 30-6 are running averages of the test results. By noting the change in results as conditions are changed, evaluations of the effect of tin alone, of the effect of chromium alone, and of the joint effect can be made. The presentation of results to manufacturing management includes statistical limits to judge whether a change is statistically significant or due to sampling variation. In this manner, specific guidance is provided for making changes in process variables that will help approach optimum conditions.

[14] *Ibid.*

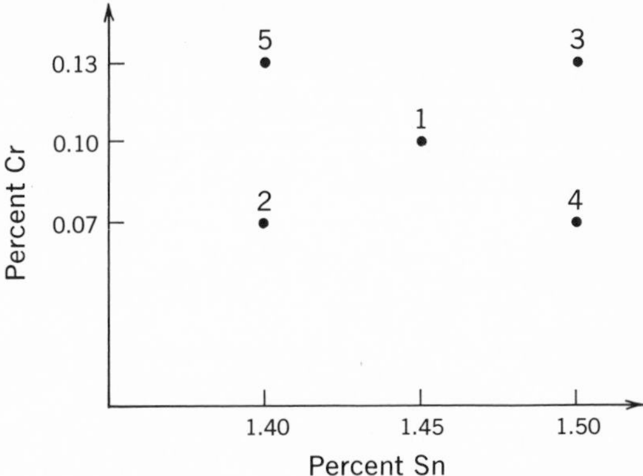

Figure 30-15 Process information board (hypothetical results—data in 10^3 psi). Last cycle completed: 1 phase: 1. Source: Richard S. Bingham, Jr., "EVOP for Systematic Process Improvement," *Industrial Quality Control,* vol. 20, no. 3, p. 17, September, 1963.

30-16 Analysis of Failures When the Specification Has Been Met

In complex products it is common to find that unfitness for use is mainly the result of causes other than poor workmanship or manufacturing nonconformance to specification. Table 30-7 summarizes the causes of

Table 30-7 SUMMARY OF CAUSES OF FAILURE*

Classification of Failures			*Percent of Total*
1 Engineering			43
Electrical	33%		
Mechanical	10%		
2 Operational			30
Abnormal or accidental conditions		12%	
Manhandling		10%	
Faulty maintenance		8%	
3 Manufacturing			20
4 Other			7
			100%

* R. M. C. Greenidge, "The Case of Reliability versus Defective Components Et Al.," *Electronic Applications Reliability Review,* no. 1, p. 12, 1953.

failure on 1,135 components which, in turn, resulted in 850 electronic equipment failures.

It should be noted that manufacturing accounts for only 20 percent of the failures.

Table 30-8 shows the results of an analysis of 2,555 component failures from seven space programs. Here, note the wide spread of functional responsibility.

These two surveys illustrate the wide range of responsibility for failures and also the inadequacy of inspection results alone as a measure of quality. One of the authors recalls a situation on a complex product which *passed all inspection tests* but failed during customer usage. An analysis of field data made it clear that a certain component had to be redesigned. The component was a vendor designed item and the vendor was extremely slow in accomplishing the redesign. It happened that the component was part of a complex subsystem that had a numerical reliability requirement that could not be adequately proved until a reasonable amount of field data on the total system were available. However, the extent of failures on the component conclusively indicated that the subsystem reliability requirement would not be met. (All this background is provided to set the stage for describing the action taken.)

As a last resort to the procrastination of the vendor, the prime contractor instructed his resident inspector to hang a rejection tag on every system submitted for inspection by the vendor. This was indeed a shock, because the systems *passed all inspection tests.* Thus, the design function was being told that it was *not* sufficient to get the product past final inspection. The situation was emotion packed. The systems were stacked up each with a reject tag stating unreliability as the reason for rejection.

Top management of the two companies was involved in the problem within a few *hours.* At this level, the procrastination suddenly disappeared and the appropriate direction was given to clear up the problem.

The basic point is that an unreliable design feature was used as the basis for rejecting all product even though formal inspection tests had been passed. The long-range effect of this *one* instance was to demonstrate the quality-mindedness of the management of the system contractor. Management refused to accept the units even though it was faced with serious problems on meeting delivery schedules to the ultimate user.

Diagnosing the causes of failures (when all inspection criteria have been met) can be difficult. A physical environment may have been underestimated, an incorrect design analysis made, poor maintenance procedures specified, the wrong spare part used, or the user may have applied the product incorrectly. In diagnosing these types of failures,

Table 30-8 **Summary of 2,555 Component Failures from Seven Space Programs**

Component		*Functional Responsibility for Failure*						
Type of Component	*Percent of Sample*	*Design,* %	*Manu-facturing,* %	*Test Equipment,* %	*Human Error,* %	*Specification Error,* %	*Normal Wear,* %	*Other Causes,* %
Electronic	16.0%	27.6%	25.4%	12.5%	3.7%	7.6%	0.7%	22.1%
Mechanical	1.0	34.6	15.4	15.4	23.1	—	—	11.5
Electromechanical	27.7	27.9	32.1	6.8	7.5	5.4	0.3	20.2
Explosive on one shot	7.5	16.7	47.9	7.3	1.0	7.3	—	19.3
Pneumatic	16.7	23.8	35.5	6.3	9.3	2.6	—	22.4
Sensors	13.5	13.6	25.3	4.2	6.5	6.4	—	29.5
Ground support equipment	15.8	32.5	16.6	5.0	9.4	8.4	0.5	27.6
Batteries	<1	4.8	4.8	38.1	9.5	33.3	—	9.5
Cables and harnesses	<1	8.7	39.1	8.7	4.4	—	4.4	34.8
Total	100	25.2	29.8	7.7	7.2	6.3	0.3	23.7

* Summarized from *Malfunctions in Missile and Space Equipment* by M. Wilson, General Electric Co., Re-entry and Environmental Systems Division, Mar. 15, 1961.

adequate data on the failures are critical and are often *not* available for field failures. A laboratory examination of the failed part is often essential from the viewpoint of the diagnosis of the cause of failure.

Based on the formal reports of product failures and the physical examination of the failures, the diagnosis should yield a detailed problem statement that includes

1 Summary of the *symptoms* of the product failure (backed up with copies of detailed reports)
2 Statement of the *cause* of the failure as concluded from the diagnosis activity
3 Summary of history of failures of the part involved
4 Effect of the part failure on system performance
5 Details of the technical investigation
6 Function responsible for obtaining corrective action

It is important to assign clear responsibility for discovering cause from symptoms. This may require a separate diagnosis activity. For some products, the diagnosis activity requires a high level of technical competence to determine the *causes* of product failure and to provide definitions of problems that are detailed enough to get corrective action. Nothing will undermine a quality improvement program faster than a statement sent to a design group that product X "did not operate satisfactorily on four occasions during the past week." A thorough analysis might show that the product was operating perfectly but that the test equipment was faulty. Perhaps inspection errors were made. The same applies to failures attributed to manufacturing or any other activity. Sufficient facts must be developed to gain agreement on the cause of the problem and the function responsible for correcting it. The success of the diagnostic activity depends on its technical reputation. It will be respected only if it provides the line with useful analyses of failures. (See Chapters 7 and 9 for discussion of the organizational aspects of achieving correction of interdepartmental quality problems.)

30-17 Measurement for Diagnosis

A common failing among supervisors and managers is to assume that the measuring instruments needed to judge conformance (of product to specification) are also adequate to conduct diagnosis of the causes of defects. It is not that the managers have thought about it and concluded that the acceptance instruments are adequate. Rather, it has simply not occurred to them that these instruments are not adequate for diagnosis.

In the case of the watch parts (see section 15-2) the measurement conducted using a precise laboratory gage established not only that the process was capable of holding tolerances; it established also that the shop gages were not adequate even for guiding the machines. In a study of carding of wool, the shop instruments were of no help—the sample pieces could be weighed only on precise laboratory balances. Such cases are legion.

In still other cases, the main contribution of the investigator is "cutting a new window" into the process. In the paper mill case (see section 30-11) cutting a hole in the digestor to be able to test a sample before concluding the cooking constituted such a new window. The act of X-raying the welding at several stages in the process was to the same effect (see section 30-11). In still another case, a plant had long endured a condition of defective gun barrels. The condition was discovered during air gaging of the finished barrels. It was not known whether the defect was created in the original drilling, in the redrilling, or in the reaming. No one knew because there was no air-gage head of a size appropriate to measure the product after either of the drilling operations; i.e., instrumentation had been provided only for product acceptance. When an air-gage head was made to enable check of the product after first drilling, the cause soon became evident.

The analyst should grasp fully the principle that instrumentation for diagnosis is to be determined by the needs for diagnosis and not by the needs for acceptance. Sometimes the acceptance instruments are adequate for both purposes. However, often they are not, and in that event the diagnosis cannot be completed. What is worse, the people long associated with the job may be so accustomed to living with the acceptance instruments that they are blocked from seeing the need for other instruments.

PROBLEMS

1 You are consultant to a company which manufactures electrical appliances sold to consumers through a chain of distributors. There have been two main quality problems:

a A high rate of consumer returns which have cost the company a large sum for replacement, servicing, etc.

b A rising rate of rejections at assembly. Last month they amounted to 25 percent.

A detailed study of the complaints has revealed a number of design weaknesses, which are now being studied by the engineering department. Your present concern is with the assembly rejections.

You have made a detailed study of the 40 different reasons for assembly rejections and by considering the causes of each, you have

Table 30-9

Breakdown of Assembly Rejections

Cause of Rejection	*Percent of Total Rejections*
Nonconforming components	40
Assembly errors	15
Visible defects	15
Engineering design faults	10
Miscellaneous unknown causes	20

Breakdown of Nonconforming Components

Component	*Percent of Total Rejections*
G	10.0
J	8.2
L	7.0
P	3.5
A	2.7
R	1.5
All others	7.1
	40.0%

boiled the picture down to the data given in Table 30-9. A meeting has been called to hear your report. Present are

The foreman of the assembly department
The foreman of the component-manufacturing departments (screw machine, punch press, plating, etc.) to whom the inspectors in these departments report
The chief inspector, to whom the assembly inspectors report
The chief engineer, who is responsible for design

After hearing your report, a lengthy discussion ensues as to the steps to be taken to improve the situation. The following is a condensed summary of the viewpoints expressed:

Assembly foreman: The departmental inspectors ought to catch the bad components stuff so they don't get to me.

Component foreman: They do. But lots of the specifications are ridiculous. We know that a few thousandths off doesn't make any difference. Should we throw away perfectly usable parts because they are a little outside of the specifications? If we did, our costs would skyrocket.

Assembly foreman: What about parts G, J, and L? These are causing most of the rejections. These specifications have to be met because the assembly is critical. But we are getting a lot of undersize on these all the time. We have to tear down the whole assembly to replace them; besides it slows down the assemblers when they have to work with these parts.

Component foreman: We ought to tighten the specifications on G, J, and L so you will have a factor of safety. Then, if they are a little out, you'll still be able to use them. That way, we'll keep your costs down.

Chief engineer: I agree with that. You always go outside of whatever we give you, so we'll just tighten up on those dimensions that are causing the rejections at assembly.

Chief inspector: I do not agree. I think the foremen ought to crack down on their inspectors and tell them not to pass anything outside of the specifications. Then we'll wipe out 40 percent of the assembly rejections.

Based on the facts and opinions given, what should this company do to reduce the high cost of assembly rejections, at the same time keeping down the cost of component manufacture? In answering, do not ramble. Make a specific list of objectives, numbering them, and telling for each *why* you think it is necessary and *how* (briefly) it should be done.[15]

2 A quality problem has been traced to variation in the thickness of 2-inch-wide incoming steel, and the sampling plan used to check the steel is being questioned. The steel carries a tolerance of ±0.001 inch, which must be met, since parts punched from the steel are later stacked up on one another and the cumulative error of over- or undersize pieces cannot be tolerated.

The present sampling plan is based on measuring the first lap of the selected coils. But the question arises: "Are the measurements on the first lap of a coil an adequate representation of the thickness of the coil?"

In order to answer this question, it is necessary to study the nature

[15] Course notes of Management of Quality Control Course, U.S. Air Force School of Logistics, 1959.

of the thickness variability, from coil to coil, within a coil, and across the coil. Accordingly, a study has been made by unrolling 12 coils. Each coil was measured on the first lap and at three other randomly selected spots along the length. At each such spot, three thickness measurements were made, one ⅜ inch in from each edge and one in the middle of the strip.

The measurements were recorded as follows. All data are in 0.0001 inch in excess of 0.0600 inch. Tolerance is 0.0620/0.0640. The first set of entries for each coil is the measurement on the first lap; the other three are at randomly selected spots along the length.

Coil No.	*Edge*	*Middle*	*Edge*	*Coil No.*	*Edge*	*Middle*	*Edge*
1	30	30	28	7	25	28	24
	35	32	35		27	28	29
	35	35	38		32	32	33
	39	35	39		29	29	31
2	30	29	30	8	28	27	26
	27	31	31		26	26	24
	29	28	30		39	38	42
	30	32	33		30	30	32
3	28	28	28	9	22	27	24
	30	32	30		21	27	27
	28	30	30		32	32	35
	40	39	45		29	27	30
4	29	30	29	10	26	28	28
	30	34	33		29	29	30
	27	27	30		22	23	23
	30	30	33		28	27	29
5	28	27	28	11	32	30	33
	18	18	18		37	35	37
	23	23	24		30	31	31
	24	24	30		28	27	28
6	39	40	40	12	22	22	23
	30	32	30		32	35	23
	30	30	30		31	31	31
	30	30	29		37	30	38

a Draw a Multi-Vari chart representing the data.

b Is the principal source of variation in steel thickness across coil, within coil, or coil to coil?

c Would you say that the measurement on the first lap of a coil adequately represents the coil?

d Can the proposed sampling plan be based on measurements made on first laps? If so, explain why. If not, on what basis *should* the sampling plan be set up?[16]

3 Trouble has been experienced in assembling Magnetron Tubes because of oversize cathode poles. About 5 percent of the cathode poles are too big to fit into the assembly fixtures, even though the fixtures are made to take poles 0.001 to 0.002 inch larger than the 0.9980 inch/1.0020 inch pole specification. The cathode poles are expensive and the 5 percent loss is well worth reducing.

A considerable hassle has been brewing between the assembly foreman and the inspection group in the pole making department. The foreman blames the inspectors for accepting lots with oversize poles, while the inspectors steadfastly maintain that they do no such thing, but that it is probably the subsequent plating or brazing operations that make the poles too big. The foreman retorts that there is only 0.0002 inch of plating put on and that the brazing could not possibly swell the poles.

The need to get facts is imperative; a lot of 40 cathode poles is taken as manufactured and followed through the plating and two brazing operations, measured at each stage. The measurements are shown in the table below:

Size	*As Made*	*After Plating*	*After First Braze*	*After Second Braze*
1.0042-3				1
1.0040-2				2
1.0038-9				4
1.0036-7			1	5
1.0034-5			1	8
1.0032-3			3	9
1.0030-1			6	6
1.0028-9			11	3
1.0026-7			7	1
1.0024-5			6	
1.0022-3			4	
1.0020-1			1	
1.0018-9		2		
1.0016-7		4		
1.0014-5		6		
1.0012-3	1	9		

[16] *Ibid.*

Size	*As Made*	*After Plating*	*After First Braze*	*After Second Braze*
1.0010-1	3	8		
1.0008-9	6	5		
1.0006-7	10	4		
1.0004-5	8	2		
1.0002-3	5			
1.0000-1	3			
0.9998-9	2			
0.9996-7	1			
0.9994-5				
0.9992-3				
0.9990-1				
0.9988-9				
0.9986-7				
0.9984-5				
0.9982-3				
0.9980-1				

a Based on the data, whose arguments appear to be correct?
b Who is actually responsible for the nonassemblable poles?
c Is the manufacturing process capable of meeting the specified tolerances?
d How much does each of the subsequent operations add to the dimension?
e What should be done to eliminate the 5 percent loss?[17]

4 The width of a slot is a critical dimension on an eccentric cam. The specification calls for 0.500 inch ± 0.002 inch. A large sample was measured with the following results:

Average = 0.499 inch
Standard deviation = 0.001 inch

The width is normally distributed. Suppose that the average value can be changed easily by adjusting the aim of the machine, but the variation about any given average cannot be reduced. If a change is made to minimize the percent defective, predict the percent defective that will result. Why might the present process setup be preferred to the revised setup to lower the total percent defective?

5 Hardness is an important characteristic of battery cases. A certain design has a hardness requirement of 60 to 70 as measured by

[17] *Ibid.*

a type D durometer. Twenty samples of five cases each were collected for a control chart:

Sample Number	$\overline{X}$	*R*	*Sample Number*	$\overline{X}$	*R*
1	64	5	11	62	11
2	63	6	12	63	8
3	59	9	13	61	6
4	59	7	14	62	7
5	62	9	15	62	5
6	62	11	16	62	6
7	61	8	17	59	9
8	63	4	18	61	8
9	59	7	19	63	8
10	61	6	20	64	11

An important process control characteristic is cure time. The average cure time for the above cases was 7 minutes and 20 seconds as compared to the 7 minutes and 30 seconds required. Operators were instructed to more carefully control the cure time. Twenty new samples of five each yielded the following results:

Sample Number	$\overline{X}$	*R*	*Sample Number*	$\overline{X}$	*R*
1	65	6	11	64	10
2	66	9	12	67	6
3	63	6	13	65	7
4	63	8	14	64	5
5	69	5	15	66	2
6	68	5	16	66	2
7	67	5	17	61	3
8	66	6	18	63	6
9	67	5	19	61	6
10	67	5	20	63	2

The average cure time for the second set of samples was 7 minutes and 37 seconds. Analyze these data in whatever way seems appropriate and draw conclusions.

6 Visit a local plant and find two examples of each of the following: (*a*) a dissectable characteristic, (*b*) a nondissectable characteristic, and (*c*) a quantity of product coming from several streams. Report your findings.

7 Draw an Ishikawa diagram for one of the following: (*a*) the quality level of a specific activity at a university, bank, or automobile repair shop, and (*b*) the quality level of *one* important characteristic of a product at a local plant.

Base the diagram on discussions with the organization involved.

31 Quality Control Engineering

31-1 The Nature of "Staff"

The term "staff" is used by industry in a wide variety of meanings. Industrial activities can be classified as:

1 Operational, covering those activities without which the enterprise cannot exist, i.e., purchasing, production, sales.
2 Staff, consisting of:
a Utility services for the operating departments. Examples are transport, plant maintenance, insurance.
b Coordinating, i.e., exerting a planning and coordinating influence on the work of other activities. Examples are production planning and inventory control
c Advisory, i.e., possessing specialized knowledge and advising other activities regarding this specialty. Examples are legal, market research, public relations, and statistics.
d Control, to see that operational and other activities adhere to prearranged plans. Examples are inspection, budgetary control, and internal audit.

The foregoing is stated in terms of "activities," not departments. In large companies, special departments are created to man these various activities and are given names descriptive of their role. This is also true of staff quality control activities; i.e., they are called quality control engineering department, reliability engineering department, etc.

Any staff department usually performs every single one of the staff activities listed, i.e., utility, coordinating, advisory, and control duties. The quality control staff departments are no exception.

31-2 Usual Staff Quality Functions

Numerous lists have been prepared to define the staff quality functions.[1] The following is a classification of the functions usually mentioned:

1 *Functions common to all stages of progression of the product.*
a Coordinate the preparation and execution of the quality plan including the tasks required during design, purchasing, manufacturing, and usage of the product.
b Design procedures for measuring quality and publish reports for all levels.
c Prepare general quality manuals, systems, and procedures.
d Recommend standards for sensory qualities.
e Design means for measuring accuracy and quantity of inspection work; design incentive plans for inspectors.
f Investigate to determine disposition of nonconforming material.
g Prepare job specifications and standards for selection and training of personnel for quality matters.

2 *Functions special to new products.*
a Review performance of similar products to identify chronic difficulties in manufacture, test, and usage.
b Recommend new quality objectives (e.g., quantitative reliability and maintainability objectives) based on customers' needs and competitive situation. Define method of evaluating design against objectives before release to production.
c Review new and changed product designs. Prepare proposals for preventing failures, for reducing defects, cost of inspection, and customer complaints, and for optimizing tolerances.
d Design tests to evaluate ability of product to meet all quality objectives.
e Identify vital few critical components and special quality efforts required during design, purchasing, manufacturing, or inspection.
f Prepare estimates of the quality costs for new product designs.

3 *Functions special to vendor relations.*
a Draft a policy on vendor quality relations.
b Draft a vendor relations plan, including procedures for proving in new sources of supply, approval and disposition of incoming shipments, vendor evaluation or rating, quality certification, and related record systems.
c Evaluate vendor quality capability.
d Provide assistance to vendors to clarify specifications, to classify importance of characteristics, to standardize measurements and ac-

[1] See, for example, *The Basic Work Elements of Quality Control Engineering,* American Society for Quality Control, September, 1961. Also, A. V. Feigenbaum, *Total Quality Control: Engineering and Management,* McGraw-Hill Book Company, New York, 1961.

ceptance criteria, and to feed back performance data.

e Draft the quality section of the vendor relations manual.

4 *Functions special to process control.*

a Prepare the inspection plan including definition of control stations, classification of characteristics, inspection and test criteria, measuring instruments, disposition of nonconforming material, and data recording systems.

b Determine quality capabilities of machines and processes and furnish results to design, manufacturing planning, and others.

c Prepare process specifications.

d Design data feedback systems for production, including chart control as appropriate.

e Design process surveillance systems as appropriate.

f Design means for rating quality performance of production operators.

g Initiate and participate in design of plans to raise level of quality-mindedness.

h Investigate causes of sporadic defects.

5 *Functions specially associated with measurement.*

a Design gages, instruments, and test equipment.

b Design and install systems for maintaining precision of measuring devices and production tools.

c Design equipment for analyzing and processing data.

d Standardize test procedures.

e Conduct or supervise laboratory testing services.

f Provide special measurement service.

g Maintain accuracy of laboratory standards and masters.

6 *Functions specially associated with customer relations.*

a Analyze customer quality complaints and returns.

b Provide liaison with customers' representatives (e.g., material review boards).

c Design plans for appraising competitors' quality, and participate in appraisal.

d Participate in surveys of customer reaction, customer costs, and quality problems as seen by customers.

e Design customer quality certification plans.

f Recommend economic level of quality guarantee to customers.

g Design plans for check inspection and rating of quality of outgoing product.

h Design quality controls for packaging and shipping operations.

i Conduct studies and recommend changes on product reliability, ease of operation, maintenance, service, length of life, and environmental conditions of use.

7 *Special analysis, audit, and consulting functions.*

a Draft a company policy on quality.

b Draft the company's major quality objectives.

c Analyze quality costs and develop the economic models needed to establish a balance between cost and value of quality.

d Study and recommend changes in the company's organization for quality.
e Determine the principal causes of quality losses.
f Diagnose causes of chronic defects and propose programs for prevention.
g Conduct quality audits and surveys.
h Design and publish executive reports on quality performance.
i Design and conduct training programs for all functions and levels.
j Provide consultation service.
k Develop new methods when required, and aid in installation.

These functions are executed throughout the spiral of quality (Figure 1-1).

31-3 Assigning the Staff Quality Functions

All the above functions need *not* be assigned to people called "quality control engineers," or to a department called "quality control engineering." In no company are all or even most of these functions in fact performed by people so labeled. On the contrary, many of these functions are performed, for example, by design engineers, manufacturing planning engineers, industrial engineers, reliability engineers, production supervisors, or inspection supervisors.

Whether to distribute these functions among operational departments or to create a staff department is always a tailor-made problem for each company. (See Chapter 8 for a discussion of some alternate organizational forms.) There is no "right" way to organize, any more than there is a "right" biological organism. Instead, there is a "right" way for the particular environment, history, culture, skills, and other local characteristics of the company. The criteria which are decisive in creating quality control staff departments include:

1 The amount of staff work to be done. This is mainly a function of the size of the company and of the complexity of the quality problems. As size and complexity grow, the staff quality control departments bud and develop.
2 The extent to which the operational departments are trained to use the tools of quality control in their day-to-day duties. In a country like Japan, where this training has been intense, the use for separate staff quality control specialists is distinctly lower than in the United States.
3 The traditional approach to separation of planning from execution. The American tradition, following the Frederick W. Taylor revolution, has tended strongly to separate planning from execution.

When the quality function grew in size and importance, the same tradition was followed.

4 The extent to which management has confidence in existing operational or staff departments to accept and execute new functions. Executives (in common with all humanity) tend to load work, including new functions, onto men who have a proved record of getting things done. As a corollary, executives will avoid assigning new functions to men whose record is mediocre, no matter how "logical" such assignment would be on the organization chart. If necessary, functions will be split, new departments will be created, and organization charts will be distorted rather than to give important new duties to men who have a record of poor performance. (These distortions may be the alternative to demotion or dismissal.)

There are other variables, of course. However, to a degree more extensive than in any other major industrial country, the United States has created staff quality control departments and has assigned to them many of these staff quality control functions. An index of the extent of this is the membership of the American Society for Quality Control which, by the late 1960s, exceeded 23,000.

While there are justifications for organization structure to give the appearance of being illogical, there is no justification for being unclear. It is most helpful to define functions clearly so that men know what is expected of them. The need for clear definition is greatest in the case of new functions and departments (see below).

31-4 Creation of the Staff Quality Control Department

The creation of a new department should start with the preparation of a statement of objectives and the plan for meeting the objectives. This should be done with the participation of other departments because the quality spiral extends throughout the organization. Some quality control departments have failed because they lacked agreement upon objectives and a plan.

In one company,[2] the air was cleared considerably when it was announced that

> It will be the job of the Quality Control Department: To discover what are the principal defects which cause customer returns and factory losses.

[2] Robert G. Mitchell, address to the American Society for Quality Control, Atlanta section, May 29, 1956.

To analyze existing data, and to generate new data, to discover what are the causes of the principal defects.

To present, to the various responsible departments, data on causes of defects and to stimulate corrective action.

To establish scoreboards so as to measure progress and help maintain the results achieved.

To study the effectiveness of inspection, and to present the results of such study to aid in improving the effectiveness of inspection.

To propose means for formalizing inspection procedures, workmanship standards, visual standards and production procedures where these affect quality.

To conduct such training as may be required in furtherance of the program.

It will *not* be the job of the Quality Control Department to:

Get involved in deciding whether the product is acceptable or not.

Attempt to foist its opinions or tell any Department what to do. (It is the facts which must supply conviction.)

Get involved in day-to-day decisions. (It should work only on long-range problems.)

Become responsible for establishing specifications, standards, methods, procedures, etc.

Particularly note the definition of what the department will *not* do. Although the negative approach is unusual in formally defining objectives, the "not do" items listed are the ones that cause much misunderstanding, and specifically highlighting them can greatly clarify the role of the department.

When it is decided to set up a staff department, organization within the department can be by job function, by product, by stage or progression of the product, by departments served, or by a combination of these. Chapter 7 discusses some alternate forms of organization.

31-5 Transferring Staff Functions to the Line

There is nothing static about staff functions. New ones are ever being created and old ones are ever being "abolished" through transfer to the line. The method of transfer is seen in the history of measuring instruments.

Early in the century, instruments such as the optical projector and the Rockwell hardness tester were housed in the measurement laboratory. Anyone requiring measurements on these instruments took the specimens to the laboratory, where a technician made the measurements and wrote the report. Since those days, the instruments have been simplified and ruggedized for shop use. In addition, shop mechanics and inspectors have been trained to use such instruments on a day-to-day routine basis without need for a laboratory technician to intervene.

A similar flow downstream takes place with all specialty knowledge. Today's techniques of reliability prediction may be known only to the reliability engineer, so that he is an essential party to the use of the technique. Tomorrow the techniques will be reduced to simplified form, and design engineers and others will be trained in their use on a day-to-day routine basis, without need for a reliability engineer to intervene, except in occasional consulting situations.

Staff specialists and departments should constantly be "working themselves out of a job" by reducing their specialized knowledge to simplified form and training the line people to take over. Actually the staff people will not be out of work thereby. They will lose the jobs that have become routine and boring, but they will become free to take on new, unsolved problems which are constantly coming over the horizon.

All companies are faced with some degree of troubles; e.g., machines cannot hold tolerances, processes are "out of control," parts do not mate in assembly, final product does not meet tests.

The man-hours required to investigate these problems are commonly provided by adding troubleshooters or supervisors. However, in some companies, the "quality control engineers" have been assigned to help solve these sporadic problems. Such assignments create risks that the "vital few" chronic quality problems may never get analyzed because the "trivial many" have priority, and there is never any end to the "trivial many."

In some companies, the quality control engineers have been transferred to report directly to the production superintendents or foremen. This means that the local, departmental problems get priority. However, the main quality problems are interdepartmental (among several departments) and only a centralized quality control department can solve *inter*departmental problems. The centralized department can provide the diagnostic arm working under the guidance of a steering arm of management people from the affected departments.

There are, of course, situations in which staff people are invited into specific fire fighting problems because of special need for some instruments, methods, records, or other competence possessed by the staff. Providing this service in selected cases is a proper function of the staff so long as the basic responsibility for troubleshooting remains with the line.

Staff departments who persist in devoting their energies to day-to-day problems thereby not only miss the greater opportunities available through solving chronic problems, but they also build up resentment in the minds of the line people because of the day-to-day interference.[3]

31-6 Jurisdictional Questions with Other Staff Departments

Creation of a staff quality control department inevitably creates problems of jurisdiction with other staff departments, such as with manufacturing engineering or process development. In addition, there may arise jurisdictional problems between departments devoted primarily to the quality function, e.g., between quality control engineering and reliability engineering. Finally, in those cases where a department performs services for several clients there is competition for authority over the service department.

For example, in one company a metallurgical laboratory conducted three kinds of activities: tests to aid production in making decisions on whether to run or stop processes, tests to aid inspection in making acceptance decisions on the product, and tests to aid research in developing new products and materials.

Each department contended that it should supervise the laboratory: production on historical grounds, inspection on the grounds of responsibility for acceptance, research on the grounds of bringing technical specialties (metallurgy) into a common department.

Investigation showed that the respective usage of the laboratory was 20 percent for process control, 65 percent for product acceptance, and 15 percent for research. The decision became obvious.

When two competing staff departments fully accept each other's status (without any attempt for one to become boss over the other), the jurisdictional questions are commonly over who is to do what work. In the case of projects, it is common to thresh out individual instances until an experience pattern emerges, after which this experience pattern is generalized into the job descriptions. For example, a common jurisdictional question in the process industries is "Which staff department, process development or quality control, is to study the problems of yields of processes?" The same question arises in the mechanical industries as between manufacturing engineering and quality control. One company came up with a solution as follows:

1 Sporadic quality troubles (trouble suddenly far in excess of operating levels) to be studied by production. Other departments to enter only as invited by production.

[3] See J. M. Juran, "A Note on the Mortality of QC Departments," *Industrial Quality Control* (*IQC*), vol. 17, no. 8, pp. 22–23, February, 1961.

2 Chronic quality troubles to be diagnosed by process development if they involve (*a*) study of relative merits of new process versus existing process, (*b*) study of proposed change in existing process, and (*c*) study of relative merits of new material.
3 Chronic quality troubles to be diagnosed by quality control if they involve (*a*) investigation of variables in existing process, (*b*) comparison of two existing processes, and (*c*) study of process capability of existing processes.
4 Simultaneous presence of chronic and sporadic quality troubles. Such simultaneous presence is no reason for deferring study of what is otherwise a high priority chronic defect.
5 Borderline cases resolved by steering arm.

When the competition includes efforts at dominance, the need may be to divide the functions performed into their elements and to assign these elements by common agreement or by management fiat. (See section 9-3 for an example.)

The rise of the reliability engineering department created some severe jurisdictional strains with the quality control engineering departments, which resented creation of these new departments to do what the quality control engineers asserted they were already doing, or should have been doing. It may well be that still more such specialist groups will emerge into widespread use, e.g., maintainability, configuration control, etc. These are not "logical" in the sense that the duties of existing departments could be expanded. The logic arises from management's conviction that the existing departments missed the opportunity to deal with an important problem, and hence are not capable of handling it.

31-7 Job Specifications for Staff Quality Control Specialists

The staff quality control specialist must possess:

1 General technical skill, because the studies are often of an engineering character and involve contacts with technical personnel
2 Special skills in the use of the tools of quality planning and analysis, i.e., the subject matter of this textbook
3 Special skills in introducing and gaining acceptance of new ideas which will require changes in existing work situations

The formal job specification statement is largely a listing of the specific functions which will be assigned to the specialist. However, the specification also includes other matters:

1 *Formal training.* Degree or equivalent in engineering, preferably including courses in industrial management; manufacturing processes;

engineering economics, methods and procedures; statistics; cost control; and human relations.

2 *Special training.* Preferably in quality planning and analysis.

3 *Experience.* Three to five years in industry, dealing with several processes and products. Product design, shop troubleshooting, methods work, process engineering, field engineering, or quality control are desirable depending on the specific job assignment.

4 *Personal characteristics.* Proved integrity. Objectivity as evidenced by a firm habit of reaching conclusions from facts, not from opinions. Ability to work well with people and accept the fact that the staff has no authority except that which it earns through painstaking study, factual reporting, and patient discussion to secure acceptance of new ideas.

5 *Relationships.* Has the right, on approval of department supervisor, to conduct studies in any department where necessary to trace causes of quality problems. May ask any department for data pertinent to the problems under investigation and may convene supervisors to discuss results. Has no authority to order changes in product, processes, or procedures.

Recently, the American Society for Quality Control has taken steps to raise the professional status of quality control practitioners. As with other emerging disciplines, it is desirable to ensure that practitioners possess adequate competence and that the profession clearly establish its value to those it serves. One example of the ASQC program on professionalism[4] is the establishment of formal certification (not licensing) procedures for a quality engineer. The requirements are

1 Membership in the ASQC.

2 Eight years of experience in quality control. A bachelor's degree (in engineering or science) counts as four years experience and a master's degree (in engineering or science) as two additional years.

3 Pass a written examination consisting of two groups of questions:

a Questions on fundamentals and principles.

Fundamental concepts of probability and statistics
Statistical quality control
Design of experiments
Quality planning
Reliability and maintainability
Quality cost analysis
Metrology
Inspection and testing
Data processing

b Questions on applications of subjects which reflect modern practices in the field of quality control engineering.

[4] H. Pitt, "New Requirements Established for Quality Engineer and Technician Certification," *Quality Progress,* August, 1969, p. 30.

The certificate issued is not a license but is a formal recognition of professional competence.

31-8 Selling "Quality Control" to Management

The concept of a new approach to quality requires that higher management approve the creation of a new department and its budget. Securing this approval is a necessary though not sufficient step in getting a program adopted. This is the first step in the breakthrough sequence, i.e., making a breakthrough in attitudes.

The strategy in selling management should first recognize that management does *not* need to be sold on the importance of good products, low costs, and prompt deliveries. Management realizes this better than the staff specialist. However, management is often not aware that:

1 "Higher quality" costs less, not more. This is usually true for quality of conformance, and false for quality of design. Many managers instinctively think of "quality" as meaning quality of design and must be shown that quality of conformance is different and has large potentials for cost saving. The key here is to calculate the "gold in the mine" (see Chapter 5).

2 Staff assistance is needed for improving quality. Management must be shown that although sporadic problems can and should be handled by regular supervision, chronic problems require staff assistance because line supervision does not have the time, skills, or objectivity to diagnose chronic quality problems.

3 Staff assistance can be paid for out of reduction in quality losses. Despite the optimism of the quality control specialist, management realizes that it is all a gamble.[5] The quality control proponent should be prepared to gamble by committing himself to reductions in quality losses and using these to justify his budget request (see next section).

There is also the problem of how a technique-oriented specialist is to communicate a proposal to a results-oriented manager. Although he may consider it heresy, the specialist must accept the concept that quality is primarily a business problem rather than a technical problem. Several principles need to be observed:

1 Identify those objectives of the manager (e.g., scrap reduction, complaint reduction) which the specialist will undertake to meet. The objectives should be quantified and means should be provided for evaluating progress.

[5] See J. M. Juran, "On Selling Quality Control to Management," *IQC*, vol. 8, no. 2, p. 29, September, 1951.

2 Spell out a sensible plan which shows how these objectives are to be reached. Any details on the "how" should stem from successful in-plant case histories and *not* from the logic of a technique. It is difficult for staff people to understand that most management people will not (initially at least) accept a technique on the basis of its logic but are much more receptive to a successful case history. Managers remember the past failures of many "logical" proposals.
3 Evaluate and present the costs and the expected results in terms of "return on investment," the common language of managers.
4 Emphasize proposals for breakthrough rather than control. Top management rightly expects control-type efforts to be handled at lower-management levels.

Once management has approved the undertaking, a major milestone has been reached, but then the real work begins. Next is the problem of establishing smooth working relations in the middle- and lower-management levels.

31-9 Administering the Staff Quality Control Department

Staff quality control work is primarily of a long-range character—extending over months and even years. This means that the department must be run through long-range planning. A key tool is the annual plan.

Some staff supervisors believe that their work is so varied and complex that it cannot be planned. Managers should resist any such contentions because it is under these conditions that planning is most important and has the greatest potential. Planning cannot be precise but preparation of a plan forces a department to think out its mission and allows other departments and management to see whether the staff department is working on important problems or trivia, on yesterday's problems or tomorrow's.

The annual plan should recognize three kinds of activities: (1) *projects* for improvement, (2) *planning* for changes on products or processes, and (3) *service* to other departments. The annual plan lists the projects to be undertaken for the forthcoming year, the expected results to be achieved in each project, and the manpower and budget to be expended on each project. Choice of projects must be made in collaboration with the line departments, which actually should have the controlling voice in deciding the order of priorities.

Aside from specific projects, time must be allotted for providing service to other departments. This is usually done on the basis of past experience.

For example, one staff manager summarized his first plan and budget as in Table 31-1.

Table 31-1 EXAMPLE OF STAFF QUALITY CONTROL BUDGET

Project Description	*Expected Annual Improvement*	*Man-months of Engineering Time*	
		Needed	*Available*
Leaks in castings	$140,000	24	
Hard castings	90,000	15	
Eccentricity	80,000	8	
Plan for vendor rating	60,000	6	
.	.	.	
.	.	.	
.	.	.	
Subtotal, projects	$600,000	85	
Inspection planning	20,000	16	12
Service, consulting	?	6	
Totals	$620,000	107	12

With such a tabulation the basic facts become clear. For example, there is an estimate of 85 man-months needed to reduce the major losses due to defects. This could, in theory, be done by assigning one man for 7 years, or seven men for 1 year. In practice, no experienced manager would authorize a sudden buildup. Instead, he would appoint one or two men and let them work at it for a year. Then, based on the experience gained, and on the revision of the plan, he would decide what, if any, expansion was needed.

A few common pitfalls of running a department might be mentioned:

1 The number of projects active at any one time must be limited to the capacity of the department.

2 Engineers should be required to use logbooks and to keep them in a workmanlike manner. Dates and project numbers should be on all sheets.

3 Engineers should not be allowed to fall more than a day behind in their analysis of data.

4 Good presentation of data should be stressed so that the value of the studies is not lost through poor "packaging and salesmanship."

5 The department work load should be decided by the list of major quality problems, not by the simplicity of product or process boundaries or by the personal preference of a technique-oriented staff specialist looking for an application of one of his pet tools. The main quality problems are almost always interdepartmental in na-

ture. Their difficulty of solution is strongly related to their interdepartmental nature.

The reports of the quality control department should be infrequent—normally quarterly. The quarterly report shows progress during the last quarter and plans for the quarter ahead, and the annual report contains the next annual plan as well. Specific projects are reported on as completed.

In reporting financial results, the department should enlist the aid of the accounting department. Computations on savings are usually more objective, and receive greater credibility, when supported by the accounting department. Even on nonfinancial matters, e.g., reports on projects, it is useful to secure the support of the departments affected. Some staff people submit such reports jointly with the line department involved. A hardy few even draft reports to be submited by the line supervisor as his report. Quality control departments that furnish extensive services to line functions may be required to formally report the hours and dollars expended on a periodic basis. This, in turn, may establish a system of ratios or other indices to evaluate performance against budgets.

The matter of "who gets the credit" is an area to which the antennae of industrial man are sharply tuned. Staff people should lean over backward to give maximum credit to line supervisors.

31-10 Introducing Change

In its creative and project work the staff department is trying to change the existing order. On the face of it, the proposed changes are "technical" in character. However, beneath the surface is the "social" aspect of the change, since the habits, beliefs, attitudes, practices, traditions, status, etc., of people are affected. It is the social aspect of the change which poses the real problems of introducing change.

Social scientists have given the name "cultural pattern" to this collection of habits, beliefs, etc. They point out that every continuing "society" (a group of people associating with each other) develops a pattern of behavior. This pattern is then taught to new members who enter the society, thus perpetuating the cultural pattern.

The cultural pattern is of such great importance to the society that attempts to change it are resisted. (By definition, these attempts threaten the habits, beliefs, etc., of the people involved.) This resistance to change can easily result in rejection of a "beneficial" technical change, since the price to be paid in habits, beliefs, etc., may be too great in the minds of the people.[6] Because cultural patterns are found in all societies, without exception, the introduction of change in industry is subject to the same forms of resistance.[7] There has been no lack of this resistance in factories with respect to new techniques for quality control.

A staff proposal to a line manager is also a potential clash between two cultures. The line manager, faced with swift-moving, harsh realities, has seen enough unrealistic staff proposals to conclude that staff people do not understand these realities and hence are impractical; also that staff people are technique oriented, not results oriented. In turn, staff people have seen enough projects succeed to conclude that line managers close their minds to new ways, lack imagination to see the potentialities of modern ways, and are unable to communicate properly. In some companies this broad relationship between the two cultures has deteriorated so far that it becomes the biggest single obstacle to successful projects.

In general, neither the schools nor the industrial companies have prepared staff specialists to deal with cultural patterns. Among these specialists there is even a widespread lack of awareness of the existence of this universal human phenomenon. But the cultural pattern is always there, so that "one of the necessary tools of the expert . . . is some way of analyzing the traditional behavior, so as to be able to estimate just where the changes are going to fall, which habits are going to change, which beliefs are going to be threatened, which attitudes will have to be altered."[8]

Studies of the social scientists have evolved principles to be observed by staff experts in introducing change. Some of these principles may be restated as follows.[9]

1 The staff specialists who propose change must understand that the premises on which they base their proposals are merely products of the culture in which the expert happened to be reared. They are not necessarily universal truths.

2 The culture of the line supervisor serves him well by providing him with precedent, practices, and explanations. These things,

[6] For an impressive array of examples, see Margaret Mead, *Cultural Patterns and Technical Change,* UNESCO, 1953; also reprinted as a Mentor Book, New American Library, Inc., New York, 1955.

[7] See J. M. Juran, "Cultural Patterns and Quality Control," *IQC,* vol. 14, no. 4, pp. 8–13, October, 1957.

[8] Mead, *op. cit.*

[9] J. M. Juran, "Improving the Relationship between Staff and Line," *Personnel,* pp. 515–524, May, 1956.

however unenlightened, have the advantage of predictability and thus assure, to some degree, peace of mind. The more the staff specialist recognizes the real values this culture has for the line supervisor (instead of disparaging it as "ignorance," "stubbornness," "too old to learn," etc.), the better will he be able to prepare his case.

3 The staff specialist should examine his proposals from the viewpoint of the line supervisor, since that is what the latter is bound to do anyhow.

4 The staff specialist must avoid the temptation to deal with a localized problem through a sweeping master plan which goes far beyond immediate needs. If he urges the sweeping plan, he risks rejection of the entire proposal, including the solution to the localized problem as well.

5 Unless the line supervisor is genuinely convinced that the change should be made, he is likely to return to his old ways rather than endure the tensions of frustrations brought about by the change.

As the cultural pattern is understood, it becomes possible to make use of some "rules of the road" for introducing change. These include:[10]

1 Secure the active participation of those who will be affected, during both the planning and the execution of the change.

2 Strip off all technical cultural baggage not strictly needed for introducing the change. (Many quality control engineers have been in violation of this.)

3 Reduce the impact of the changes by weaving them into an existing broader pattern of behavior, or by letting them ride in on the back of some acceptable change.

4 Put yourself in the other fellow's place.

5 Make use of the wide variety of methods available for dealing with resistance to change. These include:

a Persuasion.

b Change of environment in a way which makes it easy for the individual to change his point of view.

c Remedy of the cause of the resistance.

d Create a social climate which favors the new habits.

e Provide sufficient time for mental changes to take place. (Many changes have failed of acceptance on this ground alone.)

f Start small and keep it fluid.

g No surprises.

6 Treat the people with dignity.

[10] *Ibid.*

There will be strikeouts. The staff specialist will not have all his recommendations accepted. What next? He can:

1 Forget it. An 80 percent average is excellent as an adoption rate for staff recommendations.
2 Discriminate between the vital and the trivial. Save the intense arguments, the appeals, the showdowns for the important cases.
3 Try again in a few months. Sometimes it is found that the recommendation may then be acceptable (or may already have been adopted) because enough time has passed for the line man to adapt himself.
4 "Surround" the line man. Sell the idea to his subordinates and to his associates in the hope that he will then question his own position.
5 Find out what is wrong with the proposal. The outward reasons for rejection are almost always stated in technical terms but the real reasons may well be related to the cultural pattern.
6 Find out if the proposal conflicts with the personal aspirations of the man who disagrees. For example, does it threaten his security, his departmental results, or other aims.[11]
7 Identify the "law of the situation"[12] which dictates the change. This is Mary Parker Follett's term for the impersonal forces (e.g., competition from other companies) which dictate the need for the action. It is a means of depersonalizing the proposal; i.e., one person does not take orders from another person, but both take their orders from the situation.
8 Agree to disagree, and agree to let the matter be decided by higher authority.

The problem of introducing change is universal. Even quality control departments resist change. When experience showed that the use of control charts on all machines was overextended, the quality control departments resisted any reduction, since they sponsored the original proposal. For awhile, some companies had two worlds—the staff world, busy maintaining the charts which no one used, and the real world, which went on about its business.

PROBLEMS

1 For one of the institutions listed below, obtain information on the current organization structure and allocate among the departments in

[11] See J. M. Juran, "Dealing with the 'Obstructionist' Superintendent," *IQC,* July, 1955, p. 24.

[12] Mary Parker Follett in H. C. Metcalf and L. Urwick (eds.), *Dynamic Administration,* Harper & Row, Publishers, Incorporated, New York, 1941.

the organization the quality control work elements that you think are necessary for the organization.

a An institution acceptable to the instructor
b A local bank
c A local telephone company
d A local utility

2 For one of the organizations in problem 1, propose a list of functions for a staff quality control department.

3 Prepare an outline for a presentation to be made to the management of an organization in problem 1 to convince it to authorize the establishment of a staff quality control department. The outline must present the specific proposals to be made.

4 Summarize the experiences in providing technical assistance to foreign countries that led to the recommendations by Margaret Mead in introducing change. (Reference given in the chapter.)

5 Coch and French[13] conducted a classical research study on the effect of participation in introducing change. Summarize this study.

6 Make a proposal to the appropriate people to gain acceptance of one of the following:

a Control charts in manufacturing
b Failure mode/failure effect analysis in a design organization (or a design class)
c Numerical reliability requirements in a design organization (or a design class)
d Sampling audit of the accuracy of prices charged by the cashier in a supermarket
e Work simplification in the kitchen (for your mother or wife)

Identify the reasons stated to you for not accepting your proposal. Also identify possible underlying reasons for the resistance to change in terms of their cultural origin (i.e., habits, beliefs, status, etc.).

7 Read the RPM case in the Appendix and answer the following problem.

You are a quality control consultant. Engblom has called you in to look at the QC engineering department.

You have gone over the history of the department and have found the following:

A few years ago, RPM's quality function consisted pretty much of its inspection department, under Schmidt, the chief inspector. When

[13] See L. Coch and J. R. P. French, Jr., "Overcoming Resistance to Change," *Human Relations*, vol. 4, no. 1, pp. 512–533, 1948.

one of the Big Three automobile companies went in for a well-publicized program of statistical quality control, RPM's general manager was interested enough to ask Schmidt to look into it. Schmidt concluded it was mostly a lot of nonsense and said so. He was overruled. He was ordered to give SQC a try. He was also ordered to change the name "inspection" to "quality control" throughout his department.

Schmidt didn't care for either of these orders, but he was an old-timer and a good soldier. He became quality control manager. He also set up a statistical quality control section. Here again there was a problem. Schmidt wanted to put one of "his own" men in charge. Instead, the general manager pressured him to hire, on a full-time basis, the consultant, Damiano, who was advocating an SQC program.

So Schmidt hired Damiano, and purposely gave him plenty of rope. Damiano built up a sizable department, conducted courses in SQC, set up a lot of sampling plans, got Shewhart charts on all the machines and walls, and otherwise made prolific use of statistical tools.

Damiano didn't make it. He was a good promoter, a good salesman, and a good statistician. But he failed for two important reasons:

a He didn't study the economics of quality. As a result, neither he nor anyone else could prove the company was saving enough to pay for his budget.

b He reckoned without the cohesiveness of the RPM career supervisors. They fought savagely with each other, but they closed ranks against outsiders.

Accordingly, when a drive on expenses came along several years later, there was no one to speak for the SQC section. Damiano left, and the SQC section was cut back. Its name was changed to QC engineering, and Carpenter, a career RPM man, was put in charge. Carpenter had come up from the test lab to the rank of QC superintendent. (He was the man Schmidt had wanted to put in charge of SQC in the first place.)

But Carpenter didn't become the real boss. By this time the various QC superintendents ganged up on QC engineering. They proposed that the QC engineers be assigned to them for spot troubleshooting. They were willing that Carpenter exercise "technical supervision," i.e., training, standards, manuals, etc. Schmidt wasn't sure this was a good idea, but his superintendents were so strong for it that he agreed to it. Carpenter opposed it, but his bargaining power was not enough to stop it.

You have found that morale in QC engineering is very low. The men do not feel that they are being challenged enough. They don't like being in a situation of dual responsibility. They don't see any future in a department that has such a shaky past and so weak a present.

You also suspect that the fire fighting being done by the QC engineers is not nearly as important as the chronic quality problems that go on year after year. You have also found that everyone respects Carpenter and realizes that he is burdened by the past reputation of his section.

You have put together an analysis of the time being spent by the section (now a department). QC engineering is set up in four sections, as follows:

a Quality planning and procedures section. Six men, engaged full time in writing manuals and general use plans and procedures (sampling, vendor relations, inspector training, etc.).

b Data analysis group. Five men, engaged in analysis of data on field returns and vendor quality plus some special studies. They also design data processing procedures for the foregoing and for the much larger problem of shop scrap and rework data. (The actual analysis of scrap and rework data is done in the offices of the various QC superintendents.)

c Quality improvement group. Thirty men, engaged in prevention studies: causes of defects, machine capability, etc. Of these thirty, all but five are assigned to specific shop areas, where they work mainly on problems selected by the area QC superintendent. These five are working on major projects which are really guided by Carpenter.

d Statistical research. Three men, engaged in advanced statistical work—analysis of variance, design of experiments, training in statistical methods, etc.

Considering the new objectives of RPM in the quality function, prepare your recommendations to Engblom on what to do respecting the problem of managing the QC engineering department.

Appendix

Metal Containers, Inc., Case

Metal Containers, Inc., (MC) makes and sells a variety of "tin cans" and other metal containers and products. This industry is highly competitive. Two giant companies dominate the manufacturing industry. Some of the large food processors and breweries have captive plants to make their standard sizes of cans. There are a few medium-sized manufacturers and numerous small ones, the latter generally being devoted to specialty work and selling in a local market. (The shipment problem has resulted, even in the large companies, in many decentralized plants, rather than in a few monsters.)

MC is one of the medium-sized companies, with sales of about $40 million per annum and a total employment of about 1,200. MC was started a few decades ago as a small company in the specialty field. The original plant, in *Cleveland,* was first devoted mainly to specialties—galvanized pails, lunch pails, trays, housewares. A few years later, it added some lines for making standard cans for fruit and vegetable packing. In addition, MC set up a plant in *Milwaukee,* mainly to make beer cans for breweries and with some other standard cans for food processors.

Within the last several years, MC has acquired two more plants by buying out the little companies which owned them. The *Detroit* plant makes large specialty cans for lubricating oil, paint, varnish, etc. The *Chicago* plant makes mainly standard cans, but has a few popular specialty items as well.

The Process

Cans are made from tinplate as a base material. The plate is highly standardized, as are many can sizes, and is available from a number of steel companies.

The plates are decorated on the outside by the can maker, usually by lithographing before forming the can. (Alternatively, the cans are shipped undecorated to the customer, who does the decorating, usually by attaching a paper label.) Lithography or other coatings must meet requirements for identity, registration, film weight, color match, appearance, etc.

The decorated or coated sheets then go through a series of operations which shear and slit the plate for the cylindrical part of the can. Dimensional control is paramount here. The end of the can is pressed to shape, and a ring of sealing compound is applied. Dimensional and laboratory control is involved.

The cut plate is fed into an automated transfer machine which forms the cylindrical part of the can, solders the side seam, attaches the end, and double seams the assembly. These machines operate at prodigious speeds—several hundred cans per *minute*. Numerous dimensional, laboratory, and visual controls are involved.

The can bodies are conveyed to ovens which apply the appropriate inner coatings, lacquer, or whatever. Laboratory and visual controls are used here. Since this is the last production operation before shipment, a detailed visual inspection may be applied here, by production department sorters.

Finally, the cans are conveyed to the shipping department to go either to storage or direct to the customer.

Manufacture of special cans, pails, boxes, etc., is not automated. Some of the operations are performed one at a time on hand-fed general use machinery. Still others are fully operator controlled. In contrast to the amazing speeds of the can machines, the specialty products are fairly slow paced. The risk of making a huge number of defective units before detection is quite different.

Organization

Until 2 years ago, MC operated on a regional basis. The manager in each city was pretty much a general manager. But 2 years ago, MC decided to set up functional vice-presidents. Mr. Wallace, who had been plant manager at the parent Cleveland plant, became manufacturing vice-president. At first, Wallace had no staff, and merely supervised the four plant managers directly. But it soon became evident to Wallace that some central staff was needed if the company were to get the benefit of size and to prepare for further expansion. His first staff appointment was a machine development manager in recognition of the important role played, in this business, by good machinery.

Within the last 2 months, Wallace made his second staff appointment—a quality control manager, in the person of Mr. Lafferty. Wallace was driven to this move because he had various quality problems at the plants, and could not personally get into them—he was too busy on other things. Wallace's organization chart now looks like this:

Vice-president for manufacture
- Plant managers
- Machine development manager
- Quality control manager

Each plant is organized somewhat as follows:

Plant manager
- Production
- Manufacturing engineering
- Production control
- Standards and cost control
- Quality control supervisor

In each plant the quality control department is responsible direct to the plant manager. Quality control has generally organized itself as follows:

Quality control supervisor
- Chief inspectors (one for each shift)
    - Inspectors
- Laboratory

This outward similarity is deceptive. Actually there are wide differences in practice at the plants, for a variety of reasons.

1 The plants differ in size. The parent Cleveland plant, with 600 employees, has a thirty-man quality control department. The Detroit plant, with 100 employees, has a four-man quality control department (the boss, a one-man laboratory, and two inspectors). The Chicago plant, also with about 100 employees, has a nine-man quality control department, including a QC engineer and a statistical clerk.

2 The type of product. The mass production lines require quite a different approach from the specialty lines.

3 The plant traditions. The Detroit and Chicago plants still carry many of the habits of the predecessor independent companies. Also, all plants are exhibiting a lot of resistance to regulation from headquarters.

The quality control supervisors delegate to their people the inspection of the product. Other duties are generally not delegated. Investigation of quality complaints is handled by a supervisor, as is much of the work of clearing shop troubles, quarantining defective work, etc.

The plant managers look to Production to prevent a recurrence of prior troubles, but they also look to the quality control supervisor to do some follow-up as well as to provide general feedback on how things are going with respect to quality.

On being appointed to his new job, Lafferty spent the first 2 months going around to the various plants to see the problems and practices. He talked with the men in various departments outside of manufacture. He also visited some customer installations to see how the product was used and to understand some of the problems faced by MC's marketing people. These 2 months are now behind Lafferty, and it is up to him to come up with his proposals and recommendations.

The RPM Case

General

RPM, Inc., is one of the few remaining independent companies making and selling passenger automobiles. The original name of the company, the name of the car division, and the name of the car are all RPM.

The RPM car has gradually been losing share of market against the Big Three, and there has been a real threat that the RPM car might become extinct, as have so many other cars. The company has, as a hedge, gotten into other lines of business over the years, so that today it has an appliance division, a railroad division, and a defense division. But the top executives have not given up on the car business. Instead, they have concluded that new leadership and ideas should be able to arrest the downward tread in share of market.

The New General Manager

Accordingly, the general manager of the RPM car division "resigned." He was replaced by Mr. Heilman, who had been general manager of the defense division. During his years in the defense division, Heilman had learned a few things about "reliability." On assuming his new duties, Heilman concluded that there was an opportunity to make reliability a major tool for marketing automobiles. He set out to improve the reliability of the RPM car, and to build a marketing strategy which would feature reliability as a major selling point.

Heilman was convinced that this improvement in reliability would increase the sales of the RPM. He was therefore willing to spend substantial sums for a reliability program. But he suspected that this investment would yield a cost reduction as well. The RPM car had just finished a year in which guarantee charges came to $32 per car, a record in the wrong direction. Heilman felt that a reliability program could cut this figure, even if there were to be a more liberal guarantee, as the rumor mill in Detroit was then saying.

Heilman looked over the quality control department in his new job and concluded that it was too hidebound in its habits to take on such an assignment. He therefore arranged to have transferred to him, from the defense division, Mr. Engblom, his former quality manager, and he then set up a divisional organization as follows:

- General manager (Heilman)
  - Chief engineer
  - Controller
  - Marketing manager
  - Materials control manager
  - Manufacturing manager
  - Personnel manager
  - Quality manager (Engblom)

The New Quality Manager

Engblom looked over his new job and found that he had inherited an organization as follows:

- Quality control manager
  - Assistant manager
    - Engine QC superintendent
    - Transmissions QC superintendent
    - Foundry QC superintendent
    - Machining QC superintendent
    - Body QC superintendent
    - Assembly QC superintendent
  - Vendor QC manager
  - QC engineering manager

Engblom made one change in the setup. He moved QC engineering out of it to report to him direct. In addition, he created a new post of reliability manager. Now, Engblom's job looked like this:

- Quality manager (Engblom)
  - Quality control manager (Schmidt)
  - QC engineering manager (Carpenter)
  - Reliability manager (Day)

The Reliability Manager

Day, the new reliability manager, was asked to work up a program for improving reliability. Day concluded that he should treat reliability like any other need, and that he should:

1 Set targets for reliability levels to be attained
2 Measure actual reliability attained versus these targets
3 Take steps to improve reliability by
   a Eliminating long-standing causes of poor reliability
   b Detecting and eliminating new causes of poor reliability as fast as possible
   c Heading off likely new threats to reliability in the design and preproduction phases

Day also concluded that he faced several serious obstacles in carrying out such a program.

1 The annual model change, which always introduced new designs and processes, some not too well proven.
2 The tradition of the chief engineer's department, which to date has not really faced any critical outside review of its design work.
3 The dubious validity of the system of measuring reliability of cars in use. This measurement depended on feedback from hundreds of dealers whose accuracy ranged from mediocre to plain untrustworthy.
4 The low state of morale and effectiveness of QC engineering. This department had not yet recovered from the loss of prestige due to going overboard on charts and such. Hence, planning data such as process capabilities, etc., were sparse and not well accepted by other departments anyway.

The Plan of Action

In the light of the foregoing, Day evolved a specific plan of action as follows:

1 Using the available field failure data, with all the known shortcomings, as a starting point, measure the actual failure rates for the car and for major subsystems, i.e., engine, transmission, body, chassis, electrical
2 Establish reliability goals which can be the basis for a more liberal guarantee and for other marketing strategy
3 Develop means for meeting these reliability goals
4 Develop ways for improving the launching of new models so as to avoid creating or perpetuating causes of failure
5 Develop ways of improving the system of measuring actual failure rates in the field, and of getting prompt feedback in the case of new models

Table A Normal Distribution Areas*

Fractional parts of the total area (1.000) under the normal curve between the mean and a perpendicular erected at various numbers of standard deviations (K) from the mean. To illustrate the use of the table, 39.065 percent of the total area under the curve will lie between the mean and a perpendicular erected at a distance of 1.23σ from the mean.

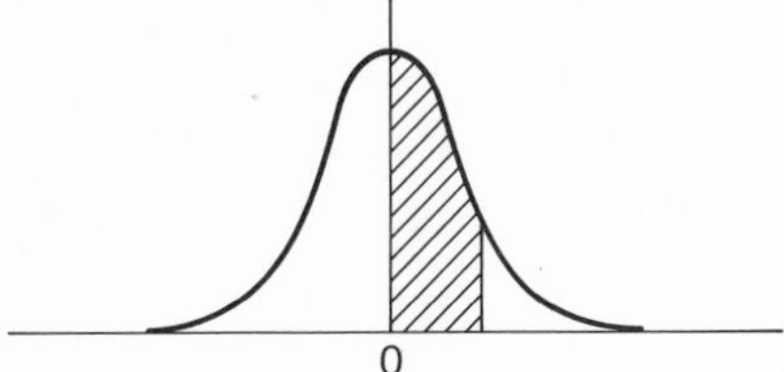

Each figure in the body of the table is preceded by a decimal point.

K	0.00	0.01	0.02	0.03	0.04	0.05	0.06	0.07	0.08	0.09
0.0	00000	00399	00798	01197	01595	01994	02392	02790	03188	03586
0.1	03983	04380	04776	05172	05567	05962	06356	06749	07142	07535
0.2	07926	08317	08706	09095	09483	09871	10257	10642	11026	11409
0.3	11791	12172	12552	12930	13307	13683	14058	14431	14803	15173
0.4	15554	15910	16276	16640	17003	17364	17724	18082	14839	18793
0.5	19146	19497	19847	20194	20540	20884	21226	21566	21904	22240
0.6	22575	22907	23237	23565	23891	24215	24537	24857	25175	25490
0.7	25804	26115	26424	26730	27035	27337	27637	27935	28230	28524
0.8	28814	29103	29389	29673	29955	30234	30511	30785	31057	31327
0.9	31594	31859	32121	32381	32639	32894	33147	33398	33646	33891
1.0	34134	34375	34614	34850	35083	35313	35543	35769	35993	36214
1.1	36433	36650	36864	37076	37286	37493	37698	37900	38100	38298
1.2	38493	38686	38877	39065	39251	39435	39617	39796	39973	40147
1.3	40320	40490	40658	40824	40988	41149	41308	41466	41621	41774
1.4	41924	42073	42220	42364	42507	42647	42786	42922	43056	43189
1.5	43319	43448	43574	43699	43822	43943	44062	44179	44295	44408
1.6	44520	44630	44738	44845	44950	45053	45154	45254	45352	45449
1.7	45543	45637	45728	45818	45907	45994	46080	46164	46246	46327
1.8	46407	46485	46562	46638	47712	46784	46856	46926	46995	47062
1.9	47128	47193	47257	47320	47381	47441	47500	47558	47615	47670
2.0	47725	47778	47831	47882	47932	47982	48030	48077	48124	48169
2.1	48214	48257	48300	48341	48382	48422	48461	48500	48537	48574
2.2	48610	48645	48679	48713	48745	48778	48809	48840	48870	48899
2.3	48928	48956	48983	49010	49036	49061	49086	49111	49134	49158
2.4	49180	49202	49224	49245	49266	49286	49305	49324	49343	49361
2.5	49379	49396	49413	49430	49446	49461	49477	49492	49506	49520
2.6	49534	49547	49560	49573	49585	49598	49609	49621	49632	49643
2.7	49653	49664	49674	49683	49693	49702	49711	49720	49728	49736
2.8	49744	49752	49760	49767	49774	49781	49788	49795	49801	49807
2.9	49813	49819	49825	49831	49836	49841	49846	49851	49856	49861
3.0	49865	49869	49874	49878	49882	49886	49889	49893	49896	49900
3.1	49903	49906	49910	49913	49915	49918	49921	49924	49926	49929
3.2	49931	49934	49936	49938	49940	49942	49944	49946	49948	49950
3.3	49952	49953	49955	49957	49958	49960	49961	49962	49964	49965

* This table has been adapted, by permission, from F. C. Kent, *Elements of Statistics*, McGraw-Hill Book Company, New York, 1924.

Table B Exponential Distribution Values of $e^{-X/\mu}$ for Various Values*

Fractional parts of the total area (1.000) under the exponential curve greater than X. To illustrate: if X/μ is 0.45, the probability of occurrence for a value greater than X is 0.6376.

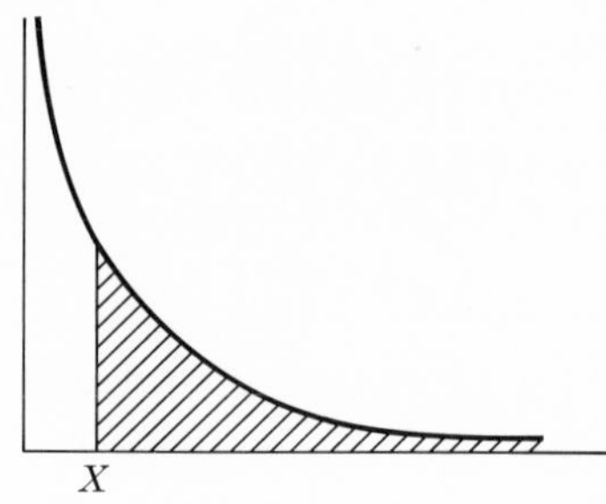

$\frac{X}{\mu}$	0.00	0.01	0.02	0.03	0.04	0.05	0.06	0.07	0.08	0.09
0.0	1.000	0.9900	0.9802	0.9704	0.9608	0.9512	0.9418	0.9324	0.9231	0.9139
0.1	0.9048	0.8958	0.8860	0.8781	0.8694	0.8607	0.8521	0.8437	0.8353	0.8270
0.2	0.8187	0.8106	0.8025	0.7945	0.7866	0.7788	0.7711	0.7634	0.7558	0.7483
0.3	0.7408	0.7334	0.7261	0.7189	0.7118	0.7047	0.6977	0.6907	0.6839	0.6771
0.4	0.6703	0.6637	0.6570	0.6505	0.6440	0.6376	0.6313	0.6250	0.6188	0.6126
0.5	0.6065	0.6005	0.5945	0.5886	0.5827	0.5769	0.5712	0.5655	0.5599	0.5543
0.6	0.5488	0.5434	0.5379	0.5326	0.5273	0.5220	0.5169	0.5117	0.5066	0.5016
0.7	0.4966	0.4916	0.4868	0.4819	0.4771	0.4724	0.4677	0.4630	0.4584	0.4538
0.8	0.4493	0.4449	0.4404	0.4360	0.4317	0.4274	0.4232	0.4190	0.4148	0.4107
0.9	0.4066	0.4025	0.3985	0.3946	0.3906	0.3867	0.3829	0.3791	0.3753	0.3716

$\frac{X}{\mu}$	0.0	0.1	0.2	0.3	0.4	0.5	0.6	0.7	0.8	0.9
1.0	0.3679	0.3329	0.3012	0.2725	0.2466	0.2231	0.2019	0.1827	0.1653	0.1496
2.0	0.1353	0.1225	0.1108	0.1003	0.0907	0.0821	0.0743	0.0672	0.0608	0.0550
3.0	0.0498	0.0450	0.0408	0.0369	0.0334	0.0302	0.0273	0.0247	0.0224	0.0202
4.0	0.0183	0.0166	0.0150	0.0130	0.0123	0.0111	0.0101	0.0091	0.0082	0.0074
5.0	0.0067	0.0061	0.0055	0.0050	0.0045	0.0041	0.0037	0.0033	0.0030	0.0027
6.0	0.0025	0.0022	0.0020	0.0018	0.0017	0.0015	0.0014	0.0012	0.0011	0.0010

* Adapted from S. M. Selby (ed.), "CRC Standard Mathematical Tables," 17th ed., The Chemical Rubber Co., 1969, pp. 201-207.

Table C Median Ranks

	Sample size = n 1	2	3	4	5	6	7	8	9	10	11	12	13	14	15	16	17	18	Sample size = n 19	20	
1	.5000	.2929	.2063	.1591	.1294	.1091	.0943	.0830	.0741	.0670	.0611	.0561	.0519	.0483	.0452	.0424	.0400	.0378	.0358	.0341	1
2		.7071	.5000	.3864	.3147	.2655	.2295	.2021	.1806	.1632	.1489	.1368	.1266	.1178	.1101	.1034	.0975	.0922	.0874	.0831	2
3			.7937	.6136	.5000	.4218	.3648	.3213	.2871	.2594	.2366	.2175	.2013	.1873	.1751	.1644	.1550	.1465	.1390	.1322	3
4				.8409	.6853	.5782	.5000	.4404	.3935	.3557	.3244	.2982	.2760	.2568	.2401	.2254	.2125	.2009	.1905	.1812	4
5					.8706	.7345	.6352	.5596	.5000	.4519	.4122	.3789	.3506	.3263	.3051	.2865	.2700	.2553	.2421	.2302	5
6						.8909	.7705	.6787	.6065	.5481	.5000	.4596	.4253	.3958	.3700	.3475	.3275	.3097	.2937	.2793	6
7							.9057	.7979	.7129	.6443	.5878	.5404	.5000	.4653	.4350	.4085	.3850	.3641	.3453	.3283	7
8								.9170	.8194	.7406	.6756	.6211	.5747	.5347	.5000	.4695	.4425	.4184	.3968	.3774	8
9									.9259	.8368	.7634	.7018	.6494	.6042	.5650	.5305	.5000	.4728	.4484	.4264	9
10										.9330	.8511	.7825	.7240	.6737	.6300	.5915	.5575	.5272	.5000	.4755	10
11											.9389	.8632	.7987	.7432	.6949	.6525	.6150	.5816	.5516	.5245	11
12												.9439	.8734	.8127	.7599	.7135	.6725	.6359	.6032	.5736	12
13													.9481	.8822	.8249	.7746	.7300	.6903	.6547	.6226	13
14														.9517	.8899	.8356	.7875	.7447	.7063	.6717	14
15															.9548	.8966	.8450	.7991	.7579	.7207	15
16																.9576	.9025	.8535	.8095	.7698	16
17																	.9600	.9078	.8610	.8188	17
18																		.9622	.9126	.8678	18
19																			.9642	.9169	19
20																				.9659	20

Table D **Distribution of** t*

Values of t corresponding to certain selected probabilities (i.e., tail areas under the curve). To illustrate: the probability is 0.05 that a sample with 20 degrees of freedom would have $t = 2.086$ or larger.

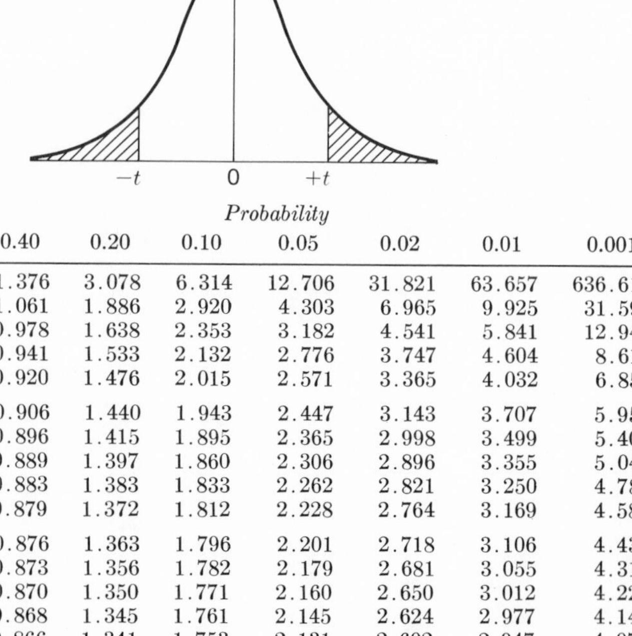

DF	*Probability* 0.80	0.40	0.20	0.10	0.05	0.02	0.01	0.001
1	0.325	1.376	3.078	6.314	12.706	31.821	63.657	636.619
2	0.289	1.061	1.886	2.920	4.303	6.965	9.925	31.598
3	0.277	0.978	1.638	2.353	3.182	4.541	5.841	12.941
4	0.271	0.941	1.533	2.132	2.776	3.747	4.604	8.610
5	0.267	0.920	1.476	2.015	2.571	3.365	4.032	6.859
6	0.265	0.906	1.440	1.943	2.447	3.143	3.707	5.959
7	0.263	0.896	1.415	1.895	2.365	2.998	3.499	5.405
8	0.262	0.889	1.397	1.860	2.306	2.896	3.355	5.041
9	0.261	0.883	1.383	1.833	2.262	2.821	3.250	4.781
10	0.260	0.879	1.372	1.812	2.228	2.764	3.169	4.587
11	0.260	0.876	1.363	1.796	2.201	2.718	3.106	4.437
12	0.259	0.873	1.356	1.782	2.179	2.681	3.055	4.318
13	0.259	0.870	1.350	1.771	2.160	2.650	3.012	4.221
14	0.258	0.868	1.345	1.761	2.145	2.624	2.977	4.140
15	0.258	0.866	1.341	1.753	2.131	2.602	2.947	4.073
16	0.258	0.865	1.337	1.746	2.120	2.583	2.921	4.015
17	0.257	0.863	1.333	1.740	2.110	2.567	2.898	3.965
18	0.257	0.862	1.330	1.734	2.101	2.552	2.878	3.922
19	0.257	0.861	1.328	1.729	2.093	2.539	2.861	3.883
20	0.257	0.860	1.325	1.725	2.086	2.528	2.845	3.850
21	0.257	0.859	1.323	1.721	2.080	2.518	2.831	3.819
22	0.256	0.858	1.321	1.717	2.074	2.508	2.819	3.792
23	0.256	0.858	1.319	1.714	2.069	2.500	2.807	3.767
24	0.256	0.857	1.318	1.711	2.064	2.492	2.797	3.745
25	0.256	0.856	1.316	1.708	2.060	2.485	2.787	3.725
26	0.256	0.856	1.315	1.706	2.056	2.479	2.779	3.707
27	0.256	0.855	1.314	1.703	2.052	2.473	2.771	3.690
28	0.256	0.855	1.313	1.701	2.048	2.467	2.763	3.674
29	0.256	0.854	1.311	1.699	2.045	2.462	2.756	3.659
30	0.256	0.854	1.310	1.697	2.042	2.457	2.750	3.646
40	0.255	0.851	1.303	1.684	2.021	2.423	2.704	3.551
60	0.254	0.848	1.296	1.671	2.000	2.390	2.660	3.460
120	0.254	0.845	1.289	1.658	1.980	2.358	2.617	3.373
∞	0.253	0.842	1.282	1.645	1.960	2.326	2.576	3.291

* Reproduced in abridged form from Table III of Fisher and Yates, *Statistical Tables for Biological, Agricultural, and Medical Research,* Oliver & Boyd Ltd., Edinburgh, by permission of the authors and publishers.

Table E **Distribution of** χ^{2}*

Values of χ^2 corresponding to certain selected probabilities (i.e., tail areas under the curve). To illustrate: the probability is 0.05 that a sample with 20 degrees of freedom, taken from a normal distribution, would have $\chi^2 =$ 31.410 or larger.

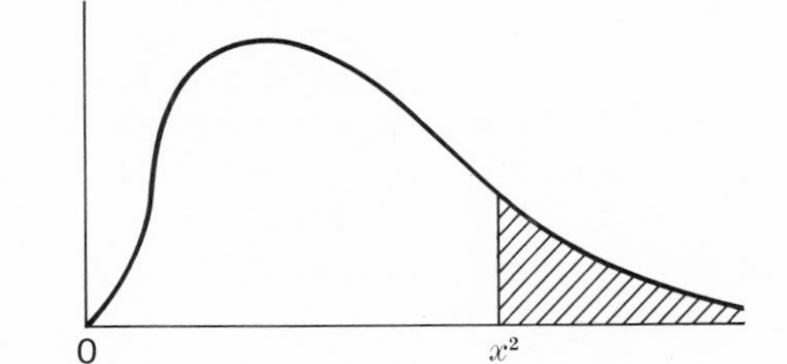

DF	*Probability* 0.99	0.98	0.95	0.90	0.80	0.20	0.10	0.05	0.02	0.01	0.001
1	0.0^3157	0.0^3628	0.00393	0.0158	0.0642	1.642	2.706	3.841	5.412	6.635	10.827
2	0.0201	0.0404	0.103	0.211	0.446	3.219	4.605	5.991	7.824	9.210	13.815
3	0.115	0.185	0.352	0.584	1.005	4.642	6.251	7.815	9.837	11.341	16.268
4	0.297	0.429	0.711	1.064	1.649	5.989	7.779	9.488	11.668	13.277	18.465
5	0.554	0.752	1.145	1.610	2.343	7.289	9.236	11.070	13.388	15.086	20.517
6	0.872	1.134	1.635	2.204	3.070	8.558	10.645	12.592	15.033	16.812	22.457
7	1.239	1.564	2.167	2.833	3.822	9.803	12.017	14.067	16.622	18.475	24.322
8	1.646	2.032	2.733	3.490	4.594	11.030	13.362	15.507	18.168	20.090	26.125
9	2.088	2.532	3.325	4.168	5.380	12.242	14.684	16.919	19.679	21.666	27.877
10	2.558	3.059	3.940	4.865	6.179	13.442	15.987	18.307	21.161	23.209	29.588

Table E DISTRIBUTION OF χ^2* (*Continued*)

	Probability										
DF	0.99	0.98	0.95	0.90	0.80	0.20	0.10	0.05	0.02	0.01	0.001
11	3.053	3.609	4.575	5.578	6.989	14.631	17.275	19.675	22.618	24.725	31.264
12	3.571	4.178	5.226	6.304	7.807	15.812	18.549	21.026	24.054	26.217	32.909
13	4.107	4.765	5.892	7.042	8.634	16.985	19.812	22.362	25.472	27.688	34.523
14	4.660	5.368	6.571	7.790	9.467	18.151	21.064	23.685	26.873	29.141	36.128
15	5.229	5.985	7.261	8.547	10.307	19.311	22.307	24.996	28.259	30.578	37.697
16	5.812	6.614	7.962	9.312	11.152	20.465	23.542	26.296	29.633	32.000	39.252
17	6.408	7.255	8.672	10.085	12.002	21.615	24.769	27.587	30.995	33.409	40.790
18	7.015	7.906	9.390	10.865	12.857	22.760	25.989	28.869	32.346	34.805	42.312
19	7.633	8.567	10.117	11.651	13.716	23.900	27.204	30.144	33.687	36.191	43.820
20	8.260	9.237	10.851	12.443	14.578	25.038	28.412	31.410	35.020	37.566	45.315
21	8.897	9.915	11.591	13.240	15.445	26.171	29.615	32.671	36.343	38.932	46.797
22	9.542	10.600	12.338	14.041	16.314	27.301	30.813	33.924	37.659	40.289	48.268
23	10.196	11.293	13.091	14.848	17.187	28.429	32.007	35.172	38.968	41.638	49.728
24	10.856	11.992	13.848	15.659	18.062	29.553	33.196	36.415	40.270	42.980	51.179
25	11.524	12.697	14.611	16.473	18.940	30.675	34.382	37.652	41.566	44.314	52.620
26	12.198	13.409	15.379	17.292	19.820	31.795	35.563	38.885	42.856	45.642	54.052
27	12.879	14.125	16.151	18.114	20.703	32.912	36.741	40.113	44.140	46.963	55.476
28	13.565	14.847	16.928	18.939	21.588	34.027	37.916	41.337	45.419	48.278	56.893
29	14.256	15.574	17.708	19.768	22.475	35.139	39.087	42.557	46.693	49.588	58.302
30	14.953	16.306	18.493	20.599	23.364	36.250	40.256	43.773	47.962	50.892	59.703

* Reproduced in abridged form from Table IV of Fisher and Yates, *Statistical Tables for Biological, Agricultural, and Medical Research*, Oliver & Boyd Ltd., Edinburgh, by permission of the authors and publishers.

Table F DISTRIBUTION OF F*

5 Percent (Roman Type) and 1 Percent (**Boldface Type**)

Values of F corresponding to two selected probabilities (i.e., tail areas under the curve). To illustrate: the probability is 0.05 that the ratio of two mean squares obtained with 20 and 10 degrees of freedom in numerator and denominator, respectively, would yield $F = 2.77$ or larger.

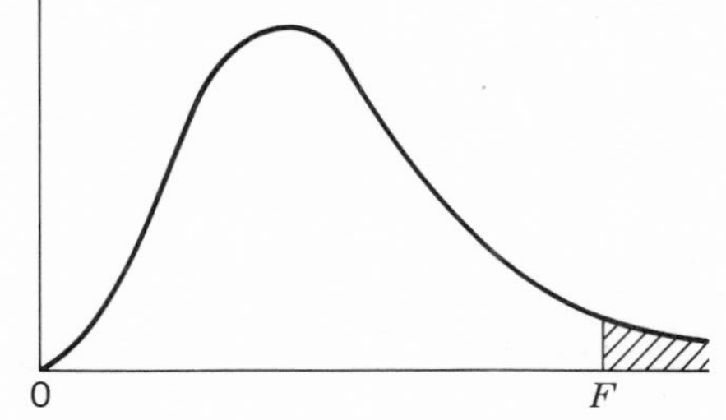

DF_2	*DF_1 Degrees of Freedom for Greater Mean Square (Placed in the Numerator)* 1	2	3	4	5	6	7	8	9	10	11	12	14	16	20	24	30	40	50	75	100	200	500	∞
1	161	200	216	225	230	234	237	239	241	242	243	244	245	246	248	249	250	251	252	253	253	254	254	254
	4,052	**4,999**	**5,403**	**5,625**	**5,764**	**5,859**	**5,928**	**5,981**	**6,022**	**6,056**	**6,082**	**6,106**	**6,142**	**6,169**	**6,208**	**6,234**	**6,258**	**6,286**	**6,302**	**6,323**	**6,334**	**6,352**	**6,361**	**6,366**
2	18.51	19.00	19.16	19.25	19.30	19.33	19.36	19.37	19.38	19.39	19.40	19.41	19.42	19.43	19.44	19.45	19.46	19.47	19.47	19.48	19.49	19.49	19.50	19.50
	98.49	**99.00**	**99.17**	**99.25**	**99.30**	**99.33**	**99.34**	**99.36**	**99.38**	**99.40**	**99.41**	**99.42**	**99.43**	**99.44**	**99.45**	**99.46**	**99.47**	**99.48**	**99.48**	**99.49**	**99.49**	**99.49**	**99.50**	**99.50**
3	10.13	9.55	9.28	9.12	9.01	8.94	8.88	8.84	8.81	8.78	8.76	8.74	8.71	8.69	8.66	8.64	8.62	8.60	8.58	8.57	8.56	8.54	8.54	8.53
	34.12	**30.82**	**29.46**	**28.71**	**28.24**	**27.91**	**27.67**	**27.49**	**27.34**	**27.23**	**27.13**	**27.05**	**26.92**	**26.83**	**26.69**	**26.60**	**26.50**	**26.41**	**26.35**	**26.27**	**26.23**	**26.18**	**26.14**	**26.12**
4	7.71	6.94	6.59	6.39	6.26	6.16	6.09	6.04	6.00	5.96	5.93	5.91	5.87	5.84	5.80	5.77	5.74	5.71	5.70	5.68	5.66	5.65	5.64	5.63
	21.20	**18.00**	**16.69**	**15.98**	**15.52**	**15.21**	**14.98**	**14.80**	**14.66**	**14.54**	**14.45**	**14.37**	**14.24**	**14.15**	**14.02**	**13.93**	**13.83**	**13.74**	**13.69**	**13.61**	**13.57**	**13.52**	**13.48**	**13.46**
5	6.61	5.79	5.41	5.19	5.05	4.95	4.88	4.82	4.78	4.74	4.70	4.68	4.64	4.60	4.56	4.53	4.50	4.46	4.44	4.42	4.40	4.38	4.37	4.36
	16.26	**13.27**	**12.06**	**11.39**	**10.97**	**10.67**	**10.45**	**10.27**	**10.15**	**10.05**	**9.96**	**9.89**	**9.77**	**9.68**	**9.55**	**9.47**	**9.38**	**9.29**	**9.24**	**9.17**	**9.13**	**9.07**	**9.04**	**9.02**

* This table is reproduced from Table 10.7 in *Statistical Methods*, 4th ed., 1946, with the permission of Prof. George W. Snedecor and the Publishers, The Iowa State University Press, Ames. Fifth edition available, 1956.

Table F Distribution of F (*Continued*)

5 Percent (Roman Type) and 1 Percent (Boldface Type) Points for the Distribution of F

DF_1 Degrees of Freedom for Greater Mean Square (Placed in the Numerator)

DF_2	1	2	3	4	5	6	7	8	9	10	11	12	14	16	20	24	30	40	50	75	100	200	500	∞
6	5.99	5.14	4.76	4.53	4.39	4.28	4.21	4.15	4.10	4.06	4.03	4.00	3.96	3.92	3.87	3.84	3.81	3.77	3.75	3.72	3.71	3.69	3.68	3.67
	13.74	**10.92**	**9.78**	**9.15**	**8.75**	**8.47**	**8.26**	**8.10**	**7.98**	**7.87**	**7.79**	**7.72**	**7.60**	**7.52**	**7.39**	**7.31**	**7.23**	**7.14**	**7.09**	**7.02**	**6.99**	**6.94**	**6.90**	**6.88**
7	5.59	4.74	4.35	4.12	3.97	3.87	3.79	3.73	3.68	3.63	3.60	3.57	3.52	3.49	3.44	3.41	3.38	3.34	3.32	3.29	3.28	3.25	3.24	3.23
	12.25	**9.55**	**8.45**	**7.85**	**7.46**	**7.19**	**7.00**	**6.84**	**6.71**	**6.62**	**6.54**	**6.47**	**6.35**	**6.27**	**6.15**	**6.07**	**5.98**	**5.90**	**5.85**	**5.78**	**5.75**	**5.70**	**5.67**	**5.65**
8	5.32	4.46	4.07	3.84	3.69	3.58	3.50	3.44	3.39	3.34	3.31	3.28	3.23	3.20	3.15	3.12	3.08	3.05	3.03	3.00	2.98	2.96	2.94	2.93
	11.26	**8.65**	**7.59**	**7.01**	**6.63**	**6.37**	**6.19**	**6.03**	**5.91**	**5.82**	**5.74**	**5.67**	**5.56**	**5.48**	**5.36**	**5.28**	**5.20**	**5.11**	**5.06**	**5.00**	**4.96**	**4.91**	**4.88**	**4.86**
9	5.12	4.26	3.86	3.63	3.48	3.37	3.29	3.23	3.18	3.13	3.10	3.07	3.02	2.98	2.93	2.90	2.86	2.82	2.80	2.77	2.76	2.73	2.72	2.71
	10.56	**8.02**	**6.99**	**6.42**	**6.06**	**5.80**	**5.62**	**5.47**	**5.35**	**5.26**	**5.18**	**5.11**	**5.00**	**4.92**	**4.80**	**4.73**	**4.64**	**4.56**	**4.51**	**4.45**	**4.41**	**4.36**	**4.33**	**4.31**
10	4.96	4.10	3.71	3.48	3.33	3.22	3.14	3.07	3.02	2.97	2.94	2.91	2.86	2.82	2.77	2.74	2.70	2.67	2.64	2.61	2.59	2.56	2.55	2.54
	10.04	**7.56**	**6.55**	**5.99**	**5.64**	**5.39**	**5.21**	**5.06**	**4.95**	**4.85**	**4.78**	**4.71**	**4.60**	**4.52**	**4.41**	**4.33**	**4.25**	**4.17**	**4.12**	**4.05**	**4.01**	**3.96**	**3.93**	**3.91**
11	4.84	3.98	3.59	3.36	3.20	3.09	3.01	2.95	2.90	2.86	2.82	2.79	2.74	2.70	2.65	2.61	2.57	2.53	2.50	2.47	2.45	2.42	2.41	2.40
	9.65	**7.20**	**6.22**	**5.67**	**5.32**	**5.07**	**4.88**	**4.74**	**4.63**	**4.54**	**4.46**	**4.40**	**4.29**	**4.21**	**4.10**	**4.02**	**3.94**	**3.86**	**3.80**	**3.74**	**3.70**	**3.66**	**3.62**	**3.60**
12	4.75	3.88	3.49	3.26	3.11	3.00	2.92	2.85	2.80	2.76	2.72	2.69	2.64	2.60	2.54	2.50	2.46	2.42	2.40	2.36	2.35	2.32	2.31	2.30
	9.33	**6.93**	**5.95**	**5.41**	**5.06**	**4.82**	**4.65**	**4.50**	**4.39**	**4.30**	**4.22**	**4.16**	**4.05**	**3.98**	**3.86**	**3.78**	**3.70**	**3.61**	**3.56**	**3.49**	**3.46**	**3.41**	**3.38**	**3.36**
13	4.67	3.80	3.41	3.18	3.02	2.92	2.84	2.77	2.72	2.67	2.63	2.60	2.55	2.51	2.46	2.42	2.38	2.34	2.32	2.28	2.26	2.24	2.22	2.21
	9.07	**6.70**	**5.74**	**5.20**	**4.86**	**4.62**	**4.44**	**4.30**	**4.19**	**4.10**	**4.02**	**3.96**	**3.85**	**3.78**	**3.67**	**3.59**	**3.51**	**3.42**	**3.37**	**3.30**	**3.27**	**3.21**	**3.18**	**3.16**
14	4.60	3.74	3.34	3.11	2.96	2.85	2.77	2.70	2.65	2.60	2.56	2.53	2.48	2.44	2.39	2.35	2.31	2.27	2.24	2.21	2.19	2.16	2.14	2.13
	8.86	**6.51**	**5.56**	**5.03**	**4.69**	**4.46**	**4.28**	**4.14**	**4.03**	**3.94**	**3.86**	**3.80**	**3.70**	**3.62**	**3.51**	**3.43**	**3.34**	**3.26**	**3.21**	**3.14**	**3.11**	**3.06**	**3.02**	**3.00**
15	4.54	3.68	3.29	3.06	2.90	2.79	2.70	2.64	2.59	2.55	2.51	2.48	2.43	2.39	2.33	2.29	2.25	2.21	2.18	2.15	2.12	2.10	2.08	2.07
	8.68	**6.36**	**5.42**	**4.89**	**4.56**	**4.32**	**4.14**	**4.00**	**3.89**	**3.80**	**3.73**	**3.67**	**3.56**	**3.48**	**3.36**	**3.29**	**3.20**	**3.12**	**3.07**	**3.00**	**2.97**	**2.92**	**2.89**	**2.87**
16	4.49	3.63	3.24	3.01	2.85	2.74	2.66	2.59	2.54	2.49	2.45	2.42	2.37	2.33	2.28	2.24	2.20	2.16	2.13	2.09	2.07	2.04	2.02	2.01
	8.53	**6.23**	**5.29**	**4.77**	**4.44**	**4.20**	**4.03**	**3.89**	**3.78**	**3.69**	**3.61**	**3.55**	**3.45**	**3.37**	**3.25**	**3.18**	**3.10**	**3.01**	**2.96**	**2.89**	**2.86**	**2.80**	**2.77**	**2.75**
17	4.45	3.59	3.20	2.96	2.81	2.70	2.62	2.55	2.50	2.45	2.41	2.38	2.33	2.29	2.23	2.19	2.15	2.11	2.08	2.04	2.02	1.99	1.97	1.96
	8.40	**6.11**	**5.18**	**4.67**	**4.34**	**4.10**	**3.93**	**3.79**	**3.68**	**3.59**	**3.52**	**3.45**	**3.35**	**3.27**	**3.16**	**3.08**	**3.00**	**2.92**	**2.86**	**2.79**	**2.76**	**2.70**	**2.67**	**2.65**
18	4.41	3.55	3.16	2.93	2.77	2.66	2.58	2.51	2.46	2.41	2.37	2.34	2.29	2.25	2.19	2.15	2.11	2.07	2.04	2.00	1.98	1.95	1.93	1.92
	8.28	**6.01**	**5.09**	**4.58**	**4.25**	**4.01**	**3.85**	**3.71**	**3.60**	**3.51**	**3.44**	**3.37**	**3.27**	**3.19**	**3.07**	**3.00**	**2.91**	**2.83**	**2.78**	**2.71**	**2.68**	**2.62**	**2.59**	**2.57**
19	4.38	3.52	3.13	2.90	2.74	2.63	2.55	2.48	2.43	2.38	2.34	2.31	2.26	2.21	2.15	2.11	2.07	2.02	2.00	1.96	1.94	1.91	1.90	1.88
	8.18	**5.93**	**5.01**	**4.50**	**4.17**	**3.94**	**3.77**	**3.63**	**3.52**	**3.43**	**3.36**	**3.30**	**3.19**	**3.12**	**3.00**	**2.92**	**2.84**	**2.76**	**2.70**	**2.63**	**2.60**	**2.54**	**2.51**	**2.49**
20	4.35	3.49	3.10	2.87	2.71	2.60	2.52	2.45	2.40	2.35	2.31	2.28	2.23	2.18	2.12	2.08	2.04	1.99	1.96	1.92	1.90	1.87	1.85	1.84
	8.10	**5.85**	**4.94**	**4.43**	**4.10**	**3.87**	**3.71**	**3.56**	**3.45**	**3.37**	**3.30**	**3.23**	**3.13**	**3.05**	**2.94**	**2.86**	**2.77**	**2.69**	**2.63**	**2.56**	**2.53**	**2.47**	**2.44**	**2.42**

21	4.32	3.47	3.07	2.84	2.68	2.57	2.49	2.42	2.37	2.32	2.28	2.25	2.20	2.15	2.09	2.05	2.00	1.96	1.93	1.89	1.87	1.84	1.82	1.81
	8.02	**5.78**	**4.87**	**4.37**	**4.04**	**3.81**	**3.65**	**3.51**	**3.40**	**3.31**	**3.24**	**3.17**	**3.07**	**2.99**	**2.88**	**2.80**	**2.72**	**2.63**	**2.58**	**2.51**	**2.47**	**2.42**	**2.38**	**2.36**
22	4.30	3.44	3.05	2.82	2.66	2.55	2.47	2.40	2.35	2.30	2.26	2.23	2.18	2.13	2.07	2.03	1.98	1.93	1.91	1.87	1.84	1.81	1.80	1.78
	7.94	**5.72**	**4.82**	**4.31**	**3.99**	**3.76**	**3.59**	**3.45**	**3.35**	**3.26**	**3.18**	**3.12**	**3.02**	**2.94**	**2.83**	**2.75**	**2.67**	**2.58**	**2.53**	**2.46**	**2.42**	**2.37**	**2.33**	**2.31**
23	4.28	3.42	3.03	2.80	2.64	2.53	2.45	2.38	2.32	2.28	2.24	2.20	2.14	2.10	2.04	2.00	1.96	1.91	1.88	1.84	1.82	1.79	1.77	1.76
	7.88	**5.66**	**4.76**	**4.26**	**3.94**	**3.71**	**3.54**	**3.41**	**3.30**	**3.21**	**3.14**	**3.07**	**2.97**	**2.89**	**2.78**	**2.70**	**2.62**	**2.53**	**2.48**	**2.41**	**2.37**	**2.32**	**2.28**	**2.26**
24	4.26	3.40	3.01	2.78	2.62	2.51	2.43	2.36	2.30	2.26	2.22	2.18	2.13	2.09	2.02	1.98	1.94	1.89	1.86	1.82	1.80	1.76	1.74	1.73
	7.82	**5.61**	**4.72**	**4.22**	**3.90**	**3.67**	**3.50**	**3.36**	**3.25**	**3.17**	**3.09**	**3.03**	**2.93**	**2.85**	**2.74**	**2.66**	**2.58**	**2.49**	**2.44**	**2.36**	**2.33**	**2.27**	**2.23**	**2.21**
25	4.24	3.38	2.99	2.76	2.60	2.49	2.41	2.34	2.28	2.24	2.20	2.16	2.11	2.06	2.00	1.96	1.92	1.87	1.84	1.80	1.77	1.74	1.72	1.71
	7.77	**5.57**	**4.68**	**4.18**	**3.86**	**3.63**	**3.46**	**3.32**	**3.21**	**3.13**	**3.05**	**2.99**	**2.89**	**2.81**	**2.70**	**2.62**	**2.54**	**2.45**	**2.40**	**2.32**	**2.29**	**2.23**	**2.19**	**2.17**
26	4.22	3.37	2.98	2.74	2.59	2.47	2.39	2.32	2.27	2.22	2.18	2.15	2.10	2.05	1.99	1.95	1.90	1.85	1.82	1.78	1.76	1.72	1.70	1.69
	7.72	**5.53**	**4.64**	**4.14**	**3.82**	**3.59**	**3.42**	**3.29**	**3.17**	**3.09**	**3.02**	**2.96**	**2.86**	**2.77**	**2.66**	**2.58**	**2.50**	**2.41**	**2.36**	**2.28**	**2.25**	**2.19**	**2.15**	**2.13**
27	4.21	3.35	2.96	2.73	2.57	2.46	2.37	2.30	2.25	2.20	2.16	2.13	2.08	2.03	1.97	1.93	1.88	1.84	1.80	1.76	1.74	1.71	1.68	1.67
	7.68	**5.49**	**4.60**	**4.11**	**3.79**	**3.56**	**3.39**	**3.26**	**3.14**	**3.06**	**2.98**	**2.93**	**2.83**	**2.74**	**2.63**	**2.55**	**2.47**	**2.38**	**2.33**	**2.25**	**2.21**	**2.16**	**2.12**	**2.10**
28	4.20	3.34	2.95	2.71	2.56	2.44	2.36	2.29	2.24	2.19	2.15	2.12	2.06	2.02	1.96	1.91	1.87	1.81	1.78	1.75	1.72	1.69	1.67	1.65
	7.64	**5.45**	**4.57**	**4.07**	**3.76**	**3.53**	**3.36**	**3.23**	**3.11**	**3.03**	**2.95**	**2.90**	**2.80**	**2.71**	**2.60**	**2.52**	**2.44**	**2.35**	**2.30**	**2.22**	**2.18**	**2.13**	**2.09**	**2.06**
29	4.18	3.33	2.93	2.70	2.54	2.43	2.35	2.28	2.22	2.18	2.14	2.10	2.05	2.00	1.94	1.90	1.85	1.80	1.77	1.73	1.71	1.68	1.65	1.64
	7.60	**5.42**	**4.54**	**4.04**	**3.73**	**3.50**	**3.33**	**3.20**	**3.08**	**3.00**	**2.92**	**2.87**	**2.77**	**2.68**	**2.57**	**2.49**	**2.41**	**2.32**	**2.27**	**2.19**	**2.15**	**2.10**	**2.06**	**2.03**
30	4.17	3.32	2.92	2.69	2.53	2.42	2.34	2.27	2.21	2.16	2.12	2.09	2.04	1.99	1.93	1.89	1.84	1.79	1.76	1.72	1.69	1.66	1.64	1.62
	7.56	**5.39**	**4.51**	**4.02**	**3.70**	**3.47**	**3.30**	**3.17**	**3.06**	**2.98**	**2.90**	**2.84**	**2.74**	**2.66**	**2.55**	**2.47**	**2.38**	**2.29**	**2.24**	**2.16**	**2.13**	**2.07**	**2.03**	**2.01**
32	4.15	3.30	2.90	2.67	2.51	2.40	2.32	2.25	2.19	2.14	2.10	2.07	2.02	1.97	1.91	1.86	1.82	1.76	1.74	1.69	1.67	1.64	1.61	1.59
	7.50	**5.34**	**4.46**	**3.97**	**3.66**	**3.42**	**3.25**	**3.12**	**3.01**	**2.94**	**2.86**	**2.80**	**2.70**	**2.62**	**2.51**	**2.42**	**2.34**	**2.25**	**2.20**	**2.12**	**2.08**	**2.02**	**1.98**	**1.96**
34	4.13	3.28	2.88	2.65	2.49	2.38	2.30	2.23	2.17	2.12	2.08	2.05	2.00	1.95	1.89	1.84	1.80	1.74	1.71	1.67	1.64	1.61	1.59	1.57
	7.44	**5.29**	**4.42**	**3.93**	**3.61**	**3.38**	**3.21**	**3.08**	**2.97**	**2.89**	**2.82**	**2.76**	**2.66**	**2.58**	**2.47**	**2.38**	**2.30**	**2.21**	**2.15**	**2.08**	**2.04**	**1.98**	**1.94**	**1.91**
36	4.11	3.26	2.86	2.63	2.48	2.36	2.28	2.21	2.15	2.10	2.06	2.03	1.98	1.93	1.87	1.82	1.78	1.72	1.69	1.65	1.62	1.59	1.56	1.59
	7.39	**5.25**	**4.38**	**3.89**	**3.58**	**3.35**	**3.18**	**3.04**	**2.94**	**2.86**	**2.78**	**2.72**	**2.62**	**2.54**	**2.43**	**2.35**	**2.26**	**2.17**	**2.12**	**2.04**	**2.00**	**1.94**	**1.90**	**1.87**
38	4.10	3.25	2.85	2.62	2.46	2.35	2.26	2.19	2.14	2.09	2.05	2.02	1.96	1.92	1.85	1.80	1.76	1.71	1.67	1.63	1.60	1.57	1.54	1.53
	7.35	**5.21**	**4.34**	**3.86**	**3.54**	**3.32**	**3.15**	**3.02**	**2.91**	**2.82**	**2.75**	**2.69**	**2.59**	**2.51**	**2.40**	**2.32**	**2.22**	**2.14**	**2.08**	**2.00**	**1.97**	**1.90**	**1.86**	**1.84**
40	4.08	3.23	2.84	2.61	2.45	2.34	2.25	2.18	2.12	2.07	2.04	2.00	1.95	1.90	1.84	1.79	1.74	1.69	1.66	1.61	1.59	1.55	1.53	1.51
	7.31	**5.18**	**4.31**	**3.83**	**3.51**	**3.29**	**3.12**	**2.99**	**2.88**	**2.80**	**2.73**	**2.66**	**2.56**	**2.49**	**2.37**	**2.29**	**2.20**	**2.11**	**2.05**	**1.97**	**1.94**	**1.88**	**1.84**	**1.89**
42	4.07	3.22	2.83	2.59	2.44	2.32	2.24	2.17	2.11	2.06	2.02	1.99	1.94	1.89	1.82	1.78	1.73	1.68	1.64	1.60	1.57	1.54	1.51	1.41
	7.27	**5.15**	**4.29**	**3.80**	**3.49**	**3.26**	**3.10**	**2.96**	**2.86**	**2.77**	**2.70**	**2.64**	**2.54**	**2.46**	**2.35**	**2.26**	**2.17**	**2.08**	**2.02**	**1.94**	**1.91**	**1.85**	**1.80**	**1.78**
44	4.06	3.21	2.82	2.58	2.43	2.31	2.23	2.16	2.10	2.05	2.01	1.98	1.92	1.88	1.81	1.76	1.72	1.66	1.63	1.58	1.56	1.52	1.50	1.48
	7.24	**5.12**	**4.26**	**3.78**	**3.46**	**3.24**	**3.07**	**2.94**	**2.84**	**2.75**	**2.68**	**2.62**	**2.52**	**2.44**	**2.32**	**2.24**	**2.15**	**2.06**	**2.00**	**1.92**	**1.88**	**1.82**	**1.78**	**1.75**
46	4.05	3.20	2.81	2.57	2.42	2.30	2.22	2.14	2.09	2.04	2.00	1.97	1.91	1.87	1.80	1.75	1.71	1.65	1.62	1.57	1.54	1.51	1.48	1.46
	7.21	**5.10**	**4.24**	**3.76**	**3.44**	**3.22**	**3.05**	**2.92**	**2.82**	**2.73**	**2.66**	**2.60**	**2.50**	**2.42**	**2.30**	**2.22**	**2.13**	**2.04**	**1.98**	**1.90**	**1.86**	**1.80**	**1.76**	**1.72**
48	4.04	3.19	2.80	2.56	2.41	2.30	2.21	2.14	2.08	2.03	1.99	1.96	1.90	1.86	1.79	1.74	1.70	1.64	1.61	1.56	1.53	1.50	1.47	1.45
	7.19	**5.08**	**4.22**	**3.74**	**3.42**	**3.20**	**3.04**	**2.90**	**2.80**	**2.71**	**2.64**	**2.58**	**2.48**	**2.40**	**2.28**	**2.20**	**2.11**	**2.02**	**1.96**	**1.88**	**1.84**	**1.78**	**1.73**	**1.70**

Table F **Distribution of** *F* (*Continued*)

5 Percent (Roman Type) and 1 Percent (Boldface Type) Points for the Distribution of *F*

	DF_1 Degrees of Freedom for Greater Mean Square (Placed in the Numerator)																							
DF_2	1	2	3	4	5	6	7	8	9	10	11	12	14	16	20	24	30	40	50	75	100	200	500	∞
50	4.03	3.18	2.79	2.56	2.40	2.29	2.20	2.13	2.07	2.02	1.98	1.95	1.90	1.85	1.78	1.74	1.69	1.63	1.60	1.55	1.52	1.48	1.46	1.44
	7.17	**5.06**	**4.20**	**3.72**	**3.41**	**3.18**	**3.02**	**2.88**	**2.78**	**2.70**	**2.62**	**2.56**	**2.46**	**2.39**	**2.26**	**2.18**	**2.10**	**2.00**	**1.94**	**1.86**	**1.82**	**1.76**	**1.71**	**1.68**
55	4.02	3.17	2.78	2.54	2.38	2.27	2.18	2.11	2.05	2.00	1.97	1.93	1.88	1.83	1.76	1.72	1.67	1.61	1.58	1.52	1.50	1.46	1.43	1.41
	7.12	**5.01**	**4.16**	**3.68**	**3.37**	**3.15**	**2.98**	**2.85**	**2.75**	**2.66**	**2.59**	**2.53**	**2.43**	**2.35**	**2.23**	**2.15**	**2.06**	**1.96**	**1.90**	**1.82**	**1.78**	**1.71**	**1.66**	**1.64**
60	4.00	3.15	2.76	2.52	2.37	2.25	2.17	2.10	2.04	1.99	1.95	1.92	1.86	1.81	1.75	1.70	1.65	1.59	1.56	1.50	1.48	1.44	1.41	1.39
	7.08	**4.98**	**4.13**	**3.65**	**3.34**	**3.12**	**2.95**	**2.82**	**2.72**	**2.63**	**2.56**	**2.50**	**2.40**	**2.32**	**2.20**	**2.12**	**2.03**	**1.93**	**1.87**	**1.79**	**1.74**	**1.68**	**1.63**	**1.60**
65	3.99	3.14	2.75	2.51	2.36	2.24	2.15	2.08	2.02	1.98	1.94	1.90	1.85	1.80	1.73	1.68	1.63	1.57	1.54	1.49	1.46	1.42	1.39	1.37
	7.04	**4.95**	**4.10**	**3.62**	**3.31**	**3.09**	**2.93**	**2.79**	**2.70**	**2.61**	**2.54**	**2.47**	**2.37**	**2.30**	**2.18**	**2.09**	**2.00**	**1.90**	**1.84**	**1.76**	**1.71**	**1.64**	**1.60**	**1.56**
70	3.98	3.13	2.74	2.50	2.35	2.23	2.14	2.07	2.01	1.97	1.93	1.89	1.84	1.79	1.72	1.67	1.62	1.56	1.53	1.47	1.45	1.40	1.37	1.35
	7.01	**4.92**	**4.08**	**3.60**	**3.29**	**3.07**	**2.91**	**2.77**	**2.67**	**2.59**	**2.51**	**2.45**	**2.35**	**2.28**	**2.15**	**2.07**	**1.98**	**1.88**	**1.82**	**1.74**	**1.69**	**1.62**	**1.56**	**1.53**
80	3.96	3.11	2.72	2.48	2.33	2.21	2.12	2.05	1.99	1.95	1.91	1.88	1.82	1.77	1.70	1.65	1.60	1.54	1.51	1.45	1.42	1.38	1.35	1.32
	6.96	**4.88**	**4.04**	**3.56**	**3.25**	**3.04**	**2.87**	**2.74**	**2.64**	**2.55**	**2.48**	**2.41**	**2.32**	**2.24**	**2.11**	**2.03**	**1.94**	**1.84**	**1.78**	**1.70**	**1.65**	**1.57**	**1.52**	**1.49**
100	3.94	3.09	2.70	2.46	2.30	2.19	2.10	2.03	1.97	1.92	1.88	1.85	1.79	1.75	1.68	1.63	1.57	1.51	1.48	1.42	1.39	1.34	1.30	1.28
	6.90	**4.82**	**3.98**	**3.51**	**3.20**	**2.99**	**2.82**	**2.69**	**2.59**	**2.51**	**2.43**	**2.36**	**2.26**	**2.19**	**2.06**	**1.98**	**1.89**	**1.79**	**1.73**	**1.64**	**1.59**	**1.51**	**1.46**	**1.43**
125	3.92	3.07	2.68	2.44	2.29	2.17	2.08	2.01	1.95	1.90	1.86	1.83	1.77	1.72	1.65	1.60	1.55	1.49	1.45	1.39	1.36	1.31	1.27	1.25
	6.84	**4.78**	**3.94**	**3.47**	**3.17**	**2.95**	**2.79**	**2.65**	**2.56**	**2.47**	**2.40**	**2.33**	**2.23**	**2.15**	**2.03**	**1.94**	**1.85**	**1.75**	**1.68**	**1.59**	**1.54**	**1.46**	**1.40**	**1.37**
150	3.91	3.06	2.67	2.43	2.27	2.16	2.07	2.00	1.94	1.89	1.85	1.82	1.76	1.71	1.64	1.59	1.54	1.47	1.44	1.37	1.34	1.29	1.25	1.22
	6.81	**4.75**	**3.91**	**3.44**	**3.14**	**2.92**	**2.76**	**2.62**	**2.53**	**2.44**	**2.37**	**2.30**	**2.20**	**2.12**	**2.00**	**1.91**	**1.83**	**1.72**	**1.66**	**1.56**	**1.51**	**1.43**	**1.37**	**1.33**
200	3.89	3.04	2.65	2.41	2.26	2.14	2.05	1.98	1.92	1.87	1.83	1.80	1.74	1.69	1.62	1.57	1.52	1.45	1.42	1.35	1.32	1.26	1.22	1.19
	6.76	**4.71**	**3.88**	**3.41**	**3.11**	**2.90**	**2.73**	**2.60**	**2.50**	**2.41**	**2.34**	**2.28**	**2.17**	**2.09**	**1.97**	**1.88**	**1.79**	**1.69**	**1.62**	**1.53**	**1.48**	**1.39**	**1.33**	**1.28**
400	3.86	3.02	2.62	2.39	2.23	2.12	2.03	1.96	1.90	1.85	1.81	1.78	1.72	1.67	1.60	1.54	1.49	1.42	1.38	1.32	1.28	1.22	1.16	1.13
	6.70	**4.66**	**3.83**	**3.36**	**3.06**	**2.85**	**2.69**	**2.55**	**2.46**	**2.37**	**2.29**	**2.23**	**2.12**	**2.04**	**1.92**	**1.84**	**1.74**	**1.64**	**1.57**	**1.47**	**1.42**	**1.32**	**1.24**	**1.19**
1,000	3.85	3.00	2.61	2.38	2.22	2.10	2.02	1.95	1.89	1.84	1.80	1.76	1.70	1.65	1.58	1.53	1.47	1.41	1.36	1.30	1.26	1.19	1.13	1.08
	6.66	**4.62**	**3.80**	**3.34**	**3.04**	**2.82**	**2.66**	**2.53**	**2.43**	**2.34**	**2.26**	**2.20**	**2.09**	**2.01**	**1.89**	**1.81**	**1.71**	**1.61**	**1.54**	**1.44**	**1.38**	**1.28**	**1.19**	**1.11**
∞	3.84	2.99	2.60	2.37	2.21	2.09	2.01	1.94	1.88	1.83	1.79	1.75	1.69	1.64	1.57	1.52	1.46	1.40	1.35	1.28	1.24	1.17	1.11	1.00
	6.64	**4.60**	**3.78**	**3.32**	**3.02**	**2.80**	**2.64**	**2.51**	**2.41**	**2.32**	**2.24**	**2.18**	**2.07**	**1.99**	**1.87**	**1.79**	**1.69**	**1.59**	**1.52**	**1.41**	**1.36**	**1.25**	**1.15**	**1.00**

Table G Tolerance Factors for Normal Distributions (Two-sided)*

N \ P	γ = 0.75					γ = 0.90					γ = 0.95					γ = 0.99				
	0.75	0.90	0.95	0.99	0.999	0.75	0.90	0.95	0.99	0.999	0.75	0.90	0.95	0.99	0.999	0.75	0.90	0.95	0.99	0.999
2	4.498	6.301	7.414	9.531	11.920	11.407	15.978	18.800	24.167	30.227	22.858	32.019	37.674	48.430	60.573	114.363	160.193	188.491	242.300	303.054
3	2.501	3.538	4.187	5.431	6.844	4.132	5.847	6.919	8.974	11.309	5.922	8.380	9.916	12.861	16.208	13.378	18.930	22.401	29.055	36.616
4	2.035	2.892	3.431	4.471	5.657	2.932	4.166	4.943	6.440	8.149	3.779	5.369	6.370	8.299	10.502	6.614	9.398	11.150	14.527	18.383
5	1.825	2.599	3.088	4.033	5.117	2.454	3.494	4.152	5.423	6.879	3.002	4.275	5.079	6.634	8.415	4.643	6.612	7.855	10.260	13.015
6	1.704	2.429	2.889	3.779	4.802	2.196	3.131	3.723	4.870	6.188	2.604	3.712	4.414	5.775	7.337	3.743	5.337	6.345	8.301	10.548
7	1.624	2.318	2.757	3.611	4.593	2.034	2.902	3.452	4.521	5.750	2.361	3.369	4.007	5.248	6.676	3.233	4.613	5.488	7.187	9.142
8	1.568	2.238	2.663	3.491	4.444	1.921	2.743	3.264	4.278	5.446	2.197	3.136	3.732	4.891	6.226	2.905	4.147	4.936	6.468	8.234
9	1.525	2.178	2.593	3.400	4.330	1.839	2.626	3.125	4.098	5.220	2.078	2.967	3.532	4.631	5.899	2.677	3.822	4.550	5.966	7.600
10	1.492	2.131	2.537	3.328	4.241	1.775	2.535	3.018	3.959	5.046	1.987	2.839	3.379	4.433	5.649	2.508	3.582	4.265	5.594	7.129
11	1.465	2.093	2.493	3.271	4.169	1.724	2.463	2.933	3.849	4.906	1.916	2.737	3.259	4.277	5.452	2.378	3.397	4.045	5.308	6.766
12	1.443	2.062	2.456	3.223	4.110	1.683	2.404	2.863	3.758	4.792	1.858	2.655	3.162	4.150	5.291	2.274	3.250	3.870	5.079	6.477
13	1.425	2.036	2.424	3.183	4.059	1.648	2.355	2.805	3.682	4.697	1.810	2.587	3.081	4.044	5.158	2.190	3.130	3.727	4.893	6.240
14	1.409	2.013	2.398	3.148	4.016	1.619	2.314	2.756	3.618	4.615	1.770	2.529	3.012	3.955	5.045	2.120	3.029	3.608	4.737	6.043
15	1.395	1.994	2.375	3.118	3.979	1.594	2.278	2.713	3.562	4.545	1.735	2.480	2.954	3.878	4.949	2.060	2.945	3.507	4.605	5.876
16	1.383	1.977	2.355	3.092	3.946	1.572	2.246	2.676	3.514	4.484	1.705	2.437	2.903	3.812	4.865	2.009	2.872	3.421	4.492	5.732
17	1.372	1.962	2.337	3.069	3.917	1.552	2.219	2.643	3.471	4.430	1.679	2.400	2.858	3.754	4.791	1.965	2.808	3.345	4.393	5.607
18	1.363	1.948	2.321	3.048	3.891	1.535	2.194	2.614	3.433	4.382	1.655	2.366	2.819	3.702	4.725	1.926	2.753	3.279	4.307	5.497
19	1.355	1.936	2.307	3.030	3.867	1.520	2.172	2.588	3.399	4.339	1.635	2.337	2.784	3.656	4.667	1.891	2.703	3.221	4.230	5.399
20	1.347	1.925	2.294	3.013	3.846	1.506	2.152	2.564	3.368	4.300	1.616	2.310	2.752	3.615	4.614	1.860	2.659	3.168	4.161	5.312
21	1.340	1.915	2.282	2.998	3.827	1.493	2.135	2.543	3.340	4.264	1.599	2.286	2.723	3.577	4.567	1.833	2.620	2.121	4.100	5.234
22	1.334	1.906	2.271	2.984	3.809	1.482	2.118	2.524	3.315	4.232	1.584	2.264	2.697	3.543	4.523	1.808	2.584	3.078	4.044	5.163
23	1.328	1.898	2.261	2.971	3.793	1.471	2.103	2.506	3.292	4.203	1.570	2.244	2.673	3.512	4.484	1.785	2.551	3.040	3.993	5.098
24	1.322	1.891	2.252	2.950	3.778	1.462	2.089	2.480	3.270	4.176	1.557	2.225	2.651	3.483	4.447	1.764	2.522	3.004	3.947	5.039
25	1.317	1.883	2.244	2.948	3.764	1.453	2.077	2.474	3.251	4.151	1.545	2.208	2.631	3.457	4.413	1.745	2.494	2.972	3.904	4.985
26	1.313	1.877	2.236	2.938	3.751	1.444	2.065	2.460	3.232	4.127	1.534	2.193	2.612	3.432	4.382	1.727	2.460	2.941	3.865	4.935
27	1.309	1.871	2.229	2.929	3.740	1.437	2.054	2.447	3.215	4.106	1.523	2.178	2.595	3.409	4.353	1.711	2.446	2.914	3.828	4.888
30	1.297	1.855	2.210	2.904	3.708	1.417	2.025	2.413	3.170	4.049	1.497	2.140	2.549	3.350	4.278	1.668	2.385	2.841	3.733	4.768
35	1.283	1.834	2.185	2.871	3.667	1.390	1.988	2.368	3.112	3.974	1.462	2.090	2.490	3.272	4.179	1.613	2.306	2.748	3.611	4.611
40	1.271	1.818	2.166	2.846	3.635	1.370	1.959	2.334	3.066	3.917	1.435	2.052	2.445	3.213	4.104	1.571	2.247	2.677	3.518	4.493
45	1.262	1.805	2.150	2.826	3.609	1.354	1.935	2.306	3.030	3.871	1.414	2.021	2.408	3.165	4.042	1.539	2.200	2.621	3.444	4.399
50	1.255	1.794	2.138	2.809	3.588	1.340	1.916	2.284	3.001	3.833	1.396	1.996	2.379	3.126	3.993	1.512	2.162	2.576	3.385	4.323
55	1.249	1.785	2.127	2.795	3.571	1.329	1.901	2.265	2.976	3.801	1.382	1.976	2.354	3.094	3.951	1.490	2.130	2.538	3.335	4.260
60	1.243	1.778	2.118	2.784	3.556	1.320	1.887	2.248	2.955	3.774	1.369	1.958	2.333	3.066	3.916	1.471	2.103	2.506	3.293	4.206

Table G Tolerance Factors for Normal Distributions (Two-sided)* (*Continued*)

P	$\gamma = 0.75$					$\gamma = 0.90$					$\gamma = 0.95$					$\gamma = 0.99$				
N	0.75	0.90	0.95	0.99	0.999	0.75	0.90	0.95	0.99	0.999	0.75	0.90	0.95	0.99	0.999	0.75	0.90	0.95	0.99	0.999
65	1.239	1.771	2.110	2.773	3.543	1.312	1.875	2.235	2.937	3.751	1.359	1.943	2.315	3.042	3.886	1.455	2.080	2.478	3.257	4.160
70	1.235	1.765	2.104	2.764	3.531	1.304	1.865	2.222	2.920	3.730	1.349	1.929	2.299	3.021	3.859	1.440	2.060	2.454	3.225	4.120
75	1.231	1.760	2.098	2.757	3.521	1.298	1.856	2.211	2.906	3.712	1.341	1.917	2.285	3.002	3.835	1.428	2.042	2.433	3.197	4.084
80	1.228	1.756	2.092	2.749	3.512	1.292	1.848	2.202	2.894	3.696	1.334	1.907	2.272	2.986	3.814	1.417	2.026	2.414	3.173	4.053
85	1.225	1.752	2.087	2.743	3.504	1.287	1.841	2.193	2.882	3.682	1.327	1.897	2.261	2.971	3.795	1.407	2.012	2.397	3.150	4.024
90	1.223	1.748	2.083	2.737	3.497	1.283	1.834	2.185	2.872	3.669	1.321	1.889	2.251	2.958	3.778	1.398	1.999	2.382	3.130	3.999
95	1.220	1.745	2.079	2.732	3.490	1.278	1.828	2.178	2.863	3.657	1.315	1.881	2.241	2.945	3.763	1.390	1.987	2.368	3.112	3.976
100	1.218	1.742	2.075	2.727	3.484	1.275	1.822	2.172	2.854	3.646	1.311	1.874	2.233	2.934	3.748	1.383	1.977	2.355	3.096	3.954
110	1.214	1.736	2.069	2.719	3.473	1.268	1.813	2.160	2.839	3.626	1.302	1.861	2.218	2.915	3.723	1.369	1.958	2.333	3.066	3.917
120	1.211	1.732	2.063	2.712	3.464	1.262	1.804	2.150	2.826	3.610	1.294	1.850	2.205	2.898	3.702	1.358	1.942	2.314	3.041	3.885
130	1.208	1.728	2.059	2.705	3.456	1.257	1.797	2.141	2.814	3.595	1.288	1.841	2.194	2.883	3.683	1.349	1.928	2.298	3.019	3.857
140	1.206	1.724	2.054	2.700	3.449	1.252	1.791	2.134	2.804	3.582	1.282	1.833	2.184	2.870	3.666	1.340	1.916	2.283	3.000	3.833
150	1.204	1.721	2.051	2.695	3.443	1.248	1.785	2.127	2.795	3.571	1.277	1.825	2.175	2.859	3.652	1.332	1.905	2.270	2.983	3.811
160	1.202	1.718	2.047	2.691	3.437	1.245	1.780	2.121	2.787	3.561	1.272	1.819	2.167	2.848	3.638	1.326	1.896	2.259	2.968	3.792
170	1.200	1.716	2.044	2.687	3.432	1.242	1.775	2.116	2.780	3.552	1.268	1.813	2.160	2.839	3.627	1.320	1.887	2.248	2.955	3.774
180	1.198	1.713	2.042	2.683	3.427	1.239	1.771	2.111	2.774	3.543	1.264	1.808	2.154	2.831	3.616	1.314	1.879	2.239	2.942	3.759
190	1.197	1.711	2.039	2.680	3.423	1.236	1.767	2.106	2.768	3.536	1.261	1.803	2.148	2.823	3.606	1.309	1.872	2.230	2.931	3.744
200	1.195	1.709	2.037	2.677	3.419	1.234	1.764	2.102	2.762	3.529	1.258	1.798	2.143	2.816	3.597	1.304	1.865	2.222	2.921	3.731
250	1.190	1.702	2.028	2.665	3.404	1.224	1.750	2.085	2.740	3.501	1.245	1.780	2.121	2.788	3.561	1.286	1.839	2.191	2.880	3.678
300	1.186	1.696	2.021	2.656	3.393	1.217	1.740	2.073	2.725	3.481	1.236	1.767	2.106	2.767	3.535	1.273	1.820	2.169	2.850	3.641
400	1.181	1.688	2.012	2.644	3.378	1.207	1.726	2.057	2.703	3.452	1.223	1.749	2.084	2.739	3.499	1.255	1.794	2.138	2.809	3.589
500	1.177	1.683	2.006	2.636	3.368	1.201	1.717	2.046	2.689	3.434	1.215	1.737	2.070	2.721	3.475	1.243	1.777	2.117	2.783	3.555
600	1.175	1.680	2.002	2.631	3.360	1.196	1.710	2.038	2.678	3.421	1.209	1.729	2.060	2.707	3.458	1.234	1.764	2.102	2.763	3.530
700	1.173	1.677	1.998	2.626	3.355	1.192	1.705	2.032	2.670	3.411	1.204	1.722	2.052	2.697	3.445	1.227	1.755	2.091	2.748	3.511
800	1.171	1.675	1.996	2.623	3.350	1.189	1.701	2.027	2.663	3.402	1.201	1.717	2.046	2.688	3.434	1.222	1.747	2.082	2.736	3.495
900	1.170	1.673	1.993	2.620	3.347	1.187	1.697	2.023	2.658	3.396	1.198	1.712	2.040	2.682	3.426	1.218	1.741	2.075	2.726	3.483
1,000	1.169	1.671	1.992	2.617	3.344	1.185	1.695	2.019	2.654	3.390	1.195	1.709	2.036	2.676	3.418	1.214	1.736	2.068	2.718	3.472
∞	1.150	1.645	1.960	2.576	3.291	1.150	1.645	1.960	2.576	3.291	1.150	1.645	1.960	2.576	3.291	1.150	1.645	1.960	2.576	3.291

* From C. Eisenhart, M. W. Hastay, and W. A. Wallis, "Selected Techniques of Statistical Analysis," McGraw-Hill Book Company, New York, 1947. Used by permission.

Table H FACTORS FOR $\overline{X}$ AND R CONTROL CHARTS—TRIAL CONTROL LIMITS*

$$\begin{cases} \text{Upper control limit for } \overline{X} = UCL_{\overline{X}} = \overline{\overline{X}} + A_2\bar{R} \\ \text{Lower control limit for } \overline{X} = LCL_{\overline{X}} = \overline{\overline{X}} - A_2\bar{R} \end{cases}$$

$$\begin{cases} \text{Upper control limit for } R = UCL_R = D_4\bar{R} \\ \text{Lower control limit for } R = LCL_R = D_3\bar{R} \end{cases}$$

Number of Observations in Sample	A_2	D_3	D_4
2	1.880	0	3.268
3	1.023	0	2.574
4	0.729	0	2.282
5	0.577	0	2.114
6	0.483	0	2.004
7	0.419	0.076	1.924
8	0.373	0.136	1.864
9	0.337	0.184	1.816
10	0.308	0.223	1.777
11	0.285	0.256	1.744
12	0.266	0.284	1.717
13	0.249	0.308	1.692
14	0.235	0.329	1.671
15	0.223	0.348	1.652

* Factors reproduced from *1950 ASTM Manual on Quality Control of Materials* by permission of the American Society for Testing Materials, Philadelphia. All factors in Table H are based on a normal distribution.

Table I FACTORS FOR ESTIMATING s FROM $\bar{R}$*

Estimate of $s = \bar{R}/d_2$. These factors assume sampling from a normal universe.

Number of Observations in Sample n	*Factor for Estimate from* $\bar{R}$ $d_2 = \bar{R}/s$
2	1.128
3	1.693
4	2.059
5	2.326
6	2.534
7	2.704
8	2.847
9	2.970
10	3.078
11	3.173
12	3.258
13	3.336
14	3.407
15	3.472

* Reproduced by permission from *ASTM Manual on Presentation of Data*, American Society for Testing Materials, Philadelphia, 1945.

Table J **Poisson Chart***

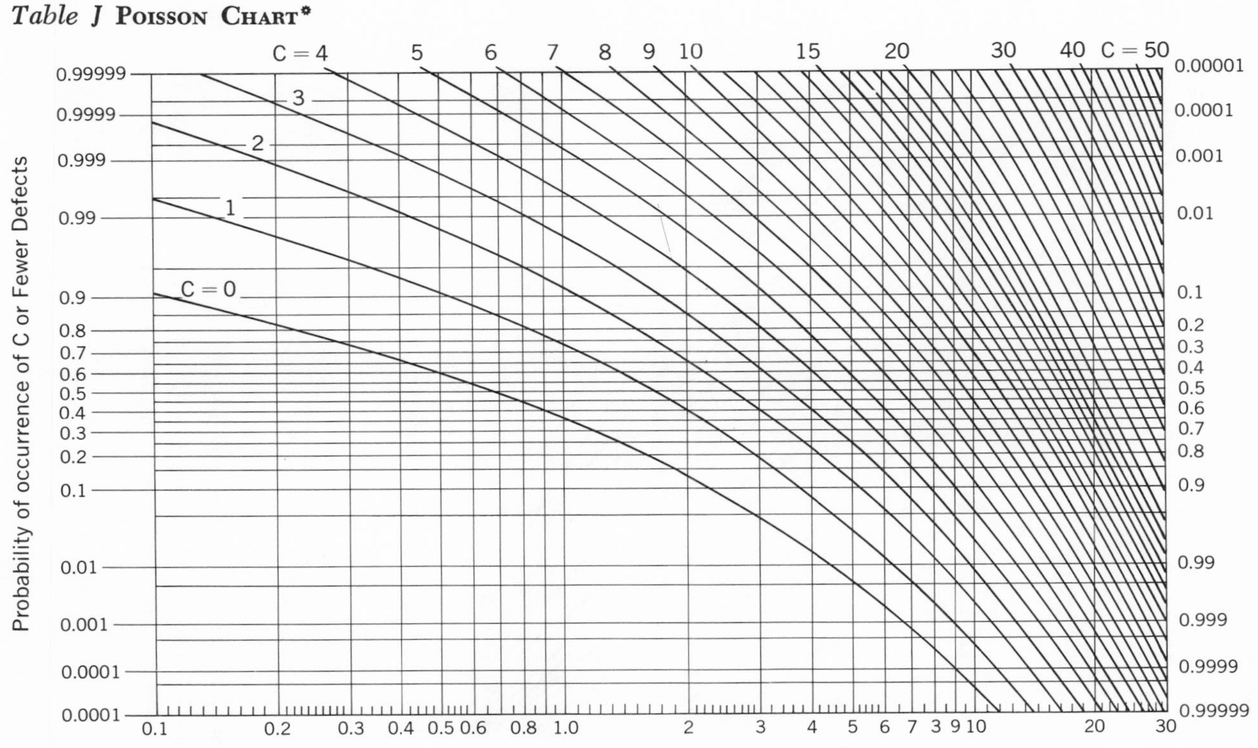

* Reproduced by permission from H. F. Dodge and H. G. Romig, *Sampling Inspection Tables*, 2d ed., John Wiley & Sons, Inc., New York, 1959.

Table K NINETY-FIVE PERCENT CONFIDENCE BELTS FOR POPULATION PROPORTION*

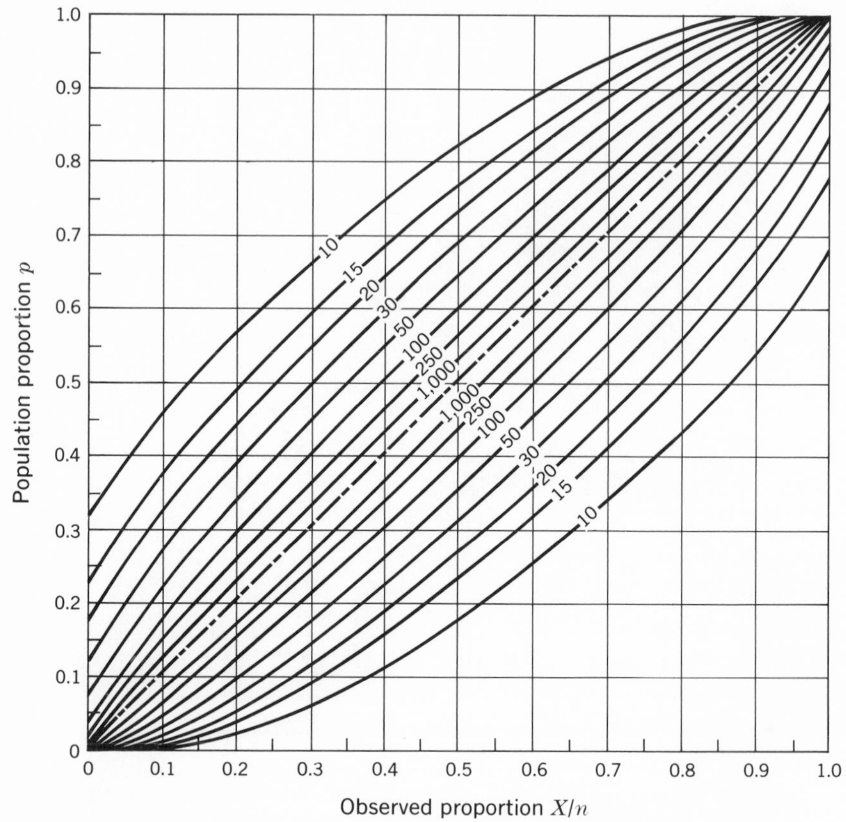

* SOURCE: C. Eisenhart, M. W. Hastay, and W. A. Wallis, *Selected Techniques of Statistical Analysis*—OSRD, McGraw-Hill Book Company, New York, 1947.

Example In a sample of 10 items, 8 were defective ($x/n = 8/10$). The 95% confidence limits on the population proportion defective are read from the two curves (for $n = 10$) as 0.43 and 0.98.

Index